Translations of
MATHEMATICAL
MONOGRAPHS

Volume 244

Arakelov Geometry

Atsushi Moriwaki

Translated by Atsushi Moriwaki

American Mathematical Society
Providence, Rhode Island

ARAKEROFU KIKA (ARAKELOV GEOMETRY)

アラケロフ幾何

by Atsushi Moriwaki

森脇 淳

Translated by Atsushi Moriwaki

First published in 2008 by Iwanami Shoten, Publishers, Tokyo.
This English language edition published in 2014
by the American Mathematical Society, Providence
by arrangement with the author c/o Iwanami Shoten, Publishers, Tokyo.

2010 *Mathematics Subject Classification.* Primary 14G40, 11G50, 37P30.

For additional information and updates on this book, visit
www.ams.org/bookpages/mmono-244

Library of Congress Cataloging-in-Publication Data

Moriwaki, Atsushi, 1960–
[Arakerofu kika. English]
Arakelov geometry / Atsushi Moriwaki ; translated by the author.
pages cm. — (Translation of mathematical monographs ; v. 244)
In English.
Originally published in Japanese: Arakerofu kika (Tokyo : Iwanami Shoten, 2008).
Includes bibliographical references and index.
ISBN 978-1-4704-1074-2 (alk. paper)
1. Arakelov theory. 2. Geometry, Algebraic. I. Title.

QA242.6.M6713 2014
516.3′5—dc23 2014021155

∞ The paper used in this book is acid-free and falls within the guidelines
established to ensure permanence and durability.
Visit the AMS home page at `http://www.ams.org/`

10 9 8 7 6 5 4 3 2 1 19 18 17 16 15 14

Contents

Preface

Arakelov geometry is one of the branches in arithmetic geometry. Nevertheless, its research area is very widespread. Algebra, geometry and analysis are all part of it. It is also closely related to theoretical physics. To give full expositions of Arakelov geometry in one book is almost impossible. Fortunately, the arithmetic Riemann-Roch theorem, which is one of the most significant theorems in Arakelov geometry, is treated in the books of Soulé [**64**] and Faltings [**20**], so this book is mainly devoted to a field of the so-called "birational Arakelov geometry". First, let us recall a brief history of Arakelov geometry.

Diophantus at Alexandria might be the founder of Arakelov geometry, as its source is Diophantine geometry. However, its concrete form has appeared in Arakelov's paper [**2**] in 1974. The intersection theory and the Riemann-Roch theorem may be the most important tools in the theory of algebraic surfaces. Due to them, Arakelov actually made significant progress on the problems of rational points over function fields. In this sense, it was very natural to seek an analogue of the intersection theory and the Riemann-Roch formula on arithmetic surfaces. A prototype of his idea is the geometry of numbers due to Minkowski (see Chapter 2). An important point of Arakelov's idea is to treat a height function in terms of intersection numbers (see Section 9.1). In this sense, the Mordell-Weil theorem from the viewpoint of the height function is also a prototype of his plan. In the paper [**2**], Arakelov had partially succeeded his dream, but it was not complete. The full treatment was done in Faltings' paper [**18**]. In his paper, Faltings established the fundamental results on arithmetic surfaces, such as the arithmetic Riemann-Roch formula, the arithmetic Noether formula and so on.

The next problem is a generalization to a higher dimensional arithmetic variety. The first one is the intersection theory in a higher dimensional case. Gillet-Soulé [**23**] solved it by considering Green currents instead of Green functions. The second one is the arithmetic Riemann-Roch formula including the arithmetic Noether formula. As treated in this book (see Section 4), the arithmetic Riemann-Roch formula on arithmetic surfaces due to Faltings, however elementary, is very complicated, so that it seemed to be difficult to generalize it to the higher dimensional case. Despite this, Deligne [**14**] reported a treatment of the arithmetic Riemann-Roch formula in terms of the Quillen metrics of determinant bundles (it was also pointed out in Faltings' paper [**18**]) and he built it up on arithmetic surfaces. The key for his proof is the usage of the anomaly formula in string theory. Soon after the generalization of intersection theory, Gillet-Soulé [**26**] succeeded again in proving the arithmetic Riemann-Roch theorem on a higher dimensional arithmetic variety. In this way, the fundamental tools for Arakelov geometry had been established.

A typical application of the arithmetic Riemann-Roch theorem is the existence of small sections, that is, sections whose norm is less than or equal to 1. It was used in the solution of the Mordell conjecture due to Vojta [**71**]. Moreover, Ullmo [**70**] and Zhang [**77**] gave an affirmative answer to the Bogomolov conjecture using Arakelov geometry. The existence of small sections was also crucial for their solution. Why are the small sections important? In algebraic geometry, it is a very important problem to find global sections of an invertible sheaf, or to determine the dimension of the vector space consisting of global sections. In this sense, it is significant to require an arithmetic analogue of global sections. As explained in Remark 5.2, a section with small norm on an arithmetic variety means an arithmetic analogue of a global section and the logarithm of the number of small sections is nothing more than the dimension in the arithmetic sense. In the geometric case, it is sufficient to consider non-archimedean norms, but archimedean norms turn out to be important on an arithmetic variety. Of course, an archimedean norm is not determined uniquely. In Arakelov geometry, choosing a different norm is viewed as a different compactification, so that we always fix a metric of an invertible sheaf. For this reason, complex geometry plays an important role in Arakelov geometry. In this book many pages are devoted to the arguments of complex geometry.

In algebraic geometry, to study the asymptotic behaviour of powers of an invertible sheaf is nothing more than the study of birational geometry. Similarly the purpose of birational Arakelov geometry is to consider an arithmetic analogue of them. In this book, we try to include the recent results of birational Arakelov geometry, like the continuity of the arithmetic volumes, the generalized Hodge index theorem, and so on.

The following is the flowchart of this book.

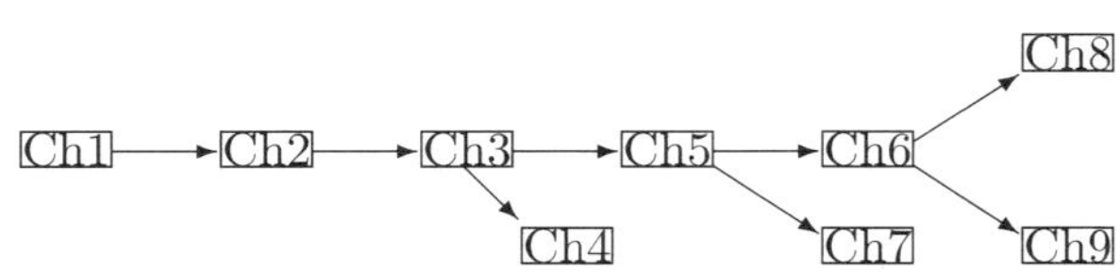

Chapter 1 is devoted to preliminaries for all chapters. In Chapter 2, the fundamental results of geometry of numbers are discussed. In Chapter 3, we give expositions of Arakelov geometry on arithmetic curves, which will be an introduction to Arakelov geometry. In Chapter 4, we consider Arakelov geometry on arithmetic surfaces. This chapter is intended for beginners of Arakelov geometry and its context is not referred to in later chapters, so the reader can skip this chapter. Chapter 5 is devoted to the expositions of the basic results of Arakelov geometry on a general arithmetic variety. In Chapter 6, the recent result of birational Arakelov geometry on the continuity of the arithmetic volumes is discussed. In Chapter 7, we treat the arithmetic Nakai-Moishezon criterion, which is also the fundamental theorem of birational Arakelov geometry. Chapter 8 is devoted to the observation of the arithmetic Bogomolov inequality. In Chapter 9, we consider the Lang-Bogomolov conjecture as an application of birational Arakelov geometry. In this translation, the following points are different from the the original Japanese book:

(1) I adopt a systematic approach to establish the arithmetic Nakai-Moishezon criterion.
(2) I omit the postscripts at the end of each chapter because their contexts are out of date.

I expect the reader to be familiar with the basic algebraic geometry including the scheme theory. The prerequisite for reading this book is, for example, the context of Hartshorne's book [**29**] and the first half of the Griffiths-Harris book [**27**] on complex geometry. I hope the reader will see the attraction of Arakelov geometry throughout this book.

Acknowledgments. I thank Professor Miyaoka, who encouraged me to write this book. I also thank the referee and Professor Kawaguchi, who read the draft of the book carefully and gave me valuable advice. Thanks also goes to Professor Iwanari who helped with the proofreading. I am grateful to all of them for their valuable contributions.

Atsushi Moriwaki
Kyoto 2014

CHAPTER 1

Preliminaries

In this chapter, we will give expositions of elementary results from several areas. These will be used throughout this book. You don't need to read them in turn. You may come back here if it is necessary.

1.1. Frequently used notation and conventions

1.1.1. The ring of rational integers, the field of rational numbers, the field of real numbers, and the field of complex numbers are denoted by $\mathbb{Z}$, $\mathbb{Q}$, $\mathbb{R}$ and $\mathbb{C}$, respectively. Let $\mathbb{R}_{\geq 0}$ (resp. $\mathbb{R}_{>0}$) denote the set of non-negative real numbers (resp. the set of positive real numbers). Similarly we use the symbols $\mathbb{Z}_{\geq 0}$, $\mathbb{Z}_{>0}$, $\mathbb{Q}_{\geq 0}$ and $\mathbb{Q}_{>0}$. Moreover, the absolute value of an element x of $\mathbb{Z}$, $\mathbb{Q}$, $\mathbb{R}$ and $\mathbb{C}$ is denoted by $|x|$.

1.1.2. In this book, a ring means a commutative ring with the unity 1. For rings A and B, a map $f : A \to B$ is called a ring homomorphism if $f(x + y) = f(x) + f(y)$ $(x, y \in A)$, $f(xy) = f(x)f(y)$ $(x, y \in A)$ and $f(1) = 1$.

1.1.3. Let A be a commutative ring with the unity. Let $\operatorname{Spec}(A)$ and $\operatorname{m-Spec}(A)$ denote the set of all prime ideals of A and the set of all maximal ideals of A, respectively.

1.1.4. Let X be either a scheme or a complex space. The structure sheaf of X is denoted by $\mathscr{O}_X$ and the maximal ideal of the local ring $\mathscr{O}_{X,x}$ at a point x of X is denoted by m_x. Moreover, let $\kappa(x)$ denote the residue field $\mathscr{O}_{X,x}/m_x$ at x. For invertible sheaves L and M on X, we often use the additive notation $L + M$ to denote the tensor product $L \otimes_{\mathscr{O}_X} M$.

1.2. Normed finite-dimensional vector space

Let $\mathbb{K}$ be either $\mathbb{R}$ or $\mathbb{C}$. Let V be a finite-dimensional vector space over $\mathbb{K}$. A map $\sigma : V \to \mathbb{R}_{\geq 0}$ is called a *seminorm* if (1) $\sigma(ax) = |a|\sigma(x)$ $(a \in \mathbb{K},\ x \in V)$ and (2) $\sigma(x + y) \leq \sigma(x) + \sigma(y)$ $(x, y \in V)$ hold. Moreover, a seminorm σ is called a *norm* if further (3) "$\sigma(x) = 0 \iff x = 0$" holds. For $x \in V$, the value $\sigma(x)$ is often denoted by either $|x|$ or $\|x\|$.

Let us fix a norm σ of V. Let $\phi : V' \to V$ be an injective homomorphism of finite-dimensional vector spaces over $\mathbb{K}$. If $\sigma' : V' \to \mathbb{R}$ is the map given by $\sigma'(x') = \sigma(\phi(x'))$, then σ' forms a norm on V'. This norm is called the *subnorm* of V' induced by the injective homomorphism $\phi : V' \to V$ and the norm σ. It is denoted by $\sigma_{V' \hookrightarrow V}$. Next, let $\psi : V \to V''$ be a surjective homomorphism of finite-dimensional vector spaces over $\mathbb{K}$. Let $\sigma' : V'' \to \mathbb{R}$ be the map given by

$$\sigma''(x'') = \inf \{\sigma(x) \mid x \in V,\ \psi(x) = x''\}.$$

Then σ'' is a norm on V''. This norm is called the *quotient norm* of V'' induced by the surjective homomorphism $\psi : V \to V''$ and the norm σ. It is denoted by $\sigma_{V \twoheadrightarrow V''}$.

Let h be a $\mathbb{K}$-inner product on V, that is, it is a usual inner product (resp. a hermitian inner product) if $\mathbb{K} = \mathbb{R}$ (resp. if $\mathbb{K} = \mathbb{C}$). In this book, a $\mathbb{K}$-inner product is often called a $\mathbb{K}$-*metric* or a *metric* for simplicity. If a norm σ is given by $\sigma(x) = \sqrt{h(x,x)}$, then σ is called an L^2-*norm*. As before, let $\phi : V' \to V$ be an injective homomorphism of finite-dimensional vector spaces over $\mathbb{K}$. If we set

$$h'(x', y') = h(\phi(x'), \phi(y')),$$

then h' is a $\mathbb{K}$-metric of V'. It is called the *submetric* of V' induced by the injective homomorphism $\phi : V' \to V$ and the metric h. It is denoted by $h_{V' \hookrightarrow V}$. Clearly if $\sigma(x) = \sqrt{h(x,x)}$, then $\sigma_{V' \hookrightarrow V}(x') = \sqrt{h'(x', x')}$. Thus, if σ is an L^2-norm, then $\sigma_{V' \hookrightarrow V}$ is also an L^2-norm. Next let $\psi : V \to V''$ be a surjective homomorphism of finite-dimensional vector spaces over $\mathbb{K}$. Let $\operatorname{Ker}(\psi)^{\perp}$ be the orthogonal complement of $\operatorname{Ker}(\psi)$ with respect to h. Then $\psi|_{\operatorname{Ker}(\psi)^{\perp}} : \operatorname{Ker}(\psi)^{\perp} \to V''$ is an isomorphism, and hence there is a unique $\mathbb{K}$-metric h'' of V'' such that

$$h(w, w') = h''(\psi(w), \psi(w')) \quad (w, w' \in \operatorname{Ker}(\psi)^{\perp}).$$

It is called the *quotient metric* of V'' induced by the surjective homomorphism $\psi : V \to V''$ and the metric h. It is denoted by $h_{V \twoheadrightarrow V''}$. If $\sigma(x) = \sqrt{h(x,x)}$, then it is easy to see that $\sigma_{V \twoheadrightarrow V''}(x'') = \sqrt{h''(x'', x'')}$. Thus, if σ is an L^2-norm, then $\sigma_{V \twoheadrightarrow V''}$ is also an L^2-norm.

In the case where $\dim_{\mathbb{K}} V = 1$, every norm is an L^2-norm, that is, for a norm σ of V, if we set

$$h(a\omega, b\omega) = a\bar{b}\sigma(\omega)^2 \quad (a, b \in \mathbb{K},\ \omega \in V),$$

then h is a $\mathbb{K}$-inner product and $\sqrt{h(\cdot,\cdot)} = \sigma(\cdot)$.

Let us begin with the following lemma.

LEMMA 1.1. *Let $0 \to V' \xrightarrow{\phi} V \xrightarrow{\psi} V'' \to 0$ be an exact sequence of finite-dimensional vector spaces over $\mathbb{K}$. Fix a $\mathbb{K}$-metric h of V. Let h' be the submetric of V' induced by $V' \to V$ and h, and let h'' be the quotient metric of V'' induced by $V \to V''$ and h. Let $\{e'_1, \ldots, e'_m\}$, $\{e_1, \ldots, e_n\}$ and $\{e''_1, \ldots, e''_{n-m}\}$ be bases of V', V and V'', respectively, such that*

$$\phi(e'_i) = e_i \quad (i = 1, \ldots, m) \quad \text{and} \quad \psi(e_{m+j}) = e''_j \quad (j = 1, \ldots, n-m).$$

Then

$$\det(h(e_i, e_j)) = \det(h'(e'_i, e'_j)) \det(h''(e''_i, e''_j)).$$

PROOF. Let $\operatorname{Ker}(\psi)^{\perp}$ be the orthogonal complement of $\operatorname{Ker}(\psi) = \phi(V')$. We decompose e_i as follows:

$$e_i = x_i + y_i \qquad (x_i \in \operatorname{Ker}(\psi), y_i \in \operatorname{Ker}(\psi)^{\perp}).$$

Then $x_1, \ldots, x_m, y_{m+1}, \ldots, y_n$ form a basis of V and

$$\begin{pmatrix} e_1 \\ \vdots \\ e_n \end{pmatrix} = \left(\begin{array}{c|c} I_m & 0 \\ \hline * & I_{n-m} \end{array} \right) \begin{pmatrix} x_1 \\ \vdots \\ x_m \\ y_{m+1} \\ \vdots \\ y_n \end{pmatrix},$$

where I_l is the $l \times l$ identity matrix. We denote the matrix on the right side of the above equation by A. Note that

$$(h(e_i, e_j)) = A \left(\begin{array}{c|c} h(x_i, x_j) & 0 \\ \hline 0 & h(y_i, y_j) \end{array} \right) {}^t\bar{A}$$

and $\det(A) = 1$. Thus

$$\det(h(e_i, e_j)) = \det(h(x_i, x_j)) \det(h(y_i, y_j)).$$

On the other hand, $h'(e'_i, e'_j) = h(x_i, x_j)$ $(i, j = 1, \ldots, m)$ and $h''(e''_i, e''_j) = h(y_{m+i}, y_{m+j})$ $(i, j = 1, \ldots, n-m)$, as required. □

Next we consider the following lemma.

LEMMA 1.2. (1) *Let $f : V \to W$ and $g : W \to U$ be surjective homomorphisms of finite-dimensional vector spaces over $\mathbb{K}$. For a norm σ on V, $(\sigma_{V \twoheadrightarrow W})_{W \twoheadrightarrow U} = \sigma_{V \twoheadrightarrow U}$ holds as norms on U.*

(2) *Let*

$$\begin{array}{ccc} W & \xrightarrow{\ f\ } & V \\ {\scriptstyle g'}\downarrow & & \downarrow{\scriptstyle g} \\ P & \xrightarrow{\ f'\ } & Q \end{array}$$

be a commutative diagram of finite-dimensional vector spaces over $\mathbb{K}$ such that f and f' are injective, and that g and g' are surjective. For a norm σ on V, the inequality

$$(\sigma_{W \hookrightarrow V})_{W \twoheadrightarrow P} \geq (\sigma_{V \twoheadrightarrow Q})_{P \hookrightarrow Q}$$

holds as norms on P. Moreover, if $\ker(g) \subseteq f(W)$, then the equality

$$(\sigma_{W \hookrightarrow V})_{W \twoheadrightarrow P} = (\sigma_{V \twoheadrightarrow Q})_{P \hookrightarrow Q}$$

holds.

PROOF. (1) Fix an element u of U. Note that

$$\sigma(v) \geq \sigma_{V \twoheadrightarrow W}(f(v)) \geq (\sigma_{V \twoheadrightarrow W})_{W \twoheadrightarrow U}(u)$$

for $v \in (g \circ f)^{-1}(u)$. Thus $\sigma_{V \twoheadrightarrow U}(u) \geq (\sigma_{V \twoheadrightarrow W})_{W \twoheadrightarrow U}(u)$.

On the other hand, we choose $v_0 \in (g \circ f)^{-1}(u)$ with $\sigma_{V \twoheadrightarrow U}(u) = \sigma(v_0)$. Then, since $f^{-1}(w) \subseteq (g \circ f)^{-1}(u)$ for any $w \in g^{-1}(u)$, we get $\sigma(v_0) \leq \sigma_{V \twoheadrightarrow W}(w)$. Therefore,

$$\sigma_{V \twoheadrightarrow U}(u) = \sigma(v_0) \leq (\sigma_{V \twoheadrightarrow W})_{W \twoheadrightarrow U}(u).$$

(2) As $f(\ker(g')) = f(W) \cap \ker(g)$, the following hold for $w \in W$:

$$\begin{cases} (\sigma_{W \hookrightarrow V})_{W \twoheadrightarrow P}(g'(w)) = \inf\{\sigma(x) \mid x \in f(w) + f(W) \cap \ker(g)\}, \\ (\sigma_{V \twoheadrightarrow Q})_{P \hookrightarrow Q}(g'(w)) = \inf\{\sigma(x) \mid x \in f(w) + \ker(g)\}. \end{cases}$$

Therefore, $(\sigma_{W \hookrightarrow V})_{W \twoheadrightarrow P}(g'(w)) \geq (\sigma_{V \twoheadrightarrow Q})_{P \hookrightarrow Q}(g'(w))$. Moreover, if

$$\ker(g) \subseteq f(W),$$

that is, $f(W) \cap \ker(g) = \ker(g)$, then $(\sigma_{W \hookrightarrow V})_{W \twoheadrightarrow P} = (\sigma_{V \twoheadrightarrow Q})_{P \hookrightarrow Q}$. □

COROLLARY 1.3. *Let V be a finite-dimensional vector space over $\mathbb{K}$, and let $T \subseteq U \subseteq W \subseteq V$ be vector subspaces of V. For a norm σ of V,*

$$(\sigma_{W\hookrightarrow V})_{W\twoheadrightarrow W/U} = ((\sigma_{V\twoheadrightarrow V/T})_{W/T\hookrightarrow V/T})_{W/T\twoheadrightarrow W/U}$$

holds on W/U.

PROOF. Let us consider the following commutative diagram:

$$\begin{array}{ccc} W & \hookrightarrow & V \\ \downarrow & & \downarrow \\ W/U & \hookrightarrow & V/U. \end{array}$$

Then, by (2) in Lemma 1.2, we have $(\sigma_{W\hookrightarrow V})_{W\twoheadrightarrow W/U} = (\sigma_{V\twoheadrightarrow V/U})_{W/U\hookrightarrow V/U}$. Moreover, considering the following commutative diagram:

$$\begin{array}{ccc} W/T & \hookrightarrow & V/T \\ \downarrow & & \downarrow \\ W/U & \hookrightarrow & V/U, \end{array}$$

if we set $\sigma' = \sigma_{V\twoheadrightarrow V/T}$, then $(\sigma'_{V/T\twoheadrightarrow V/U})_{W/U\hookrightarrow V/U} = (\sigma'_{W/T\hookrightarrow V/T})_{W/T\twoheadrightarrow W/U}$. Thus the corollary follows because $\sigma_{V\twoheadrightarrow V/U} = \sigma'_{V/T\twoheadrightarrow V/U}$ by (1) in Lemma 1.2. □

Let $\overline{V} = (V, \sigma_V)$ and $\overline{W} = (W, \sigma_W)$ be finite-dimensional vector spaces over $\mathbb{K}$ with norms σ_V and σ_W. For a homomorphism $\phi : V \to W$, the *norm of the homomorphism* $\sigma_{\mathrm{Hom}(\overline{V},\overline{W})}(\phi)$ of ϕ is given by

$$\sigma_{\mathrm{Hom}(\overline{V},\overline{W})}(\phi) = \sup\left\{ \frac{\sigma_W(\phi(v))}{\sigma_V(v)} \;\middle|\; v \in V \setminus \{0\} \right\}.$$

For simplicity, it is often denoted by $\sigma(\overline{V} \to \overline{W})$ or $\sigma(\phi)$. Clearly,

$$\sigma_{\mathrm{Hom}(\overline{V},\overline{W})}(\phi) = \sup\left\{ \sigma_W(\phi(v)) \mid v \in V \text{ and } \sigma_V(v) = 1 \right\}.$$

In the case where $W = \mathbb{K}$ and σ_W is the usual absolute value of $\mathbb{K}$, $V^\vee = \mathrm{Hom}_{\mathbb{K}}(V, \mathbb{K})$ possesses the natural norm induced by σ as above. This norm is called the *dual norm* and it is denoted by $\sigma^\vee$.

LEMMA 1.4. *Let V be a finite-dimensional vector space over $\mathbb{K}$ and let h be a $\mathbb{K}$-metric of V. Then we have the following.*

(1) *Let $e_1, \ldots, e_n$ be an orthonormal basis of V with respect to h, and let $e_1^\vee, \ldots, e_n^\vee$ be the dual basis of V. Let h' be the $\mathbb{K}$-metric of $V^\vee$ such that $e_1^\vee, \ldots, e_n^\vee$ is an orthonormal basis of $V^\vee$ with respect to h', that is,*

$$h'(x_1e_1^\vee + \cdots + x_ne_n^\vee, y_1e_1^\vee + \cdots + y_ne_n^\vee) = x_1\bar{y}_1 + \cdots + x_n\bar{y}_n.$$

Then h' does not depend on the choice of the orthonormal basis $e_1, \ldots, e_n$. Its $\mathbb{K}$-metric of $V^\vee$ is called the dual metric *of $V^\vee$ and it is denoted by $h^\vee$.*

(2)

$$h^\vee(\phi,\phi) = \sup\left\{ \frac{|\phi(x)|^2}{h(x,x)} \;\middle|\; x \in V \setminus \{0\} \right\}$$

holds for $\phi \in V^\vee$. *Thus, if* σ *is the norm given by* $\sigma(x) = \sqrt{h(x,x)}$, *then*

$$\sigma^\vee(\phi) = \sqrt{h^\vee(\phi,\phi)}.$$

(3) *Let* $x_1, \ldots, x_n$ *be a basis of* V *and let* $x_1^\vee, \ldots, x_n^\vee$ *be the dual basis of* V. *Then*

$$(h(x_i,x_j))\, {}^t(h^\vee(x_i^\vee, x_j^\vee)) = I_n.$$

PROOF. Before starting the proof of the lemma, let us consider the following claim.

CLAIM 1. *Let* $x_1, \ldots, x_n$ *and* $y_1, \ldots, y_n$ *be bases of* V, *and let* $x_1^\vee, \ldots, x_n^\vee$ *and* $y_1^\vee, \ldots, y_n^\vee$ *be the dual bases of* $x_1, \ldots, x_n$ *and* $y_1, \ldots, y_n$, *respectively. Let* $A = (a_{ij})$ *and* $B = (b_{ij})$ *be the matrices given by* $y_i = \sum_{j=1}^n a_{ij}x_j$ *and* $x_i^\vee = \sum_{j=1}^n b_{ij}y_j^\vee$. *Then* $B = {}^tA$.

PROOF. Let us calculate $x_i^\vee(y_j)$ in two ways. First,

$$x_i^\vee(y_j) = x_i^\vee\left(\sum_{k=1}^n a_{jk}x_k\right) = a_{ji}.$$

Moreover,

$$x_i^\vee(y_j) = \left(\sum_{k=1}^n b_{ik}y_k^\vee\right)(y_j) = b_{ij}.$$

Thus $b_{ij} = a_{ji}$, as required. □

(1) Let $e_1', \ldots, e_n'$ be another orthonormal basis of V. Let $A = (a_{ij})$ be the matrix given by $e_i' = \sum_{j=1}^n a_{ij}e_j$. Then A is an orthogonal matrix (resp. a unitary matrix) if $\mathbb{K} = \mathbb{R}$ (resp. if $\mathbb{K} = \mathbb{C}$). We set $B = {}^tA$ and $B = (b_{ij})$. By the previous claim, $e_i^\vee = \sum_{j=1}^n b_{ij}e_j'^\vee$. Therefore, $e_1'^\vee, \ldots, e_n'^\vee$ is an orthonormal basis with respect to h'.

(2) If $\phi = 0$, then the assertion is obvious, so that we assume $\phi \neq 0$. If we set $\phi = \sum \phi_i e_i^\vee$ and $x = \sum x_i e_i$, then

$$\begin{cases} \phi(x) = \sum \phi_i x_i, \\ h(x,x) = \sum |x_i|^2, \\ h^\vee(\varphi,\phi) = \sum |\phi_i|^2. \end{cases}$$

Note that $|\sum \phi_i x_i|^2 \leq \left(\sum |\phi_i|^2\right)\left(\sum |x_i|^2\right)$ by the Cauchy-Schwarz inequality, which implies

$$\frac{|\phi(x)|^2}{h(x,x)} \leq h^\vee(\phi,\phi).$$

On the other hand, if we set $x_i = \bar{\varphi}_i$ $(i = 1, \ldots, n)$, then the equality holds in the above inequality.

(3) Let $C = (c_{ij})$ be the matrix given by $x_i = \sum_{j=1}^n c_{ij}e_j$ and let $D = {}^tC = (d_{ij})$. Then, by the previous claim, $e_i^\vee = \sum_{j=1}^n d_{ij}x_j^\vee$. Note that

$$(h(x_i,x_j)) = C\,{}^t\bar{C} = {}^tD\bar{D}$$

and $I_n = (h^\vee(e_i^\vee, e_j^\vee)) = D(h^\vee(x_i^\vee, x_j^\vee))^t\bar{D}$. Thus we have

$$^t(h^\vee(x_i^\vee, x_j^\vee)) = \bar{D}^{-1}\,{}^tD^{-1}.$$

Therefore,

$$(h(x_i, x_j))\,{}^t\big(h^\vee(x_i^\vee, x_j^\vee)\big) = {}^tD\bar{D}\bar{D}^{-1}\,{}^tD^{-1} = I_n.$$ □

Finally, we consider an inequality on norms of homomorphisms.

LEMMA 1.5. *Let*

$$\begin{array}{ccc} W & \xrightarrow{f} & V \\ {\scriptstyle g'}\big\downarrow & & \big\downarrow{\scriptstyle g} \\ P & \xrightarrow{f'} & Q \end{array}$$

be a commutative diagram of finite-dimensional vector spaces over $\mathbb{K}$ *such that* g *and* g' *are surjective. Let* σ_W, σ_V, σ_P *and* σ_Q *be norms of* W, V, P *and* Q, *respectively, such that* $\sigma_P = (\sigma_W)_{W \twoheadrightarrow P}$ *and* $\sigma_Q = (\sigma_V)_{V \twoheadrightarrow Q}$. *Then an inequality*

$$\sigma_{\mathrm{Hom}(\overline{P},\overline{Q})}(f') \le \sigma_{\mathrm{Hom}(\overline{W},\overline{V})}(f)$$

holds, where $\overline{W} = (W, \sigma_W)$, $\overline{V} = (V, \sigma_V)$, $\overline{P} = (P, \sigma_P)$ *and* $\overline{Q} = (Q, \sigma_Q)$.

PROOF. Let us fix $y \in P \setminus \{0\}$. Let $x \in W$ with $g'(x) = y$. Then, as $g(f(x)) = f'(y)$, we have

$$\sigma_{\mathrm{Hom}(\overline{W},\overline{V})}(f) \ge \frac{\sigma_V(f(x))}{\sigma_W(x)} \ge \frac{\sigma_Q(g(f(x)))}{\sigma_W(x)} = \frac{\sigma_Q(f'(y))}{\sigma_W(x)},$$

so that

$$\sigma_{\mathrm{Hom}(\overline{W},\overline{V})}(f) \ge \frac{\sigma_Q(f'(y))}{\sigma_P(y)},$$

and hence $\sigma_{\mathrm{Hom}(\overline{W},\overline{V})}(f) \ge \sigma_{\mathrm{Hom}(\overline{P},\overline{Q})}(f')$. □

1.3. Lemmas on the length of modules

In this section, we will discuss several lemmas on the length of modules over a commutative ring. First, let us recall basic definitions. Let A be a commutative ring and let M be an A-module. An element a of A is called an *M-regular element* if the a-multiplication map $M \xrightarrow{a\cdot} M$ $(m \mapsto am)$ is injective. An A-regular element is simply called a *regular element of* A. The support of M is given by $\mathrm{Supp}(M) = \{P \in \mathrm{Spec}(A) \mid M_P \neq \{0\}\}$. We say that M has *composition series* if there is a sequence

$$\{0\} = M_0 \subsetneq M_1 \subsetneq \cdots \subsetneq M_{l-1} \subsetneq M_l = M_l$$

of A-submodules of M such that M_i/M_{i-1} is simple, that is, an A-submodule of M_i/M_{i-1} is either M_i/M_{i-1} or $\{0\}$ for each $i = 1, \ldots, l$. Note that the length l of composite series does not depend on the choice of series, so that it is called the *length of* M and is denoted by $\mathrm{length}_A(M)$. If M has no composite series, then the length of M is defined to be ∞. By definition, $M = \{0\}$ if and only if $\mathrm{length}_A(M) = 0$. Moreover, if A is a field, then $\mathrm{length}_A(M)$ is nothing more than the dimension of M over A. Let us begin with the following lemma.

LEMMA 1.6. *Let* A *be a noetherian ring and let* M *be a finitely generated* A-*module. Then we have the following.*

(1) *If* $\operatorname{Supp}(M) \subseteq \text{m-Spec}(A)$, *then* $\operatorname{Supp}(M)$ *is finite, where* m-Spec(A) *denotes the set of all maximal ideals of* A.
(2) M *has composite series if and only if* $\operatorname{Supp}(M) \subseteq \text{m-Spec}(A)$.
(3) *If* $\operatorname{Supp}(M) \subseteq \text{m-Spec}(A)$, *then the natural homomorphism*

$$\alpha_M : M \to \bigoplus_{m \in \operatorname{Supp}(M)} M_m$$

is an isomorphism. Moreover, in this case,

$$\operatorname{length}_A(M) = \sum_{m \in \operatorname{Supp}(M)} \operatorname{length}_{A_m}(M_m).$$

PROOF. First let us see the following claim.

CLAIM 2. *Let* A *and* M *be as in the lemma. Then we have the following.*
(a) *If* $\operatorname{length}_A(M) = 1$, *then there is a maximal ideal* m *of* A *with* $M \simeq A/m$. *In particular,* $\operatorname{Supp}(M) = \{m\}$.
(b) *If there is a maximal ideal* m *with* $\operatorname{Supp}(M) = \{m\}$, *then the length of* A *as an* A*-module is finite. Moreover,* M *is an* A_m*-module in the natural way.*

PROOF. (a) As M is finitely generated, there are $x_1, \ldots, x_l \in M$ with $M = Ax_1 + \cdots + Ax_l$. Choose a non-zero x_i. Note that $\{0\} \subsetneq Ax_i \subseteq M$, and hence $M = Ax_i$. If I is the kernel of a homomorphism $A \to M$ given by $a \mapsto ax_i$, then $A/I \simeq M$. Let m be a maximal ideal of A with $I \subseteq m$. As $m/I \subsetneq A/I$, we have $m = I$, which shows (a).

(b) Here we use the noetherian condition. As $\operatorname{Supp}(M) = V(\operatorname{ann}(M))$, there is a positive integer N with $m^N M = 0$. Consider the sequence

$$0 = m^N M \subseteq m^{N-1} M \subseteq \cdots \subseteq mM \subseteq M.$$

Since $m^i M / m^{i+1} M$ $(i = 0, \ldots, N-1)$ is a finitely generated A/m-module and A/m is a field, the length of $m^i M/m^{i+1} M$ as an A-module is finite, and hence $\operatorname{length}_A(M) < \infty$.

Next let us check that M is naturally an A_m-module. For $a \in A$, let $\phi_a : M \to M$ be a homomorphism of A-modules given by $\phi_a(x) = ax$. In order to see the last assertion of (b), it is sufficient to see that ϕ_a is an isomorphism if $a \in A \setminus m$. As $\phi_{ab} = \phi_a \circ \phi_b = \phi_b \circ \phi_a$ for $a, b \in A \setminus m$, if ϕ_{ab} is an isomorphism, then ϕ_a and ϕ_b are also isomorphisms. On the the hand, there is $b \in A \setminus m$ with $ab \equiv 1 \mod m$ because A/m is a field. Thus we may assume that $a \equiv 1 \mod m$, that is, we can put $a = 1 - t$ $(t \in m)$. Therefore, if we set $b = 1 + t + \cdots + t^{N-1}$, then $ab = 1 - t^N$. Thus, as $m^N \cdot M = \{0\}$, we have $\phi_{ab} = \mathrm{id}_M$, which shows that ϕ_a is an isomorphism. $\square$

(1) Since $\operatorname{Supp}(M) \subseteq \text{m-Spec}(A)$, every prime ideal in $\operatorname{Supp}(M)$ is minimal in $\operatorname{Supp}(M)$. Note that the minimal elements in $\operatorname{Supp}(M)$ are finite. Thus $\operatorname{Supp}(M)$ is finite.

(2) If the length of M is finite, then, by (a) in Claim 2, $\operatorname{Supp}(M) \subseteq \text{m-Spec}(A)$. Conversely we assume that $\operatorname{Supp}(M) \subseteq \text{m-Spec}(A)$. We put

$$\operatorname{Supp}(M) = \{m_1, \ldots, m_l\} \qquad (m_1, \ldots, m_l \in \text{m-Spec}(A)).$$

We prove $\operatorname{length}_A(M) < \infty$ by induction on l. If $l = 1$, then the assertion holds by (b) in Claim 2. We assume that $l > 1$. Note that $V(\operatorname{ann}(M)) = \operatorname{Supp}(M)$. Thus there is a positive integer N with $(m_1 \cdots m_l)^N M = \{0\}$. As $(m_1 \cdots m_{l-1})^N m_l^N M = \{0\}$ and $m_l^N (M/m_l^N M) = \{0\}$, we have

$$\operatorname{Supp}(m_l^N M) \subseteq \{m_1, \ldots, m_{l-1}\} \quad \text{and} \quad \operatorname{Supp}(M/m_l^N M) \subseteq \{m_l\}.$$

Therefore, by the hypothesis of induction,

$$\operatorname{length}_A(m_l^N M) < \infty \quad \text{and} \quad \operatorname{length}_A(M/m_l^N M) < \infty,$$

which implies that the length of M is finite.

(3) For an exact sequence $0 \to N' \to N \to N'' \to 0$ of finitely generated A-modules, we have the following commutative diagram:

$$\begin{array}{ccccccccc}
0 & \longrightarrow & N' & \longrightarrow & N & \longrightarrow & N'' & \longrightarrow & 0 \\
 & & \big\downarrow \alpha_{N'} & & \big\downarrow \alpha_N & & \big\downarrow \alpha_{N''} & & \\
0 & \longrightarrow & \bigoplus_{m \in \operatorname{Supp}(N')} N'_m & \longrightarrow & \bigoplus_{m \in \operatorname{Supp}(N)} N_m & \longrightarrow & \bigoplus_{m \in \operatorname{Supp}(N'')} N''_m & \longrightarrow & 0,
\end{array}$$

where the bottom sequence is also exact. Therefore, if $\alpha_{N'}$ and $\alpha_{N''}$ are bijective, then so is α_N. On the other hand, as the length of M is finite by (2), there is a sequence $0 = M_0 \subset M_1 \subset \cdots \subset M_l = M$ of A-submodules of M such that $M_i/M_i \simeq A/m_i$ and $m_i \in \text{m-Spec}(A)$ for all i. Moreover, if $N = A/m_i$, then α_N is an isomorphism. Therefore, the first assertion of (3) follows by induction on l.

By virtue of the isomorphism $\alpha_M : M \to \bigoplus_{m \in \operatorname{Supp}(M)} M_m$, we have

$$\operatorname{length}_A(M) = \sum_{m \in \operatorname{Supp}(M)} \operatorname{length}_A(M_m).$$

On the other hand, by (b) in Claim 2, $\operatorname{length}_A(M_m) = \operatorname{length}_{A_m}(M_m)$. Thus the last assertion follows. □

According to Appendix A of Fulton's book [**22**], we introduce the invariant $e_A(\phi, M)$ and consider its several properties. Let A be a noetherian ring and let M be a finitely generated A-module. Let $\phi : M \to M$ be a homomorphism of A-modules. We say that $e_A(\phi, M)$ is definable if $\operatorname{length}_A(\operatorname{Ker}(\phi)) < \infty$ and $\operatorname{length}_A(\operatorname{Coker}(\phi)) < \infty$, and we set

$$e_A(\phi, M) := \operatorname{length}_A(\operatorname{Coker}(\phi)) - \operatorname{length}_A(\operatorname{Ker}(\phi)).$$

In the case where ϕ is a multiplication map by an element a of A ($m \mapsto a \cdot m$), $e_A(\phi, M)$ is denoted by $e_A(a, M)$. The invariant $e_A(\phi, M)$ has the following properties.

LEMMA 1.7. (1) *If M is of finite length, that is,* $\operatorname{length}_A(M) < \infty$, *then* $e_A(\phi, M) = 0$.

(2) *If A is a noetherian ring of dimension ≤ 1 and ϕ is either injective or surjective, then $e_A(\phi, M)$ is definable.*

(3) *Let* $0 \to M' \to M \to M'' \to 0$ *be an exact sequence of finitely generated* A*-modules. Let* $\phi' : M' \to M'$, $\phi : M \to M$ *and* $\phi'' : M'' \to M''$ *be homomorphisms of* A*-modules such that the following diagram is commutative:*

$$\begin{array}{ccccccccc} 0 & \longrightarrow & M' & \longrightarrow & M & \longrightarrow & M'' & \longrightarrow & 0 \\ & & \big\downarrow{\scriptstyle \phi'} & & \big\downarrow{\scriptstyle \phi} & & \big\downarrow{\scriptstyle \phi''} & & \\ 0 & \longrightarrow & M' & \longrightarrow & M & \longrightarrow & M'' & \longrightarrow & 0 \end{array}$$

If two of $e_A(\phi', M')$, $e_A(\phi, M)$ *and* $e_A(\phi'', M'')$ *are definable, then the remaining is also definable and*

$$e_A(\phi, M) = e_A(\phi', M') + e_A(\phi'', M'')$$

holds.

(4) *Let* ϕ *and* ψ *be* A*-endomorphisms of* M*. If* $e_A(\phi, M)$ *and* $e_A(\psi, M)$ *are definable, then* $e_A(\phi\psi, M)$ *is also definable and*

$$e_A(\phi\psi, M) = e_A(\phi, M) + e_A(\psi, M)$$

holds.

(5) *Let* ϕ *and* ψ *be* A*-endomorphisms of* M *such that* ϕ *and* ψ *are injective and* $\phi\psi = \psi\phi$*. Let* $\bar{\phi} : \mathrm{Coker}(\psi) \to \mathrm{Coker}(\psi)$ *and* $\bar{\psi} : \mathrm{Coker}(\phi) \to \mathrm{Coker}(\phi)$ *be homomorphisms induced by* ϕ *and* ψ*. Then* $e_A(\bar{\phi}, \mathrm{Coker}(\psi))$ *is definable if and only if* $e_A(\bar{\psi}, \mathrm{Coker}(\phi))$ *is definable. Moreover, in this case, we have*

$$e_A(\bar{\phi}, \mathrm{Coker}(\psi)) = e_A(\bar{\psi}, \mathrm{Coker}(\phi)).$$

(6) *Let* A *be a noetherian* 1*-dimensional local ring and* $P_1, \dots, P_l$ *the minimal primes of* A*. If* $a \in A \setminus P_1 \cup \dots \cup P_l$*, then* $e_A(a, M)$ *is definable and*

$$\begin{aligned} e_A(a, M) &= \sum_{i=1}^{l} \mathrm{length}_{A_{P_i}}(M_{P_i}) \cdot e_A(a, A/P_i) \\ &= \sum_{i=1}^{l} \mathrm{length}_{A_{P_i}}(M_{P_i}) \cdot \mathrm{length}_A(A/(P_i + aA)). \end{aligned}$$

PROOF. (1) It follows from the exact sequence $0 \to \mathrm{Ker}(\phi) \to M \xrightarrow{\phi} M \to \mathrm{Coker}(\phi) \to 0$.

(2) Let P be a minimal prime ideal of A. As A_P is artinian, $\phi_P : M_P \to M_P$ is an isomorphism by (1). Therefore,

$$\mathrm{Supp}(\mathrm{Ker}(\phi)),\ \mathrm{Supp}(\mathrm{Coker}(\phi)) \subseteq \text{m-Spec}(A).$$

Thus $e_A(\phi, M)$ is definable.

(3) Applying the Snake lemma to the diagram in (3) of the lemma, we have the exact sequence

$$\begin{aligned} 0 \to \mathrm{Ker}(\phi') \to \mathrm{Ker}(\phi) \to \mathrm{Ker}(\phi'') \to \mathrm{Coker}(\phi') & \\ \to \mathrm{Coker}(\phi) \to \mathrm{Coker}(\phi'') \to 0&, \end{aligned}$$

which yields (3).

(4) This follows from the natural exact sequence

$$0 \to \mathrm{Ker}(\psi) \to \mathrm{Ker}(\phi\psi) \xrightarrow{\psi} \mathrm{Ker}(\phi) \to \mathrm{Coker}(\psi) \xrightarrow{\phi} \mathrm{Coker}(\phi\psi) \to \mathrm{Coker}(\phi) \to 0.$$

(5) Applying the Snake lemma to the following commutative diagram

$$\begin{array}{ccccccccc} 0 & \longrightarrow & M & \xrightarrow{\phi} & M & \longrightarrow & \mathrm{Coker}(\phi) & \longrightarrow & 0 \\ & & \downarrow{\scriptstyle\psi} & & \downarrow{\scriptstyle\psi} & & \downarrow{\scriptstyle\bar{\psi}} & & \\ 0 & \longrightarrow & M & \xrightarrow{\phi} & M & \longrightarrow & \mathrm{Coker}(\phi) & \longrightarrow & 0, \end{array}$$

we have an exact sequence

$$0 \to \mathrm{Ker}(\bar{\psi}) \to \mathrm{Coker}(\psi) \xrightarrow{\bar{\phi}} \mathrm{Coker}(\psi) \to \mathrm{Coker}(\bar{\psi}) \to 0.$$

Thus (5) follows.

(6) Note that the a-multiplication map $a\cdot : A/P_i \to A/P_i$ is injective. Thus, by (2), we have $e_A(a, A/P_i) = \mathrm{length}_A(A/(P_i + aA)) < \infty$. Hence it is enough to see the definability of $e_A(a, M)$ and the first equation. As each side of the equation is additive with respect to an exact sequence by using (3), we may assume that $M = A/P$ ($P \in \mathrm{Spec}(A)$). First we consider the case where P is maximal. Then $\mathrm{length}_{A_{P_i}}(M_{P_i}) = 0$ ($i = 1, \ldots, l$). On the other hand, by (1), $e_A(a, M) = 0$, as required. Next we consider the case where $P = P_j$. In this case,

$$\mathrm{length}_{A_{P_i}}(M_{P_i}) = \begin{cases} 1 & \text{if } i = j, \\ 0 & \text{if } i \neq j. \end{cases}$$

Thus the assertion holds. □

Finally, let us consider the definition of the order function and its basic properties. Let A be a 1-dimensional noetherian ring and x a regular element of A. Then the x-multiplication map $x\cdot : A \to A$ ($a \mapsto xa$) is injective, so that, by (2) in Lemma 1.7, $e_{A_m}(x, A_m)$ is definable for a maximal ideal m of A. We define $\mathrm{ord}_m(x)$ by

$$\mathrm{ord}_m(x) = e_{A_m}(x, A_m) = \mathrm{length}_{A_m}(A_m/xA_m).$$

By (4) in Lemma 1.7,

$$\mathrm{ord}_m(xy) = \mathrm{ord}_m(x) + \mathrm{ord}_m(y)$$

for regular elements x, y of A. Let K be the total quotient field (i.e. the localization with respect to the set of regular elements). For $z \in K^{\times}$, we put $z = x/y$ (x, y are regular elements of A). Then

$$\mathrm{ord}_m(z) = \mathrm{ord}_m(x) - \mathrm{ord}_m(y)$$

is well defined by the above formula. Moreover, we have

$$\mathrm{ord}_m(xy) = \mathrm{ord}_m(x) + \mathrm{ord}_m(y)$$

for all $x, y \in K^{\times}$.

1.4. Image of a homomorphism and its determinant

Let A be a noetherian ring and let M be a finitely generated flat A-module. As M is flat, we can find $f_1, \ldots, f_n \in A$ such that M_{f_i} is a free A_{f_i}-module for every i and $\bigcup_{i=1}^n \operatorname{Spec}(A_{f_i}) = \operatorname{Spec}(A)$. Let $\phi : M \to M$ be an endomorphism of M over A. Since ϕ gives rise to an endmorphism $\phi_i : M_{f_i} \to M_{f_i}$ and M_{f_i} is a free A_{f_i}-module, we have the determinant $\det(\phi_i) \in A_{f_i}$ of ϕ_i. Clearly, $\det(\phi_i)$ is equal to $\det(\phi_j)$ in $A_{f_i f_j}$. Thus there is a unique $d \in A$ such that d coincides with $\det(\phi_i)$ in A_{f_i} for every i. This d is denoted by $\det(\phi)$. The purpose of this section is to prove the following theorem.

THEOREM 1.8. *The definability of $e_A(\phi, M)$ is equivalent to the definability of $e_A(\det(\phi), A)$. Moreover, in this case, we have*

$$e_A(\phi, M) = e_A(\det(\phi), A).$$

Before starting the proof of Theorem 1.8, we need to prepare several facts. For a commutative ring A, the ring of all $n \times n$-matrices whose entries belong to A is denoted by $M_{n\times n}(A)$. Moreover, the subgroup consisting of invertible matrices in $M_{n\times n}(A)$ is denoted by $\operatorname{GL}(n, A)$. Let I_n be the identity matrix and let E_{ij} be the matrix such that the (i,j)-entry is 1 and other entries are 0. For $a \in A$, we set

$$T_{ij}(a) = \begin{cases} I_n + aE_{ij} & \text{if } i \neq j, \\ I_n & \text{if } i = j. \end{cases}$$

An upper triangle matrix whose diagonal entries are 1 is called a *unipotent upper triangle matrix*, and, similarly, a lower triangle matrix whose diagonal entries are 1 is called a *unipotent lower triangle matrix*. Note that

$$T_{ij}(a) = \begin{cases} \text{unipotent upper triangle matrix} & \text{if } i \leq j, \\ \text{unipotent lower triangle matrix} & \text{if } i \geq j. \end{cases}$$

If $B \in M_{n\times n}(A)$ and $i \neq j$, then $BT_{ij}(a)$ is a matrix obtained by adding the a-multiplication of the i-th column of B to the j-th column of B, and $T_{ij}(a)B$ is a matrix obtained by adding the a-multiplication of the j-th row of B to the i-th row of B. In particular, the $(1,1)$-entry of $BT_{1i}(-1)T_{i1}(1)$ is a_{1i}. Let us begin with the following lemma.

LEMMA 1.9. *We assume that A is an artinian ring. For $B \in \operatorname{GL}(n, A)$, there are a unipotent upper triangle matrix C and a unipotent lower triangle matrix D such that the $(1,1)$-entry of BCD is a unit element.*

PROOF. First we assume that A is local. Let $a_{11}, \ldots, a_{1n}$ be the first column of B. By considering the expansion of $\det(B)$ along the first column, if $a_{11}, \ldots, a_{1n}$ belong to the maximal ideal of A, then $\det(B)$ belongs to the maximal ideal. Thus there is i such that a_{1i} is a unit element. The $(1,1)$-th entry of $BT_{1i}(-1)T_{i1}(1)$ is a_{1i}, as required.

Next we consider a general case. Let $P_1, \ldots, P_r$ be all maximal ideals of A. Then, by Lemma 1.6, we have the natural isomorphism

$$\alpha : A \to \bigoplus_{l=1}^r A_{P_l}.$$

We denote the natural homomorphism $A \to A_{P_l}$ by α_l. Moreover, for $E = (e_{ij}) \in M_{n\times n}(A)$, $(\alpha_l(e_{ij})) \in M_{n\times n}(A_{P_l})$ is denoted by $\alpha_l(E)$. By the previous case, there are a unipotent upper triangle matrix C_l and a unipotent lower triangle matrix D_l in $\mathrm{GL}(n, A_{P_l})$ such that the $(1,1)$-entry of $\alpha_l(B)C_lD_l$ is a unit element of A_{P_l}. Here we choose $C, D \in M_{n\times n}(A)$ such that $\alpha_l(C) = C_l$ and $\alpha_l(D) = D_l$ for all $l = 1, \ldots, r$. Then C is a unipotent upper triangle matrix and D is a unipotent lower triangle matrix. Moreover, the $(1,1)$-entry of $\alpha_l(BCD) = \alpha_l(B)C_lD_l$ is a unit element of A_{P_l}, so that the $(1,1)$-entry of BCD is a unit. □

The following theorem is well known if A is a field. However, we need a number of arguments for a general artinian ring.

THEOREM 1.10. *Let A be an artinian ring. Then $\mathrm{GL}(n, A)$ is generated by unipotent upper triangle matrices, unipotent lower triangle matrices and diagonal matrices whose diagonal elements are units.*

PROOF. We prove this theorem by induction on n. If $n = 1$, then the assertion is obvious. We assume that $n > 1$. Let $B \in \mathrm{GL}(n, A)$. By the previous lemma, we may assume that the $(1,1)$-entry of B is a unit. By multiplying matrices of type $T_{ij}(a)$ to the left and right, we may further assume that B has a form of the type

$$B = \begin{pmatrix} u & 0 \\ 0 & B' \end{pmatrix},$$

where $B' \in \mathrm{GL}(n-1, A)$ and $u \in A^{\times}$. Thus, by induction of hypothesis, we have our assertion. □

THE PROOF OF THEOREM 1.8. Let us start the proof of Theorem 1.8. The first half of the theorem is a consequence of the following three facts, which can be shown by using (2) in Lemma 1.6.

(a) $e_A(\phi, M)$ is definable $\iff$ there is a finite set S consisting of maximal ideals of A such that $\phi_P : M_P \to M_P$ is an isomorphism for every $P \in \mathrm{Spec}(A) \setminus S$.

(b) $e_A(\det(\phi), A)$ is definable $\iff$ there is a finite set S' consisting of maximal ideals of A such that $\det(\phi)_P \in A_P^{\times}$ holds for every $P \in \mathrm{Spec}(A) \setminus S'$.

(c) $\phi_P : M_P \to M_P$ is an isomorphism for $P \in \mathrm{Spec}(A) \iff \det(\phi)_P \in A_P^{\times}$.

From now on, we assume that $e_A(\phi, M)$ is definable and we consider the equation in the theorem. First we suppose that A is a local ring. Let m be the maximal ideal of A. Clearly we may assume that $\det(\phi)$ is not a unit. Then $\dim A \leq 1$ because $A/\det(\phi)A$ has the finite length. If $\dim A = 0$, then our assertion is obvious by (1) of Lemma 1.7, so that we may assume that $\dim A = 1$.

We consider the case where $m \in \mathrm{Ass}(A)$. We put

$$I = \bigcup_{n>0} \mathrm{Ann}(m^n) = \bigcup_{n>0} \{a \in A \mid m^n \cdot a = 0\}.$$

I is an ideal of A and $I \neq \{0\}$ because $m \in \mathrm{Ass}(A)$. Moreover, as

$$\mathrm{Ann}(m) \subseteq \mathrm{Ann}(m^2) \subseteq \cdots,$$

there is $N > 0$ with $I = \mathrm{Ann}(m^N)$. We set $\bar{A} = A/I$ and $\bar{\phi} = \phi \otimes \mathrm{id}_{\bar{A}}$. Then we have the following two commutative diagrams:

$$\begin{array}{ccccccccc} 0 & \longrightarrow & IM & \longrightarrow & M & \longrightarrow & M\otimes_A \bar{A} & \longrightarrow & 0 \\ & & \downarrow{\scriptstyle \phi} & & \downarrow{\scriptstyle \phi} & & \downarrow{\scriptstyle \bar{\phi}} & & \\ 0 & \longrightarrow & IM & \longrightarrow & M & \longrightarrow & M\otimes_A \bar{A} & \longrightarrow & 0 \end{array} \qquad \begin{array}{ccccccccc} 0 & \longrightarrow & I & \longrightarrow & A & \longrightarrow & \bar{A} & \longrightarrow & 0 \\ & & \downarrow{\scriptstyle \det(\phi)\cdot} & & \downarrow{\scriptstyle \det(\phi)\cdot} & & \downarrow{\scriptstyle \det(\bar{\phi})\cdot} & & \\ 0 & \longrightarrow & I & \longrightarrow & A & \longrightarrow & \bar{A} & \longrightarrow & 0 \end{array}$$

Note that I and IM are A/m^N-modules because $m^N \cdot I = \{0\}$. Therefore, by (1) in Lemma 1.7, $e_A(\phi, IM) = e_A(\det(\phi), I) = 0$. Thus, by (3) in Lemma 1.7,

$$e_A(\phi, M) = e_{\bar{A}}(\bar{\phi}, M \otimes_A \bar{A}) \quad \text{and} \quad e_A(\det(\phi), A) = e_{\bar{A}}(\det(\bar{\phi}), \bar{A})$$

hold. Let $\bar{m}$ be the image of m in $\bar{A}$ Then $\bigcup_{n>0} \operatorname{Ann}(\bar{m}^n) = \{0\}$. Indeed,

$$\begin{aligned} \bar{a} \in \operatorname{Ann}(\bar{m}^n) &\Longrightarrow m^{N+n} \cdot a \subseteq m^N \cdot I = \{0\} \qquad (\because m^n \cdot a \subseteq I) \\ &\Longrightarrow a \in \operatorname{Ann}(m^{N+n}) = I \\ &\Longrightarrow \bar{a} = 0 \text{ in } \bar{A}. \end{aligned}$$

In particular, $\bar{m} \notin \operatorname{Ass}(\bar{A})$. Hence we can reduce the assertion of the theorem to the case where $m \notin \operatorname{Ass}(A)$.

Next we consider the case where $m \notin \operatorname{Ass}(A)$. We set $\operatorname{Ass}(A) = \{P_1, \ldots, P_l\}$. By our assumption, $P_1, \ldots, P_l$ are minimal primes of A. Let S be the set of all regular elements of A. Then $S \cap m \neq \emptyset$ because $A \setminus S = P_1 \cup \cdots \cup P_l$. Therefore, the total quotient ring A_S of A is artinian. Since $e(\phi, A)$ is definable, we have

$$\operatorname{Supp}(\operatorname{Ker}(\phi)),\ \operatorname{Supp}(\operatorname{Coker}(\phi)) \subseteq \{m\},$$

that is, $\phi_S : M_S \to M_S$ is an isomorphism.

Here let us construct a homomorphism

$$\gamma : \mathrm{GL}(n, A_S) \to \mathbb{Z}.$$

For $B \in \mathrm{GL}(n, A_S)$, we choose $s \in S$ with $sB \in M_{n\times n}(A)$. We would like to define $\gamma(B)$ to be

$$\gamma(B) = e_A(sB, A^n) - e_A(\det(sB), A).$$

We need to see that the above quantity does not depend on the choice of s. For this purpose, it is sufficient to show that the quantity by s coincides with the quantity by ss' for any $s' \in S$. By using (4) in Lemma 1.7,

$$\begin{aligned} e_A(ss'B, A^n) &= e_A(s'I_n, A^n) + e(sB, A^n) = ne_A(s', A) + e_A(sB, A^n) \\ e_A(\det(ss'B), A) &= e_A(s'^n \det(sB), A) = ne_A(s', A) + e_A(\det(sB), A). \end{aligned}$$

Therefore,

$$e_A(ss'B, A^n) - e_A(\det(ss'B), A) = e_A(sB, A^n) - e_A(\det(sB), A).$$

Next let us see that γ is a homomorphism. Let $B, B' \in \mathrm{GL}(n, A_S)$ and $s, s' \in S$ with $sB, s'B' \in M_{n\times n}(A)$. Then, by (4) in Lemma 1.7,

$$e_A((sB)(s'B'), A^n) = e_A(sB, A^n) + e_A(s'B', A^n)$$

and

$$\begin{aligned} e_A(\det((sB)(s'B')), A) &= e_A(\det(sB)\det(s'B'), A) \\ &= e_A(\det(sB), A) + e_A(\det(s'B'), A), \end{aligned}$$

which show that γ is a homomorphism.

As A is a local ring, M is a free A-module. Thus ϕ can be expressed by a matrix whose entries belong to A. Moreover, $\phi_S : M_S \to M_S$ is an isomorphism. Thus, in order to show the equality, it is enough to show $\gamma(B) = 0$ for all $B \in \mathrm{GL}(n, A_S)$. As γ is a homomorphism, by Theorem 1.10, we may assume that B is either an upper triangle matrix whose diagonal entries are units, or a lower triangle matrix whose diagonal entries are units. Let B be an upper triangle matrix whose diagonal entries are units. Let $s_1, \ldots, s_n$ be the diagonal entries of B in this order. Replacing

B with sB ($s \in S$), we may assume that every entry of B belongs to A. Here we would like to show

$$e_A(B, A^n) = \sum_{i=1}^{n} e_A(s_i, A).$$

We prove this by induction on n. Let $i : A \to A^n$ and $p : A^n \to A^{n-1}$ be homomorphisms given by $i(x) = {}^t(x, 0, \ldots, 0)$ and $p({}^t(x_1, \ldots, x_n)) = {}^t(x_2, \ldots, x_n)$. Let B' be an $(n-1) \times (n-1)$-matrix obtained by extracting the first row and the first column from B. Then the following diagram is commutative:

$$\begin{array}{ccccccccc} 0 & \longrightarrow & A & \xrightarrow{i} & A^n & \xrightarrow{p} & A^{n-1} & \longrightarrow & 0 \\ & & \downarrow{\scriptstyle s_1} & & \downarrow{\scriptstyle B} & & \downarrow{\scriptstyle B'} & & \\ 0 & \longrightarrow & A & \xrightarrow{i} & A^n & \xrightarrow{p} & A^{n-1} & \longrightarrow & 0. \end{array}$$

By (3) in Lemma 1.7 and the hypothesis of induction, we have our desired equation as above. On the other hand, since $\det(B) = s_1 \cdots s_n$, we have

$$e_A(\det(B), A) = \sum_{i=1}^{n} e_A(s_i, A).$$

Therefore, $\gamma(B) = 0$. The same argument works in the case where B is a lower triangle matrix whose diagonal entries are units.

Finally, we consider a general case (i.e. A is not necessarily a local ring). As $e_A(\phi, M)$ is definable, ϕ_m is an isomorphism except finitely many maximal ideals m. Thus, by (3) in Lemma 1.6,

$$e_A(\phi, M) = \sum_{m \in \text{m-Spec}(A)} e_{A_m}(\phi_m, M_m)$$

$$e_A(\det(\phi), A) = \sum_{m \in \text{m-Spec}(A)} e_{A_m}(\det(\phi_m), A_m)$$

hold. Therefore, the general case follows from the case where A is a local ring. □

Corollary 1.11. *Let (A, m) be a noetherian local domain of dimension ≤ 1 and let K be the quotient field. Let M be a finitely generated A-module and let $\phi : M \to M$ be a homomorphism of A-modules. We denote the homomorphism $\phi \otimes_A \mathrm{id}_K : M \otimes_A K \to M \otimes_A K$ by ϕ_K. If $\det(\phi_K) \neq 0$, then $e_A(\phi, M) = \mathrm{ord}_m(\det(\phi_K))$.*

Proof. We set $d = \dim_K(M \otimes_A K)$. There are $e_1, \ldots, e_d \in M$ such that $e_1 \otimes 1, \ldots, e_d \otimes 1$ forms a basis of $M \otimes_A K$. Let $F = Ae_1 + \cdots + Ae_d$. We can see that F is a free A-module because $e_1 \otimes 1, \ldots, e_d \otimes 1$ forms a basis. Moreover, M/F is a torsion module because $F \otimes_A K = M \otimes_A K$. Therefore, there is $a \in A \setminus \{0\}$ with $aM \subseteq F$. In particular, $a\phi(F) \subseteq aM \subseteq F$. As M/F has the finite length, by (1) and (3) in Lemma 1.7, $e_A(a\phi, M) = e_A(a\phi, F)$. Moreover, by (1), (3) and (4) in Lemma 1.7,

$$e_A(a\phi, M) = e_A(a, M) + e_A(\phi, M) = e_A(a, F) + e_A(\phi, M) = d\,\mathrm{ord}_m(a) + e_A(\phi, M).$$

On the other hand, by Theorem 1.8,

$$\begin{aligned} e_A(a\phi, F) = e_A(\det(a\phi), A) = \mathrm{ord}_m(\det(a\phi)) &= \mathrm{ord}_m(\det(a\phi_K)) \\ &= d\,\mathrm{ord}_m(a) + \mathrm{ord}_m(\det(\phi_K)). \end{aligned}$$

Hence the corollary follows. □

1.5. Norm of flat and finite homomorphisms

Let $A \to B$ be a ring homomorphism of noetherian rings such that B is flat and finitely generated as an A-module. For an element b of B, the b-multiplication map $b\cdot : B \to B$ is an A-homomorphism. The determinant $\det(b\cdot)$ as defined in Section 1.4 is called the *norm of b with respect $A \to B$* and it is denoted by $N_{B/A}(b)$. For $b, b' \in B$, it is easy to see that $N_{B/A}(bb') = N_{B/A}(b)N_{B/A}(b')$. Thus, if A and B are integral domains, and K and L are the quotient fields of A and B, respectively, then $N_{L/K}(x) := N_{B/A}(b)/N_{B/A}(b')$ for $x = b/b' \in L$ is well defined. This is nothing more than the norm treated in the field theory. Here let us show the basic lemma on norm.

LEMMA 1.12. *Let (A, m) be a 1-dimensional noetherian local ring and B an A-algebra such that B is finitely generated as an A-module.*

(1) *If B is flat over A, then*

$$\sum_{n \in \text{m-Spec}(B)} \operatorname{ord}_n(b)[B/n : A/m] = \operatorname{ord}_m(N_{B/A}(b))$$

for a regular element b of B.

(2) *We assume that A and B are integral domains and that A is a subring of B. Let K and L be the quotient fields of A and B, respectively. Then*

$$\sum_{n \in \text{m-Spec}(B)} \operatorname{ord}_n(b)[B/n : A/m] = \operatorname{ord}_m(N_{L/K}(b))$$

for $b \in L^{\times}$.

PROOF. (1) By Theorem 1.8, $\operatorname{length}_A(B/bB) = \operatorname{ord}_m(N_{B/A}(b))$. On the other hand, as $B/bB = \bigoplus_{n \in \text{m-Spec}(B)} B_n/bB_n$,

$$\operatorname{length}_A(B/bB) = \sum_{n \in \text{m-Spec}(B)} \operatorname{length}_A(B_n/bB_n) = \sum_{n \in \text{m-Spec}(B)} \operatorname{ord}_n(b)[B/n : A/m].$$

Thus (1) follows.

(2) We may assume that $b \in B$. By Corollary 1.11,

$$\operatorname{length}_A(B/bB) = \operatorname{ord}_m(N_{L/K}(b)).$$

In the same way as in (1),

$$\operatorname{length}_A(B/bB) = \sum_{n \in \text{m-Spec}(B)} \operatorname{ord}_n(b)[B/n : A/m],$$

as required. □

Moreover, in the case where A is a field K, the following two lemmas hold.

LEMMA 1.13. *For $b \in B$,*

$$N_{B/K}(b) = \prod_{n \in \text{m-Spec}(B)} N_{(B/n)/K}(b \bmod n)^{\operatorname{length}_{B_n}(B_n)}.$$

PROOF. We consider the following commutative diagram by Lemma 1.6:

$$\begin{array}{ccc} B & \longrightarrow & \bigoplus_{n\in\text{m-Spec}(B)} B_n \\ \big\downarrow{\scriptstyle b\cdot} & & \big\downarrow{\scriptstyle \oplus b\cdot} \\ B & \longrightarrow & \bigoplus_{n\in\text{m-Spec}(B)} B_n, \end{array}$$

which implies

$$N_{B/K}(b) = \prod_{n\in\text{m-Spec}(B)} N_{B_n/K}(b).$$

Here we claim the following.

CLAIM 3. *Let J_1 and J_2 be ideals of B_n such that $J_1 \subseteq J_2$, $\operatorname{Supp}(B_n/J_1) \subseteq \{nB_n\}$ and $J_2/J_1 \simeq B/n$ (as a B-module). Then*

$$N_{(B_n/J_1)/K}(b \bmod J_1) = N_{(B_n/J_2)/K}(b \bmod J_2) \cdot N_{(B/n)/K}(b \bmod n).$$

PROOF. This claim follows from the following commutative diagram:

$$\begin{array}{ccccccccc} 0 & \longrightarrow & J_2/J_1 & \longrightarrow & B_n/J_1 & \longrightarrow & B_n/J_2 & \longrightarrow & 0 \\ & & \big\downarrow{\scriptstyle b\cdot} & & \big\downarrow{\scriptstyle b\cdot} & & \big\downarrow{\scriptstyle b\cdot} & & \\ 0 & \longrightarrow & J_2/J_1 & \longrightarrow & B_n/J_1 & \longrightarrow & B_n/J_2 & \longrightarrow & 0. \end{array}$$

□

By the above claim, we have

$$N_{B_n/K}(b) = N_{(B/n)/K}(b \bmod n)^{\operatorname{length}_{B_n}(B_n)}.$$

Hence we obtain the lemma. □

LEMMA 1.14. *Let K' be a field and let $\sigma : K \to K'$ be a field homomorphism. Then, for $b \in B$,*

$$N_{(B\otimes_K K')/K'}(b\otimes 1) = \sigma(N_{B/K}(b)),$$

where K' is viewed as a K-module via σ.

PROOF. Let $\omega_1,\dots,\omega_n$ be a basis of B as a K-module. We write $b\omega_i = \sum \beta_{ij}\omega_j$ ($\beta_{ij} \in K$). Then $N_{B/K}(b) = \det(\beta_{ij})$. Note that

$$(b\otimes 1)(\omega_i \otimes 1) = b\omega_i \otimes 1 = \sum \beta_{ij}\omega_j \otimes 1 = \sum \omega_j \otimes \sigma(\beta_{ij}) = \sum \sigma(\beta_{ij})(\omega_j\otimes 1).$$

Therefore,

$$N_{(B\otimes_K K')/K'}(b\otimes 1) = \det(\sigma(\beta_{ij})) = \sigma(\det(\beta_{ij})) = \sigma(N_B(b)).$$ □

Let $\pi : Y \to X$ be a flat and finite morphism of noetherian schemes. Let $\{U_\lambda\}$ be an affine covering of X. We can take a ring homomorphism $A_\lambda \to B_\lambda$ of noetherian rings such that $U_\lambda = \operatorname{Spec}(A_\lambda)$, $\pi^{-1}(U_\lambda) = \operatorname{Spec}(B_\lambda)$, $\pi|_{\pi^{-1}(U_\lambda)}$ is induced by the ring homomorphism $A_\lambda \to B_\lambda$ and that B_λ is flat and finitely generated as an A_λ-module. For $\phi \in H^0(Y, \mathscr{O}_Y)$, $N_{B_\lambda/A_\lambda}(\phi)$ is an element of $H^0(U_\lambda, \mathscr{O}_X)$. Moreover, $N_{B_\lambda/A_\lambda}(\phi) = N_{B_{\lambda'}/A_{\lambda'}}(\phi)$ holds on $U_\lambda \cap U_{\lambda'}$. Thus there is a unique $N \in H^0(X, \mathscr{O}_X)$ such that N coincides with $N_{B_\lambda/A_\lambda}(\phi)$ on U_λ. This N is denoted by $N_{Y/X}(\phi)$ and it is called the *norm of ϕ with respect to $Y \to X$*.

Let L be an invertible sheaf on Y. We would like to construct an invertible sheaf $N_{Y/X}(L)$ on X. Let us begin with the following lemma.

LEMMA 1.15. *Let $\pi : Y \to X$ be a finite morphism of noetherian schemes ($\pi : Y \to X$ is not necessarily flat) and let L be an invertible sheaf on Y. Then, for $x \in X$, there are an neighborhood U of x and $l \in L|_{\pi^{-1}(U)}$ such that $L|_{\pi^{-1}(U)} = \mathscr{O}_{\pi^{-1}(U)}l$.*

PROOF. This is a local problem with respect to X. Thus we may assume that there is a ring homomorphism $f : A \to B$ of noetherian rings such that $X = \operatorname{Spec}(A)$, $Y = \operatorname{Spec}(B)$ and $\pi : Y \to X$ is induced by f. Moreover, we may assume that A is a local ring. Since π is finite, B is finitely generated as an A-module.

First we consider the case where A is a field. Note that B is an artinian ring. Let $P_1, \ldots, P_r$ be all maximal ideals of B. By Lemma 1.6, $B = \bigoplus_{i=1}^r B_{P_i}$ and $L = \bigoplus_{i=1}^r L_{P_i}$. As L_{P_i} is a free B_{P_i}-module of rank 1, there is $\omega_i \in L_{P_i}$ with $L_{P_i} = B_{P_i}\omega_i$. We set $\omega = (\omega_1, \ldots, \omega_n)$ in $L = \bigoplus_{i=1}^r L_{P_i}$. Then $L = B\omega$.

Next we discuss a general case. Let m be the maximal ideal of A. Then B/mB is a finitely generated as an A/m-module. By the previous case, there is $l \in L$ with $L/mL = (B/mB) \cdot \bar{l}$, where $\bar{l}$ is the class of l in L/mL. Thus $L = mL + Bl$. Since L is finitely generated as an A-module, by using Nakayama's lemma, we have $L = Bl$. □

Let us go back to the flat and finite morphism $\pi : Y \to X$ of noetherian schemes. Let L be an invertible sheaf on Y. Then, by the previous lemma, there is an affine covering $\{X_i\}$ of X such that $L|_{\pi^{-1}(X_i)} = \mathscr{O}_{\pi^{-1}(X_i)}\omega_i$ for some $\omega_i \in L|_{\pi^{-1}(X_i)}$. Here we write $X_{ij} = X_i \cap X_i$, $Y_i = \pi^{-1}(X_i)$ and $Y_{ij} = \pi^{-1}(X_{ij})$. Moreover, we set $\omega_j = g_{ij}\omega_i$ on Y_{ij}. Then the collection $\{N_{Y_{ij}/X_{ij}}(g_{ij})\}$ yields an invertible sheaf on X. This is denoted by $N_{Y/X}(L)$ or $N_\pi(L)$. Let s be a global section of L and we put $s = s_i\omega_i$. As $s_i = g_{ij}s_j$, we have

$$N_{Y_{ij}/X_{ij}}(s_i) = N_{Y_{ij}/X_{ij}}(g_{ij})N_{Y_{ij}/X_{ij}}(s_j).$$

Therefore, $\{N_{Y_{ij}/X_{ij}}(s_i)\}$ gives rise to a global section of $N_{X/Y}(L)$. This is denoted by $N_{X/Y}(s)$ or $N_\pi(s)$. In the case where X and Y are integral schemes and s is a rational section, we can construct a rational section $N_{X/Y}(s)$ of $N_{X/Y}(L)$ in the same way as above.

1.6. Principal divisor and Weil's reciprocity law

In this section, we discuss Weil's reciprocity law. Let K be a field and let A be an essentially finitely generated K-algebra, that is, A is a localization of a finitely generated K-algebra. For a maximal ideal m and an element a of A, we define $N_m(a)$ to be

$$N_m(a) = N_{(A/m)/K}(a \bmod m) = \det_K(a\cdot : A/m \to A/m),$$

where $\det_K$ is the determinant of a linear endomorphism over K (see Section 1.5 for notation of norm). More generally, let I be a proper ideal of A with $\operatorname{Supp}(A/I) \subseteq \operatorname{m-Spec}(A)$. Then we define $N_I(a)$ to be

$$N_I(a) = N_{(A/I)/K}(a \bmod I) = \det_K(a\cdot : A/I \to A/I).$$

By Lemma 1.13,

$$N_I(a) = \prod_{m \in \operatorname{m-Spec}(A)} N_m(a)^{\operatorname{length}_{A_m}(A_m/I_m)}.$$

First let us consider the following lemma.

LEMMA 1.16. *Let K be a field, and let A and B be essentially finitely generated K-algebras. Let $f : A \to B$ be a homomorphism as a K-algebra. We assume that B is finitely generated as an A-module. Then we have the following.*

(1) *Let n be the maximal ideal and $m = f^{-1}(n)$. Then, for $a \in A$,*

$$N_n(f(a)) = N_m(a)^{[B/n:A/m]}.$$

(2) *We assume that A is a local ring and m is the maximal ideal of A. Then, for $b \in B$,*

$$\prod_{n \in \text{m-Spec}(B)} N_n(b)^{\text{length}_{B_n}(B_n/mB_n)} = N_m(\det_{A/m}(b\cdot : B/mB \to B/mB)).$$

Moreover, if B is flat over A, then

$$\det_{A/m}(b\cdot : B/mB \to B/mB)$$

is the class of $N_{B/A}(b)$ in A/m.

PROOF. Let $K \subseteq L$ be a finite extension of fields. Let V be a finite-dimensional vector space over L and let $\phi : V \to V$ be an endomorphism over L. We denote the determinant of ϕ as an endomorphism over L (resp. K) by $\det_L(\phi)$ (resp. $\det_K(\phi)$). Then we would like to show the following equation:

$$N_{L/K}(\det_L(\phi)) = \det_K(\phi) \tag{1.1}$$

For this purpose, let us see the following claim.

CLAIM 4. (a) *Let $\psi : V \to V$ be another linear map over L. If (1.1) holds for ϕ and ψ, then (1.1) also holds for $\phi \circ \psi$.*

(b) *Let $0 \to V' \to V \to V'' \to 0$ be an exact sequence of finite-dimensional vector spaces over L. Moreover, let $\phi' : V' \to V'$, $\phi : V \to V$ and $\phi'' : V'' \to V''$ be endomorphisms over L such that the following diagram is commutative:*

$$\begin{array}{ccccccccc} 0 & \longrightarrow & V' & \longrightarrow & V & \longrightarrow & V' & \longrightarrow & 0 \\ & & {\scriptstyle \phi'}\downarrow & & {\scriptstyle \phi}\downarrow & & {\scriptstyle \phi''}\downarrow & & \\ 0 & \longrightarrow & V' & \longrightarrow & V & \longrightarrow & V'' & \longrightarrow & 0. \end{array}$$

If (1.1) holds for ϕ' and ϕ'', then (1.1) also holds for ϕ.

PROOF. (a) is obvious. (b) is also easily checked because

$$\det_L(\phi) = \det_L(\phi') \det_L(\phi'') \quad \text{and} \quad \det_K(\phi) = \det_K(\phi') \det_K(\phi''). \qquad \square$$

Let us start the proof of (1.1). If ϕ is not an isomorphism, then $\det_L(\phi) = \det_K(\phi) = 0$. Thus the assertion is trivial, so that we may assume that ϕ is an isomorphism. Let us fix a basis $x_1, \ldots, x_n$ of V over L. For a matrix $A = (a_{ij}) \in \text{GL}(n, L)$, we define by $\phi_A : V \to V$ by $\phi_A(x_i) = \sum_{j=1}^n a_{ij} x_j$. By Theorem 1.10, $\text{GL}(n, L)$ is generated by regular upper triangle matrices and regular lower triangle matrices. Thus, by using (a) of the above claim, it is sufficient to show (1.1) for ϕ_A in cases where A is either a regular upper triangle matrix or a regular lower triangle matrix. We prove this by induction on n. If $n = 1$, then the assertion is trivial. First we assume that A is a regular upper triangle matrix. We set

$$A = \left(\begin{array}{c|c} a & * \\ \hline 0 & A' \end{array} \right),$$

where $a \in L^{\times}$ and A' is an $(n-1) \times (n-1)$ regular upper triangle matrix. We set $V' = Lx_1$ and $V'' = Lx_2 + \cdots + Lx_n$. Let $\iota : V' \to V$ be the homomorphism given by $\iota(x_1) = x_1$ and $p : V \to V''$ the homomorphism given by $p(x_1) = 0$ and $p(x_i) = x_i$ $(i = 2, \ldots, n)$. Then $0 \to V' \xrightarrow{\iota} V \xrightarrow{p} V'' \to 0$ is an exact sequence and

$$\begin{array}{ccccccccc} 0 & \longrightarrow & V' & \xrightarrow{\iota} & V & \xrightarrow{p} & V' & \longrightarrow & 0 \\ & & {\scriptstyle \phi_a}\downarrow & & {\scriptstyle \phi_A}\downarrow & & {\scriptstyle \phi_{A'}}\downarrow & & \\ 0 & \longrightarrow & V' & \xrightarrow{\iota} & V & \xrightarrow{p} & V'' & \longrightarrow & 0 \end{array}$$

is a commutative diagram. Using the hypothesis of induction and (b) in the above claim, we can see (1.1) for ϕ_A. In the same way, we can check (1.1) in the case where A is a regular lower triangle matrix.

(1) Let $K \subseteq E \subseteq L$ be a sequence of finite extensions of fields. (1) means that $N_{L/K}(x) = N_{E/K}(x)^{[L:E]}$ for $x \in E$. By using (1.1), it can be checked in the following way:

$$N_{L/K}(x) = N_{E/K}(N_{L/E}(x)) = N_{E/K}(x^{[L:E]}) = N_{E/K}(x)^{[L:E]}.$$

(2) Let $b\cdot : B/mB \to B/mB$ be the b-multiplication map. Then, by (1.1),

$$N_m(\det_{A/m}(b\cdot : B/mB \to B/mB)) = \det_K(b\cdot : B/mB \to B/mB).$$

On the other hand, by Lemma 1.13,

$$\det_K(b\cdot : B/mB \to B/mB) = \prod_{n \in \text{m-Spec}(B)} N_n(b)^{\text{length}_{B_n}(B_n/mB_n)}.$$

Therefore, the first half of (2) follows. The later half of (2) is obvious because B is free as an A-module. □

Let us turn to the main topics of this section. Let X be an integral scheme essentially of finite type over a field K. For a closed point P of X and a rational function ϕ on X with $\phi \in \mathscr{O}_{X,P}$, we define $N_P(\phi)$ to be

$$N_P(\phi) := N_{\kappa(P)/K}(\phi|_P).$$

where $\phi|_P$ is the class of ϕ in the residue field $\kappa(P)$ at P and $N_{\kappa(P)/K}$ is the norm map with respect to $K \to \kappa(P)$. A 0-cycle on X means an element of the free abelian group generated by closed points of X. For a 0-cycle $D = \sum n_i P_i$ and a rational function ϕ with $\phi \in \mathscr{O}_{X,P_i}^{\times}$ for all i, we define $N_D(\phi)$ to be

$$N_D(\phi) = \prod_i N_{P_i}(\phi)^{n_i}.$$

It is easy to see that

$$\begin{cases} N_D(\phi\phi') = N_D(\phi)N_D(\phi'), \\ N_{D+D'}(\phi) = N_D(\phi)N_{D'}(\phi). \end{cases}$$

From now on, we assume that X is 1-dimensional. For a non-zero rational function ϕ on X, the *principal divisor* (ϕ) for ϕ is a 0-cycle defined by

$$(\phi) = \sum_{P \in X} \operatorname{ord}_P(\phi) P.$$

Moreover, for a 0-cycle $D = \sum n_i P_i$ on X, we define the degree of D by

$$\deg(D) = \sum n_i [\kappa(P_i) : K].$$

Let Y be another 1-dimensional integral scheme essentially of finite type over K, and let $\pi : X \to Y$ be a finite morphism. The *push-forward* $\pi_*(D)$ of a 0-cycle $D = \sum n_i P_i$ by π is defined by

$$\pi_*(D) = \sum_i n_i [\kappa(P_i) : \kappa(\pi(P_i))] \pi(P_i).$$

Finally, assuming that Y is normal, the *pull-back* $\pi^*(E)$ of a 0-cycle E on Y is defined as follows: First we consider a case where E consists of a closed point Q. Note that $\mathscr{O}_{Y,Q}$ is a discrete valuation ring. Let t_Q be a local parameter of $\mathscr{O}_{Y,Q}$. Then $\pi^*(Q)$ is defined by

$$\pi^*(Q) = \sum_{P \in \pi^{-1}(Q)} \operatorname{ord}_P(\pi^*(t_Q)) P.$$

For a general 0-cycle $E = \sum n_i Q_i$,

$$\pi^*(E) = \sum n_i \pi^*(Q_i).$$

It is easy to see that $(\pi^* \psi) = \pi^*((\psi))$ for a non-zero rational function ψ. Here let us observe the following lemma.

LEMMA 1.17. *Let ϕ be a non-zero rational function on X, ψ a non-zero rational function on Y, D a 0-cycle on X and E a 0-cycle on Y. Moreover, let L and M be the function fields of X and Y, respectively. Then the following hold:*

(1) $N_D(\pi^* \psi) = N_{\pi_*(D)}(\psi)$.
(2) $\pi_*(\phi) = (N_{M/L}(\phi))$. *In particular, if $\pi : X \to Y$ is a birational morphism, then* $\pi_*(\pi^* \psi) = (\psi)$.
(3) *If Y is normal, then* $N_{\pi^*(E)}(\phi) = N_E(N_{M/L}(\phi))$.
(4) $\deg(D) = \deg(\pi_*(D))$.

PROOF. (1) Each side is linear with respect to D. Thus we may assume that $D = P$ for some closed point P of X. Hence (1) follows from (1) in Lemma 1.16.

(2) This is a local problem with respect to Y. Thus it follows from (2) in Lemma 1.12.

(3) Each side is linear with respect to E. Thus we may assume that $E = Q$ for some closed point Q of Y. Moreover, it is a local problem with respect to Y. Hence, by (2) in Lemma 1.16, we obtain (3).

(4) Since each side is linear with respect to D, we may assume that $D = P$ for some closed point P of X. Then

$$\deg(\pi_*(P)) = [\kappa(P) : \kappa(\pi(P))][\kappa(\pi(P)) : K] = [\kappa(P) : K] = \deg(P). \qquad \square$$

Let X be a 1-dimensional projective integral scheme over a field K. We denote the function field of X by $\operatorname{Rat}(X)$. Then we have the following two propositions.

PROPOSITION 1.18. $\deg((\phi)) = 0$ *for* $\phi \in \operatorname{Rat}(X)^{\times}$.

PROPOSITION 1.19 (Weil's reciprocity law). *Let* $\phi, \psi \in \operatorname{Rat}(X)^{\times}$ *such that* (ϕ) *and* (ψ) *have no common component. Then*

$$N_{(\phi)}(\psi) = N_{(\psi)}(\phi)$$

holds.

PROOF. We will prove two propositions simultaneously. First we assume that X is normal. Let K' be the field consisting of algebraic elements of $\operatorname{Rat}(X)$ over K. Since X is normal, we have $K' \subseteq H^0(X, \mathscr{O}_X)$. We consider two cases.

(a) The case where ϕ is algebraic over K: Then $\phi, \phi^{-1} \in K'$. Thus $(\phi) = 0$.
(b) The case where ϕ is transcendental over K: Let $K(t)$ be the quotient field of the ring of polynomials $K[t]$ and let $K(t) \to \operatorname{Rat}(X)$ be the field homomorphism over K such that t maps to ϕ. This yields the scheme morphism $X \to \mathbb{P}^1_K$, which is denoted by f. Note that f is finite and flat. Moreover, let $(x : y)$ be a homogeneous coordinate of $\mathbb{P}^1_K$ with $t = x/y$. Then $f^*(x/y) = \phi$.

Here we give the proofs of Proposition 1.18 and Proposition 1.19 for the cases (a) and (b) under the assumption that X is normal.

Proposition 1.18: The assertion is obvious in the case (a), so that we consider the case (b). In this case, since $\phi = f^*(x/y)$, by (2) and (4) in Lemma 1.17,

$$\begin{aligned}\deg((\phi)) &= \deg(f_*(\phi)) = \deg((N_{\operatorname{Rat}(X)/K(t)}(f^*(x/y)))) \\ &= \deg(((x/y)^{\deg(f)})) = \deg(f)\deg((x/y)) \\ &= \deg(f)\deg((0:1) - (1:0)) = 0.\end{aligned}$$

Proposition 1.19: First we consider the case (a). As $(\phi) = 0$, we have $N_{(\phi)}(\psi) = 1$. If we set $(\psi) = \sum n_i P_i$, then

$$\begin{aligned}N_{(\psi)}(\phi) &= \prod N_{P_i}(\phi)^{n_i} = \prod N_{K'/K}(\phi)^{n_i[\kappa(P_i):K']} \\ &= N_{K'/K}(\phi)^{\sum n_i[\kappa(P_i):K']} \\ &= N_{K'/K}(\phi)^{\deg((\psi))/[K':K]} = N_{K'/K}(\phi)^0 = 1,\end{aligned}$$

as desired.

Next we consider the case (b). Then, by (1), (2) and (3) in Lemma 1.17,

$$\begin{aligned}N_{(\psi)}(\phi) &= N_{(\psi)}(f^*(t)) = N_{(N_{\operatorname{Rat}(X)/K(t)}(\psi))}(t), \\ N_{(\phi)}(\psi) &= N_{(f^*(t))}(\psi) = N_{(t)}(N_{\operatorname{Rat}(X)/K(t)}(\psi)).\end{aligned}$$

Thus it is sufficient to show the following: Let h be a rational function on $\mathbb{P}^1_K$ such that h has no pole and no zero at 0 and ∞. Then $N_{(h)}(t) = N_{(t)}(h)$.

Since h has no pole and no zero at 0 and ∞, there are monic polynomials $\alpha(t), \beta(t) \in K[t]$ and $a \in K \setminus \{0\}$ such that

$$h = a\frac{\alpha}{\beta} \quad (\alpha(0) \neq 0, \beta(0) \neq 0, \deg(\alpha) = \deg(\beta)).$$

Let $\alpha = \alpha_1 \cdots \alpha_d$ and $\beta = \beta_1 \cdots \beta_l$ be decompositions into irreducible monic polynomials. Let P_i, Q_j be the points on $\mathbb{P}^1_K$ corresponding to α_i, β_j, respectively. Then

$$N_{(t)}(h) = N_0(h)N_\infty(h)^{-1} = a\frac{\alpha(0)}{\beta(0)} \cdot \frac{1}{a} = \frac{\alpha(0)}{\beta(0)}.$$

Thus, by using Lemma 1.20 below,

$$\begin{aligned} N_{(h)}(t) &= \frac{N_{P_1}(t) \cdots N_{P_d}(t)}{N_{Q_1}(t) \cdots N_{Q_l}(t)} \\ &= \frac{((-1)^{\deg(\alpha_1)}\alpha_1(0)) \cdots ((-1)^{\deg(\alpha_d)}\alpha_d(0))}{((-1)^{\deg(\beta_1)}\beta_1(0)) \cdots ((-1)^{\deg(\beta_l)}\beta_l(0))} \\ &= \frac{\alpha_1(0) \cdots \alpha_d(0)}{\beta_1(0) \cdots \beta_l(0)} = \frac{\alpha(0)}{\beta(0)}, \end{aligned}$$

as required.

Next we discuss a general case, that is, X is is not necessarily normal. Let $\pi : \widetilde{X} \to X$ be the normalization of X.

First we consider Proposition 1.18. Note that $\deg((\pi^*\phi)) = 0$. On the the hand, by the last half of (2) in Lemma 1.17, $\pi_*(\pi^*\phi) = (\phi)$. Thus, by (4) in Lemma 1.17,

$$\deg((\phi)) = \deg(\pi_*(\pi^*\phi)) = \deg((\pi^*\phi)) = 0.$$

Finally we consider Proposition 1.19. Note that $N_{(\pi^*\psi)}(\pi^*\phi) = N_{(\pi^*\phi)}(\pi^*\psi)$. On the other hand, by using the last half of (1) and (2) in Lemma 1.17,

$$N_{(\pi^*\psi)}(\pi^*\phi) = N_{\pi_*(\pi^*\psi)}(\phi) = N_{(\psi)}(\phi).$$

In the same way, $N_{(\pi^*\phi)}(\pi^*\psi) = N_{(\phi)}(\psi)$. Therefore, we obtain $N_{(\psi)}(\phi) = N_{(\phi)}(\psi)$. □

LEMMA 1.20. *Let K be a field and let $f = a_nT^n + \cdots + a_1T + a_0 \in K[T]$ be an irreducible polynomial of degree n. We set $F = K[T]/(f)$. Then*

$$N_{F/K}(T \mod (f)) = (-1)^n(a_0/a_n).$$

PROOF. We set $t = T \mod (f)$. Then $\{1, t, \ldots, t^{n-1}\}$ is a basis of F over K. The matrix representation of the t-multiplication map with respect to the basis is

$$\begin{pmatrix} 0 & 0 & \cdots & 0 & -a_0/a_n \\ 1 & 0 & \cdots & 0 & -a_1/a_n \\ 0 & 1 & \cdots & 0 & -a_2/a_n \\ \vdots & \vdots & \ddots & \vdots & \vdots \\ 0 & 0 & \cdots & 1 & -a_{n-1}/a_n \end{pmatrix}.$$

Thus its determinant is $(-1)^n(a_0/a_n)$. □

1.7. Existence of rational sections not passing through given points

Let X be a noetherian integral scheme and let η be the generic point of X. Let L be an invertible sheaf on X and let s be a non-zero rational section of L, that is, s is a non-zero element of L_η. For a point $x \in X$, let ω_x be a local basis of L at x. Then there is a non-zero rational function f_x with $s = f_x\omega_x$. We say that s is a unit at x if f_x is a unit. Note that this definition does not depend on the choice of ω_x. We denote the set of all codimension 1 points of X (i.e. points with

$\dim \mathscr{O}_{X,x} = 1$) by $X^{(1)}$. For $x \in X^{(1)}$, $\operatorname{ord}_x(f_x)$ defined in Section 1.3 does not depend on the choice of the local basis ω_x at x, so that it is denoted by $\operatorname{ord}_x(L,s)$. Moreover, we define $\operatorname{Supp}(L,s)$, $\operatorname{div}(L,s)$ and $\operatorname{Supp}(\operatorname{div}(L,s))$ as follows:

$$\begin{cases} \operatorname{Supp}(L,s) = \{x \in X \mid s \text{ is not a unit at } x\}, \\ \operatorname{div}(L,s) = \sum_{x\in X^{(1)}} \operatorname{ord}_x(s)\overline{\{x\}} \qquad \text{(formal sum)}, \\ \operatorname{Supp}(\operatorname{div}(L,s)) = \bigcup_{\operatorname{ord}_x(s)\neq 0,\ x\in X^{(1)}} \overline{\{x\}}. \end{cases}$$

For simplicity, we often denote $\operatorname{ord}_x(L,s)$ and $\operatorname{div}(L,s)$ by $\operatorname{ord}_x(s)$ and $\operatorname{div}(s)$, respectively. Moreover, if f is a non-zero rational function, $\operatorname{div}(\mathscr{O}_X, f)$ is called a *principal divisor* as in the previous section, and it is often denoted by (f). Then we have the following.

PROPOSITION 1.21. *$\operatorname{Supp}(L,s)$ is a closed set and* $\operatorname{Supp}(\operatorname{div}(L,s)) \subseteq \operatorname{Supp}(L,s)$.

PROOF. The first half of the proposition is a local problem, so that we may assume there is a noetherian integral domain A such that $X = \operatorname{Spec}(A)$, $L = A$, s is a non-zero element of the quotient field of A. We set $I = \{a \in A \mid as \in A\}$ and $J = Is$. Then I and J are ideals of A. Moreover, for $P \in \operatorname{Spec}(A)$, $I_P = \{a \in A_P \mid as \in A_P\}$ (for example, see [**3**, Corollary 3.15]). Therefore, we can see that

$$s \in A_P^{\times} \iff I_P = A_P \text{ and } J_P = A_P.$$

Thus $\operatorname{Supp}(L,s) = \operatorname{Supp}(A/I) \cup \operatorname{Supp}(A/J)$, which is a closed set.

Finally we show the last half of the proposition. We assume that $\operatorname{ord}_x(L,s) \neq 0$ for $x \in X^{(1)}$. Then s is not a unit at x, so that $x \in \operatorname{Supp}(L,s)$. This shows $\operatorname{Supp}(\operatorname{div}(L,s)) \subseteq \operatorname{Supp}(L,s)$. □

Next we consider the following proposition.

PROPOSITION 1.22. *Let R be a noetherian integral domain, and let $X \to \operatorname{Spec}(R)$ be a quasi-projective morphism of noetherian integral scheme. Let L be an invertible sheaf on X and let $x_1, \ldots, x_r$ be points of X. Then there is a non-zero rational section s of L with*

$$\operatorname{Supp}(L,s) \cap \{x_1, \ldots, x_s\} = \emptyset.$$

In particular, $\operatorname{Supp}(\operatorname{div}(L,s))$ *does not pass through $x_1, \ldots, x_r$.*

PROOF. Let us begin with the following lemma.

LEMMA 1.23. *Let R be a noetherian ring and let X be a quasi-projective scheme over* $\operatorname{Spec}(R)$. *Let A be a very ample invertible sheaf with respect to $X \to \operatorname{Spec}(R)$, and let Q be an invertible sheaf on X. Moreover, let $x_1, \ldots, x_r$ be points of X. Then there is n_0 such that, for each $n \geq n_0$, there is a non-zero section $s \in H^0(X, Q \otimes A^{\otimes n})$ with $s(x_i) \neq 0$ for all $i = 1, \ldots, r$, where $t(x)$ means $t \otimes 1$ in $(Q \otimes A^{\otimes n}) \otimes \kappa(x)$ for $x \in X$ and $t \in H^0(X, Q \otimes A^{\otimes n})$.*

PROOF. First we consider the case where $x_1, \ldots, x_r$ are closed points. By our assumption, there are a projective scheme X' over $\operatorname{Spec}(R)$, an open embedding $i : X \hookrightarrow X'$ over $\operatorname{Spec}(R)$ and a very ample invertible sheaf A' on X' with respect to $X' \to \operatorname{Spec}(R)$ such that $A = i^*(A')$. Moreover, there is a coherent sheaf Q' on

X' with $Q = i^*(Q')$. Let $m_1, \ldots, m_r$ be the maximal ideal sheaf corresponding to $x_1, \ldots, x_r$ and $I = m_1 \cdots m_r$. Then there is n_0 such that

$$H^1(X', IQ' \otimes A'^{\otimes n}) = 0$$

for all $n \geq n_0$. Therefore, as

$$\begin{aligned} &H^0(X', Q' \otimes A'^{\otimes n}/IQ' \otimes A'^{\otimes n}) \\ &\qquad = H^0(X', Q' \otimes A'^{\otimes n} \otimes (\mathscr{O}_{X'}/I)) = \bigoplus_{i=1}^{r} Q \otimes A^{\otimes n} \otimes \kappa(x_i), \end{aligned}$$

we obtain that

$$H^0(X', Q' \otimes A'^{\otimes n}) \to \bigoplus_{i=1}^{r} Q \otimes A^{\otimes n} \otimes \kappa(x_i)$$

is surjective. Hence there is $s' \in H^0(X', Q' \otimes A'^{\otimes n})$ with $s'(x_i) \neq 0$ for all i. Thus, if we set $s = s'|_X$, then we have the assertion.

For the general case, we choose a closed point y_i with $y_i \in \overline{\{x_i\}}$. Then there is $s \in H^0(X, Q \otimes A^{\otimes n})$ with $s(y_i) \neq 0$ $(i = 1, \ldots, r)$. Note that

$$\operatorname{Supp}(\operatorname{Coker}(\mathscr{O}_X \xrightarrow{s} Q \otimes A^{\otimes n}))$$

is a closed set. Thus $s(x_i) \neq 0$ for all i. □

Let us go back to the proof of the proposition. Let A be a very ample invertible sheaf on X with respect to $X \to \operatorname{Spec}(R)$. Then, by Lemma 1.23, there are a positive integer n, $t_1 \in H^0(Y, A^{\otimes n} \otimes L) \setminus \{0\}$ and $t_2 \in H^0(Y, A^{\otimes n}) \setminus \{0\}$ such that $t_1(x_i) \neq 0$ and $t_2(x_i) \neq 0$ for all $i = 1, \ldots, r$. In particular,

$$\begin{cases} \operatorname{Supp}(A^{\otimes n} \otimes L, t_1) \cap \{x_1, \ldots, x_r\} = \emptyset, \\ \operatorname{Supp}(A^{\otimes n}, t_2) \cap \{x_1, \ldots, x_r\} = \emptyset. \end{cases}$$

Set $s = t_1 \otimes t_2^{-1}$. Then s is a non-zero rational section of L and

$$\begin{aligned} \operatorname{Supp}(L, s) &\subseteq \operatorname{Supp}(A^{\otimes n} \otimes L, t_1) \cup \operatorname{Supp}(A^{\otimes -n}, t_2^{-1}) \\ &= \operatorname{Supp}(A^{\otimes n} \otimes L, t_1) \cup \operatorname{Supp}(A^{\otimes n}, t_2), \end{aligned}$$

as required. □

1.8. Graded modules and ample invertible sheaves

Let R be a noetherian ring and let X be a projective and flat scheme over R. Let us begin with the following proposition.

PROPOSITION 1.24. *Let A be an ample invertible sheaf on X and let F be a coherent sheaf on X. Then there are positive integers d and n_0 such that, for all $n \geq n_0$, the natural homomorphism*

$$H^0(X, A^{\otimes d}) \otimes_R H^0(X, F \otimes_{\mathscr{O}_X} A^{\otimes n}) \to H^0(X, F \otimes_{\mathscr{O}_X} A^{\otimes n+d})$$

is surjective.

PROOF. Since A is ample, there is a positive integer d such that

$$H^0(X, A^{\otimes d}) \otimes_R \mathscr{O}_X \to A^{\otimes d}$$

is surjective. Let K be the kernel of the above homomorphism. As $A^{\otimes d}$ is invertible,

$$\operatorname{Tor}_1^{\mathscr{O}_X}(A^{\otimes d}, F \otimes_{\mathscr{O}_X} A^{\otimes n}) = 0.$$

Thus the exact sequence

$$0 \to K \to H^0(X, A^{\otimes d}) \otimes_R \mathscr{O}_X \to A^{\otimes d} \to 0$$

yields the exact sequence

$$0 \to K \otimes_{\mathscr{O}_X} F \otimes_{\mathscr{O}_X} A^{\otimes n} \to H^0(X, A^{\otimes d}) \otimes_R F \otimes_{\mathscr{O}_X} A^{\otimes n} \to F \otimes_{\mathscr{O}_X} A^{\otimes n+d} \to 0$$

by applying $\otimes_{\mathscr{O}_X} F \otimes_{\mathscr{O}_X} A^{\otimes n}$. Since A is ample, there is a positive integer n_0 such that, for all $n \geq n_0$, $H^1(X, K \otimes_{\mathscr{O}_X} F \otimes_{\mathscr{O}_X} A^{\otimes n}) = \{0\}$. Therefore,

$$\begin{aligned} H^0(X, A^{\otimes d}) \otimes_R H^0(X, F \otimes_{\mathscr{O}_X} A^{\otimes n}) = H^0\left(X, H^0(X, A^{\otimes d}) \otimes_R F \otimes_{\mathscr{O}_X} A^{\otimes n}\right) \\ \longrightarrow H^0(X, F \otimes_{\mathscr{O}_X} A^{\otimes n+d}) \end{aligned}$$

is surjective. □

COROLLARY 1.25. *Let A and F be as in Proposition 1.24. For a positive integer l, we define the R-algebra B_l to be*

$$B_l := \bigoplus_{n \geq 0} H^0(X, A^{\otimes nl}).$$

Then we have the following.

(1) *$\bigoplus_{n\geq 0} H^0(X, F \otimes_{\mathscr{O}_X} A^{\otimes n})$ is finitely generated as a B_l-module.*
(2) *B_l is finitely generated as an R-algebra.*

PROOF. (1) First we consider the case where $l = 1$. By Proposition 1.24, there are positive integers d and n_0 such that, for all $n \geq n_0$, the natural homomorphism

$$H^0(X, A^{\otimes d}) \otimes_R H^0(X, F \otimes_{\mathscr{O}_X} A^{\otimes n}) \to H^0(X, F \otimes_{\mathscr{O}_X} A^{\otimes n+d})$$

is surjective. Thus, for $0 \leq r < d$ and $a \geq 0$,

$$H^0(X, A^{\otimes d})^{\otimes a} \otimes_R H^0(X, F \otimes_{\mathscr{O}_X} A^{\otimes(n_0+r)}) \to H^0(X, F \otimes_{\mathscr{O}_X} A^{\otimes ad+n_0+r})$$

is surjective. Therefore, $\bigoplus_{n\geq 0} H^0(X, F \otimes_{\mathscr{O}_X} A^{\otimes n})$ is generated by finitely generated R-modules

$$H^0(X, F), H^0(X, F \otimes_{\mathscr{O}_X} A), \ldots, H^0(X, F \otimes_{\mathscr{O}_X} A^{\otimes n_0+d-1})$$

as a B_1-module. This completes the proof in the case $l = 1$.

In the case where $l > 1$, we can decompose $\bigoplus_{n\geq 0} H^0(X, F \otimes_{\mathscr{O}_X} A^{\otimes n})$ into

$$\bigoplus_{n \geq 0} H^0(X, F \otimes_{\mathscr{O}_X} A^{\otimes n}) = \bigoplus_{i=0}^{l-1} \left(\bigoplus_{n \geq 0} H^0(X, (F \otimes_{\mathscr{O}_X} A^{\otimes i}) \otimes_{\mathscr{O}_X} (A^{\otimes l})^{\otimes n}) \right).$$

Thus, applying the previous result to the case where A and F are given by $A^{\otimes l}$ and $F \otimes_{\mathscr{O}_X} A^{\otimes i}$, respectively, we can see that $\bigoplus_{n\geq 0} H^0(X, (F \otimes_{\mathscr{O}_X} A^{\otimes i}) \otimes_{\mathscr{O}_X} (A^{\otimes l})^{\otimes n})$ is finitely generated as a B_l-module. Therefore, $\bigoplus_{n\geq 0} H^0(X, F \otimes_{\mathscr{O}_X} A^{\otimes n})$ is finitely generated as a B_l-module.

(2) It is sufficient to consider the case $l = 1$. In the same way as in (1), there are positive integers d and n_0 such that

$$H^0(X, A^{\otimes d})^{\otimes a} \otimes_R H^0(X, A^{n_0+r}) \to H^0(X, A^{\otimes ad+n_0+r})$$

is surjective for all a and r with $a \geq 0$ and $0 \leq r < d$. Therefore, B_1 is generated by finitely generated R-modules

$$H^0(X, \mathscr{O}_X), H^0(X, A), \ldots, H^0(X, A^{\otimes n_0+d-1})$$

as an R-algebra. □

1.9. Several results on separable extensions of the base field

First recall a theorem on associated primes and ring homomorphisms.

THEOREM 1.26. *Let $\phi : A \to B$ be a ring homomorphism of noetherian rings and let $\phi^a : \operatorname{Spec}(B) \to \operatorname{Spec}(A)$ be the morphism induced by ϕ. Let M be an A-module and let N be a B-module such that N is flat as an A-module. Then we have the following.*

(1) *For $p \in \operatorname{Spec}(A)$,*

$$\phi^a(\operatorname{Ass}_B(N/pN)) = \begin{cases} \{p\} & \text{if } N \neq pN, \\ \emptyset & \text{if } N = pN. \end{cases}$$

(2) $\operatorname{Ass}_B(M \otimes_A N) = \bigcup_{p \in \operatorname{Ass}_A(M)} \operatorname{Ass}_B(N/pN)$.

PROOF. For example, see [**44**, (9.B) Theorem 12]. □

As a corollary of the above theorem, we have the following.

COROLLARY 1.27. *Let $\phi : A \to B$ be a ring homomorphism of noetherian rings such that B is flat over A. Let M be an A-module. Here we assume the following.*

(1) *M has no embedded prime as an A-module.*
(2) *B/pB has no embedded prime as a B-module for all $p \in \operatorname{Ass}_A(M)$.*

Then $M \otimes_A B$ has no embedded prime as a B-module.

PROOF. We assume that $M \otimes_A B$ has an embedded prime as a B-module. Namely, there are $q, q' \in \operatorname{Ass}_B(M \otimes_A B)$ with $q \subsetneq q'$. By (2) in Theorem 1.26, there are $p, p' \in \operatorname{Ass}_A(M)$ with $q \in \operatorname{Ass}_B(B/pB)$ and $q' \in \operatorname{Ass}_B(B/p'B)$. Thus, by (1) in Theorem 1.26, $p = \phi^a(q)$, $p' = \phi^a(q')$. In particular, $p \subseteq p'$. Hence $p = p'$ because M has no embedded prime. Therefore, $q, q' \in \operatorname{Ass}_B(B/pB)$, which contradicts to the assumption that B/pB has no embedded prime. □

Let us consider the following proposition as an application of the above corollary.

PROPOSITION 1.28. *Let k be a field and let A be a noetherian k-algebra. Let K be a separable extension field (K is not necessarily algebraic) and $B = A \otimes_k K$. Moreover, let M be an A-module. If M has no embedded prime as an A-module, then $M \otimes_A B$ has no embedded prime as a B-module.*

PROOF. By Corollary 1.27, it is sufficient to show that B/pB has no embedded prime for all $p \in \operatorname{Ass}_A(M)$. Applying $\otimes_k K$ to $0 \to p \to A \to A/p \to 0$, we have an exact sequence

$$0 \to p \otimes_k K \to B \to (A/p) \otimes_k K \to 0.$$

As $p \otimes_k K = pB$, we obtain $B/pB = (A/p) \otimes_k K$. Thus, by Lemma 1.29 as below, B/pB is reduced. In particular, it has no embedded prime. □

LEMMA 1.29. *Let k be a field and let A be a noetherian k-algebra. Let K be a separable extension field of k (K is not necessarily algebraic over k). If A is reduced, then $A \otimes_k K$ is also reduced.*

PROOF. Let x be a nilpotent element of $A \otimes_k K$. We need to show that $x = 0$. We can find $x_1, \ldots, x_n \in K$ such that

$$x \in A \otimes_k k(x_1, \ldots, x_n) \subseteq A \otimes_k K.$$

Thus we may assume that K is finitely generated over k.

CLAIM 5. *If K is algebraic over k, then the lemma holds.*

PROOF. Let $P_1, \ldots, P_m$ be the minimal primes of A. As A is reduced, $P_1 \cap \cdots \cap P_m = \{0\}$. Thus

$$A \hookrightarrow A/P_1 \oplus \cdots \oplus A/P_m \hookrightarrow Q(A/P_1) \oplus \cdots \oplus Q(A/P_m),$$

where $Q(A/P_i)$ is the quotient field of A/P_i. Therefore, we may assume A is a field. Since K is algebraic and finitely generated over k, K is a finite and separable extension over k. Thus there is an irreducible polynomial $f(T)$ over k such that $K = k[T]/(f(T))$ and $f(T)$ has only simple roots. Thus $A \otimes_k K = A[T]/(f(T))$ is also reduced. □

CLAIM 6. *If K is a purely transcendental extension over k, then the lemma holds.*

PROOF. In this case, there are $t_1, \ldots, t_r \in K$ such that $t_1, \ldots, t_r \in K$ are algebraically independent over k and $K = k(t_1, \ldots, t_r)$. For a nilpotent element x of $A \otimes_k K$, we need to see $x = 0$. There is a non-zero $a \in k[t_1, \ldots, t_r]$ with $ax \in A \otimes_k k[t_1, \ldots, t_r]$. The a-multiplication map $a\cdot : K \to K$ is injective, so that $a\cdot : A \otimes_k K \to A \otimes_k K$ is also injective because A is flat over k. Thus we may assume that $x \in A \otimes_k k[t_1, \ldots, t_r]$. As $A \otimes_k k[t_1, \ldots, t_r]$ is isomorphic to a polynomial ring over A, x can be expressed by a polynomial with coefficients in A. Since x is nilpotent, all coefficients of x are nilpotent (for example, see Atiyah-Macdonald's book [**3**, Chapter 1, Exercise 2]), so that they should be zero. □

Finally we consider a general case. Since K is separable over k, there are $t_1, \ldots, t_r$ such that $t_1, \ldots, t_r$ are algebraically independent over k and K is a separable algebraic extension over $k(t_1, \ldots, t_r)$. Note that

$$A \otimes_k K = (A \otimes_k k(t_1, \ldots, t_r)) \otimes_{k(t_1, \ldots, t_r)} K.$$

Thus the lemma follows from Claim 5 and Claim 6. □

Gathering results in this section, we have the following proposition.

PROPOSITION 1.30. *Let k be a field and let K be a separable extension over k. Let X be an algebraic scheme over k, $X' = X \times_{\mathrm{Spec}(k)} \mathrm{Spec}(K)$ and $\phi : X' \to X$ the natural morphism. Moreover, let F be a coherent sheaf on X. Then we have the following.*

(1) *If F has no embedded prime, then $\phi^*(F)$ has also no embedded prime.*
(2) *If X is reduced, then X' is also reduced.*
(3) $\mathrm{Supp}(\phi^*(F)) = \phi^{-1}(\mathrm{Supp}(F))$.

PROOF. (1) and (2) are consequences of Proposition 1.28 and Lemma 1.29. (3) is easy. □

1.10. Determinant bundle

In this section, we will discuss necessary facts on determinant bundles for later use. For details, see [**36**].

Let Σ be a set. For two sequences $\Lambda = (x_1, \dots, x_l)$ and $\Lambda' = (x'_1, \dots, x'_{l'})$ of elements of Σ, we write

$$\Lambda \amalg \Lambda' = (x_1, \dots, x_l, x'_1, \dots, x'_{l'}).$$

For the empty sequence $\emptyset$, $\Lambda = \emptyset \amalg \Lambda = \Lambda \amalg \emptyset$. Moreover, if $\varphi : \Sigma \to \Sigma'$ is a map of sets, then, for a sequence $\Lambda = (x_1, \dots, x_l)$ of elements of Σ, we set $\varphi(\Lambda) = (\varphi(x_1), \dots, \varphi(x_l))$.

Let k be a field and let V be a finite-dimensional vector space over k. For a sequence $\Lambda = (x_1, \dots, x_r)$ consisting of a basis of V (i.e. $x_1, \dots, x_r$ forms a basis of V), we put $\det(\Lambda) = x_1 \wedge \cdots \wedge x_r \in \det(V)$. Let

$$0 \to V_0 \xrightarrow{f_0} V_1 \xrightarrow{f_1} \cdots \xrightarrow{f_{n-2}} V_{n-1} \xrightarrow{f_{n-1}} V_n \to 0$$

be an exact sequence of finite-dimensional vector spaces over k. For each $i = 0, \dots, n$, let Λ_i be a sequence consisting of a basis of V_i. Then $\{\Lambda_i\}_{i=0}^n$ is called a *system of bases of the exact sequence* if, for every i, there is a decomposition $\Lambda_i = \Lambda'_i \amalg \Lambda''_i$ with the following properties: (1) $\Lambda'_0 = \emptyset$, $\Lambda''_n = \emptyset$ and (2) $f_{i-1}(\Lambda''_{i-1}) = \Lambda'_i$ for $i = 1, \dots, n$. It is easy to see that there is a system of bases for any exact sequence. Moreover, an element

$$\bigotimes_{i=0}^{n} \det(\Lambda_i)^{(-1)^i}$$

of $\bigotimes_{i=0}^n \det(V_i)^{(-1)^i}$ does not depend on the choice of the system of bases. Indeed, let $\{\tilde{\Lambda}_i\}$ be another system of bases. Let us choose a matrix A_i with $\tilde{\Lambda}_i = \Lambda_i A_i$. Then we have

$$A_i = \begin{pmatrix} B'_i & C_i \\ 0 & B''_i \end{pmatrix}$$

and $B''_{i-1} = B'_i$ for $i = 1, \dots, n$. This shows that

$$\bigotimes_{i=0}^{n} \det(\Lambda_i)^{(-1)^i} = \bigotimes_{i=0}^{n} \det(\tilde{\Lambda}_i)^{(-1)^i}.$$

In this sense, $\bigotimes_{i=0}^n \det(V_i)^{\otimes(-1)^i}$ is canonically isomorphic to k.

Let S be a regular and integral noetherian scheme. Let U be an open set of S and let $i : U \hookrightarrow S$ be the inclusion map. We assume that $\operatorname{codim}(S \setminus U) \geq 2$ (i.e. for all $x \in S \setminus U$, $\dim \mathscr{O}_{S,x} \geq 2$). Then, for any invertible sheaf L on U, $i_*(L)$ gives rise to an invertible sheaf on S. Indeed, since S is regular and L is invertible, there are a divisor D and an isomorphism $L \simeq \mathscr{O}_U(D|_U)$. Therefore, we have an isomorphism $i_*(L) \simeq i_*(\mathscr{O}_U(D|_U))$. On the other hand, since $\operatorname{codim}(S \setminus U) \geq 2$, $i_*(\mathscr{O}_U(D|_U)) = \mathscr{O}_S(D)$, that is, $i_*(L)$ is an invertible sheaf on S.

Let F be a torsion free coherent sheaf on F. There is an open set U of S such that $F|_U$ is free and $\operatorname{codim}(S \setminus U) \geq 2$. Thus we define $\det(F)$ by $\det(F) :=$

$i_*(\det(F|_U))$, where i is the inclusion map $U \hookrightarrow S$ as before. Moreover, it is easy to see that $\det(F)$ does not depend on the choice of U.

Next let F be a coherent sheaf on X. Let T be the torsion part of F. Let $S^{(1)}$ be the set of all codimension 1 points of S. We define $\det(F)$ to be

$$\det(F) := \det(F/T) \otimes \mathscr{O}_S\left(\sum\nolimits_{x\in S^{(1)}} \operatorname{length}_{\mathscr{O}_{S,x}}(T_x)\overline{\{x\}}\right).$$

For an exact sequence

$$E : 0 \to F_0 \to F_1 \to \cdots \to F_{n-1} \to F_n \to 0$$

of coherent sheaves on S, $\det(E)$ is defined to be

$$\det(E) := \bigotimes\nolimits_{i=0}^{n} \det(F_i)^{\otimes(-1)^i}.$$

Let η be the generic point of S and let $K = \mathscr{O}_{S,\eta}$ be the function field of S. As described before, there is a natural isomorphism $\iota_\eta : K \xrightarrow{\sim} \det(E)_\eta$. We have the following theorem.

THEOREM 1.31. *There is a unique isomorphism* $\tilde{\iota} : \mathscr{O}_S \to \det(E)$ *such that* $\tilde{\iota}$ *coincides with* ι_η *at the generic point, that is,* $(\tilde{\iota})_\eta = \iota_\eta$.

PROOF. The uniqueness is obvious, so that we consider the existence. For a point $x \in S$, the natural homomorphisms $\mathscr{O}_{S,x} \to \mathscr{O}_{S,\eta}$ and $\det(E)_x \to \det(E)_\eta$ are denoted by α_x and β_x, respectively. Let us consider the following diagram:

$$\begin{array}{ccc} \mathscr{O}_{S,x} & & \det(E)_x \\ \big\downarrow{\scriptstyle \alpha_x} & & \big\downarrow{\scriptstyle \beta_x} \\ \mathscr{O}_{S,\eta} & \xrightarrow{\iota_\eta} & \det(E)_\eta. \end{array}$$

In order to construct $\iota_x : \mathscr{O}_{S,x} \to \det(E)_x$, it is sufficient to see that

$$\iota_\eta(\alpha_x(\mathscr{O}_{S,x})) = \beta_x(\det(E)_x)$$

for all $x \in S$. For this purpose, we may assume $x \in S^{(1)}$ because S is normal.

We set $A = \mathscr{O}_{S,x}$. For a finitely generated A-module V, we say that $x_1, \ldots, x_r \in V$ is a *quasi-basis* of V if $x_1, \ldots, x_r$ gives rise to a basis of $V \otimes_A K$. Let

$$0 \to V_0 \xrightarrow{f_0} V_1 \xrightarrow{f_1} \cdots \xrightarrow{f_{n-2}} V_{n-1} \xrightarrow{f_{n-1}} V_n \to 0$$

be an exact sequence of finitely generated A-modules. For each $i = 0, \ldots, n$, let Λ_i be a sequence consisting of a quasi-basis of V_i. Then $\{\Lambda_i\}$ is called a *system of quasi-bases of the exact sequence* if there are decompositions $\Lambda_i = \Lambda_i' \amalg \Lambda_i''$ for all i such that $\Lambda_0' = \emptyset$, $\Lambda_n'' = \emptyset$ and that $f_{i-1}(\Lambda_{i-1}'') = \Lambda_i'$ holds for all $i = 1, \ldots, n$. It is easy that such a system of quasi-bases of the exact sequence exists.

Let N_i be the torsion part of F_i and $\bar{F}_i = F_i/N_i$. Let t be a local parameter of A. Let $\{\Lambda_i\}$ be a system of quasi-bases of the exact sequence

$$E_x : 0 \to F_{0,x} \to F_{1,x} \to \cdots \to F_{n-1,x} \to F_{n,x} \to 0.$$

We set $\Lambda_i = (a_{i1}, \ldots, a_{ir_i})$ and $\bar{\Lambda}_i = (\bar{a}_{i1}, \ldots, \bar{a}_{ir_i})$, where $\bar{a}_{ij}$ is the image of a_{ij} by the natural homomorphism $F_{i,x} \to \bar{F}_{i,x}$. By abuse of notation, we set

$$A\Lambda_i := Aa_{i1} + \cdots + Aa_{ir_i} \quad \text{and} \quad A\bar{\Lambda}_i := A\bar{a}_{i1} + \cdots + A\bar{a}_{ir_i}.$$

Note that Λ_i yields a system of bases over K. It is easy to see that

$$0 \to F_{0,x}/A\Lambda_0 \to F_{1,x}/A\Lambda_1 \to \cdots \to F_{n-1,x}/A\Lambda_{n-1} \to F_{n,x}/A\Lambda_n \to 0$$

and

$$0 \to N_{i,x} \to F_{i,x}/A\Lambda_i \to \bar{F}_{i,x}/A\bar{\Lambda}_i \to 0$$

are exact. Therefore, if we set

$$e_i = \mathrm{length}_A(N_{i,x}), \quad f_i = \mathrm{length}_A(F_{i,x}/A\Lambda_i) \quad \text{and} \quad g_i = \mathrm{length}_A(\bar{F}_{i,x}/A\bar{\Lambda}_i),$$

then we have

$$\sum_{i=0}^{n}(-1)^i f_i = 0 \quad \text{and} \quad f_i = e_i + g_i \ (i = 0, \dots, n).$$

We put $\omega_i = \bar{a}_{i1} \wedge \cdots \wedge \bar{a}_{ir_i}$. Then $\omega_i \in \det(\bar{F}_{i,x})$ and, by Theorem 1.8, we obtain

$$\mathrm{length}_A(\bar{F}_{i,x}/A\bar{\Lambda}_i) = \mathrm{length}_A(\det(\bar{F}_{i,x})/A\omega_i).$$

In particular, $t^{-g_i}\omega_i$ forms a basis of $\det(\bar{F}_{i,x})$. Therefore, $t^{-g_i}\omega_i \otimes t^{-e_i}$ yields a basis of $\det(F_{i,x})$, that is, $\otimes_{i=0}^{n}(t^{-g_i}\omega_i \otimes t^{-e_i})^{(-1)^i}$ is a basis of $\det(E)_x$. Thus, if we denote the natural homomorphism $\det(F_i)_x \to \det(F_i)_\eta$ by $\beta_{i,x}$, then

$$\begin{aligned} \iota_\eta(\alpha_x(1)) = \otimes_{i=0}^{n}\omega_i^{(-1)^i} &= \otimes_{i=0}^{n}(t^{-f_i}\omega_i)^{(-1)^i} \\ &= \otimes_{i=0}^{n}\beta_{i,x}(t^{-g_i}\omega_i \otimes t^{-e_i})^{(-1)^i} \\ &= \beta_x\left(\otimes_{i=0}^{n}(t^{-g_i}\omega_i \otimes t^{-e_i})^{(-1)^i}\right), \end{aligned}$$

and hence the theorem follows. □

Let $\pi : X \to S$ be a surjective and proper morphism of noetherian schemes such that S is regular and integral. Let L be an invertible sheaf on X. Then we define $\det \mathbf{R}\pi_*(L)$ to be

$$\det \mathbf{R}\pi_*(L) := \bigotimes_{i \geq 0} \det(R^i\pi_*(L))^{(-1)^i}.$$

This is called the *determinant bundle* of L with respect to $\pi : X \to S$.

Proposition 1.32. (1) *Let $\nu : S' \to S$ be a morphism of regular and integral schemes such that either* (i) *ν is flat and finite, or,* (ii) *S' is an open set of S and ν is the inclusion map. We set the induced morphism as follows:*

$$\begin{array}{ccc} X \times_S S' & \xrightarrow{\nu'} & X \\ \downarrow{\scriptstyle \pi'} & & \downarrow{\scriptstyle \pi} \\ S' & \xrightarrow{\nu} & S. \end{array}$$

Then there is a natural isomorphism

$$\nu^*(\det \mathbf{R}\pi_*(L)) \xrightarrow{\sim} \det \mathbf{R}\pi'_*\left(\nu'^*(L)\right).$$

(2) *If M is an invertible sheaf on S, then there is a natural isomorphism*

$$\det \mathbf{R}\pi_*(L \otimes \pi^* M) \xrightarrow{\sim} \det \mathbf{R}\pi_*(L) \otimes M^{\chi},$$

where χ is $\chi(X_\eta, L_\eta)$ on the generic fiber.

PROOF. (1) By the base change theorem of cohomologies in terms of a flat morphism, there is the natural isomorphism

$$\nu^*(R^i\pi_*(L)) \xrightarrow{\sim} R^i\pi'_*(\nu'^*(L)).$$

Therefore, it is sufficient to show that $\nu^*(\det(F)) = \det(\nu^*(F))$ for a coherent sheaf F on S.

First we consider the case where F is torsion free. By Theorem 1.26, $\nu^*(F)$ is also torsion free. Let U be an open set of S such that $F|_U$ is locally free and $\operatorname{codim}(S \setminus U) \geq 2$. Then $\nu^*(F)|_{\nu^{-1}(U)}$ is locally free and $\operatorname{codim}(S' \setminus \nu^{-1}(U)) \geq 2$. Therefore, $\det(\nu^*(F)) = \nu^*(\det(F))$.

Next we consider the case where F is a torsion sheaf. Let $x' \in S'^{(1)}$ and $x = \nu(x')$. Then $x \in S^{(1)}$. Thus it is sufficient to show

$$\operatorname{length}_{\mathscr{O}_{S',x'}}(\nu^*(F)_{x'}) = e_{x'} \operatorname{length}_{\mathscr{O}_{S,x}}(F_x),$$

where $e_{x'}$ is the ramification index of $\mathscr{O}_{S,x} \to \mathscr{O}_{S',x'}$, that is, if t' and t are local parameters of $\mathscr{O}_{S',x'}$ and $\mathscr{O}_{S,x}$, then $e_{x'}$ is defined by $t\mathscr{O}_{S',x'} = t'^{e_{x'}}\mathscr{O}_{S',x'}$. By elementary divisor theory, we can set

$$F_x \simeq (\mathscr{O}_{S,x}/t^{r_1}\mathscr{O}_{S,x}) \oplus \cdots \oplus (\mathscr{O}_{S,x}/t^{r_n}\mathscr{O}_{S,x}).$$

Note that $\operatorname{length}_{\mathscr{O}_{S,x}}(F_x) = r_1 + \cdots + r_n$. On the other hand,

$$\begin{aligned}\nu^*(F)_{x'} = F_x \otimes_{\mathscr{O}_{S,x}} \mathscr{O}_{S',x'} &\simeq (\mathscr{O}_{S',x'}/t^{r_1}\mathscr{O}_{S',x'}) \oplus \cdots \oplus (\mathscr{O}_{S',x'}/t^{r_n}\mathscr{O}_{S',x'}) \\ &= (\mathscr{O}_{S',x'}/t'^{e_{x'}r_1}\mathscr{O}_{S',x'}) \oplus \cdots \oplus (\mathscr{O}_{S',x'}/t'^{e_{x'}r_n}\mathscr{O}_{S',x'}).\end{aligned}$$

Thus $\operatorname{length}_{\mathscr{O}_{S',x'}}(\nu^*(F)_{x'}) = e_{x'}(r_1 + \cdots + r_n)$.

Finally we consider a general case. Let T be the torsion part of F and $\bar{F} = F/T$. An exact sequence $0 \to T \to F \to \bar{F} \to 0$ yields an exact sequence $0 \to \nu^*(T) \to \nu^*(F) \to \nu^*(\bar{F}) \to 0$ because ν is flat. Thus the assertion follows from the previous two cases.

(2) By the projection formula, we obtain $R^i\pi_*(L \otimes \pi^*M) = R^i\pi_*(L) \otimes M$. Thus, if we set

$$\chi_i = \dim_K R^i\pi_*(L)_\eta = h^i(X_\eta, L_\eta),$$

then

$$\det(R^i\pi_*(L \otimes \pi^*M)) = \det(R^i\pi_*(L)) \otimes M^{\chi_i}.$$

Therefore,

$$\det \mathbf{R}\pi_*(L \otimes \pi^*M) = \det \mathbf{R}\pi_*(L) \otimes M^{\chi(X_\eta, L_\eta)}.$$

□

1.11. Complex manifold and Hodge theory

Let X be an n-dimensional complex manifold. Let $\mathscr{A}_X^{p,q}$ be the sheaf consisting of C^∞ (p,q)-forms on X. We define $A^{p,q}(X)$ to be

$$A^{p,q}(X) = \Gamma(X, \mathscr{A}_X^{p,q}).$$

Let $ds^2 = \sum_{ij} h_{ij} dz_i \otimes d\bar{z}_j$ be a hermitian metric of X. We set

$$\Omega := \frac{\sqrt{-1}}{2}\sum_{ij} h_{ij} dz_i \wedge d\bar{z}_j.$$

It is called the *fundamental form* of X associated with ds^2. Let

$$* : A^{p,q}(X) \to A^{n-q,n-p}(X)$$

be the $*$-*operator* of Hodge arising from ds^2 (for example, see [**37**]. The $*$-operator in [**27**] is the complex conjugate of our $*$-operator). We define d^*, ∂^* and $\bar{\partial}^*$ to be

$$d^* = -*d*, \quad \partial^* = -*\bar{\partial}*, \quad \bar{\partial}^* = -*\partial*.$$

In the case where X is compact, we put

$$(\alpha, \beta) = \int_X \alpha \wedge \overline{*\beta}$$

for $\alpha, \beta \in A^{p,q}(X)$. Then $(\ , \ )$ yields a $\mathbb{C}$-inner product of $A^{p,q}(X)$ and d^*, ∂^* and $\bar{\partial}^*$ gives rise to the adjoint operators of d, ∂ and $\bar{\partial}$ with respect to the $\mathbb{C}$-inner product $(\ , \ )$. We define the *d-Laplacian* Δ_d, the *∂-Laplacian* Δ_∂ and the *$\bar{\partial}$-Laplacian* $\Delta_{\bar{\partial}}$ to be

$$\Delta_d = dd^* + d^*d, \quad \Delta_\partial = \partial\partial^* + \partial^*\partial, \quad \text{and} \quad \Delta_{\bar{\partial}} = \bar{\partial}\bar{\partial}^* + \bar{\partial}^*\bar{\partial}.$$

The space $\mathbf{H}^{p,q}_d(X)$ of *d-harmonic forms* on X, the space $\mathbf{H}^{p,q}_\partial(X)$ of *∂-harmonic forms* on X and the space $\mathbf{H}^{p,q}_{\bar{\partial}}(X)$ of *$\bar{\partial}$-harmonic forms* on X are defined by

$$\begin{cases} \mathbf{H}^{p,q}_d(X) = \{\eta \in A^{p,q}(X) \mid \Delta_d(\eta) = 0\}, \\ \mathbf{H}^{p,q}_\partial(X) = \{\eta \in A^{p,q}(X) \mid \Delta_\partial(\eta) = 0\}, \\ \mathbf{H}^{p,q}_{\bar{\partial}}(X) = \{\eta \in A^{p,q}(X) \mid \Delta_{\bar{\partial}}(\eta) = 0\}. \end{cases}$$

If X is compact, then

$$\begin{cases} (\Delta_d\eta, \eta) = \|d\eta\|^2 + \|d^*\eta\|^2, \\ (\Delta_\partial\eta, \eta) = \|\partial\eta\|^2 + \|\partial^*\eta\|^2, \\ (\Delta_{\bar{\partial}}\eta, \eta) = \|\bar{\partial}\eta\|^2 + \|\bar{\partial}^*\eta\|^2. \end{cases}$$

Thus

$$\begin{cases} \mathbf{H}^{p,q}_d(X) = \{\eta \in A^{p,q}(X) \mid d(\eta) = d^*(\eta) = 0\}, \\ \mathbf{H}^{p,q}_\partial(X) = \{\eta \in A^{p,q}(X) \mid \partial(\eta) = \partial^*(\eta) = 0\}, \\ \mathbf{H}^{p,q}_{\bar{\partial}}(X) = \{\eta \in A^{p,q}(X) \mid \bar{\partial}(\eta) = \bar{\partial}^*(\eta) = 0\}. \end{cases}$$

THEOREM 1.33 (Hodge's theorem). *If X is compact, then the following hold.*

(1) *$\mathbf{H}^{p,q}_{\bar{\partial}}(X)$ is a finite-dimensional complex vector space for any p, q. In particular, there is the orthogonal projection $H_{\bar{\partial}} : A^{p,q}(X) \to \mathbf{H}^{p,q}_{\bar{\partial}}(X)$.*

(2) *There is a unique Green operator $G_{\bar{\partial}} : A^{p,q}(X) \to A^{p,q}(X)$ with the following properties.*
  (2.1) *$G_{\bar{\partial}}(\mathbf{H}^{p,q}_{\bar{\partial}}(X)) = \{0\}$, $\bar{\partial}G_{\bar{\partial}} = G_{\bar{\partial}}\bar{\partial}$ and $\bar{\partial}^*G_{\bar{\partial}} = G_{\bar{\partial}}\bar{\partial}^*$.*
  (2.2) *$H_{\bar{\partial}} + \Delta_{\bar{\partial}}G_{\bar{\partial}} = \mathrm{id}_{A^{p,q}(X)}$.*

PROOF. For example, see [**27**, Chapter 0, Section 6]. □

COROLLARY 1.34. *We assume that X is connected and compact. If*

$$\int_X f\Omega^n = 0$$

for a C^∞-function f on X, then there is a C^∞-function g on X with $f = \Delta_{\bar{\partial}}(g)$.

PROOF. The orthogonal projection $H_{\bar\partial} : A^{0,0}(X) \to \mathbf{H}^{0,0}_{\bar\partial}(X) = \mathbb{C}$ is given by

$$H_{\bar\partial}(f) = \frac{\displaystyle\int_X f\Omega^n}{\displaystyle\int_X \Omega^n}.$$

Thus the corollary follows from Theorem 1.33. □

If $d(\Omega) = 0$, then we say that X is a *Kähler manifold* and Ω is the *Kähler form* of X. Let $L : A^{p,q}(X) \to A^{p+1,q+1}(X)$ be the operator given by $L(\eta) = \Omega \wedge \eta$ and $\Lambda = *^{-1}L*$. If X is compact, then Λ is the adjoint operator of L with respect to the $\mathbb{C}$-inner product $(\ ,\ )$. If X is a compact Kähler manifold, then it is well known that *the Hodge identities*:

$$\Lambda\bar\partial - \bar\partial\Lambda = -\sqrt{-1}\partial^* \quad \text{and} \quad \Lambda\partial - \partial\Lambda = \sqrt{-1}\bar\partial^*$$

hold (for example, see [**27**, Chapter 0, Section 7]). By using the Hodge identities, we have

$$2\Delta_\partial = 2\Delta_{\bar\partial} = \Delta_d.$$

Thus we have the following as a corollary of Theorem 1.33.

COROLLARY 1.35. *If X is a compact Kähler manifold, then*

$$\mathbf{H}^{p,q}_d(X) = \mathbf{H}^{p,q}_\partial(X) = \mathbf{H}^{p,q}_{\bar\partial}(X)$$

and the orthogonal decompositions

$$\begin{aligned} A^{p,q}(X) &= \mathbf{H}^{p,q}_{\bar\partial}(X) \oplus^\perp \bar\partial(A^{p,q-1}(X)) \oplus^\perp \bar\partial^*(A^{p,q+1}(X)) \\ &= \mathbf{H}^{p,q}_\partial(X) \oplus^\perp \partial(A^{p-1,q}(X)) \oplus^\perp \partial^*(A^{p+1,q}(X)) \end{aligned}$$

hold.

As an application of the above corollary, we have the following $\partial\bar\partial$-lemma.

LEMMA 1.36 ($\partial\bar\partial$-lemma). *We assume that X is a compact Kähler manifold. For a C^∞ (p,q)-form η on X, the following (1) ~ (3) are equivalent.*

(1) *η is d-exact, that is, there is a C^∞-form ξ with $\eta = d(\xi)$.*
(2) *$d(\eta) = 0$ and there are $\xi_1 \in A^{p-1,q}(X)$ and $\xi_2 \in A^{p,q-1}(X)$ with $\eta = \partial(\xi_1) + \bar\partial(\xi_2)$.*
(3) *There is $\mu \in A^{(p-1,q-1)}(X)$ with $\eta = \sqrt{-1}\partial\bar\partial(\mu)$.*

Moreover, if the above conditions hold, $p = q$ and η is real, then there is a real $\mu \in A^{(p-1,q-1)}(X)$ with $\eta = \sqrt{-1}\partial\bar\partial(\mu)$.

PROOF. (1) $\Longrightarrow$ (2) is obvious.
(2) $\Longrightarrow$ (3) : Applying Corollary 1.35 to ξ_1 and ξ_2, there are

$$h_1 \in \mathbf{H}^{p-1,q}_\partial(X), \quad a_1 \in A^{p-1,q-1}(X), \quad b_1 \in A^{p-1,q+1}(X)$$

and

$$h_2 \in \mathbf{H}^{p,q-1}_\partial(X), \quad a_2 \in A^{p-1,q-1}(X), \quad b_2 \in A^{p+1,q-1}(X)$$

such that

$$\xi_1 = h_1 + \bar\partial(a_1) + \bar\partial^*(b_1), \quad \xi_2 = h_2 + \partial(a_2) + \partial^*(b_2).$$

Thus $\eta = \partial\bar{\partial}(a_1) + \partial\bar{\partial}^*(b_1) + \bar{\partial}\partial(a_2) + \bar{\partial}\partial^*(b_2)$. Since $d(\eta) = 0$, we have $\partial(\eta) = \bar{\partial}(\eta) = 0$. Therefore, $\bar{\partial}\partial\bar{\partial}^*(b_1) = \partial\bar{\partial}\partial^*(b_2) = 0$. On the other hand, by the Hodge identities, we can see $\partial\bar{\partial}^* = -\bar{\partial}^*\partial$ and $\bar{\partial}\partial^* = -\partial^*\bar{\partial}$, which implies

$$\|\partial\bar{\partial}^*(b_1)\|^2 = -(\partial\bar{\partial}^*(b_1), \bar{\partial}^*\partial(b_1)) = -(\bar{\partial}\partial\bar{\partial}^*(b_1), \partial(b_1)) = 0,$$
$$\|\bar{\partial}\partial^*(b_2)\|^2 = -(\bar{\partial}\partial^*(b_2), \partial^*\bar{\partial}(b_2)) = -(\partial\bar{\partial}\partial^*(b_2), \bar{\partial}(b_2)) = 0.$$

Therefore, $\partial\bar{\partial}^*(b_1) = \bar{\partial}\partial^*(b_2) = 0$. Hence

$$\eta = \partial\bar{\partial}(a_1) + \bar{\partial}\partial(a_2) = \sqrt{-1}\partial\bar{\partial}(\sqrt{-1}(a_2 - a_1)).$$

(3) $\Longrightarrow$ (1) : If we set $\xi = (\sqrt{-1}/2)\bar{\partial}(\mu) - (\sqrt{-1}/2)\partial(\mu)$, then $d(\xi) = \eta$.

Finally we consider the last assertion. Let us take $\mu \in A^{p-1,p-1}(X)$ with $\eta = \sqrt{-1}\partial\bar{\partial}(\mu)$. Note that

$$\eta = \bar{\eta} = \overline{\sqrt{-1}\partial\bar{\partial}(\mu)} = \sqrt{-1}\partial\bar{\partial}(\bar{\mu}).$$

If we set $\mu' = (\mu + \bar{\mu})/2$, then μ' is real and $\eta = \sqrt{-1}\partial\bar{\partial}(\mu')$. □

Next we observe an expression of the $\bar{\partial}$-Laplacian in terms of $\partial\bar{\partial}$.

PROPOSITION 1.37. *We assume that X is a compact Kähler manifold. Then*

$$\Delta_{\bar{\partial}}(f)\Omega^n = -n\sqrt{-1}\partial\bar{\partial}(f) \wedge \Omega^{n-1}$$

for $f \in A^{0,0}(X)$.

PROOF. First we consider the following.

CLAIM 7. *For a positive integer a,*

$$L^a\Lambda = \Lambda L^a + a(p + q - n + a - 1)L^{a-1}$$

holds on $A^{p,q}(X)$.

PROOF. If $a = 1$, then $L\Lambda = \Lambda L + (p + q - n)\,\mathrm{id}$ is the well-known formula (for example, see [**27**, Chapter 0, Section 7]). In general, it can be easily proved by induction on a. □

As $L^n = 0$ on $A^{1,1}(X)$, by the above claim,

$$L^n\Lambda = nL^{n-1} \quad \text{on } A^{1,1}(X). \tag{1.2}$$

Next we have the following formula:

$$\Lambda\partial\bar{\partial} = \sqrt{-1}\Delta_{\bar{\partial}} \quad \text{on } A^{0,0}(X). \tag{1.3}$$

Note that $\Lambda = 0$ on $A^{0,1}(X)$ and $\Delta_{\bar{\partial}} = \bar{\partial}^*\bar{\partial}$ on $A^{0,0}(X)$. Thus, using Hodge's identities,

$$\Lambda\partial\bar{\partial} = (\partial\Lambda + \sqrt{-1}\bar{\partial}^*)\bar{\partial} = \sqrt{-1}\Delta_{\bar{\partial}}$$

on $A^{0,0}(X)$.

By (1.2) and (1.3), we have

$$\begin{aligned} -n\sqrt{-1}\partial\bar{\partial}(f) \wedge \Omega^{n-1} &= -nL^{n-1}(\sqrt{-1}\partial\bar{\partial}(f)) = -L^n\Lambda(\sqrt{-1}\partial\bar{\partial}(f)) \\ &= -\sqrt{-1}L^n(\sqrt{-1}\Delta_{\bar{\partial}}(f)) = \Delta_{\bar{\partial}}(f)\Omega^n. \end{aligned}$$ □

Finally we consider Künneth's formula in terms of harmonic forms.

THEOREM 1.38 (Künneth's formula in terms of harmonic forms). *Let X and Y be compact Kähler manifolds, and let Ω_X and Ω_Y be Kähler forms on X and Y, respectively. Let $p : X \times Y \to X$ and $q : X \times Y \to Y$ be the projections to the first factor and the second factor, respectively. Let $\Omega_{X\times Y}$ be the Kähler form on $X \times Y$ given by $\Omega_{X\times Y} = p^*(\Omega_X) + q^*(\Omega_Y)$. Let Δ_X, Δ_Y and $\Delta_{X\times Y}$ be the $\bar{\partial}$-Laplacians induced by Ω_X, Ω_Y and $\Omega_{X\times Y}$, respectively. Then we have the following.*

(1) $\Delta_{X\times Y} = \Delta_X + \Delta_Y$. *More precisely, for $\psi \in A^{p,q}(X)$ and $\eta \in A^{p',q'}(Y)$,*
$$\Delta_{X\times Y}(p^*(\psi) \wedge q^*(\eta)) = p^*(\Delta_X(\psi)) \wedge q^*(\eta) + p^*(\psi) \wedge q^*(\Delta_Y(\eta)).$$
In particular, If $\psi \in \mathbf{H}^{p,q}(X)$ and $\eta \in \mathbf{H}^{p',q'}(Y)$, then
$$p^*(\psi) \wedge q^*(\eta) \in \mathbf{H}^{p+p',q+q'}(X \times Y).$$

(2) *Let $\mathbf{H}^{p,q}(X) \otimes \mathbf{H}^{p',q'}(Y) \to \mathbf{H}^{p+p',q+q'}(X \times Y)$ be the natural homomorphism by* (1). *In terms of the above homomorphisms,*
$$\mathbf{H}^{r,s}(X \times Y) = \bigoplus_{\substack{r=p+p'\\ s=q+q'}} \mathbf{H}^{p,q}(X) \otimes \mathbf{H}^{p',q'}(Y)$$
holds.

PROOF. For example, see [**27**, Chapter 0, Section 6]. □

1.12. Connection and curvature

Let X be an n-dimensional complex manifold and let E be a locally free coherent sheaf on X. As before, let $\mathscr{A}_X^{p,q}$ be the sheaf consisting of C^∞ (p,q)-forms on X. We define $A^{p,q}(E)$ to be
$$A^{p,q}(E) = \Gamma(X, \mathscr{A}_X^{p,q} \otimes_{\mathscr{O}_X} E).$$
In the case where $E = \mathscr{O}_X$, $A^{p,q}(\mathscr{O}_X)$ is denoted by $A^{p,q}(X)$ as in the previous section. Moreover, we set
$$A^k(E) = \bigoplus_{p+q=k} A^{p,q}(E).$$
Note that $A^k(X)$ means $A^k(\mathscr{O}_X)$ as before. Since $\bar{\partial} : \mathscr{A}_X^{p,q} \to \mathscr{A}_X^{p,q+1}$ is $\mathscr{O}_X$-linear, we can define the homomorphism
$$\mathscr{A}_X^{p,q} \otimes_{\mathscr{O}_X} E \to \mathscr{A}_X^{p,q+1} \otimes_{\mathscr{O}_X} E$$
given by $\eta \otimes s \mapsto \bar{\partial}(\eta) \otimes s$. It is called the *Cauchy-Riemann operator* and it is denoted by $\bar{\partial}_E$, or, $\bar{\partial}$ for simplicity.

A *connection* ∇ of E is a $\mathbb{C}$-linear map
$$\nabla : A^0(E) \to A^1(E)$$
such that
$$\nabla(fs) = df \otimes s + f\nabla(s) \quad (f \in A^0(X),\ s \in A^0(E))$$
holds. The decomposition $A^1(E) = A^{1,0}(E) \oplus A^{0,1}(E)$ induces
$$\nabla = \nabla^{1,0} + \nabla^{0,1}$$
on the connection accordingly.

Using Leibniz's rule,
$$\nabla(\eta \otimes s) = d\eta \otimes s + (-1)^k \eta \wedge \nabla(s) \quad (\eta \in A^k(X),\ s \in A^0(E)),$$

the connection ∇ extends to $\nabla : A^k(E) \to A^{k+1}(E)$. The *curvature* of the connection ∇ is defined by ∇^2, that is, the composition

$$A^0(E) \xrightarrow{\nabla} A^1(E) \xrightarrow{\nabla} A^2(E).$$

We can easily check that ∇^2 is $A^0(X)$-linear. Thus ∇^2 can be viewed as an element of $A^0(E \otimes E^\vee) = A^0(\mathscr{H}om(E, E))$.

Let $\overline{E} = (E, h)$ be a pair consisting of a locally free coherent sheaf E on X and a C^∞-hermitian metric h of E. It is called a *C^∞-hermitian locally free sheaf.* If E is invertible, then $\overline{E}$ is called a *C^∞-hermitian invertible sheaf.* It is well known (for example, see [**37**, Proposition 4.9]) that there is a unique connection $\nabla_{\overline{E}}$ of E with the following properties.

(a) $\nabla_{\overline{E}}^{0,1} = \bar{\partial}_E$.

(b) $d(h(s,t)) = h(\nabla_{\overline{E}}(s), t) + h(s, \nabla_{\overline{E}}(t))$ holds for all $s, t \in A^0(E)$.

This connection is called the *h-connection.* By the above properties, we can see that the curvature $\nabla_{\overline{E}}^2$ of the h-connection is an element of $A^{1,1}(E \otimes E^\vee)$. The curvature $\nabla_{\overline{E}}^2$ is denoted by $\Theta_{\overline{E}}$. From now on, we consider only the h-connection after fixing a hermitian metric.

In the case where E is an invertible sheaf L, the curvature is an element of $A^{1,1}(X)$ because $L \otimes L^\vee$ is naturally isomorphic to $\mathscr{O}_X$. If L has a non-zero meromorphic section s, then the curvature $\Theta_{\overline{L}}$ of the h-connection $\nabla_{\overline{L}}$ is actually given by

$$\Theta_{\overline{L}} = \partial\bar{\partial}(-\log(h(s,s))).$$

$\dfrac{\sqrt{-1}}{2\pi}\Theta_{\overline{L}}$ is called the *first Chern form* of $\overline{L}$ and it is denoted by $c_1(\overline{L})$. Note that $c_1(\overline{L})$ is a real $(1,1)$-form.

Let η be a real $(1,1)$-form on X. We write

$$\eta = \sqrt{-1}\sum_{i,j} h_{ij}(z) dz_i \wedge d\bar{z}_j$$

using a local coordinate $(z_1, \ldots, z_n)$. We say that η is *positive* if the hermitian matrix $(h_{ij}(z))$ is positive definite at each point z. Moreover, if $(h_{ij}(z))$ is positive semidefinite at each z, then η is said to be *semipositive.* For a C^∞-hermitian invertible sheaf $\overline{L}$ on X, $\overline{L}$ is said to be *positive* (resp. *semipositive*) if $c_1(\overline{L})$ is positive (resp. semipositive).

1.13. Poincaré-Lelong formula

In this section, we discuss the Poincaré-Lelong formula which will be used later.

THEOREM 1.39 (Poincaré-Lelong formula). *Let X be a d-dimensional non-singular complex quasi-projective variety and let (L, h) be a C^∞-hermitian invertible sheaf on X. Let s be a rational section of L such that s is non-zero on each connected component of X. Then the following formula*

$$\int_X \log(h(s,s))\partial\bar{\partial}(\eta) + 2\pi\sqrt{-1}\int_{\operatorname{div}(s)} \eta = \int_X \partial\bar{\partial}(\log(h(s,s))) \wedge \eta$$

holds for all C^∞ $(d-1, d-1)$-form η on X with compact support.

PROOF. Clearly we may assume that X is connected. First we consider the case where $\mathrm{Supp}(\mathrm{div}(s))$ is a normal crossing divisor. Then there is a locally finite open covering $\{U_\lambda\}_{\lambda\in\Lambda}$ with the following property: there are a local coordinate $(z_1^\lambda,\ldots,z_d^\lambda)$ and a local basis ω_λ of L on U_λ such that $s|_{U_\lambda} = (z_1^\lambda)^{a_1^\lambda}\cdots(z_d^\lambda)^{a_d^\lambda}\omega_\lambda$, where $a_1^\lambda,\ldots,a_d^\lambda$ are integers. Let $\{\rho_\lambda\}$ be a partition of unity subordinate to $\{U_\lambda\}$ and let $\eta_\lambda = \rho_\lambda\eta$. As $\eta = \sum_\lambda \eta_\lambda$, it is sufficient to see the formula for η_λ. Using the coordinate $(z_1^\lambda,\ldots,z_d^\lambda)$, U_λ can be identified with an open set of $\mathbb{C}^d$. Let f_λ be a C^∞-function on $\mathbb{C}^d$ such that f_λ coincides with $\log h(\omega_\lambda,\omega_\lambda)$ on $\mathrm{Supp}(\eta_\lambda)$. Then the formula

$$\int_X \log(h(s,s))\partial\bar\partial(\eta_\lambda) + 2\pi\sqrt{-1}\int_{\mathrm{div}(s)}\eta_\lambda = \int_X \partial\bar\partial(\log(h(s,s)))\wedge\eta_\lambda$$

for η_λ is equivalent to the formula

$$\sum_{i=1}^d a_i^\lambda\left(\int_{\mathbb{C}^d}\log(|z_i^\lambda|^2)\partial\bar\partial(\eta_\lambda) + 2\pi\sqrt{-1}\int_{z_i^\lambda=0}\eta_\lambda\right) + \int_{\mathbb{C}^d} f_\lambda\partial\bar\partial(\eta_\lambda) = \int_{\mathbb{C}^d}\partial\bar\partial(f_\lambda)\wedge\eta_\lambda.$$

Therefore the theorem in the case where $\mathrm{Supp}(\mathrm{div}(s))$ is a normal crossing divisor follows from Lemma 1.40 below.

Finally, we consider a general case. By virtue of Hironaka's theorem (see [**30**]), there is a projective birational morphism $\nu: Y \to X$ of non-singular complex quasi-projective varieties such that $\nu^{-1}(\mathrm{Supp}(\mathrm{div}(s)))$ is a normal crossing divisor. Then, by the previous result,

$$\begin{aligned}\int_Y \log(\nu^*(h)(\nu^*s,\nu^*s))\partial\bar\partial(\nu^*\eta) + 2\pi\sqrt{-1}&\int_{\mathrm{div}(\nu^*s)}\nu^*\eta \\ &= \int_Y \partial\bar\partial(\log(\nu^*(h)(\nu^*s,\nu^*s)))\wedge\nu^*\eta\end{aligned}$$

holds. Clearly we have

$$\int_Y \log(\nu^*(h)(\nu^*s,\nu^*s))\partial\bar\partial(\nu^*\eta) = \int_X \log(h(s,s))\partial\bar\partial(\eta)$$

and

$$\int_Y \partial\bar\partial(\log(\nu^*(h)(\nu^*s,\nu^*s)))\wedge\nu^*\eta = \int_X \partial\bar\partial(\log(h(s,s)))\wedge\eta.$$

On the other hand, by the projection formula,

$$\int_{\mathrm{div}(\nu^*s)}\nu^*\eta = \int_{\mathrm{div}(s)}\eta.$$

Thus the theorem follows. □

LEMMA 1.40. *Let f be a C^∞-function on $\mathbb{C}^d$ and let η be a C^∞ $(d-1,d-1)$-form on $\mathbb{C}^d$ with compact support. Then we have the following.*

(1) $\displaystyle\int_{\mathbb{C}^d} f\partial\bar\partial(\eta) = \int_{\mathbb{C}^d}\partial\bar\partial(f)\wedge\eta.$

(2) $\displaystyle\int_{\mathbb{C}^d}\log(|z_i|^2)\partial\bar\partial(\eta) + 2\pi\sqrt{-1}\int_{z_i=0}\eta = 0 \quad (i=1,\ldots,d).$

PROOF. (1) First, since $f\bar{\partial}(\eta)$ is a $(d-1,d)$-form and $\partial(f)\wedge\eta$ is a $(d,d-1)$-form, we have

$$\begin{aligned} d(f\bar{\partial}(\eta)) &= \partial(f\bar{\partial}(\eta)) = \partial(f)\wedge\bar{\partial}(\eta) + f\partial\bar{\partial}(\eta), \\ d(\partial(f)\wedge\eta) &= \bar{\partial}(\partial(f)\wedge\eta) = -\partial\bar{\partial}(f)\wedge\eta - \partial(f)\wedge\bar{\partial}(\eta). \end{aligned}$$

By Stokes' theorem,

$$\int_{\mathbb{C}^d} f\partial\bar{\partial}(\eta) = -\int_{\mathbb{C}^d} \partial(f)\wedge\bar{\partial}(\eta) = \int_{\mathbb{C}^d} \partial\bar{\partial}(f)\wedge\eta.$$

(2) Changing the order of the coordinate, we may assume $i=1$ without loss of generality. Let us see the following claim.

CLAIM 8. (a) *For a C^∞ $(2d-1)$-form ω with compact support,*

$$\lim_{\epsilon\to 0}\int_{|z_1|=\epsilon} \log|z_1|^2\omega = 0.$$

(b) *For a C^∞ $(d-1,d-1)$-form η with compact support,*

$$\begin{aligned} \lim_{\epsilon\to 0}\int_{|z_1|=\epsilon} \frac{dz_1}{z_1}\wedge\eta &= 2\pi\sqrt{-1}\int_{z_1=0}\eta, \\ \lim_{\epsilon\to 0}\int_{|z_1|=\epsilon} \frac{d\bar{z}_1}{\bar{z}_1}\wedge\eta &= -2\pi\sqrt{-1}\int_{z_1=0}\eta. \end{aligned}$$

PROOF. (a) We consider the case where ω is a $(d-1,d)$-form. The case where ω is a $(d,d-1)$-form is similar. We put

$$\omega = \sum_{i=1}^d \alpha_i(z_1,\ldots,z_d)(dz_1\wedge d\bar{z}_1)\wedge\cdots\wedge dz_i\wedge\cdots\wedge(dz_d\wedge d\bar{z}_d).$$

If we set $z_1=\epsilon\exp(\sqrt{-1}\theta)$, then

$$\begin{cases} dz_1 = \sqrt{-1}\epsilon\exp(\sqrt{-1}\theta)d\theta, \\ d\bar{z}_1 = -\sqrt{-1}\epsilon\exp(-\sqrt{-1}\theta)d\theta \end{cases}$$

on $|z_1|=\epsilon$. Thus

$$\begin{aligned} \log(|z_1|^2)\omega\big|_{|z_1|=\epsilon} = \epsilon\log(\epsilon^2)\sqrt{-1}\exp(\sqrt{-1}\theta)\alpha_1(\epsilon\exp(\sqrt{-1}\theta), z_2,\ldots,z_d) \\ d\theta\wedge(dz_2\wedge d\bar{z}_2)\wedge\cdots\wedge(dz_d\wedge d\bar{z}_d). \end{aligned}$$

Here $\sqrt{-1}\exp(\sqrt{-1}\theta)\alpha_1(\epsilon\exp(\sqrt{-1}\theta), z_2,\ldots,z_d)$ is bounded. Hence (a) follows.

(b) We put

$$\eta = \sum_{i,j}\beta_{ij}(z_1,\ldots,z_d)(dz_1\wedge\cdots\wedge\widehat{dz_i}\wedge\cdots\wedge d\bar{z}_d)\wedge(d\bar{z}_1\wedge\cdots\wedge\widehat{d\bar{z}_j}\wedge\cdots\wedge d\bar{z}_d).$$

If we set $z_1=\epsilon\exp(\sqrt{-1}\theta)$, then, as in (a),

$$\begin{aligned} \frac{dz_1}{z_1}\wedge\eta\bigg|_{|z_1|=\epsilon} = \sqrt{-1}\beta_{11}(\epsilon\exp(\sqrt{-1}\theta), z_2,\ldots,z_d) \\ d\theta\wedge dz_2\wedge\cdots\wedge dz_d\wedge d\bar{z}_2\wedge\cdots\wedge d\bar{z}_d. \end{aligned}$$

Therefore,

$$\int_{|z_1|=\epsilon} \frac{dz_1}{z_1} \wedge \eta = \int_{\mathbb{C}^{d-1}} \left(\int_0^{2\pi} \sqrt{-1}\beta_{11}(\epsilon \exp(\sqrt{-1}\theta), z_2, \ldots, z_d) d\theta \right) dz_2 \wedge \cdots \wedge dz_d \wedge d\bar{z}_2 \wedge \cdots \wedge d\bar{z}_d.$$

Note that

$$\lim_{\epsilon \to 0} \int_0^{2\pi} \sqrt{-1}\beta_{11}(\epsilon \exp(\sqrt{-1}\theta), z_2, \ldots, z_d) d\theta = 2\pi\sqrt{-1}\beta_{11}(0, z_2, \ldots, z_d)$$

and $\eta|_{z_1=0} = \beta_{11}(0, z_2, \ldots, z_d) dz_2 \wedge \cdots \wedge dz_d \wedge d\bar{z}_2 \wedge \cdots \wedge \cdots \wedge d\bar{z}_d$. Thus the first equation holds. The second equation is similar. □

Let us go back to the proof of (2). As $\log(|z_1|^2)\bar{\partial}(\eta)$ is a $(d-1, d)$-form and $(dz_1/z_1) \wedge \eta$ is a $(d, d-1)$-form, we have

$$d(\log(|z_1|^2)\bar{\partial}(\eta)) = \partial(\log(|z_1|^2)\bar{\partial}(\eta)) = \frac{dz_1}{z_1} \wedge \bar{\partial}(\eta) + \log(|z_1|^2)\partial\bar{\partial}(\eta)$$

and

$$d\left(\frac{dz_1}{z_1} \wedge \eta\right) = \bar{\partial}\left(\frac{dz_1}{z_1} \wedge \eta\right) = -\frac{dz_1}{z_1} \wedge \bar{\partial}(\eta).$$

Thus, by Stokes' theorem,

$$\begin{aligned}
\int_{|z_1|\geq\epsilon} \log(|z_1|^2)\partial\bar{\partial}(\eta) &= \int_{|z_1|\geq\epsilon} d(\log(|z_1|^2)\bar{\partial}(\eta)) - \int_{|z_1|\geq\epsilon} \frac{dz_1}{z_1} \wedge \bar{\partial}(\eta) \\
&= -\int_{|z_1|=\epsilon} \log(|z_1|^2)\bar{\partial}(\eta) + \int_{|z_1|\geq\epsilon} d\left(\frac{dz_1}{z_1} \wedge \eta\right) \\
&= -\int_{|z_1|=\epsilon} \log(|z_1|^2)\bar{\partial}(\eta) - \int_{|z_1|=\epsilon} \frac{dz_1}{z_1} \wedge \eta.
\end{aligned}$$

Therefore, (2) follows from the previous claim. □

REMARK 1.1. Theorem 1.39 can be derived from the Poincaré-Lelong formula for a meromorphic function (for example, [**27**, p. 388, Lemma (Poincaré-Lelong Equation)] or [**58**, Theorem 5.1.13]) by reducing it to a local problem with help of a partition of unity. In this case, Hironaka's theorem is not necessary.

1.14. C^∞ on reduced complex space

Let T be a reduced complex space. A complex valued continuous function f on T is called a *C^∞-function* if, for any analytic map $\phi : M \to T$ from any complex manifold M to T, $f \circ \phi$ is a C^∞-function on M. Let $\psi : T' \to T$ be an analytic map of reduced complex spaces and let f be a C^∞ complex valued function on T. Then $f \circ \psi$ is a C^∞ complex valued function on T'. Next let $g : T' \to T$ be a continuous map of reduced complex spaces. It is called a *C^∞-map* if, for any $t' \in T'$ and any C^∞-function h around $g(t')$, $h \circ g$ is a C^∞-function around t. It is easy to see that if $g' : T'' \to T'$ and $g : T' \to T$ are C^∞ maps of reduced complex spaces, then $g \circ g'$ is also a C^∞ map.

Let F be a locally free coherent sheaf on T and let $h = \{h_t\}_{t \in T}$ be a family of hermitian metrics h_t of $F(t) = F_t \otimes_{\mathcal{O}_{T,t}} \kappa(t)$ at all points $t \in T$. We say that h is a

C^∞*-hermitian metric* if, for any open set U of T and any $s, s' \in F(U)$, $h_t(s(t), s'(t))$ is a C^∞-function on U, where $s(t)$ means $s \otimes 1$ in $F_t \otimes_{\mathscr{O}_{T,t}} \kappa(t)$.

We assume that the rank of F is 1. Let h be a C^∞-hermitian metric of F. We say that (F, h) is *semipositive* if, for any analytic map $\phi : M \to T$ from any complex manifold M to T, $c_1(\phi^*(F), \phi^*(h))$ is a semipositive $(1,1)$-form on M. Note that T is not necessarily non-singular, so that $c_1(F, h)$ may be not defined on T. However, we say that $c_1(F, h)$ is semipositive by abuse of notation if (F, h) is semipositive. Moreover, (F, h) is said to be *positive* if, for any C^∞ real-valued function f with compact support, there is a positive real number λ_0 such that $(F, \exp(-\lambda f)h)$ is semipositive for any real number λ with $|\lambda| \leq \lambda_0$. In the same way as before, we say that $c_1(F, h)$ is positive if (F, h) is positive. If T is non-singular, then the semipositivity and the positivity in this section coincides with the semipositivity and the positivity in Section 1.12.

Let (F', h') be another pair of an invertible sheaf F' on T and a C^∞-hermitian metric h'. Then $c_1(F, h) = c_1(F', h')$ means that, for any analytic map $\phi : M \to T$ from any complex manifold M to T, $c_1(\phi^*(F), \phi^*(h)) = c_1(\phi^*(F'), \phi^*(h'))$ holds.

CHAPTER 2

Geometry of Numbers

In this chapter, we discuss several topics on geometry of numbers, such as Minkowski's theorem, Mahler's inequality, the Brunn-Minkowski theorem and a theorem due to Gillet-Soulé. In particular, an estimate of the number of points in a convex lattice by Gillet-Soulé is very important for later use. It can be viewed as a Riemann-Roch type theorem for normed $\mathbb{Z}$-modules and it will be a fundamental tool for an asymptotic estimate of small sections.

2.1. Convex set and Minkowski's theorem

First, let us recall basic materials on convex sets. For details, see [**72**]. Let V be an n-dimensional real vector space. A subset K of V is called a *convex set* if $tx+(1-t)y \in K$ for any $x, y \in K$ and any real number t with $0 \leq t \leq 1$. Moreover, K is said to be *symmetric* if $x \in K$ implies $-x \in K$. Note that a convex set K has an interior point if and only if K is not contained in a proper affine subspace of V (cf. [**72**, Corollary 2.3.2]). It is easy to see that if K is a symmetric convex set with an interior point, then the origin is an internal point of K. Further, a compact convex set with an interior point is called a *convex body*.

Let h be an inner product of V. There is a unique Haar measure vol_h such that, for any basis $x_1, \ldots, x_n$ of V,

$$\mathrm{vol}_h\left(\{\lambda_1 x_1 + \cdots + \lambda_n x_n \mid 0 \leq \lambda_i \leq 1\ (\forall i)\}\right) = \left(\det(h(x_i, x_j))\right)^{1/2}.$$

It is called *the Haar measure induced by h*. The Haar measure of $\mathbb{R}^n$ induced by the standard inner product of $\mathbb{R}^n$ is denoted by vol_n. Here we give a special notation: in the case where $V = \{0\}$, for a subset X of V,

$$\mathrm{vol}(X) = \begin{cases} 1 & \text{if } X = \{0\}, \\ 0 & \text{if } X = \emptyset. \end{cases}$$

Note that a bounded convex set K is Jordan measurable (cf. [**72**, Theorem 6.2.11]). In particular, if we denote the set of interior points of K and the topological closure of K by K° and $\overline{K}$, respectively, then

$$\mathrm{vol}_h(K^\circ) = \mathrm{vol}_h(K) = \mathrm{vol}_h(\overline{K})$$

(cf. [**72**, Theorem 6.2.5]).

Here we consider Minkowski's theorem due to Corput.

THEOREM 2.1 (Minkowski's theorem). *Let K be a bounded and symmetric convex set in $\mathbb{R}^n$. Then we have the following:*

(1) *If k is an integer with $\mathrm{vol}_n(K) > 2^n k$, then $\#(\mathbb{Z}^n \cap K) \geq 2k+1$.*
(2) *We assume that K is closed. If k is an integer with $\mathrm{vol}_n(K) \geq 2^n k$, then $\#(\mathbb{Z}^n \cap K) \geq 2k+1$.*

PROOF. (1) If $k \le 0$, then the assertion is obvious, so that we may assume that $k > 0$. We set $K' = (1/2)K = \{x/2 \mid x \in K\}$. Then $\mathrm{vol}_n(K') = 2^{-n}\,\mathrm{vol}_n(K) > k$. Since K' is Jordan measurable, we have

$$\lim_{r\to\infty} \# \left(K' \cap \frac{1}{r}\mathbb{Z}^n \right) \frac{1}{r^n} = \mathrm{vol}_n(K').$$

In particular, for a sufficiently large integer r,

$$\# \left(K' \cap \frac{1}{r}\mathbb{Z}^n \right) > kr^n$$

holds. Let $\pi : (1/r)\mathbb{Z}^n \to (1/r)\mathbb{Z}^n/\mathbb{Z}^n$ be the natural homomorphism. As

$$\#((1/r)\mathbb{Z}^n/\mathbb{Z}^n) = r^n,$$

using the pigeon hole principle, there are distinct $k+1$ elements

$$u_1, \dots, u_{k+1} \in K' \cap \frac{1}{r}\mathbb{Z}^n$$

such that $\pi(u_1) = \cdots = \pi(u_{k+1})$, that is, $u_i - u_j \in \mathbb{Z}^n$ $(\forall i, j)$. Renumbering $u_1, \dots, u_{k+1}$, we may assume that $u_1, \dots, u_{k+1}$ is arranged according to the lexicographic order. Here we put $v_i = u_{i+1} - u_1$ $(i = 1, \dots, k)$. If we choose $x_i \in K$ with $u_i = x_i/2$, then

$$v_i = \frac{x_{i+1} - x_1}{2} = \frac{x_{i+1} + (-x_1)}{2} \in K$$

because K is symmetric. Moreover, as $u_1, \dots, u_{k+1}$ is arranged according to the lexicographic order, the first non-zero entry of v_i is positive. Thus

$$v_1, \dots, v_k, -v_1, \dots, -v_k \in K \cap \mathbb{Z}^n$$

are non-zero distinct elements. Therefore, counting the origin 0, (1) follows.

(2) For a positive integer m, we set $A_m = ((1 + (1/m))K) \cap \mathbb{Z}^n$. Then

$$A_1 \supseteq A_2 \supseteq \cdots \supseteq A_m \supseteq A_{m+1} \supseteq \cdots$$

and A_m is a finite set. Therefore, there are a positive integer m_0 and a finite subset A of $\mathbb{Z}^n$ such that $A_m = A$ for all $m \ge m_0$. Since

$$\mathrm{vol}_n((1 + (1/m))K) > \mathrm{vol}_n(K) \ge 2^n k,$$

we have $\#(A_m) \ge 2k + 1$ by (1). In particular, $\#(A) \ge 2k + 1$. Thus it is sufficient to see that $A \subseteq K \cap \mathbb{Z}^n$. Let a be an element of A. For $m \ge m_0$, as $a \in A = A_m \subseteq (1 + (1/m))K$, there is $x_m \in K$ such that $a = (1 + (1/m))x_m$, and hence $\lim_{m\to\infty} x_m = a$. This implies $a \in K$ because K is closed. □

Let V be an n-dimensional vector space. A $\mathbb{Z}$-submodule Λ of V is called a *lattice* of V if the natural homomorphism $\Lambda \otimes_{\mathbb{Z}} \mathbb{R} \to V$ is an isomorphism. Let us fix a Haar measure vol of V. A quantity

$$\mathrm{vol}\left(\{\lambda_1\omega_1 + \cdots + \lambda_n\omega_n \mid 0 \le \lambda_i \le 1\ (\forall i)\}\right)$$

does not depend on the choice of a free basis $\omega_1, \dots, \omega_n$ of Λ, so that it is denoted by $\mathrm{vol}(V/\Lambda)$. If V has an inner product h, then

$$\mathrm{vol}_h(V/\Lambda) = \sqrt{\det(h(\omega_i, \omega_j))},$$

where vol_h is the Haar measure induced by h. For a measurable subset X of V, the quantity $\log(\mathrm{vol}(X)/\mathrm{vol}(V/\Lambda))$ does not depend on the choice of the Haar measure, so that it is denoted by $\widehat{\chi}(X;\Lambda)$, that is, $\exp(\widehat{\chi}(X;\Lambda)) = \mathrm{vol}(X)/\mathrm{vol}(V/\Lambda)$.

Let K be a convex set in V. The set $K \cap \Lambda$ is called a *convex lattice of* V. Minkowski's theorem asserts a good lower bound of the number of points in a convex lattice.

COROLLARY 2.2. *If K is bounded and symmetric, then we have the following:*

(1) $\#(K \cap \Lambda) \geq 2^{-n}\exp(\widehat{\chi}(K;\Lambda))$.

(2) *If K is closed, then* $\#(K \cap \Lambda) > 2^{-n}\exp(\widehat{\chi}(K;\Lambda))$.

PROOF. Let $\omega_1, \ldots, \omega_n$ be a free basis of Λ. Let us consider an isomorphism $\phi : V \to \mathbb{R}^n$ given by $\phi(\omega_i) = e_i$ $(i = 1, \ldots, n)$, where $\{e_1, \ldots, e_n\}$ is the standard basis of $\mathbb{R}^n$. Then, since $\#(K \cap \Lambda) = \#(\phi(K) \cap \mathbb{Z}^n)$ and $\widehat{\chi}(K;\Lambda) = \widehat{\chi}(\phi(K);\mathbb{Z}^n)$, we may assume that $V = \mathbb{R}^n$ and $\Lambda = \mathbb{Z}^n$.

Let vol_n be the standard Haar measure of $\mathbb{R}^n$. Note that $\exp(\widehat{\chi}(K;\Lambda)) = \mathrm{vol}_n(K)$. If $\mathrm{vol}_n(K) < 2^n$, then $2^{-n}\mathrm{vol}_n(K) < 1$. Thus, in this case, (1) and (2) are obvious. Next we assume that $\mathrm{vol}_n(K) = 2^n$. Then (1) is also obvious. Moreover, if K is closed, then, by (2) in Theorem 2.1, $\#(K \cap \Lambda) \geq 3$, so that (2) follows. Finally we suppose that $\mathrm{vol}_n(K) > 2^n$. Let us take a positive integer k and a real number ϵ with

$$2^{-n}\mathrm{vol}_n(K) = k + \epsilon \quad (0 < \epsilon \leq 1).$$

By Theorem 2.1, we have

$$\#(K \cap \mathbb{Z}^n) \geq 2k + 1.$$

As $2^{-n}\mathrm{vol}_n(K) > 1$, we have

$$\#(K \cap \mathbb{Z}^n) \geq 2(2^{-n}\mathrm{vol}_n(K) - \epsilon) + 1 \geq 2(2^{-n}\mathrm{vol}_n(K)) - 1 > 2^{-n}\mathrm{vol}_n(K).$$

□

Finally we consider the following proposition as an application of Minkowski's theorem.

PROPOSITION 2.3. *Let V be an n-dimensional real vector space and let q be a quadratic form on V. If there is a lattice Λ of V such that*

$$\inf\{q(\lambda) \mid \lambda \in \Lambda \setminus \{0\}\} > 0,$$

then q is positive definite.

PROOF. Let us choose a basis $e_1, \ldots, e_n$ of V such that

$$q(x_1e_1 + \cdots + x_ne_n) = x_1^2 + \cdots + x_s^2 - x_{s+1}^2 - \cdots - x_{s+t}^2.$$

We set $\epsilon = \inf\{q(\lambda) \mid \lambda \in \Lambda \setminus \{0\}\} > 0$. For a positive real number a, we put

$$K(a) = \{x_1e_1 + \cdots + x_ne_n \in V \mid x_1^2 + \cdots + x_s^2 \leq \epsilon/2,\ x_{s+1}^2 + \cdots + x_n^2 \leq a\}.$$

Note that $q \leq \epsilon/2$ on $K(a)$. Therefore, $K(a) \cap \Lambda = \{0\}$. We assume that $s < n$. Then, for a sufficiently large a, $2^{-n}\exp(\widehat{\chi}(K(a);\Lambda)) \geq 1$. Hence, as $K(a)$ is a bounded and symmetric closed convex set, by Corollary 2.2,

$$\#(K(a) \cap \Lambda) > 2^{-n}\exp(\widehat{\chi}(K(a);\Lambda)) \geq 1.$$

This is a contradiction. Thus $s = n$, that is, q is positive definite. □

2.2. Polar dual set and Mahler's inequality

Let (V, h) be a pair of an n-dimensional real vector space V and an inner product h of V. For a subset X of V, the *polar dual set* X^* of X is defined by

$$X^* = \{x \in V \mid h(x, y) \leq 1 \quad (\forall y \in X)\}.$$

It is easy to see that X^* is a closed convex set of V and $X \subseteq (X^*)^*$. Moreover, if X is a convex set with the origin, then $\overline{X} = (X^*)^*$ (cf. [**72**, Theorem 2.8.3]). Note that if X is symmetric, that is, $-x \in X$ for all $x \in X$, then

$$X^* = \{x \in V \mid |h(x, y)| \leq 1 \quad (\forall y \in X)\}.$$

Further, if X is a bounded and symmetric convex set with an interior point, then so is X^* (cf. [**72**, Theorem 2.8.4]). Let us begin with the following lemma.

LEMMA 2.4. *Let $\langle , \rangle$ be the standard inner product of $\mathbb{R}^n$. Let X be a subset of $\mathbb{R}^n$ and let A be an $n \times n$ regular real matrix. Then $(AX)^* = {}^tA^{-1}X^*$.*

PROOF. Since $\langle x, Ay\rangle = \langle {}^tAx, y\rangle$ in general, we have

$$\begin{aligned} x \in (AX)^* &\iff \langle x, Ay\rangle \leq 1 \ (\forall y \in X) \\ &\iff \langle {}^tAx, y\rangle \leq 1 \ (\forall y \in X) \\ &\iff {}^tAx \in X^* \\ &\iff x \in {}^tA^{-1}X^*, \end{aligned}$$

as required. □

The following theorem is important for this book.

THEOREM 2.5 (Mahler's inequality). *Let (V, h) be a pair of an n-dimensional real vector space V and an inner product h of V. For a bounded and symmetric convex set K of V with an interior point, the following two inequalities hold:*

$$\frac{4^n}{(n!)^2} \leq \operatorname{vol}_h(K) \operatorname{vol}_h(K^*) \leq 4^n.$$

PROOF. Considering an orthonormal basis of V with respect to h, we may assume that $V = \mathbb{R}^n$ and h is the standard inner product of $\mathbb{R}^n$. Clearly we may also assume that K is closed. Let $\phi : K^n \to \mathbb{R}$ be the continuous map given by $\phi(v_1, \ldots, v_n) = |\det(v_1, \ldots, v_n)|$, where $\det(v_1, \ldots, v_n)$ is the determinant of the matrix $(v_1, \ldots, v_n)$ obtained by arranging the vectors $v_1, \ldots, v_n$ as column vectors. Since K is compact, there are $u_1, \ldots, u_n \in K$ such that ϕ takes the maximal value at $(u_1, \ldots, u_n)$. Note that $u_1, \ldots, u_n$ are linearly independent because $\phi(u_1, \ldots, u_n) > 0$. Thus there is a regular $n \times n$ real matrix A with $Au_i = e_i$ $(i = 1, \ldots, n)$, where $e_1, \ldots, e_n$ is the standard basis of $\mathbb{R}^n$. As

$$\begin{cases} \phi(Av_1, \ldots, Av_n) = |\det A|\phi(v_1, \ldots, v_n), \\ \operatorname{vol}_n(K)\operatorname{vol}_n(K^*) = \operatorname{vol}_n(AK)\operatorname{vol}_n((AK)^*) \quad (\because \text{Lemma 2.4}), \end{cases}$$

we may further assume that $u_i = e_i$ $(i = 1, \ldots, n)$. We set

$$W = \{(x_1, \ldots, x_n) \in \mathbb{R}^n \mid |x_i| \leq 1 \ (\forall i)\}.$$

Then it is easy to see that $W^* = \{(x_1, \ldots, x_n) \in \mathbb{R}^n \mid |x_1| + \cdots + |x_n| \leq 1\}$, $\operatorname{vol}_n(W) = 2^n$ and $\operatorname{vol}_n(W^*) = 2^n/n!$. Moreover, we can see that

$$W^* \subseteq K \subseteq W. \tag{2.1}$$

Indeed, as $\pm e_i \in K$, it is obvious that $W^* \subseteq K$. Further, if $x = (x_1, \ldots, x_n) \in K$, then, for any i,

$$|x_i| = \phi(e_1, \ldots, e_{i-1}, x, e_{i+1}, \ldots, e_n) \leq \phi(e_1, \ldots, e_n) = 1.$$

Thus $x \in W$.

By (2.1),

$$\frac{2^n}{n!} \leq \mathrm{vol}_n(K) \leq 2^n.$$

On the other hand, considering the polar dual of (2.1), we obtain

$$W^* \subseteq K^* \subseteq (W^*)^* = W,$$

which implies

$$\frac{2^n}{n!} \leq \mathrm{vol}_n(K^*) \leq 2^n.$$

Hence the theorem follows. □

REMARK 2.1. By virtue of Blaschke and Santaló, we know the inequality

$$\mathrm{vol}_h(K)\,\mathrm{vol}_h(K^*) \leq \mathrm{vol}_h\left(\{x \in V \mid h(x,x) \leq 1\}\right)^2.$$

Moreover, it is expected that

$$\frac{4^n}{n!} \leq \mathrm{vol}_h(K)\,\mathrm{vol}_h(K^*)$$

holds (Mahler's conjecture), and the above theorem is sufficient for this book.

2.3. The Brunn-Minkowski theorem

In this section, we prove the Brunn-Minkowski theorem, which is one of the fundamental theorems in the geometry of convex sets.

THEOREM 2.3 (Brunn-Minkowski theorem). *Let (V, h) be a pair of an n-dimensional real vector space V and an inner product h of V. For compact sets X and Y of V, the inequality*

$$\mathrm{vol}_h(X+Y)^{1/n} \geq \mathrm{vol}_h(X)^{1/n} + \mathrm{vol}_h(Y)^{1/n}$$

holds, where $X + Y = \{x + y \mid x \in X,\ y \in Y\}$ (note that $X + Y$ is a compact set).

PROOF. Taking an orthonormal basis of V, we may assume that $V = \mathbb{R}^n$ and that h is the standard inner product. Here we introduce several definitions. A subset B of $\mathbb{R}^n$ is called a *compact interval* if there are compact intervals $I_1, \ldots, I_n$ of $\mathbb{R}$ such that $B = I_1 \times \cdots \times I_n$. Moreover, a subset X of $\mathbb{R}^n$ is called a *compact elementary set* if there are finitely many compact intervals $B_1, \ldots, B_r$ such that $X = B_1 \cup \cdots \cup B_r$ and that $B_i \cap B_j$ has no interior points for all $i \neq j$.

Since X and Y are compact, there are sequences $\{X_l\}$ and $\{Y_l\}$ of compact elementary sets such that

$$\begin{cases} X_1 \supseteq X_2 \supseteq \cdots \supseteq X_l \supseteq \cdots, & X = \bigcap_{l \geq 0} X_l, \\ Y_1 \supseteq Y_2 \supseteq \cdots \supseteq Y_l \supseteq \cdots, & Y = \bigcap_{l \geq 0} Y_l. \end{cases}$$

Then we have the following:

CLAIM 1. $X + Y = \bigcap_{l \geq 0}(X_l + Y_l)$.

PROOF. It is obvious that $X + Y \subseteq \bigcap_{l\geq 0}(X_l + Y_l)$. Conversely we assume $z \in \bigcap_{l\geq 0}(X_l + Y_l)$. Then we can set $z = x_l + y_l$ ($x_l \in X_l,\ y_l \in Y_l$) for all l. Since $x_l \in X_1$ ($\forall l$) and X_1 is compact, we can find a convergent subsequence $\{x_{l_i}\}$ of $\{x_l\}$. As $y_{l_i} = z - x_{l_i}$, $\{y_{l_i}\}$ is also a convergent sequence. Here we set

$$x = \lim_{i\to\infty} x_{l_i} \quad \text{and} \quad y = \lim_{i\to\infty} y_{l_i}.$$

We assume that $x \notin X$. Then there is l with $x \notin X_l$. Since $\mathbb{R}^n \setminus X_l$ is an open set, we can find i_0 such that, for all $i \geq i_0$, $x_{l_i} \notin X_l$. Here we choose i with $l_i \geq l$ and $i \geq i_0$. Then $x_{l_i} \in X_{l_i} \subseteq X_l$, which is a contradiction. Therefore $x \in X$. In the same way, we can check $y \in Y$. Hence $z = x + y \in X + Y$, as required. □

By the above claim, we have

$$\begin{cases} \mathrm{vol}_n(X) = \lim_{l\to\infty} \mathrm{vol}_n(X_l), \\ \mathrm{vol}_n(Y) = \lim_{l\to\infty} \mathrm{vol}_n(Y_l), \\ \mathrm{vol}_n(X+Y) = \lim_{l\to\infty} \mathrm{vol}_n(X_l + Y_l). \end{cases}$$

Therefore we may assume that X and Y are compact elementary sets in order to show the theorem. Moreover, as

$$\mathrm{vol}_n(X+Y) \geq \max\{\mathrm{vol}_n(X), \mathrm{vol}_n(Y)\},$$

the assertion of the theorem is trivial if either $\mathrm{vol}_n(X) = 0$ or $\mathrm{vol}_n(Y) = 0$. Hence we may further assume that $\mathrm{vol}_n(X) > 0$ and $\mathrm{vol}_n(Y) > 0$.

For a compact elementary set Z, there is no unique way to express Z as a union of compact intervals without sharing an interior point. If the number of compact intervals is minimal among all expressions of Z as above, then such an expression is called a minimal expression of Z and the number of compact intervals in a minimal expression is denoted by $\sigma(Z)$.

First we consider the case where $\sigma(X) + \sigma(Y) = 2$, that is, $\sigma(X) = \sigma(Y) = 1$. In this case, there are compact intervals $I_1, \ldots, I_n, J_1, \ldots, J_n$ in $\mathbb{R}$ such that $X = I_1 \times \cdots \times I_n$ and $Y = J_1 \times \cdots \times J_n$. Then $X + Y = (I_1 + J_1) \times \cdots \times (I_n + J_n)$. If we set

$$a_1 = \mathrm{vol}_1(I_1), \ldots, a_n = \mathrm{vol}_1(I_n), b_1 = \mathrm{vol}_1(J_1), \ldots, b_n = \mathrm{vol}_1(J_n),$$

then

$$\begin{cases} \mathrm{vol}_n(X) = a_1 \cdots a_n, \\ \mathrm{vol}_n(Y) = b_1 \cdots b_n, \\ \mathrm{vol}_n(X+Y) = (a_1 + b_1) \cdots (a_n + b_n). \end{cases}$$

As the arithmetic mean is greater than or equal to the geometric mean, we have

$$\begin{aligned} &\left(\frac{\mathrm{vol}_n(X)}{\mathrm{vol}_n(X+Y)}\right)^{1/n} + \left(\frac{\mathrm{vol}_n(Y)}{\mathrm{vol}_n(X+Y)}\right)^{1/n} \\ &\qquad = \left(\frac{a_1}{a_1+b_1} \cdots \frac{a_n}{a_n+b_n}\right)^{1/n} + \left(\frac{b_1}{a_1+b_1} \cdots \frac{b_n}{a_n+b_n}\right)^{1/n} \\ &\qquad \leq \frac{1}{n}\left(\frac{a_1}{a_1+b_1} + \cdots + \frac{a_n}{a_n+b_n}\right) + \frac{1}{n}\left(\frac{b_1}{a_1+b_1} + \cdots + \frac{b_n}{a_n+b_n}\right) = 1, \end{aligned}$$

as required.

Next we consider the case where $\sigma(X)+\sigma(Y) \geq 3$. Clearly we may assume that $\sigma(X) \geq 2$. Then there are two compact intervals B and B' in a minimal expression of X.

CLAIM 2. *There are a real number c and an i-th coordinate x_i of $\mathbb{R}^n$ such that B and B' are separated by a hyperplane $\{x_i = c\}$.*

PROOF. We choose compact intervals $T_1, \ldots, T_n, T'_1, \ldots, T'_n$ in $\mathbb{R}$ such that $B = T_1 \times \cdots \times T_n$, $B' = T'_1 \times \cdots \times T'_n$. Then $B \cap B' = (T_1 \cap T'_1) \times \cdots \times (T_n \cap T'_n)$. Since $B \cap B'$ has no interior point, there is i such that $T_i \cap T'_i$ is either empty or a point. Thus our claim follows. □

We set $X_+ = X \cap \{x_i \geq c\}$ and $X_- = X \cap \{x_i \leq c\}$. Then X_+ and X_- are compact elementary sets. Moreover, as B and B' are separated by a hyperplane $\{x_i = c\}$, we can see $\sigma(X_+) \leq \sigma(X) - 1$ and $\sigma(X_-) \leq \sigma(X) - 1$. Here we consider the function $\phi(t)$ given by $\phi(t) = \operatorname{vol}_n(Y \cap \{x_i \leq t\})$. It is easy to see that ϕ is continuous. Thus there is a real number d such that

$$\frac{\operatorname{vol}_n(X_-)}{\operatorname{vol}_n(X)} = \frac{\operatorname{vol}_n(Y_-)}{\operatorname{vol}_n(Y)},$$

where $Y_+ = Y \cap \{x_i \geq d\}$ and $Y_- = Y \cap \{x_i \leq d\}$. Namely, we have

$$\operatorname{vol}_n(X) : \operatorname{vol}_n(Y) = \operatorname{vol}_n(X_-) : \operatorname{vol}_n(Y_-) = \operatorname{vol}_n(X_+) : \operatorname{vol}_n(Y_+).$$

Moreover, we can see that

$$\begin{cases} X_- + Y_- \subseteq (X+Y) \cap \{x_i \leq c+d\}, \\ X_+ + Y_+ \subseteq (X+Y) \cap \{x_i \geq c+d\} \end{cases}$$

and

$$\begin{cases} \sigma(X_-) + \sigma(Y_-) \leq \sigma(X) + \sigma(Y) - 1, \\ \sigma(X_+) + \sigma(Y_+) \leq \sigma(X) + \sigma(Y) - 1. \end{cases}$$

Therefore, using the hypothesis of induction, we obtain

$$\begin{aligned} \operatorname{vol}_n(X+Y) &\geq \operatorname{vol}_n((X_- + Y_-) \cup (X_+ + Y_+)) \\ &= \operatorname{vol}_n(X_- + Y_-) + \operatorname{vol}_n(X_+ + Y_+) \\ &\geq \left(\operatorname{vol}_n(X_-)^{1/n} + \operatorname{vol}(Y_-)^{1/n}\right)^n + \left(\operatorname{vol}_n(X_+)^{1/n} + \operatorname{vol}(Y_+)^{1/n}\right)^n \\ &= \operatorname{vol}_n(X_-)\left(1 + \frac{\operatorname{vol}_n(Y)^{1/n}}{\operatorname{vol}_n(X)^{1/n}}\right)^n + \operatorname{vol}_n(X_+)\left(1 + \frac{\operatorname{vol}_n(Y)^{1/n}}{\operatorname{vol}_n(X)^{1/n}}\right)^n \\ &= \operatorname{vol}_n(X)\left(1 + \frac{\operatorname{vol}_n(Y)^{1/n}}{\operatorname{vol}_n(X)^{1/n}}\right)^n = \left(\operatorname{vol}_n(X)^{1/n} + \operatorname{vol}_n(Y)^{1/n}\right)^n. \end{aligned}$$

□

COROLLARY 2.7. *For bounded convex sets K_1 and K_2 of V, we have*

$$\operatorname{vol}_h(K_1 + K_2)^{1/n} \geq \operatorname{vol}_h(K_1)^{1/n} + \operatorname{vol}_h(K_2)^{1/n}.$$

PROOF. Note that K_1, K_2 and $K_1 + K_2$ are bounded convex sets. Moreover, $\overline{K}_1$ and $\overline{K}_2$ are compact, and $\overline{K}_1 + \overline{K}_2 \subseteq \overline{K_1 + K_2}$. By Theorem 2.6, we have

$$\begin{aligned}\operatorname{vol}_h(K_1 + K_2)^{1/n} &= \operatorname{vol}_h(\overline{K_1 + K_2})^{1/n} \geq \operatorname{vol}_h(\overline{K}_1 + \overline{K}_2)^{1/n} \\ &\geq \operatorname{vol}_h(\overline{K}_1)^{1/n} + \operatorname{vol}_h(\overline{K}_2)^{1/n} \\ &= \operatorname{vol}_h(K_1)^{1/n} + \operatorname{vol}_h(K_2)^{1/n}.\end{aligned}$$

□

COROLLARY 2.8. *Let L be a subspace of V and let K be a bounded and symmetric convex set of V. Let h_L be the subinner product of L induced by h. Then, for any $x \in V$,*

$$\operatorname{vol}_{h_L}(K \cap L) \geq \operatorname{vol}_{h_L}(K \cap (L + x)),$$

where the volume of $K \cap (L + x)$ is measured on L by the natural identification $L \xrightarrow{\sim} L + x$ $(l \mapsto l + x)$.

PROOF. Let $\pi : V \to L$ be the orthogonal projection with respect to h. We set

$$K_1 = \pi(K \cap (L + x)) \quad \text{and} \quad K_2 = \pi(K \cap (L - x)).$$

For a subset X of V, we put $-X = \{-x \mid x \in X\}$. Then, since

$$-K_2 = \pi(-(K \cap (L - x))) = \pi(K \cap (L + x)) = K_1,$$

$\operatorname{vol}_{h_L}(K_1) = \operatorname{vol}_{h_L}(K_2)$. Therefore, by Corollary 2.7, if we set $l = \dim L$, then

$$\begin{aligned}\operatorname{vol}_{h_L}((1/2)K_1 + (1/2)K_2) &\geq \left(\operatorname{vol}_{h_L}((1/2)K_1)^{1/l} + \operatorname{vol}_{h_L}((1/2)K_2)^{1/l}\right)^l \\ &= \left((1/2)\operatorname{vol}_{h_L}(K_1)^{1/l} + (1/2)\operatorname{vol}_{h_L}(K_2)^{1/l}\right)^l \\ &= \operatorname{vol}_{h_L}(K_1).\end{aligned}$$

On the other hand,

$$\frac{1}{2}K_1 + \frac{1}{2}K_2 \subseteq \pi\left(\frac{1}{2}(K \cap (L + x)) + \frac{1}{2}(K \cap (L - x))\right) \subseteq K \cap L,$$

and hence

$$\begin{aligned}\operatorname{vol}_{h_L}(K \cap L) &\geq \operatorname{vol}_{h_L}((1/2)K_1 + (1/2)K_2) \\ &\geq \operatorname{vol}_{h_L}(K_1) = \operatorname{vol}_{h_L}(K \cap (L + x)).\end{aligned}$$

□

2.4. Estimate of the number of points in a convex lattice

Let (V, h) be a pair of an n-dimensional real vector space V and an inner product h of V. Let Λ be a lattice of V. In this section, we prove the following estimate of the number of points in a convex lattice due to Gillet-Soulé [**25**].

THEOREM 2.9. *We assume that (V, h, Λ) is unimodular, that is, $h(x, y) \in \mathbb{Z}$ for all $x, y \in \Lambda$, and, for a free basis $\{\omega_1, \ldots, \omega_n\}$ of Λ, $(h(\omega_i, \omega_j)) \in \mathrm{GL}(n, \mathbb{Z})$. Then, for a bounded and symmetric convex set K with an interior point, we have the following inequalities:*

$$6^{-n} \leq \frac{\#(K \cap \Lambda)}{\#(K^* \cap \Lambda)\operatorname{vol}_h(K)} \leq \left(\frac{3}{2}\right)^n (n!)^2.$$

For the proof of Theorem 2.9, we need to prepare several lemmas. Let us begin with the following lemma.

LEMMA 2.10. *Let $\phi : V \to V'$ be a surjective homomorphism of finite-dimensional real vector spaces, and let $W = \operatorname{Ker}(\phi)$. Let h_W be the subinner product of W induced by h, and let h' be the quotient inner product of V' induced by ϕ and h. Then, for a bounded and symmetric convex set K with an interior point, we have the following.*

(1) $\operatorname{vol}_h(K) \leq \operatorname{vol}_{h'}(\phi(K)) \operatorname{vol}_{h_W}(K \cap W)$.
(2) $\operatorname{vol}_h(K^*) \leq \operatorname{vol}_{h'}(\phi(K)^*) \operatorname{vol}_{h_W}((K \cap W)^*)$.

PROOF. (1) Let $W^\perp$ be the orthogonal complement of W with respect to h, and let $p : V \to W^\perp$ be the orthogonal projection to $W^\perp$ with respect to h. Moreover, let $h_{W^\perp}$ be the subinner product of $W^\perp$ induced by h. Then

$$\operatorname{vol}_h(K) = \int_{p(K)} \operatorname{vol}_{h_W}(p^{-1}(x) \cap K) d\operatorname{vol}_{h_{W^\perp}}.$$

Note that

$$\operatorname{vol}_{h_W}(p^{-1}(x) \cap K) \leq \operatorname{vol}_{h_W}(W \cap K)$$

by Corollary 2.8, and hence

$$\operatorname{vol}_h(K) \leq \operatorname{vol}_{h_{W^\perp}}(p(K)) \operatorname{vol}_{h_W}(K \cap W).$$

As the following diagram is commutative:

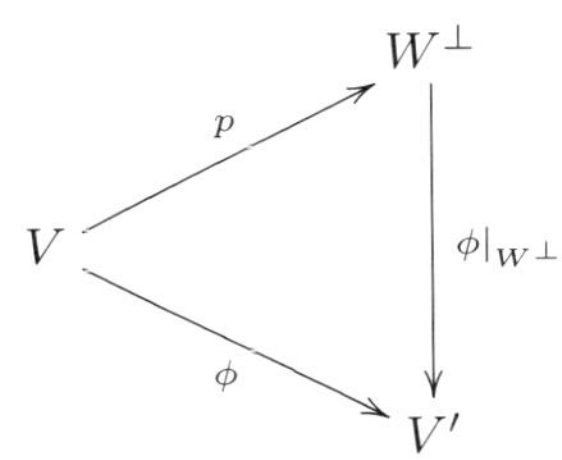

and $\phi|_{W^\perp}$ gives rise to an isometry $(W^\perp, h_{W^\perp}) \xrightarrow{\sim} (V', h')$, we obtain

$$\operatorname{vol}_{h_{W^\perp}}(p(K)) = \operatorname{vol}_{h'}(\phi(K)).$$

Therefore, (1) follows.

(2) Let $q : V \to W$ be the orthogonal projection to W with respect to h. Here we claim the following:

CLAIM 3. (a) $\phi(K^* \cap W^\perp) \subseteq \phi(K)^*$.
(b) $q(K^*) \subseteq (K \cap W)^*$.

PROOF. Let $x \in K^* \cap W^\perp$ and $y \in K$. We put

$$y = y' + y'' \quad (y' \in W^\perp, y'' \in W).$$

Then $h(x, y) = h(x, y') = h'(\phi(x), \phi(y))$ On the other hand, as $x \in K^*$ and $y \in K$, we obtain $|h(x, y)| \leq 1$. Thus $|h'(\phi(x), \phi(y))| \leq 1$, which shows (a).

Next let $x \in K^*$ and $y \in K \cap W$. We set

$$x = x' + x'' \quad (x' \in W, x'' \in W^\perp).$$

Then $h(x, y) = h(x', y) = h_W(q(x), y)$. On the other hand, as $x \in K^*$ and $y \in K$, we have $|h(x, y)| \leq 1$. Thus $|h_W(q(x), y)| \leq 1$, which shows (b). □

Let us go back to the proof of (2). Note that $\operatorname{Ker}(q) = W^{\perp}$ and that h_W coincides with the quotient inner product of W induced by $q : V \to W$ and h. Thus, applying (1) to K^* and $q : V \to W$, we obtain

$$\begin{aligned}\operatorname{vol}_h(K^*) &\leq \operatorname{vol}_{h_{W^{\perp}}}(K^* \cap W^{\perp}) \operatorname{vol}_{h_W}(q(K^*)) \\ &= \operatorname{vol}_{h'}(\phi(K^* \cap W^{\perp})) \operatorname{vol}_{h_W}(q(K^*)).\end{aligned}$$

Therefore, (2) follows from the above claim. □

Next we consider the following lemma.

LEMMA 2.11. (1) *Let* $0 \to V' \xrightarrow{\phi} V \xrightarrow{\psi} V'' \to 0$ *be an exact sequence of finite-dimensional real vector spaces. Let* h *be an inner product of* V. *Let* h' *be the subinner product of* V' *induced by* $\phi : V' \to V$ *and* h, *and let* h'' *be the quotient inner product of* V'' *induced by* $\psi : V \to V''$ *and* h. *Let* Λ', Λ *and* Λ'' *be lattices of* V', V *and* V'', *respectively. We assume that the above exact sequence gives rise to an exact sequence* $0 \to \Lambda' \to \Lambda \to \Lambda'' \to 0$ *as* $\mathbb{Z}$-*modules. Then we have*

$$\operatorname{vol}_h(V/\Lambda) = \operatorname{vol}_{h'}(V'/\Lambda') \operatorname{vol}_{h''}(V''/\Lambda'').$$

(2) *Let* V *be a finite-dimensional real vector space,* h *an inner product of* V *and* $h^{\vee}$ *the dual inner product of* h. *Let* Λ *be a lattice of* V *and* $\Lambda^{\vee} = \operatorname{Hom}_{\mathbb{Z}}(\Lambda, \mathbb{Z})$. *Then* $\Lambda^{\vee}$ *is a lattice of* $V^{\vee}$ *and* $\operatorname{vol}_{h^{\vee}}(V^{\vee}/\Lambda^{\vee}) = \operatorname{vol}_h(V/\Lambda)^{-1}$.

PROOF. (1) Let $\{e'_1, \ldots, e'_m\}$, $\{e_1, \ldots, e_n\}$ and $\{e''_1, \ldots, e''_{n-m}\}$ be free bases of Λ', Λ and Λ'' such that

$$\phi(e'_i) = e_i \quad (i = 1, \ldots, m) \quad \text{and} \quad \psi(e_{m+j}) = e''_j \quad (j = 1, \ldots, n-m).$$

Then, by Lemma 1.1,

$$\det(h(e_i, e_j)) = \det(h'(e'_i, e'_j)) \det(h''(e''_i, e''_j)).$$

Thus we obtain (1).

(2) Let $e_1, \ldots, e_n$ be a free basis of Λ, and let $e_1^{\vee}, \ldots, e_n^{\vee}$ be the dual basis of $e_1, \ldots, e_n$ in $V^{\vee}$. Then $e_1^{\vee}, \ldots, e_n^{\vee}$ form a free basis of $\Lambda^{\vee}$. On the other hand, by Lemma 1.4, $(h(e_i, e_j))(h^{\vee}(e_i^{\vee}, e_j^{\vee})) = I_n$, and hence (2) follows. □

From now on, let Λ be a lattice of V.

LEMMA 2.12. *Let* K *be a bounded and symmetric convex set with an interior point. If* $K \cap \Lambda$ *generates* V *as a real vector space, then*

$$\#(K \cap \Lambda) \operatorname{vol}_h(K^*) \operatorname{vol}_h(V/\Lambda) \leq 6^n.$$

Note that we do not assume that (V, h, Λ) *is unimodular.*

PROOF. First, let us recall the primitivity of a lattice point. For $x \in \Lambda \setminus \{0\}$, x is said to be *primitive* if there are no integer $l \geq 2$ and $y \in \Lambda \setminus \{0\}$ such that $x = ly$. Note that x is primitive if and only if $\Lambda/\mathbb{Z}x$ has no torsion. Moreover, as K is a symmetric convex set, we can find a primitive lattice point in K if $K \cap \Lambda \neq \{0\}$.

Let Λ' be a $\mathbb{Z}$-module generated by $K \cap \Lambda$. Then $\Lambda' \subseteq \Lambda$ and $\operatorname{rk} \Lambda' = \operatorname{rk} \Lambda$. Thus $\operatorname{vol}_h(V/\Lambda') \geq \operatorname{vol}_h(V/\Lambda)$. Note that $\#(K \cap \Lambda') = \#(K \cap \Lambda)$. Therefore, if the assertion of the lemma holds for Λ', then it also holds for Λ, so that we may assume that $K \cap \Lambda$ generates Λ as a $\mathbb{Z}$-module.

We prove the lemma by induction on n under the assumption that $K \cap \Lambda$ generates Λ as a $\mathbb{Z}$-module. First we have the case where $n = 1$. Clearly we may assume that $V = \mathbb{R}$, h is the standard inner product, $\Lambda = \mathbb{Z}a$ ($a > 0$) and K is a symmetric interval. It is easy to see the following claim.

CLAIM 4. *Let a be a positive real number and let I be an interval in $\mathbb{R}$. Then*

$$\#(I \cap \mathbb{Z}a) \leq \frac{\mathrm{vol}_1(I)}{a} + 1.$$

Using the above claim, $\#(K \cap \Lambda) \leq \mathrm{vol}_1(K)/a + 1$. Moreover, since $[-a, a] \subseteq K$, $2a \leq \mathrm{vol}_1(K)$. On the other hand, $\mathrm{vol}_1(K)\,\mathrm{vol}_1(K^*) = 4$. Thus we obtain

$$\#(K \cap \Lambda)\,\mathrm{vol}_1(K^*)\,\mathrm{vol}_1(\mathbb{R}/\Lambda) \leq 4 + \frac{4a}{\mathrm{vol}_1(K)} \leq 6.$$

Next we assume that $n \geq 2$. We choose a primitive lattice point ω in K. Let $V'' = V/\mathbb{R}\omega$, $\psi : V \to V''$ the natural homomorphism, $\Lambda'' = \psi(\Lambda)$ and $K'' = \psi(K)$. Moreover, we set $V' = \mathbb{R}\omega$, $\Lambda' = \mathbb{Z}\omega$ and $K' = K \cap V'$. Let h'' be the quotient inner product of V'' induced by $\psi : V \to V''$ and h, and let h' be the subinner product of V' induced by h. Then, as $K'' \cap \Lambda''$ generates Λ'' as a $\mathbb{Z}$-module, by using the hypothesis of induction,

$$\#(K'' \cap \Lambda'')\,\mathrm{vol}_{h''}(K''^*)\,\mathrm{vol}_{h''}(V''/\Lambda'') \leq 6^{n-1} \tag{2.2}$$

holds. Moreover, we can see

$$\#(K \cap \Lambda) \leq \#(K'' \cap \Lambda'') \cdot \sup_{y \in \Lambda''} \#(\psi^{-1}(y) \cap K \cap \Lambda).$$

For $y \in \Lambda''$, let us fix $x \in \Lambda$ with $\psi(x) = y$. Since

$$\#(\psi^{-1}(y) \cap K \cap \Lambda) = \#(\psi^{-1}(0) \cap (K - x) \cap \Lambda)$$

and

$$\mathrm{vol}_{h'}(\psi^{-1}(y) \cap K) = \mathrm{vol}_{h'}(\psi^{-1}(0) \cap (K - x)),$$

using Claim 4, we have

$$\#(\psi^{-1}(y) \cap K \cap \Lambda) \leq \frac{\mathrm{vol}_{h'}(\psi^{-1}(y) \cap K)}{\mathrm{vol}_{h'}(V'/\Lambda')} + 1.$$

On the other hand, by Corollary 2.8,

$$\mathrm{vol}_{h'}(\psi^{-1}(y) \cap K) \leq \mathrm{vol}_{h'}(\psi^{-1}(0) \cap K) = \mathrm{vol}_{h'}(K').$$

Note that $\pm\omega \in K'$. Thus $\mathrm{vol}_{h'}(K')/\mathrm{vol}_{h'}(V'/\Lambda') \geq 2$, so that

$$\frac{\mathrm{vol}_{h'}(K')}{\mathrm{vol}_{h'}(V'/\Lambda')} + 1 \leq (3/2)\frac{\mathrm{vol}_{h'}(K')}{\mathrm{vol}_{h'}(V'/\Lambda')}.$$

Hence

$$\#(K \cap \Lambda) \leq \#(K'' \cap \Lambda'')(3/2)\frac{\mathrm{vol}_{h'}(K')}{\mathrm{vol}_{h'}(V'/\Lambda')}. \tag{2.3}$$

Moreover, by Lemma 2.10,

$$\mathrm{vol}_h(K^*) \leq \mathrm{vol}_{h'}(K'^*)\,\mathrm{vol}_{h''}(K''^*). \tag{2.4}$$

Therefore, by (2.2), (2.3) and (2.4), we obtain

$$\begin{aligned}
&\#(K \cap \Lambda) \operatorname{vol}_h(K^*) \operatorname{vol}_h(V/\Lambda) \\
&\qquad \leq 6^n \frac{\operatorname{vol}_{h'}(K') \operatorname{vol}_{h'}(K'^*)}{4} \frac{\operatorname{vol}_h(V/\Lambda)}{\operatorname{vol}_{h'}(V'/\Lambda') \operatorname{vol}_{h''}(V''/\Lambda'')}.
\end{aligned}$$

Note that $\operatorname{vol}_{h'}(K') \operatorname{vol}_{h'}(K'^*) = 4$ ($\because \dim V' = 1$), and that

$$\operatorname{vol}_h(V/\Lambda) = \operatorname{vol}_{h'}(V'/\Lambda') \operatorname{vol}_{h''}(V''/\Lambda'')$$

by Lemma 2.11. Hence the lemma follows. □

LEMMA 2.13. *We assume that (V, h, Λ) is unimodular. Let T be a saturated $\mathbb{Z}$-submodule of Λ, that is, T is a $\mathbb{Z}$-submodule of Λ and Λ/T has no torsion. We set*

$$T^\perp = \{x \in \Lambda \mid h(x, y) = 0 \text{ for all } y \in T\}.$$

Let W be a vector subspace of V generated by T, and let $W^\perp$ be the orthogonal complement of W. Then we have the following:

(1) *$T^\perp$ is a lattice of $W^\perp$.*

(2) *Let h_W and $h_{W^\perp}$ be the subinner products of W and $W^\perp$ induced by h, respectively. Then*

$$\operatorname{vol}_{h_W}(W/T) = \operatorname{vol}_{h_{W^\perp}}(W^\perp/T^\perp).$$

PROOF. Let us consider a homomorphism

$$\alpha : T^\perp \to \operatorname{Hom}_{\mathbb{Z}}(\Lambda/T, \mathbb{Z})$$

given by $\alpha(x) = h(x, \cdot)$. First let us see that α is an isomorphism. Clearly α is injective. Let us choose $\phi \in \operatorname{Hom}_{\mathbb{Z}}(\Lambda/T, \mathbb{Z})$. A homomorphism $\Lambda \to \Lambda/T \xrightarrow{\phi} \mathbb{Z}$ is denoted by $\widetilde{\phi}$. Since (V, h, Λ) is unimodular, there is $x \in \Lambda$ such that $\widetilde{\phi}(\cdot) = h(x, \cdot)$. Note that $\widetilde{\phi}(y) = 0$ for all $y \in T$. Thus $x \in T^\perp$, which show the surjectivity of α.

(1) Since α is bijective,

$$\operatorname{rk} T^\perp = \dim_{\mathbb{R}} V - \dim_{\mathbb{R}} W = \dim_{\mathbb{R}} W^\perp.$$

On the other hand, as $T^\perp \subseteq W^\perp$, $T^\perp$ is a lattice of $W^\perp$.

(2) Let $h_{V/W}$ be the quotient inner product of V/W induced by h, and let $h^\vee_{V/W}$ be the dual inner product of $\operatorname{Hom}_{\mathbb{R}}(V/W, \mathbb{R})$ induced by $h_{V/W}$. Let $x_1, \ldots, x_n$ be an orthonormal basis of V such that $x_1, \ldots, x_l \in W^\perp$ and $x_{l+1}, \ldots, x_n \in W$. Moreover, let $x_1^\vee, \ldots, x_n^\vee$ be the dual basis of $x_1, \ldots, x_n$. Then $x_1^\vee, \ldots, x_l^\vee$ give rise to an orthonormal basis of $\operatorname{Hom}_{\mathbb{R}}(V/W, \mathbb{R})$ with respect to $h^\vee_{V/W}$. Let $\alpha_{\mathbb{R}} : W^\perp \to \operatorname{Hom}_{\mathbb{R}}(V/W, \mathbb{R})$ be a homomorphism given by $\alpha_{\mathbb{R}}(x) = h(x, \cdot)$. Then we can see that $\alpha_{\mathbb{R}}(x_i) = x_i^\vee$ $(i = 1, \ldots, l)$. Thus $\alpha_{\mathbb{R}}$ yields an isometry

$$\alpha_{\mathbb{R}} : (W^\perp, h_{W^\perp}) \to \left(\operatorname{Hom}_{\mathbb{R}}(V/W, \mathbb{R}), h^\vee_{V/W}\right).$$

Therefore,

$$\operatorname{vol}_{h_{W^\perp}}(W^\perp/T^\perp) = \operatorname{vol}_{h^\vee_{V/W}}(\operatorname{Hom}_{\mathbb{R}}(V/W, \mathbb{R})/\operatorname{Hom}_{\mathbb{Z}}(\Lambda/T, \mathbb{Z})),$$

which implies that

$$\operatorname{vol}_{h_{W^\perp}}(W^\perp/T^\perp) = \operatorname{vol}_{h_{V/W}}((V/W)/(\Lambda/T))^{-1}$$

by (2) in Lemma 2.11. On the other hand, by using (1) in Lemma 2.11,

$$\mathrm{vol}_h(V/\Lambda) = \mathrm{vol}_{h_W}(W/T)\,\mathrm{vol}_{h_{V/W}}((V/W)/(\Lambda/T)).$$

Note that $\mathrm{vol}_h(V/\Lambda) = 1$ because (V, h, Λ) is unimodular. Thus (2) follows. □

PROOF OF THEOREM 2.9. First let us show the inequality

$$6^{-n} \leq \frac{\#(K \cap \Lambda)}{\#(K^* \cap \Lambda)\,\mathrm{vol}_h(K)}, \tag{2.5}$$

which is the left side inequality in Theorem 2.9. Let T be the saturation of a $\mathbb{Z}$-submodule of Λ generated by $K^* \cap \Lambda$, that is, T is the minimal saturated $\mathbb{Z}$-submodule of Λ containing $K^* \cap \Lambda$. Let W be the vector subspace generated by T, and let $W^\perp$ be the orthonormal complement of W with respect to h. Moreover, let vol_W be the Haar measure of W induced by $h|_{W \times W}$. Then, since $\#(K^* \cap \Lambda) = \#(K^* \cap T)$, by Lemma 2.12,

$$\#(K^* \cap \Lambda) = \#(K^* \cap T) \leq \frac{6^{n_0}}{\mathrm{vol}_W((K^* \cap W)^*)\,\mathrm{vol}_W(W/T)}, \tag{2.6}$$

where $n_0 = \dim W$. Let $p : V \to W$ be the orthogonal projection to W with respect to h. Let us see that $p(K) \subseteq (K^* \cap W)^*$. Indeed, let $x \in K$ and $y \in K^* \cap W$. We set $x = x' + x''$ ($x' \in W$, $x'' \in W^\perp$). Then

$$h(x, y) = h(x', y) = h(p(x), y).$$

On the other hand, as $x \in K$ and $y \in K^*$, we have $|h(x, y)| \leq 1$. Thus $|h(p(x), y)| \leq 1$, which shows that $p(K) \subseteq (K^* \cap W)^*$. Therefore, (2.6) implies

$$\#(K^* \cap \Lambda) \leq \frac{6^{n_0}}{\mathrm{vol}_W(p(K))\,\mathrm{vol}_W(W/T)}. \tag{2.7}$$

CLAIM 5. *Let K° be the set of interior points of K. Then, for all $x \in K^\circ$ and $y \in K^*$, $|h(x, y)| < 1$.*

PROOF. We assume the contrary. Since K^* is symmetric, there are $x \in K^\circ$ and $y \in K^*$ with $h(x, y) = 1$. In particular, $y \neq 0$. As x is an interior point of K, for a sufficiently small positive real number ϵ, $x + \epsilon y \in K$. Therefore, $h(x + \epsilon y, y) \leq 1$. On the other hand, $h(x + \epsilon y, y) = 1 + \epsilon h(y, y) > 1$. This is a contradiction. □

For $x \in K^\circ \cap \Lambda$ and $y \in K^* \cap \Lambda$, $h(x, y) \in \mathbb{Z}$ and $|h(x, y)| < 1$ by the above claim. Thus $h(x, y) = 0$. This means that $K^\circ \cap \Lambda \subseteq W^\perp$. Therefore,

$$\#(K \cap \Lambda) \geq \#(K^\circ \cap \Lambda) = \#(K^\circ \cap W^\perp \cap \Lambda).$$

Here we set

$$T^\perp = \{x \in \Lambda \mid h(x, y) = 0 \text{ for all } y \in T\},$$

that is, $W^\perp \cap \Lambda = T^\perp$. By Lemma 2.13, $T^\perp$ is a lattice of $W^\perp$. By Corollary 2.2,

$$\#(K^\circ \cap T^\perp) \geq 2^{n_0 - n} \frac{\mathrm{vol}_{W^\perp}(K^\circ \cap W^\perp)}{\mathrm{vol}_{W^\perp}(W^\perp/T^\perp)},$$

where $\mathrm{vol}_{W^\perp}$ is the Haar measure induced by $h|_{W^\perp \times W^\perp}$. Therefore, as $2^{n_0-n} \geq 6^{n_0-n}$,

$$\#(K^\circ \cap \Lambda) \geq 6^{n_0 - n} \frac{\mathrm{vol}_{W^\perp}(K^\circ \cap W^\perp)}{\mathrm{vol}_{W^\perp}(W^\perp/T^\perp)}. \tag{2.8}$$

By the inequalities (2.7), (2.8), $K^\circ \subseteq K$ and Lemma 2.13, we obtain

$$\frac{\#(K \cap \Lambda)}{\#(K^* \cap \Lambda)} \geq 6^{-n} \operatorname{vol}_{W^\perp}(K^\circ \cap W^\perp) \operatorname{vol}_W(p(K^\circ)).$$

Therefore, by virtue of Lemma 2.10, we have

$$\frac{\#(K \cap \Lambda)}{\#(K^* \cap \Lambda)} \geq 6^{-n} \operatorname{vol}_h(K^\circ) = 6^{-n} \operatorname{vol}_h(K).$$

Hence the inequality (2.5) follows.

Next let us consider the right side inequality in Theorem 2.9:

$$\frac{\#(K \cap \Lambda)}{\#(K^* \cap \Lambda) \operatorname{vol}_h(K)} \leq \left(\frac{3}{2}\right)^n (n!)^2. \tag{2.9}$$

Applying the inequality (2.5) to K^*, we have

$$6^{-n} \leq \frac{\#(K^* \cap \Lambda)}{\#((K^*)^* \cap \Lambda) \operatorname{vol}(K^*)}.$$

On the other hand, $K \subseteq (K^*)^*$ and $4^n(n!)^{-2} \leq \operatorname{vol}(K)\operatorname{vol}(K^*)$ by Mahler's inequality. Therefore,

$$\begin{aligned} 6^{-n} &\leq \frac{\#(K^* \cap \Lambda)}{\#((K^*)^* \cap \Lambda) \operatorname{vol}(K^*)} \leq \frac{\#(K^* \cap \Lambda)}{\#(K \cap \Lambda) \operatorname{vol}(K^*)} \\ &\leq \frac{(n!)^2 \#(K^* \cap \Lambda) \operatorname{vol}(K)}{4^n \#(K \cap \Lambda)}, \end{aligned}$$

which shows (2.9). □

As an application of Theorem 2.9, we can see that

$$\#((aK) \cap \Lambda) \leq a^n 9^n (n!)^2 \#(K \cap \Lambda)$$

for a real number a with $a \geq 1$. Indeed, as $(aK)^* \subseteq K^*$ and $\operatorname{vol}(aK) = a^n \operatorname{vol}(K)$, applying Theorem 2.9 to aK,

$$\begin{aligned} \#((aK) \cap \Lambda) &\leq \left(\frac{3}{2}\right)^n (n!)^2 \#((aK)^* \cap \Lambda) \operatorname{vol}(aK) \\ &\leq a^n \left(\frac{3}{2}\right)^n (n!)^2 \#(K^* \cap \Lambda) \operatorname{vol}(K). \end{aligned}$$

Moreover, by Theorem 2.9,

$$\#(K^* \cap \Lambda) \operatorname{vol}(K) \leq 6^n \#(K \cap \Lambda),$$

as required.

Using Yuan's idea [**74**] or [**53**], we have the following better estimation.

LEMMA 2.14. *Let V be a finite-dimensional real vector space and K a bounded symmetric convex set in V (note that K does not necessarily generate V). Let Λ be a lattice of V, Λ' a $\mathbb{Z}$-submodule of Λ and $r : \Lambda \to \Lambda/\Lambda'$ the natural homomorphism. Then we have the following estimation:*

$$\log \#r(K \cap \Lambda) \geq \log \#(K \cap \Lambda) - \log \#(2K \cap \Lambda'), \tag{2.10}$$

$$\log \#r(K \cap \Lambda) \leq \log \#(2K \cap \Lambda) - \log \#(K \cap \Lambda'). \tag{2.11}$$

Moreover, if a is a real number with $a \geq 1$,

$$0 \leq \log \#(aK \cap \Lambda) - \log \#(K \cap \Lambda) \leq \log(\lceil 2a \rceil) \dim V. \tag{2.12}$$

PROOF. (2.10) Let $t \in r(K \cap \Lambda)$ and fix $s_0 \in K \cap \Lambda$ with $r(s_0) = t$. Then, for any $s \in r^{-1}(t) \cap K \cap \Lambda$,

$$s - s_0 = s + (-s_0) \in 2K \cap \Lambda'.$$

Thus

$$\#(r^{-1}(t) \cap K \cap \Lambda) \leq \#(2K \cap \Lambda').$$

Therefore,

$$\#(K \cap \Lambda) = \sum_{t \in r(K \cap \Lambda)} \#(r^{-1}(t) \cap K \cap \Lambda) \leq \#(r(K \cap \Lambda))\#(2K \cap \Lambda'),$$

as required.

(2.11) We set $S = K \cap \Lambda + K \cap \Lambda'$. Then $r(S) = r(K \cap \Lambda)$ and $S \subseteq 2K \cap \Lambda$. Moreover, for all $t \in r(S)$,

$$\#(K \cap \Lambda') \leq \#(S \cap r^{-1}(t)).$$

Indeed, if we choose $s_0 \in K \cap \Lambda$ with $r(s_0) = t$, then

$$s_0 + K \cap \Lambda' \subseteq S \cap r^{-1}(t).$$

Therefore,

$$\begin{aligned} \#(2K \cap \Lambda) \geq \#(S) = \sum_{t \in r(S)} \#(r^{-1}(t) \cap S) &\geq \#(r(S))\#(K \cap \Lambda') \\ &= \#(r(K \cap \Lambda))\#(K \cap \Lambda') \end{aligned}$$

as required.

(2.12) We set $n = \lceil 2a \rceil$. Applying (2.10) to the case where $(n/2)K$ and $\Lambda' = n\Lambda$, we have

$$\log \#((n/2)K \cap \Lambda) - \log \#(nK \cap n\Lambda) \leq \log \#\Lambda/n\Lambda = \log(n) \dim V.$$

Note that $a \leq n/2$ and $\#(nK \cap n\Lambda) = \#(K \cap \Lambda)$. Hence we obtain

$$\begin{aligned} 0 &\leq \log \#(aK \cap \Lambda) - \log \#(K \cap \Lambda) \\ &\leq \log \#((n/2)K \cap \Lambda) - \log \#(nK \cap n\Lambda) \leq \log(n) \dim V. \end{aligned}$$

□

2.5. Normed finitely generated $\mathbb{Z}$-module

A pair $(M, \|\cdot\|)$ is called a *normed finitely generated $\mathbb{Z}$-module* if M is a finitely generated $\mathbb{Z}$-module and $\|\cdot\|$ is a norm of $M_{\mathbb{R}} = M \otimes_{\mathbb{Z}} \mathbb{R}$. For a normed finitely generated $\mathbb{Z}$-module $(M, \|\cdot\|)$, we define $\hat{H}^0(M, \|\cdot\|)$ to be

$$\hat{H}^0(M, \|\cdot\|) := \{x \in M \mid \|x\| \leq 1\},$$

and the logarithm of the number of $\hat{H}^0(M, \|\cdot\|)$ is denoted by $\hat{h}^0(M, \|\cdot\|)$, that is,

$$\hat{h}^0(M, \|\cdot\|) := \log \#\hat{H}^0(M, \|\cdot\|).$$

Moreover, we set

$$\hat{H}^0_{<1}(M, \|\cdot\|) := \{x \in M \mid \|x\| < 1\} \quad \text{and} \quad \hat{h}^0_{<1}(M, \|\cdot\|) := \log \#\hat{H}^0_{<1}(M, \|\cdot\|).$$

Let M_{tor} be the torsion part of M. By its definition, it is easy to see that

$$\hat{h}^0(M, \|\cdot\|) = \hat{h}^0(M/M_{tor}, \|\cdot\|) + \log \#(M_{tor}).$$

Note that M/M_{tor} is a lattice of $M_{\mathbb{R}}$. We define $\widehat{\chi}(M, \|\cdot\|)$ to be

$$\widehat{\chi}(M, \|\cdot\|) := \log\left(\frac{\mathrm{vol}(B(M, \|\cdot\|))}{\mathrm{vol}(M_{\mathbb{R}}/(M/M_{tor}))}\right) + \log \#(M_{tor}),$$

where $B(M, \|\cdot\|) = \{x \in M_{\mathbb{R}} \mid \|x\| \leq 1\}$. Note that $\widehat{\chi}(M, \|\cdot\|)$ does not depend on the choice of a Haar measure of $M_{\mathbb{R}}$. In terms of the notation in Section 2.1, if M is free, then

$$\widehat{\chi}(M, \|\cdot\|) = \widehat{\chi}(B(M, \|\cdot\|); M).$$

Let $M^{\vee}$ be the dual space of M over $\mathbb{Z}$, that is, $M^{\vee} = \mathrm{Hom}_{\mathbb{Z}}(M, \mathbb{Z})$. Note that $M^{\vee}$ has no torsion elements. As $(M^{\vee})_{\mathbb{R}}$ is naturally isomorphic to $(M_{\mathbb{R}})^{\vee}$, we denote $(M^{\vee})_{\mathbb{R}}$ by $M_{\mathbb{R}}^{\vee}$. The dual norm of the norm $M_{\mathbb{R}}$ of $\|\cdot\|$ is denoted by $\|\cdot\|^{\vee}$ (cf. Section 1.2). Then $\hat{H}^1(M, \|\cdot\|)$ and $\hat{h}^1(M, \|\cdot\|)$ are defined by

$$\hat{H}^1(M, \|\cdot\|) := \hat{H}^0(M^{\vee}, \|\cdot\|^{\vee}) \quad \text{and} \quad \hat{h}^1(M, \|\cdot\|) := \hat{h}^0(M^{\vee}, \|\cdot\|^{\vee}).$$

If $M = \{0\}$, then $\hat{h}^0(M, \|\cdot\|)$, $\hat{h}^1(M, \|\cdot\|)$ and $\widehat{\chi}(M, \|\cdot\|)$ are defined to be 0.

Let $\Sigma = \{e_1, \ldots, e_r\}$ be a free basis of M/M_{tor}, and let $\langle\ ,\ \rangle_{\Sigma}$ be the standard inner product with respect to Σ, that is, for $x = a_1e_1, + \cdots + a_re_r, y = b_1e_1 + \cdots + b_re_r \in M/M_{tor}$,

$$\langle x, y\rangle_{\Sigma} = a_1b_1 + \cdots + a_rb_r.$$

Then we have the following.

LEMMA 2.15. *We can calculate $\hat{h}^1(M, \|\cdot\|)$ by the following formula:*

$$\hat{h}^1(M, \|\cdot\|) = \log \# \left\{x \in M/M_{tor} \mid |\langle x, y\rangle_{\Sigma}| \leq 1 \quad (\forall y \in B(M, \|\cdot\|)\right\}.$$

PROOF. Let $e_1^{\vee}, \ldots, e_r^{\vee}$ be the dual basis of $e_1, \ldots, e_r$. For $\phi \in \mathrm{Hom}_{\mathbb{Z}}(M, \mathbb{Z})$, there are unique $a_1, \ldots, a_r \in \mathbb{Z}$ with $\phi = a_1e_1^{\vee} + \cdots + a_re_r^{\vee}$. Since

$$\phi(y) = a_1b_1 + \cdots + a_rb_r = \langle a_1e_1 + \cdots + a_re_r, y\rangle_{\Sigma}$$

for $y = b_1e_1 + \cdots + b_re_r \in M_{\mathbb{R}}$,

$$\begin{aligned} \|\phi\|^{\vee} \leq 1 \quad &\iff \quad |\phi(y)| \leq 1 \ (\forall y \in B(M, \|\cdot\|)) \\ &\iff \quad |\langle a_1e_1 + \cdots + a_re_r, y\rangle_{\Sigma}| \leq 1 \ (\forall y \in B(M, \|\cdot\|)). \end{aligned}$$

Thus the lemma follows. □

Here let us consider the following two propositions. The second one is an important tool to estimate $\hat{h}^0$ of normed finitely generated $\mathbb{Z}$-modules.

PROPOSITION 2.16. *For a finitely generated $\mathbb{Z}$-module M, the following hold.*

(1) *Let $\|\cdot\|_1$ and $\|\cdot\|_2$ be norms of $M_{\mathbb{R}}$ with $\|\cdot\|_1 \leq \|\cdot\|_2$. Then*

$$\widehat{\chi}(M, \|\cdot\|_1) \geq \widehat{\chi}(M, \|\cdot\|_2).$$

(2) *Let $\|\cdot\|$ be a norm of M. For a real number λ,*

$$\widehat{\chi}(M, \exp(-\lambda)\|\cdot\|) = \lambda\,\mathrm{rk}(M) + \widehat{\chi}(M, \|\cdot\|).$$

PROOF. (1) is obvious because $B(M, \|\cdot\|_1) \supseteq B(M, \|\cdot\|_2)$. (2) follows from the fact that $B(M, \exp(-\lambda)\|\cdot\|) = \exp(\lambda)B(M, \|\cdot\|)$. □

PROPOSITION 2.17. *The following (1)–(5) hold.*

(1) *For a normed finitely generated* $\mathbb{Z}$*-module* $(M, \|\cdot\|)$,

$$-\log(6)\operatorname{rk} M \leq \hat{h}^0(M, \|\cdot\|) - \hat{h}^1(M, \|\cdot\|) - \widehat{\chi}(M, \|\cdot\|) \\ \leq \log(3/2)\operatorname{rk} M + 2\log((\operatorname{rk} M)!)$$

(2) *Let* $\|\cdot\|_1$ *and* $\|\cdot\|_2$ *be norms of a finitely generated* $\mathbb{Z}$*-module* M *with* $\|\cdot\|_1 \leq \|\cdot\|_2$. *Then we have*

$$\hat{h}^0(M, \|\cdot\|_1) \geq \hat{h}^0(M, \|\cdot\|_2) \quad \text{and} \quad \hat{h}^1(M, \|\cdot\|_1) \leq \hat{h}^1(M, \|\cdot\|_2).$$

Moreover,

$$\widehat{\chi}(M, \|\cdot\|_2) - \widehat{\chi}(M, \|\cdot\|_1) \leq \hat{h}^0(M, \|\cdot\|_2) - \hat{h}^0(M, \|\cdot\|_1) \\ + \log(9)\operatorname{rk} M + 2\log((\operatorname{rk} M)!).$$

(3) *For a non-negative real number* λ,

$$0 \leq \hat{h}^0(M, \exp(-\lambda)\|\cdot\|) - \hat{h}^0(M, \|\cdot\|) \leq \lambda \operatorname{rk} M + \log(3)\operatorname{rk} M.$$

(4) *Let*

$$0 \to (M', \|\cdot\|') \xrightarrow{f} (M, \|\cdot\|) \xrightarrow{g} (M'', \|\cdot\|'') \to 0$$

be an exact sequence of normed finitely generated $\mathbb{Z}$*-modules, that is,*

$$0 \to M' \xrightarrow{f} M \xrightarrow{g} M'' \to 0$$

is an exact sequence of finitely generated $\mathbb{Z}$*-modules,* $\|\cdot\|'$ *is the subnorm induced by* $M'_{\mathbb{R}} \to M_{\mathbb{R}}$ *and* $\|\cdot\|$, *and* $\|\cdot\|''$ *is the quotient norm induced by* $M_{\mathbb{R}} \to M''_{\mathbb{R}}$ *and* $\|\cdot\|$. *Then*

$$\hat{h}^0(M, \|\cdot\|) \leq \hat{h}^0(M', \|\cdot\|') + \hat{h}^0(M'', \|\cdot\|'') + \log(6)\operatorname{rk} M'.$$

(5) *If there is a free basis* $\{e_1, \ldots, e_{\operatorname{rk} M}\}$ *of* M/M_{tor} *with* $\|e_i\| \leq 1$ *for all* i, *then*

$$\hat{h}^1(M, \|\cdot\|) \leq \log(3)\operatorname{rk} M.$$

PROOF. (1) Since

$$\hat{h}^0(M, \|\cdot\|) - \hat{h}^1(M, \|\cdot\|) - \widehat{\chi}(M, \|\cdot\|) \\ = \hat{h}^0(M/M_{tor}, \|\cdot\|) - \hat{h}^1(M/M_{tor}, \|\cdot\|) - \widehat{\chi}(M/M_{tor}, \|\cdot\|),$$

we may assume that M is torsion free. If we put $K = \{x \in M_{\mathbb{R}} \mid \|x\| \leq 1\}$, then $\hat{h}^0(M, \|\cdot\|) = \log \#(K \cap M)$ and $\hat{h}^1(M, \|\cdot\|) = \log \#(K^* \cap M)$ by Lemma 2.15. Moreover, if h is a metric of $M_{\mathbb{R}}$ such that a free basis of M is an orthogonal basis with respect to h, then $\widehat{\chi}(M, \|\cdot\|) = \log \operatorname{vol}_h(K)$. Thus (1) follows from Theorem 2.9.

(2) The inequalities

$$\hat{h}^0(M, \|\cdot\|_1) \geq \hat{h}^0(M, \|\cdot\|_2) \quad \text{and} \quad \hat{h}^1(M, \|\cdot\|_1) \leq \hat{h}^1(M, \|\cdot\|_2)$$

are obvious. The third inequality follows from (1).

(3) Since

$$\hat{h}^0(M, \exp(-\lambda)\|\cdot\|) - \hat{h}^0(M, \|\cdot\|) = \hat{h}^0(M/M_{tor}, \exp(-\lambda)\|\cdot\|) - \hat{h}^0(M/M_{tor}, \|\cdot\|),$$

we may assume that M is torsion free. Note that $\lceil 2a \rceil \leq 3a$ for $a \geq 1$. Thus (3) follows from Lemma 2.14.

(4) We may assume that M' is a submodule of M. We choose $x_1, \dots, x_l \in M$ as follows:

(i) $\|x_i\| \leq 1$ $(\forall i)$.
(ii) $g(x_i) \neq g(x_j)$ $(\forall i \neq j)$.
(iii) For any $x \in M$ with $\|x\| \leq 1$, there is x_i with $g(x) = g(x_i)$.

Using (ii) and (iii), for $x \in M$ with $\|x\| \leq 1$, there is a unique x_i with $g(x) = g(x_i)$. Moreover, $x - x_i \in M'$ and $\|x - x_i\| \leq 2$. On the other hand, $\log(l) \leq \hat{h}^0(M'', \|\cdot\|'')$ because $\|g(x_i)\|'' \leq 1$ for all $i = 1, \dots, l$. Therefore, we have

$$\hat{h}^0(M, \|\cdot\|) \leq \hat{h}^0(M'', \|\cdot\|'') + \log \#\{x' \in M' \mid \|x'\| \leq 2\}.$$

Thus (4) is a consequence of (3).

(5) Let $\langle\ ,\ \rangle$ be the standard inner product with respect to the basis $\{e_1, \dots, e_{\mathrm{rk}\, M}\}$. Then, for $x = a_1 e_1 + \cdots + a_{\mathrm{rk}\, M} e_{\mathrm{rk}\, M}$, if $|\langle x, e_i \rangle| \leq 1$ holds for every i, then $|a_i| \leq 1$ $(\forall i)$. Thus (5) follows from Lemma 2.15. □

2.6. $\lambda_{\mathbb{Q}}$ and $\lambda_{\mathbb{Z}}$

In this section, we consider basic properties of $\lambda_{\mathbb{Q}}$ and $\lambda_{\mathbb{Z}}$ treated in [**75**], [**76**] and [**54**].

For a normed finitely generated $\mathbb{Z}$-module $(M, \|\cdot\|)$, we introduce $\lambda_{\mathbb{Q}}(M, \|\cdot\|)$ and $\lambda_{\mathbb{Z}}(M, \|\cdot\|)$ as follows: If M is not a torsion module, then

$$\lambda_{\mathbb{Q}}(M, \|\cdot\|) = \min \left\{ \lambda \in \mathbb{R}_{\geq 0} \ \middle|\ \begin{array}{l} \exists\ x_1, \dots, x_n \in M \text{ such that } \{x_1, \dots, x_n\} \text{ gives} \\ \text{a basis of } M_{\mathbb{Q}} = M \otimes_{\mathbb{Z}} \mathbb{Q} \text{ and } \max_i\{\|x_i\|\} \leq \lambda \end{array} \right\}$$

and

$$\lambda_{\mathbb{Z}}(M, \|\cdot\|) = \min \left\{ \lambda' \in \mathbb{R}_{\geq 0} \ \middle|\ \begin{array}{l} \exists\ x_1, \dots, x_n \in M \text{ such that } \{x_1, \dots, x_n\} \text{ gives} \\ \text{a free basis of } M/M_{tor} \text{ and } \max_i\{\|x_i\|\} \leq \lambda' \end{array} \right\}.$$

In the case where M is a torsion module,

$$\lambda_{\mathbb{Q}}(M, \|\cdot\|) = \lambda_{\mathbb{Z}}(M, \|\cdot\|) = 0.$$

We have the following inequalities on $\lambda_{\mathbb{Q}}(M, \|\cdot\|)$ and $\lambda_{\mathbb{Z}}(M, \|\cdot\|)$.

LEMMA 2.18.

$$\lambda_{\mathbb{Q}}(M, \|\cdot\|) \leq \lambda_{\mathbb{Z}}(M, \|\cdot\|) \leq \mathrm{rk}(M) \lambda_{\mathbb{Q}}(M, \|\cdot\|).$$

PROOF. Considering M/M_{tor}, we may assume that M is free. The inequality $\lambda_{\mathbb{Q}}(M, \|\cdot\|) \leq \lambda_{\mathbb{Z}}(M, \|\cdot\|)$ is obvious, so that we only need to show the inequality

$$\lambda_{\mathbb{Z}}(M, \|\cdot\|) \leq \mathrm{rk}(M) \lambda_{\mathbb{Q}}(M, \|\cdot\|).$$

For this purpose, let us claim the following.

CLAIM 6. *Let $s_1, \dots, s_n \in M$ such that $s_1, \dots, s_n$ give rise to a basis of $M_{\mathbb{Q}}$. Then there is an $n \times n$-matrix (a_{ij}) such that a_{ij} is a rational number with $|a_{ij}| \leq 1$ for all i, j, and that if we set $e_i = \sum_j a_{ij} s_j$, then $e_1, \dots, e_n \in M$ and $e_1, \dots, e_n$ form a free basis of M.*

PROOF. We prove it by induction on n. If $n = 0$, the assertion is obvious. Let us choose a positive integer a such that if we set $e_n = (1/a) s_n$, then $e_n \in M$ and $M' = M/\mathbb{Z} e_n$ has no torsion. Let $s'_1, \dots, s'_{n-1}$ be the image of $s_1, \dots, s_{n-1}$ in M'. Then, by the hypothesis of induction, there is an $(n-1) \times (n-1)$-matrix

(a'_{ij}) such that a'_{ij} is a rational number with $|a'_{ij}| \leq 1$ for all i, j, and that if we set $e'_i = \sum_j a'_{ij} s'_j$, then $e'_1, \ldots, e'_{n-1} \in M'$ and $e'_1, \ldots, e'_{n-1}$ form a free basis of M'. Let us choose $f_1, \ldots, f_{n-1} \in M$ such that the images of $f_1, \ldots, f_{n-1}$ in M' are $e'_1, \ldots, e'_{n-1}$, respectively. Since $\sum_j a'_{ij} s_j - f_i \in \mathbb{Q} e_n$, there is a rational number c_i with $\sum_j a'_{ij} s_j - f_i = c_i e_n$. Here we put $e_i = f_i + \lceil c_i \rceil e_n$ $(i = 1, \ldots, n-1)$, where $\lceil x \rceil = \min\{b \in \mathbb{Z} \mid x \leq b\}$ for $x \in \mathbb{R}$. Then $e_1, \ldots, e_{n-1}, e_n$ yield a free basis of M. Moreover,

$$e_i = \sum\nolimits_j a'_{ij} s_j + \frac{(\lceil c_i \rceil - c_i)}{a} s_n \ (1 \leq i \leq n-1), \quad e_n = (1/a) s_n.$$

Thus we get the claim. □

By the above claim,

$$\|e_i\| \leq \mathrm{rk}(M) \max\{\|s_1\|, \ldots, \|s_n\|\}$$

holds, and hence the lemma follows. □

Moreover, the following two lemmas hold for $\lambda_{\mathbb{Q}}$.

LEMMA 2.19. *Let $(M_1, \|\cdot\|_1)$ and $(M_2, \|\cdot\|_2)$ be normed $\mathbb{Z}$-modules, and let $\phi : M_1 \to M_2$ be a homomorphism such that ϕ yields an isomorphism over $\mathbb{Q}$. Then we have the following:*

(1) $\lambda_{\mathbb{Q}}(M_2, \|\cdot\|_2) \leq \|\phi\| \lambda_{\mathbb{Q}}(M_1, \|\cdot\|_1)$.

(2) *We further assume that ϕ is surjective and that ϕ yields an isometry*

$$((M_1)_{\mathbb{R}}, \|\cdot\|_1) \xrightarrow{\sim} ((M_2)_{\mathbb{R}}, \|\cdot\|_2).$$

Then $\lambda_{\mathbb{Q}}(M_2, \|\cdot\|_2) = \lambda_{\mathbb{Q}}(M_1, \|\cdot\|_1)$.

PROOF. (1) Let $e_1, \ldots, e_n \in M_1$ such that $e_1, \ldots, e_n$ form a basis of M_1 over $\mathbb{Q}$ and $\max\{\|e_1\|_1, \ldots, \|e_n\|_1\} = \lambda_{\mathbb{Q}}(M_1, \|\cdot\|_1)$. Note that $\phi(e_1), \ldots, \phi(e_n)$ form a basis of M_2 over $\mathbb{Q}$ and

$$\|\phi(e_i)\|_2 \leq \|\phi\| \cdot \|e_i\|_1 \leq \|\phi\| \lambda_{\mathbb{Q}}(M_1, \|\cdot\|_1)$$

for all i, as required.

(2) By (1), we have $\lambda_{\mathbb{Q}}(M_2, \|\cdot\|_2) \leq \lambda_{\mathbb{Q}}(M_1, \|\cdot\|_1)$. On the other hand, let us choose $y_1, \ldots, y_n \in M_2$ such that $y_1, \ldots, y_n$ form a basis of M_2 over $\mathbb{Q}$ and

$$\max\{\|y_1\|_2, \ldots, \|y_n\|_2\} = \lambda_{\mathbb{Q}}(M_2, \|\cdot\|_2).$$

By our assumption, for each i, we can choose $x_i \in M_1$ with $\phi(x_i) = y_i$. Then $\|x_i\|_1 = \|y_i\|_2$ for all i. Thus $\lambda_{\mathbb{Q}}(M_1, \|\cdot\|_1) \leq \lambda_{\mathbb{Q}}(M_2, \|\cdot\|_2)$. □

LEMMA 2.20. *Let $(M_1, \|\cdot\|_1), \ldots, (M_n, \|\cdot\|_n)$ be normed $\mathbb{Z}$-modules. Let*

$$M_1 \xrightarrow{\alpha_2} M_2 \xrightarrow{\alpha_3} M_3 \longrightarrow \cdots \longrightarrow M_{i-1} \xrightarrow{\alpha_i} M_i \longrightarrow \cdots \longrightarrow M_{n-1} \xrightarrow{\alpha_n} M_n$$

be a sequence of homomorphisms such that $\alpha_i : M_{i-1} \to M_i$ yields an injective homomorphism over $\mathbb{Q}$ for each i with $2 \leq i \leq n$. If we set $\phi_i = \alpha_n \circ \cdots \circ \alpha_{i+1} : M_i \to M_n$ for $i = 1, \ldots, n-1$, and

$$Q_i = \begin{cases} \mathrm{Coker}(\alpha_i : M_{i-1} \to M_i) & \textit{if } i \geq 2, \\ M_1 & \textit{if } i = 1, \end{cases}$$

for $i = 1, \ldots, n$, then we have

$$\lambda_{\mathbb{Q}}(M_n, \|\cdot\|_n) \leq \lambda_{\mathbb{Q}}(Q_n, \|\cdot\|_{n,M_n\twoheadrightarrow Q_n}) + \sum_{i=1}^{n-1} \|\phi_i\| \lambda_{\mathbb{Q}}(Q_i, \|\cdot\|_{i,M_i\twoheadrightarrow Q_i}) \operatorname{rk} Q_i.$$

PROOF. We prove this lemma by induction on n. First we consider the case where $n = 2$. We need to show that

$$\lambda_{\mathbb{Q}}(M_2, \|\cdot\|_2) \leq \lambda_{\mathbb{Q}}(Q_2, \|\cdot\|_{2,M_2\twoheadrightarrow Q_2}) + \|\alpha_2\| \lambda_{\mathbb{Q}}(M_1, \|\cdot\|_1) \operatorname{rk} M_1.$$

Let $e_1, \ldots, e_s \in M_1$ and $f_1, \ldots, f_t \in Q_2$ such that $e_1, \ldots, e_s$ and $f_1, \ldots, f_t$ form bases of M_1 and Q_2 over $\mathbb{Q}$, respectively, and that

$$\begin{cases} \lambda_{\mathbb{Q}}(M_1, \|\cdot\|_1) = \max\left\{\|e_1\|_1, \ldots, \|e_s\|_1\right\}, \\ \lambda_{\mathbb{Q}}(Q_2, \|\cdot\|_{2,M_2\twoheadrightarrow Q_2}) = \max\left\{\|f_1\|_{2,M_2\twoheadrightarrow Q_2}, \ldots, \|f_t\|_{2,M_2\twoheadrightarrow Q_2}\right\}. \end{cases}$$

We choose $f'_j \in M_2$ and $f''_j \in (M_2)_{\mathbb{R}}$ such that $f'_j = f_j$ on Q_2, $f''_j = f_j$ on $(Q_2)_{\mathbb{R}}$ and that $\|f''_j\|_2 = \|f_j\|_{2,M_2\twoheadrightarrow Q_2}$. Since $f'_j \otimes 1 - f''_j \in \alpha_2(M_1)_{\mathbb{R}}$, there are $a_{ji} \in \mathbb{R}$ such that $f'_j \otimes 1 - f''_j = \sum_i a_{ji}(\alpha_2(e_i) \otimes 1)$. If we set $g_j = f'_j - \sum_i \lfloor a_{ji} \rfloor \alpha_2(e_i)$, then $\alpha_2(e_1), \ldots, \alpha_2(e_s), g_1, \ldots, g_t \in M_2$ form a basis of M_2 over $\mathbb{Q}$, where

$$\lfloor x \rfloor = \max\{a \in \mathbb{Z} \mid a \leq x\}$$

for $x \in \mathbb{R}$. Moreover, as $g_j \otimes 1 = f''_j + \sum_i (a_{ji} - \lfloor a_{ji} \rfloor)(\alpha_2(e_i) \otimes 1)$, we have

$$\begin{aligned} \|g_j\|_2 &\leq \|f''_j\|_2 + \sum_i \|\alpha_2(e_i)\|_2 \leq \|f''_j\|_2 + \sum_i \|\alpha_2\| \cdot \|e_i\|_1 \\ &\leq \lambda_{\mathbb{Q}}(Q_2, \|\cdot\|_{2,M_2\twoheadrightarrow Q_2}) + \|\alpha_2\| \lambda_{\mathbb{Q}}(M_1, \|\cdot\|_1) \operatorname{rk} M_1, \end{aligned}$$

which implies the assertion for $n = 2$ because $\|\alpha_2(e_i)\|_2 \leq \|\alpha_2\| \lambda_{\mathbb{Q}}(M_1, \|\cdot\|_1)$.

Here we assume that $n \geq 3$. We set

$$\widetilde{M}_i = \operatorname{Coker}(\alpha_i \circ \cdots \circ \alpha_2 : M_1 \longrightarrow M_i)$$

for $i = 2, \ldots, n$. Let $\tilde{\alpha}_i : \widetilde{M}_{i-1} \to \widetilde{M}_i$ be the induced homomorphism by α_i for $i = 3, \ldots, n$. Note that $\tilde{\alpha}_i$ yields an injective homomorphism over $\mathbb{Q}$. Here we put $\tilde{\phi}_i = \tilde{\alpha}_n \circ \cdots \circ \tilde{\alpha}_{i+1} : \widetilde{M}_i \to \widetilde{M}_n$ for $i = 2, \ldots, n-1$, and

$$\widetilde{Q}_i = \begin{cases} \operatorname{Coker}(\tilde{\alpha}_i : \widetilde{M}_{i-1} \to \widetilde{M}_i) & \text{if } i \geq 3, \\ \widetilde{M}_2 & \text{if } i = 2. \end{cases}$$

Then, by the hypothesis of induction, we have

$$\begin{aligned} \lambda_{\mathbb{Q}}(\widetilde{M}_n, \|\cdot\|_{n,M_n\twoheadrightarrow\widetilde{M}_n}) \leq{}& \lambda_{\mathbb{Q}}\left(\widetilde{Q}_n, \left(\|\cdot\|_{n,M_n\twoheadrightarrow\widetilde{M}_n}\right)_{\widetilde{M}_n\twoheadrightarrow\widetilde{Q}_n}\right) \\ &+ \sum_{i=2}^{n-1} \|\tilde{\phi}_i\| \lambda_{\mathbb{Q}}\left(\widetilde{Q}_i, \left(\|\cdot\|_{i,M_i\twoheadrightarrow\widetilde{M}_i}\right)_{\widetilde{M}_i\twoheadrightarrow\widetilde{Q}_i}\right) \operatorname{rk} \widetilde{Q}_i. \end{aligned} \tag{2.13}$$

By (1) in Lemma 1.2, $\left(\|\cdot\|_{i,M_i\twoheadrightarrow\widetilde{M}_i}\right)_{\widetilde{M}_i\twoheadrightarrow\widetilde{Q}_i} = \|\cdot\|_{i,M_i\twoheadrightarrow\widetilde{Q}_i}$ for $i = 2, \ldots, n$. Moreover, the natural homomorphism $h_i : Q_i \to \widetilde{Q}_i$ is surjective and $(Q_i)_{\mathbb{Q}} \to (\widetilde{Q}_i)_{\mathbb{Q}}$ is an isomorphism. Therefore, by (2) in Lemma 2.19,

$$\lambda_{\mathbb{Q}}\left(\widetilde{Q}_i, \left(\|\cdot\|_{i,M_i\twoheadrightarrow\widetilde{M}_i}\right)_{\widetilde{M}_i\twoheadrightarrow\widetilde{Q}_i}\right) = \lambda_{\mathbb{Q}}\left(Q_i, \|\cdot\|_{i,M_i\twoheadrightarrow Q_i}\right)$$

for $i = 2, \ldots, n$. On the other hand, by Lemma 1.5, $\|\tilde{\phi}_i\| \leq \|\phi_i\|$ for $i = 2, \ldots, n-1$. Thus, (2.13) implies that

$$\begin{aligned} \lambda_{\mathbb{Q}}\left(\widetilde{M}_n, \|\cdot\|_{n, M_n \twoheadrightarrow \widetilde{M}_n}\right) \leq \lambda_{\mathbb{Q}}\left(Q_n, \|\cdot\|_{n, M_n \twoheadrightarrow Q_n}\right) \\ + \sum_{i=2}^{n-1} \|\phi_i\| \lambda_{\mathbb{Q}}\left(Q_i, \|\cdot\|_{i, M_i \twoheadrightarrow Q_i}\right) \operatorname{rk} Q_i. \end{aligned} \tag{2.14}$$

Further, applying the induction hypothesis to $M_1 \xrightarrow{\phi_1} M_n$, we obtain

$$\lambda_{\mathbb{Q}}(M_n, \|\cdot\|_n) \leq \lambda_{\mathbb{Q}}\left(\widetilde{M}_n, \|\cdot\|_{n, M_n \twoheadrightarrow \widetilde{M}_n}\right) + \|\phi_1\| \lambda_{\mathbb{Q}}(M_1, \|\cdot\|_1) \operatorname{rk} M_1. \tag{2.15}$$

Therefore, the assertion of the lemma follows from (2.14) and (2.15). $\square$

CHAPTER 3

Arakelov Geometry on Arithmetic Curves

In this chapter, we discuss Arakelov geometry over a reduced order, which gives an introduction to further developments of Arakelov geometry. Since every argument is algebraic, it is not difficult to read this chapter. However, there are several important ideas of Arakelov geometry, so that if you come to a deadlock in later chapters, you may come back to this chapter.

3.1. Orders

In this section, let us recall several facts concerning orders. First, let us begin with the following proposition.

PROPOSITION 3.1. *Let R be a commutative ring. We assume that R is a finitely generated free $\mathbb{Z}$-module. Let n be the rank of R as a $\mathbb{Z}$-module. Then we have the following.*

(1) *For any $a \in R$, there are $c_0, \ldots, c_{n-1} \in \mathbb{Z}$ such that*
$$a^n + c_{n-1}a^{n-1} + \cdots + c_1 a + c_0 = 0.$$
Moreover, if a is a regular element, then we can take c_0 as $c_0 \neq 0$.

(2) *R is integral over $\mathbb{Z}$.*

(3) *For any regular element a of R, there is a regular element b of R with $ab \in \mathbb{Z} \setminus \{0\}$.*

(4) *Let K be the total quotient ring of R. Then K is the localization of R with respect to the multiplicative set $\mathbb{Z} \setminus \{0\}$, namely, $K = R \otimes_{\mathbb{Z}} \mathbb{Q}$.*

(5) *The total quotient ring K is an Artinian ring. If we set $\operatorname{Spec}(K) = \{m_1, \ldots, m_l\}$, then the natural homomorphism $K \to \bigoplus_{i=1}^{l} K_{m_i}$ is an isomorphism. Moreover, if we set $Q_i = R \cap m_i$ $(i = 1, \ldots, l)$, then $Q_1, \ldots, Q_l$ are all minimal prime ideals of R and $K_{m_i} = R_{Q_i}$. Further, if R is reduced (i.e. $\sqrt{0_R} = \{0\}$), then K_{m_i} is an algebraic number field (i.e. a finite extension field over $\mathbb{Q}$).*

(6) *We assume that R is reduced. Then, by (5), $K = K_1 \oplus \cdots \oplus K_l$ (K_i's are algebraic number fields). Let $K(\mathbb{C})$ be the set of all ring homomorphisms from K to $\mathbb{C}$. In the same way, let $K_i(\mathbb{C})$ be the set of all ring homomorphisms from K_i to $\mathbb{C}$. For $\tau \in K_i(\mathbb{C})$, let $\tilde{\tau} \in K(\mathbb{C})$ given by $\tilde{\tau}(x_1, \ldots, x_l) = \tau(x_i)$. Then a map*
$$\coprod_{i=1}^{l} K_i(\mathbb{C}) \to K(\mathbb{C}) \qquad (\tau \mapsto \tilde{\tau})$$
is bijective. In particular, $\#(K(\mathbb{C})) = n = \dim_{\mathbb{Q}} K$.

PROOF. (1) Let $\omega_1, \ldots, \omega_n$ be a free basis of R as a $\mathbb{Z}$-module. We choose $c_{ij} \in \mathbb{Z}$ such that $a\omega_i = \sum_j c_{ij}\omega_j$. Let us consider a matrix A given by $A = (a\delta_{ij} - c_{ij})$. Then

$$A \begin{pmatrix} \omega_1 \\ \vdots \\ \omega_n \end{pmatrix} = \begin{pmatrix} 0 \\ \vdots \\ 0 \end{pmatrix}.$$

Let $\widetilde{A}$ be the cofactor matrix of A. As $\widetilde{A}A = A\widetilde{A} = \det(A)I_n$,

$$\det(A) \begin{pmatrix} \omega_1 \\ \vdots \\ \omega_n \end{pmatrix} = \widetilde{A}A \begin{pmatrix} \omega_1 \\ \vdots \\ \omega_n \end{pmatrix} = \begin{pmatrix} 0 \\ \vdots \\ 0 \end{pmatrix}.$$

Therefore, we have $\det(A) = 0$. Expanding this determinant, we have

$$a^n + c_{n-1}a^{n-1} + \cdots + c_1 a + c_0 = 0$$

and $c_0 = (-1)^n \det((c_{ij}))$. If a is regular, then the multiplication map by a is injective, so that $c_0 \neq 0$.

(2) is a consequence of (1). For (3), if we set $b = a^{n-1} + c_{n-1}a^{n-2} + \cdots + c_1$ in the proof of (1), then $ab + c_0 = 0$, which shows (3). (4) is a consequence of (3).

(5) Since R is flat over $\mathbb{Z}$, it is easy to see that, for $Q \in \operatorname{Spec}(R)$, Q is a minimal prime if and only if $Q \cap \mathbb{Z} = \{0\}$. Thus, by using (4), K is an Artinian ring. Therefore, by (3) in Lemma 1.6, we obtain (5) except the final assertion. If R is reduced, then K is also reduced. Thus K_{m_i} $(i = 1, \ldots, l)$ is an algebraic number field.

(6) For $\tau \in K_i(\mathbb{C})$, $\tilde{\tau}(K_j) = 0$ $(j \neq i)$ and $\tilde{\tau}|_{K_i} = \tau$. Thus the map $\tau \mapsto \tilde{\tau}$ is injective. Let us see its surjectivity. Let e_i be the element of K such that the i-th entry is 1 and other entries are 0. Let σ be an arbitrary element of $K(\mathbb{C})$. As $e_i^2 = e_i$, $\sigma(e_i)^2 = \sigma(e_i)$. Thus $\sigma(e_i)$ is either 0 or 1. If $\sigma(e_i) = 0$ for all i, then $\sigma = 0$, which contradicts $\sigma(1) = 1$. Therefore, there is i with $\sigma(e_i) = 1$. We assume $\sigma(e_j) = 1$ for some j with $j \neq i$. Then, as $e_i e_j = 0$,

$$0 = \sigma(e_i e_j) = \sigma(e_i)\sigma(e_j) = 1.$$

This is a contradiction. Thus there is a unique i with $\sigma(e_i) = 1$. Hence, if we set $\tau = \sigma|_{K_i}$, then $\tilde{\tau} = \sigma$. □

In this book, a commutative ring R is called an *order* if R is a finitely generated free module as a $\mathbb{Z}$-module. Moreover, an order R is called an *integral order* if R is an integral domain. For an algebraic number field K, we set

$$O_K = \{x \in K \mid x \text{ is integral over } \mathbb{Z}\},$$

which is called the *maximal order in* K or the *ring of integers of* K. Note that O_K is an integral order by using [**3**, Proposition 5.17] for example. As a corollary of Proposition 3.1, we have the following.

COROLLARY 3.2. *For an order R, we have the following.*

(1) *R is reduced if and only if there are algebraic number fields $K_1, \ldots, K_l$ and an injective ring homomorphism $\phi : R \to O_{K_1} \oplus \cdots \oplus O_{K_l}$.*
(2) *R is normal (i.e. the localization of R at every prime ideal of R is an integrally closed domain) if and only if there are algebraic number fields $K_1, \ldots, K_l$ and a ring isomorphism $\phi : R \xrightarrow{\sim} O_{K_1} \oplus \cdots \oplus O_{K_l}$.*

PROOF. (1) follows from Proposition 3.1. Let us consider (2). If there are algebraic number fields $K_1, \ldots, K_l$ and a ring isomorphism $\phi : R \xrightarrow{\sim} O_{K_1} \oplus \cdots \oplus O_{K_l}$, then clearly R is normal. Conversely we assume that R is normal. In particular, R is reduced. Let $Q_1, \ldots, Q_l$ be all minimal primes of R and let K be the total quotient ring of R. Then, by Proposition 3.1, if we set $K_i = R_{Q_i}$, then K_i's are algebraic number fields and $K = K_1 \oplus \cdots \oplus K_l$. A natural injective ring homomorphism $R \to K$ gives rise to an injective homomorphism $\phi : R \to O_{K_1} \oplus \cdots \oplus O_{K_l}$. In order to see that ϕ is an isomorphism, it is sufficient to show that $\phi_P : R_P \to (O_{K_1} \oplus \cdots \oplus O_{K_l})_P$ is an isomorphism for all prime ideal P of R. Since R_P is an integral domain, R_P has the unique minimal ideal, which means that there is a unique Q_i with $Q_i \subseteq P$. Therefore, for $j \neq i$, $(K_j)_P = \{0\}$ because $(R \setminus P) \cap Q_j \neq \emptyset$ and every element of Q_j is zero in $K_j = R_{Q_j}$. Thus ϕ_P is nothing more than $R_P \to (O_{K_i})_P$. Hence ϕ_P is an isomorphism because R_P in integrally closed. □

THEOREM 3.3 (Product formula). *Let R is a reduced order and let K be the total quotient ring. Then, for $x \in K^{\times}$,*

$$\prod_{P \in \text{m-Spec}(R)} \#(R/P)^{-\operatorname{ord}_P(x)} \cdot \prod_{\sigma \in K(\mathbb{C})} |\sigma(x)| = 1,$$

where ord_P is defined in the last part of Section 1.3 and $K(\mathbb{C})$ is the set of all ring homomorphisms from K to $\mathbb{C}$.

PROOF. Let us begin with the following lemma.

LEMMA 3.4. *Let M be a finitely generated free $\mathbb{Z}$-module and let ϕ be an injective homomorphism. Then $\#(M/\phi(M)) = |\det(\phi)|$.*

PROOF. This lemma is a consequence of Theorem 1.8. However, we would like to give an elementary proof. Let $\{e_1, \ldots, e_r\}$ be a free basis of M. We set $\phi(e_i) = \sum_{j=1}^{r} b_{ij} e_j$ and $B = (b_{ij})$. Then, by using elementary divisor theory, there are $P, Q \in \mathrm{GL}(r, \mathbb{Z})$ and $c_1, \ldots, c_r \in \mathbb{Z}$ such that

$$C = PBQ = \begin{pmatrix} c_1 & & 0 \\ & \ddots & \\ 0 & & c_r \end{pmatrix}.$$

Let f_P, f_Q and f_C be endomorphisms of M given by P, Q and C with respect to the basis $\{e_1, \ldots, e_r\}$, respectively. Then the diagram

$$\begin{array}{ccc} M & \xrightarrow[f_Q]{\sim} & M \\ {\scriptstyle f_C}\big\downarrow & & \big\downarrow{\scriptstyle \phi} \\ M & \xleftarrow[f_P]{\sim} & M \end{array}$$

is commutative. Thus

$$\#(M/\phi(M)) = \#(M/f_C(M)) = |c_1 \cdots c_r| = |\det(f_C)|.$$

Note that $\det(f_P), \det(f_Q) \in \{\pm 1\}$. Hence the lemma follows. □

Let us go back to the proof of Theorem 3.3. Clearly we may assume that x is a regular element of R. Let us consider a homomorphism $\phi_x : R \to R$ given by $a \mapsto x \cdot a$. Then, by Lemma 3.4,

$$\#(R/xR) = |\det(\phi_x)|.$$

By Proposition 3.1, we can set $K = K_1 \oplus \cdots \oplus K_l$ (K_i's are algebraic number fields). We put $x = (x_1, \ldots, x_l)$. Then, since $\phi_x(a_1, \ldots, a_l) = (x_1 a_1, \ldots, x_l a_l)$, if $\phi_i : K_i \to K_i$ is a homomorphism given by $\phi_i(a_i) = x_i a_i$, then $\det(\phi_x) = \det(\phi_1) \cdots \det(\phi_l)$. It is well known as a standard field theory that

$$\det(\phi_i) = \prod_{\tau \in K_i(\mathbb{C})} \tau(x_i),$$

and hence

$$|\det(\phi_i)| = \prod_{\tau \in K_i(\mathbb{C})} |\tau(x_i)|.$$

For $\tau \in K_i(\mathbb{C})$, we define $\tilde{\tau} \in K(\mathbb{C})$ by $\tilde{\tau}(x_1, \ldots, x_l) = \tau(x_i)$. By Proposition 3.1, a map $\tau \mapsto \tilde{\tau}$ gives rise to a bijection $\coprod_{i=1}^{l} K_i(\mathbb{C}) \xrightarrow{\sim} K(\mathbb{C})$. Therefore,

$$\begin{aligned} |\det(\phi_x)| &= \prod_{i=1}^{l} |\det(\phi_i)| = \prod_{i=1}^{l} \prod_{\tau \in K_i(\mathbb{C})} |\tau(x_i)| = \prod_{i=1}^{l} \prod_{\tau \in K_i(\mathbb{C})} |\tilde{\tau}(x)| \\ &= \prod_{\sigma \in K(\mathbb{C})} |\sigma(x)|. \end{aligned}$$

Hence, it is sufficient to show

$$\#(R/xR) = \prod_{P \in \text{m-Spec}(R)} \#(R/P)^{\operatorname{ord}_P(x)}.$$

We set $\operatorname{Supp}(R/xR) = \{P_1, \ldots, P_n\}$. Then, by Lemma 1.6,

$$R/xR \simeq R_{P_1}/xR_{P_1} \oplus \cdots \oplus R_{P_n}/xR_{P_n}.$$

Thus

$$\#(R/xR) = \prod_i \#(R_{P_i}/xR_{P_i}).$$

On the other hand, for an Artinian R_{P_i}-module M, if

$$\{0\} = M_0 \subsetneq M_1 \subsetneq \cdots \subsetneq M_l = M$$

is a composite series of M as an R_{P_i}-module, then

$$M_j/M_{j-1} \simeq R_{P_i}/P_i R_{P_i} \simeq R/P_i \quad (\forall j).$$

Thus we can see that $\#(M) = \#(R/P_i)^l$. Therefore,

$$\#(R_{P_i}/xR_{P_i}) = \#(R/P_i)^{\operatorname{ord}_{P_i}(x)}.$$

Hence we complete the proof of the theorem. □

Let R be a reduced order. For a non-negative integer r, a finitely generated R-module H is said to be *of pure rank* r if $r = \dim_{R_Q} H_Q$ for all minimal prime ideals Q. This r is called the rank of H.

PROPOSITION 3.5. (1) *The following are equivalent for a finitely generated R-module H.*

(a) *H is a torsion module as a $\mathbb{Z}$-module.*

(b) *For all minimal prime ideals Q of R, $H_Q = \{0\}$.*
(c) *H is an R-module of pure rank 0.*

(2) *We assume that H is of pure rank r. For $x_1, \ldots, x_r \in H$, if $H/(Rx_1 + \cdots + Rx_r)$ is a torsion module as a $\mathbb{Z}$-module, then $Rx_1 + \cdots + Rx_r$ is a free R-module of rank r.*

(3) *H is of pure rank r if and only if there are $x_1, \ldots, x_r \in H$ such that*

$$H/(Rx_1 + \cdots + Rx_r)$$

is a torsion module as a $\mathbb{Z}$-module and $Rx_1 + \cdots + Rx_r$ is a free R-module of rank r.

PROOF. (1) (a) $\Longrightarrow$ (b): This is obvious because $Q \cap \mathbb{Z} = \{0\}$.

(b) $\Longrightarrow$ (c): This is trivial.

(c) $\Longrightarrow$ (a): Let $Q_1, \ldots, Q_l$ be all minimal prime ideals of R. Since $K = R_{Q_1} \oplus \cdots \oplus R_{Q_l}$ by Proposition 3.1, $H \otimes_R K = \{0\}$. On the other hand, by Proposition 3.1, K is the localization of R with respect to $\mathbb{Z} \setminus \{0\}$. Thus $H \otimes_R K$ is also the localization of H with respect to $\mathbb{Z} \setminus \{0\}$. Thus H is a torsion module as a $\mathbb{Z}$-module.

(2) We assume that $a_1x_1 + \cdots + a_rx_r = 0$. Since $H_{Q_i} = R_{Q_i}x_1 + \cdots + R_{Q_i}x_r$ and $\dim_{R_{Q_i}} H_{Q_i} = r$, $x_1, \ldots, x_r$ form a basis of H_{Q_i}. Thus $a_1 = \cdots = a_r = 0$ in R_{Q_i} for all i. Therefore $a_1 = \cdots = a_r = 0$ in R.

(3) Let $Q_1, \ldots, Q_l$ be all minimal prime ideals of R. First we assume that H is of pure rank r. By Proposition 3.1, $K = R_{Q_1} \oplus \cdots \oplus R_{Q_l}$. Thus

$$H \otimes_R K = \bigoplus_{i=1}^{l} H \otimes_R R_{Q_i} = \bigoplus_{i=1}^{l} H_{Q_i}.$$

Let $y_{i1}, \ldots, y_{ir}$ be a basis of H_{Q_i} as an R_{Q_i}-vector space. We set

$$y_j = (y_{1j}, \ldots, y_{lj}) \in \bigoplus_{i=1}^{l} H_{Q_i} = H \otimes_R K.$$

Since $H \otimes_R K = H \otimes_{\mathbb{Z}} \mathbb{Q}$, there are a non-zero integer n and $x_1, \ldots, x_r \in H$ such that $ny_j = x_j \otimes 1$ for all i. Then $H/Rx_1 + \cdots + Rx_r$ is a torsion module as a $\mathbb{Z}$-module. Therefore, by (2), $Rx_1 + \cdots + Rx_r$ is a free R-module.

Conversely we assume that there are $x_1, \ldots, x_r \in H$ satisfying the properties of (3). Then $H_{Q_i} = R_{Q_i}x_1 + \cdots + R_{Q_i}x_r$, which shows that H_{Q_i} is a free R_{Q_i}-module of rank r. Thus H is an R-module of pure rank r. □

Finally we consider the following two lemmas which will be used later.

LEMMA 3.6. *Let K and K' be a reduced $\mathbb{Q}$-algebra such that $\mathbb{Q} \subseteq K \subseteq K'$, and that K and K' are finite-dimensional over $\mathbb{Q}$. Let $K(\mathbb{C})$ (resp. $K'(\mathbb{C})$) be the set of ring homomorphisms from K to $\mathbb{C}$ (resp. from K' to $\mathbb{C}$). For $\sigma \in K(\mathbb{C})$, we set $K'(\mathbb{C})_\sigma = \{\sigma' \in K'(\mathbb{C}) \mid \sigma'|_K = \sigma\}$. For $x' \in K'$, the norm of x' over K (the determinant of the K-homomorphism $K' \to K'$ given by $a \mapsto ax$) is denoted by $N_{K'/K}(x')$. Then*

$$\sigma(N_{K'/K}(x')) = \prod_{\sigma' \in K'(\mathbb{C})_\sigma} \sigma'(x').$$

PROOF. First let us consider the case where K' is an algebraic number field, so that K is also an algebraic number field. We fix $\sigma_1 \in K'(\mathbb{C})_\sigma$. We set $\sigma_1(K) = K_1$ and $\sigma_1(K') = K_2$. Let $(K_2/K_1)(\mathbb{C})$ be the set of all ring homomorphisms from K_2 to $\mathbb{C}$ over K_1. Then a map $(K_2/K_1)(\mathbb{C}) \to K'(\mathbb{C})_\sigma$ given by $\tau \mapsto \tau \circ \sigma_1$ is bijective. On the other hand, it is well known as a standard fact of field theory that

$$\prod_{\tau \in (K_2/K_1)(\mathbb{C})} \tau(\sigma_1(x')) = N_{K_2/K_1}(\sigma_1(x')).$$

Moreover, it is easy to see that $N_{K_2/K_1}(\sigma_1(x')) = \sigma(N_{K'/K}(x'))$. Thus the lemma follows in this case.

Next we assume that K is an algebraic number field. Then K' is the product of algebraic number fields, that is, $K' = K'_1 \oplus \cdots \oplus K'_l$ ($K'_1, \ldots, K'_l$ are algebraic number fields over K). We set $x' = (x'_1, \ldots, x'_l)$. Then $N_{K'/K}(x') = N_{K'_1/K}(x'_1) \cdots N_{K'_l/K}(x'_l)$. In the same way as in Proposition 3.1, for $\tau \in K'_i(\mathbb{C})_\sigma$, let us define $\tilde{\tau} : K' \to \mathbb{C}$ by $\tilde{\tau}(x'_1, \ldots, x'_l) = \tau(x'_i)$. Then we can see that a map $\coprod_{i=1}^l K'_i(\mathbb{C})_\sigma \to K'(\mathbb{C})_\sigma$ given by $\tau \mapsto \tilde{\tau}$ is bijective. Therefore,

$$\prod_{\sigma' \in K'(\mathbb{C})_\sigma} \sigma'(x') = \prod_{i=1}^l \prod_{\tau \in K'_i(\mathbb{C})_\sigma} \tau(x'_i) = \prod_{i=1}^l \sigma(N_{K'_i/K}(x'_i)) = \sigma(N_{K'/K}(x')),$$

as desired.

In general, $K = K_1 \oplus \cdots \oplus K_r$ for some algebraic number fields $K_1, \ldots, K_r$. Let P_j be the minimal prime ideal of K such that $K_{P_j} = K_j$, that is, $P_j = \{(x_1, \ldots, x_r) \in K \mid x_j = 0\}$. If we set $K'_j = K'_{P_j}$, then

$$K' = K'_1 \oplus \cdots \oplus K'_r \quad \text{and} \quad N_{K'/K}(x) = (N_{K'_1/K_1}(x'_1), \ldots, N_{K'_r/K_r}(x'_r)),$$

where $x' = (x'_1, \ldots, x'_r) \in K' = K'_1 \oplus \cdots \oplus K'_r$. By Proposition 3.1, there is $\sigma_j \in K_j(\mathbb{C})$ such that $\sigma(x_1, \ldots, x_r) = \sigma_j(x_j)$ for $(x_1, \ldots, x_r) \in K$. In particular, $\sigma(N_{K'/K}(x)) = \sigma_j(N_{K'_j/K_j}(x'_j))$. Moreover, it is easy to see that $K'(\mathbb{C})_\sigma = K'_j(\mathbb{C})_{\sigma_j}$. Therefore, by using the previous case,

$$\prod_{\sigma' \in K'(\mathbb{C})_\sigma} \sigma'(x') = \prod_{\delta \in K'_j(\mathbb{C})_{\sigma_j}} \delta(x'_j) = \sigma_j(N_{K'_j/K_j}(x'_j)) = \sigma(N_{K'/K}(x')). \qquad \square$$

LEMMA 3.7. *Let $p_1, \ldots, p_l$ be distinct prime numbers and let $n_1, \ldots, n_l$ be positive integers. Then there is an algebraic number field K such that*

$$[O_K/P : \mathbb{Z}/p_i\mathbb{Z}] \geq n_i$$

for $i = 1, \ldots, l$ and all prime ideals P of O_K with $P \cap \mathbb{Z} = p_i\mathbb{Z}$.

PROOF. Let $f_i(X)$ be an irreducible monic polynomial of degree n_i in $(\mathbb{Z}/p_i\mathbb{Z})[X]$. Let $F_i(X)$ be a monic polynomial of degree n_i in $\mathbb{Z}[X]$ such that the image of $F_i(X)$ via the natural homomorphism $\mathbb{Z}[X] \to (\mathbb{Z}/p_i\mathbb{Z})[X]$ is $f_i(X)$. Let $F(X) = F_1(X) \cdots F_l(X)$ and let K be the minimal splitting field of $F(X)$. Then a root α_i of $F_i(X)$ is an element of O_K. Let P be a prime ideal of O_K with $P \cap \mathbb{Z} = p_i\mathbb{Z}$. Let $\bar{\alpha}_i$ be the class of α_i in O_K/P. Then $(\mathbb{Z}/p_i\mathbb{Z})(\bar{\alpha}_i) \subseteq O_K/P$. As $f_i(\bar{\alpha}_i) = 0$ and f_i is irreducible, f_i is the minimal polynomial of $\bar{\alpha}_i$ over $\mathbb{Z}/p_i\mathbb{Z}$. Therefore,

$$[O_K/P : \mathbb{Z}/p_i\mathbb{Z}] \geq [(\mathbb{Z}/p_i\mathbb{Z})(\bar{\alpha}_i) : \mathbb{Z}/p_i\mathbb{Z}] = n_i. \qquad \square$$

3.2. Arithmetic Chow group over a reduced order

Let R be a reduced order, that is, R is a reduced commutative ring such that R is a finitely generated free $\mathbb{Z}$-module. Let K be the total quotient ring and let $K(\mathbb{C})$ be the set of all ring homomorphisms from K to $\mathbb{C}$. The space $\widehat{Z}^1(R)$ of *arithmetic cycles of codimension one* of R is defined to be the product of a free abelian group generated by maximal ideals of R and an $\mathbb{R}$-vector space generated by $K(\mathbb{C})$, that is,

$$\widehat{Z}^1(R) := \left(\bigoplus_{P \in \text{m-Spec}(R)} \mathbb{Z}[P] \right) \times \left(\bigoplus_{\sigma \in K(\mathbb{C})} \mathbb{R}[\sigma] \right).$$

For $x \in K^{\times}$, the *arithmetic principal divisor* $\widehat{(x)}$ is defined by

$$\widehat{(x)} = \left(\sum_P \operatorname{ord}_P(x)[P], \ \sum_\sigma -\log|\sigma(x)|^2[\sigma] \right).$$

Since the set $\{\widehat{(x)} \mid x \in K^{\times}\}$ of all arithmetic principal divisors is a subgroup of $\widehat{Z}^1(R)$, the *arithmetic Chow group* $\widehat{\mathrm{CH}}^1(R)$ is defined to be

$$\widehat{\mathrm{CH}}^1(R) := \frac{\widehat{Z}^1(R)}{\left\{ \widehat{(x)} \;\middle|\; x \in K^{\times} \right\}}.$$

For an element $z = (\sum_P a_P[P], \ \sum_\sigma g_\sigma[\sigma])$ of $\widehat{Z}^1(R)$, the *arithmetic degree* is defined by

$$\widehat{\deg}(z) = \sum_P a_P \log \#(R/P) + \frac{1}{2} \sum_\sigma g_\sigma.$$

It is easy to see that $\widehat{\deg} : \widehat{Z}^1(R) \to \mathbb{R}$ is a homomorphism. By using the product formula (Theorem 3.3), we have

$$\begin{aligned} \widehat{\deg}\left(\widehat{(x)}\right) &= \sum_P \operatorname{ord}_P(x) \log \#(R/P) - \sum_\sigma \log|\sigma(x)| \\ &= -\log \left(\prod_P \#(R/P)^{-\operatorname{ord}_P(x)} \cdot \prod_\sigma |\sigma(x)| \right) = 0. \end{aligned}$$

Thus there is a homomorphism $\widehat{\deg}' : \widehat{\mathrm{CH}}^1(R) \to \mathbb{R}$ such that the following diagram is commutative:

$$\begin{array}{ccc} \widehat{Z}^1(R) & \longrightarrow & \widehat{\mathrm{CH}}^1(R) \\ & \searrow^{\widehat{\deg}} & \downarrow {\scriptstyle \widehat{\deg}'} \\ & & \mathbb{R} \end{array}$$

where $\widehat{Z}^1(R) \to \widehat{\mathrm{CH}}^1(R)$ is the natural homomorphism. By abuse of notation, we denote $\widehat{\deg}'$ by $\widehat{\deg}$.

3.3. Hermitian R-module

Let R be a reduced order and let H be a finitely generated R-module. Let K be the total quotient ring of R. For $\sigma \in K(\mathbb{C})$, the tensor product $H \otimes_R \mathbb{C}$ with respect to the ring homomorphism $R \to K \xrightarrow{\sigma} \mathbb{C}$ is denoted by $H \otimes_R^\sigma \mathbb{C}$ or H_σ. An element $x \otimes \alpha$ of $H \otimes_R^\sigma \mathbb{C}$ $(x \in H, \alpha \in \mathbb{C})$ is often denoted by $x \otimes^\sigma \alpha$ in order to distinguish it from the tensor product by other elements of $K(\mathbb{C})$. For $\sigma \in K(\mathbb{C})$, we define $\bar{\sigma} \in K(\mathbb{C})$ to be $\bar{\sigma}(x) = \overline{\sigma(x)}$, where $^-$ is the complex conjugation. Then we have an anti-$\mathbb{C}$-linear map $\iota_{\sigma,\bar{\sigma}} : H_\sigma \to H_{\bar{\sigma}}$ given by $h \otimes^\sigma \alpha \mapsto h \otimes^{\bar{\sigma}} \bar{\alpha}$, where a map $\psi : V \to W$ of complex vector spaces is called an *anti-$\mathbb{C}$-linear map* if the following conditions are satisfied:

$$\begin{cases} \psi(v + v') = \psi(v) + \psi(v') & (\forall v, v' \in V), \\ \psi(\alpha v) = \bar{\alpha}\psi(v) & (\forall v \in V, \alpha \in \mathbb{C}). \end{cases}$$

Let us begin with the following proposition.

PROPOSITION 3.8. *The natural homomorphism*

$$\phi : H \otimes_{\mathbb{Z}} \mathbb{C} \to \bigoplus_{\sigma \in K(\mathbb{C})} H_\sigma$$

is bijective. Moreover, if we set $F_\infty : H \otimes_{\mathbb{Z}} \mathbb{C} \to H \otimes_{\mathbb{Z}} \mathbb{C}$ by $F_\infty(h \otimes \alpha) = h \otimes \bar{\alpha}$ and $F'_\infty : \bigoplus_{\sigma \in K(\mathbb{C})} H_\sigma \to \bigoplus_{\sigma \in K(\mathbb{C})} H_\sigma$ by $\bigoplus_{\sigma \in K(\mathbb{C})} \iota_{\sigma,\bar{\sigma}}$, then $\phi \circ F_\infty = F'_\infty \circ \phi$. In particular,

$$\phi(H \otimes_{\mathbb{Z}} \mathbb{R}) = \left\{ x \in \bigoplus_{\sigma \in K(\mathbb{C})} H_\sigma \;\middle|\; F'_\infty(x) = x \right\}.$$

PROOF. The formula $\phi \circ F_\infty = F'_\infty \circ \phi$ is obvious by these definitions. If the first assertion holds for the case where $H = R$, that is,

$$R \otimes_{\mathbb{Z}} \mathbb{C} \to \bigoplus_{\sigma \in K(\mathbb{C})} R \otimes_R^\sigma \mathbb{C}$$

is bijective, then

$$H \otimes_R (R \otimes_{\mathbb{Z}} \mathbb{C}) \to H \otimes_R \left(\bigoplus_{\sigma \in K(\mathbb{C})} R \otimes_R^\sigma \mathbb{C} \right)$$

is also bijective. This is nothing more than the first assertion of the proposition in general. Therefore, it is sufficient to prove the bijectivity for the case where $H = R$, which is equivalent to the bijectivity of the homomorphism

$$K \otimes_{\mathbb{Q}} \mathbb{C} \to \bigoplus\nolimits_{\sigma \in K(\mathbb{C})} K \otimes_K^\sigma \mathbb{C}$$

because $R \otimes_{\mathbb{Z}} \mathbb{C} = K \otimes_{\mathbb{Q}} \mathbb{C}$ and $R \otimes_R^\sigma \mathbb{C} = K \otimes_K^\sigma \mathbb{C}$ for all $\sigma \in K(\mathbb{C})$. Note that each side has the same dimension over $\mathbb{C}$. Thus it is enough to show the injectivity of the homomorphism. We set $A = K \otimes_{\mathbb{Q}} \mathbb{C}$. We denote the set of all ring homomorphisms from A to $\mathbb{C}$ over $\mathbb{C}$ by $(A/\mathbb{C})(\mathbb{C})$. For $\sigma \in K(\mathbb{C})$, the natural ring homomorphism $A \to K \otimes_K^\sigma \mathbb{C} = \mathbb{C}$ is denoted by $\sigma_{\mathbb{C}}$. Then $\sigma_{\mathbb{C}} \in (A/\mathbb{C})(\mathbb{C})$. It is easy to see that a map given by $\sigma \mapsto \sigma_{\mathbb{C}}$ yields one-to-one correspondence between $K(\mathbb{C})$ and $(A/\mathbb{C})(\mathbb{C})$. Moreover, since A is a finite-dimensional Artinian algebra over $\mathbb{C}$ and

$\mathbb{C}$ is algebraically closed, we have $\{\mathrm{Ker}(\tau) \mid \tau \in (A/\mathbb{C})(\mathbb{C})\} = \mathrm{Spec}(A)$. Therefore, we need to see that

$$A \to \bigoplus_{m \in \mathrm{Spec}(A)} A/m$$

is injective. On the other hand, since $K \otimes_{\mathbb{Q}} \mathbb{C}$ is reduced by Lemma 1.29,

$$\bigcap\nolimits_{m \in \mathrm{Spec}(A)} m = \{0\},$$

which means that the above homomorphism is injective. □

Each H_σ is a complex vector space, so that we may consider a hermitian inner product h_σ of H_σ. In this way, we have a collection $\left(H, \{h_\sigma\}_{\sigma \in K(\mathbb{C})}\right)$ of a finitely generated R-module H and a hermitian inner product h_σ of each H_σ ($\sigma \in K(\mathbb{C})$). This collection is called a *hermitian R-module*. For simplicity, $\left(H, \{h_\sigma\}_{\sigma \in K(\mathbb{C})}\right)$ is often denoted by (H, h) or $\overline{H}$. Moreover, if H is of pure rank r (i.e. the rank of H at each minimal prime ideal is r), then it is called a *hermitian R-module of pure rank r*.

A hermitian R-module (H, h) is said to be *of real type* if

$$h_{\bar{\sigma}}(x \otimes^{\bar{\sigma}} 1, y \otimes^{\bar{\sigma}} 1) = \overline{h_\sigma(x \otimes^{\sigma} 1, y \otimes^{\sigma} 1)}$$

holds for all $\sigma \in K(\mathbb{C})$ and $x, y \in H$.

For example, R has the canonical hermitian metric $h^{can} = \{h^{can}_\sigma\}$ of real type given by $h^{can}_\sigma(x \otimes 1, y \otimes 1) = \sigma(x) \cdot \overline{\sigma(y)}$.

Let M be a finitely generated $\mathbb{Z}$-module and let $\langle\ ,\ \rangle : M \times M \to \mathbb{C}$ be a bilinear form. Clearly, $\langle x, y\rangle = \langle y, x\rangle = 0$ hold for all $x \in M_{tor}$ and $y \in M$. We say that $\langle\ ,\ \rangle : M \times M \to \mathbb{C}$ is a *hermitian form* if $\langle x, y\rangle = \overline{\langle y, x\rangle}$ holds for all $x, y \in M$. Then $\langle\ ,\ \rangle : M \times M \to \mathbb{C}$ naturally extends to a hermitian form on $M \otimes_{\mathbb{Z}} \mathbb{C}$ by setting $\langle x \otimes a, y \otimes b\rangle = a\bar{b}\langle x, y\rangle$. If this hermitian form on $M \otimes_{\mathbb{Z}} \mathbb{C}$ is positive definite, then the hermitian form $\langle\ ,\ \rangle$ on M is said to be *positive definite*. If M is a torsion module, then $\langle\ ,\ \rangle$ is zero. For our convenience, it is also said to be positive definite.

For a hermitian R-module (H, h), we set

$$\langle x, y\rangle_h = \sum_{\sigma \in K(\mathbb{C})} h_\sigma(x \otimes 1, y \otimes 1) \quad (x, y \in H).$$

Then we have the following.

PROPOSITION 3.9. *The above $\langle\ ,\ \rangle_h$ gives a positive definite hermitian form on H as a $\mathbb{Z}$-module.*

PROOF. It is obvious that $\langle\ ,\ \rangle_h$ is a hermitian form. Let us show that it is positive definite. By Proposition 3.8, the natural homomorphism

$$\phi : H \otimes_{\mathbb{Z}} \mathbb{C} \to \bigoplus_{\sigma \in K(\mathbb{C})} H_\sigma$$

is an isomorphism. The complex vector space $\bigoplus_{\sigma \in K(\mathbb{C})} H_\sigma$ has the hermitian inner product $h' = \bigoplus^{\perp}_{\sigma \in K(\mathbb{C})} h_\sigma$, that is, $h'(H_\sigma, H_{\sigma'}) = 0$ for all $\sigma \neq \sigma'$ and $h'|_{H_\sigma \times H_\sigma} = h_\sigma$ for all σ. The above isomorphism and h' induces a hermitian inner product $h_{\mathbb{C}}$ on $H \otimes_{\mathbb{Z}} \mathbb{C}$, which is nothing more than the extension of $\langle\ ,\ \rangle_h$. Thus $\langle\ ,\ \rangle_h$ is positive definite. □

REMARK 3.1. If h is of real type, then it is easy to see that $\langle x, y\rangle_h \in \mathbb{R}$ for all $x, y \in H$. Thus $\langle\ ,\ \rangle_h$ yields an inner product on $H \otimes_{\mathbb{Z}} \mathbb{R}$.

Let T be a finitely generated R-module such that T is a torsion module as a $\mathbb{Z}$-module, that is, $nT = \{0\}$ for some non-zero integer n. Then we can associate T with a codimension 1 cycle $[T]$ as follows:

$$[T] := \sum_{P \in \text{m-Spec}(R)} \text{length}_{R_P}(T_P)[P].$$

Since $T \simeq \bigoplus_{P \in \text{m-Spec}(R)} T_P$ by Lemma 1.6, we have

$$\log \#(T) = \sum_{P \in \text{m-Spec}(R)} \text{length}_{R_P}(T_P) \log \#(R/P). \tag{3.1}$$

In order to define the arithmetic first Chern class of a hermitian R-module of pure rank, let us consider the following proposition.

PROPOSITION 3.10. *Let (H, h) be a hermitian R-module of pure rank r. As in Proposition 3.5, let $s_1, \ldots, s_r \in H$ such that $H/(Rs_1 + \cdots + Rs_r)$ is a torsion module as a $\mathbb{Z}$-module. Let us consider an element $z(s_1, \ldots, s_r) \in \widehat{Z}^1(R)$ given by*

$$z(s_1, \ldots, s_r) := \left([H/(Rs_1 + \cdots + Rs_r)], \right.$$
$$\left. \sum_{\sigma} - \log \det \left((h_\sigma(s_i \otimes^\sigma 1, s_j \otimes^\sigma 1))_{\substack{1 \le i \le r \\ 1 \le j \le r}} \right) [\sigma] \right),$$

where z means $([H], 0)$ in the case where $r = 0$. Then $z(s_1, \ldots, s_r)$ does not depend on the choice of $s_1, \ldots, s_r$ as an element of $\widehat{\mathrm{CH}}^1(R)$. In this way, the class of the arithmetic cycle $z(s_1, \ldots, s_r)$ in $\widehat{\mathrm{CH}}^1(R)$ is called the arithmetic first Chern class *of (H, h) and it is denoted by $\widehat{c}_1(H, h)$.*

PROOF. In the case where $r = 0$, the assertion of the proposition is trivial, so that we may assume $r > 0$. Let $s'_1, \ldots, s'_r$ be another set of elements of H such that $H/Rs'_1 + \cdots + Rs'_r$ is a torsion module as a $\mathbb{Z}$-module. First we assume that $s'_1, \ldots, s'_r \in Rs_1 + \cdots + Rs_r$. Then we can choose $a_{ij} \in R$ with $s'_i = \sum_{j=1}^r a_{ij} s_j$. We denote a matrix (a_{ij}) by A. Let us see that $\det(A)$ is regular and

$$z(s'_1, \ldots, s'_r) = z(s_1, \ldots, s_r) + (\widehat{\det(A)}).$$

Using an exact sequence

$$0 \to \frac{Rs_1 + \cdots + Rs_r}{Rs'_1 + \cdots + Rs'_r} \to \frac{H}{Rs'_1 + \cdots + Rs'_r} \to \frac{H}{Rs_1 + \cdots + Rs_r} \to 0,$$

we have

$$\left[\frac{H}{Rs'_1 + \cdots + Rs'_r}\right] = \left[\frac{H}{Rs_1 + \cdots + Rs_r}\right] + \left[\frac{Rs_1 + \cdots + Rs_r}{Rs'_1 + \cdots + Rs'_r}\right].$$

On the other hand, if we set $M = Rs_1 + \cdots + Rs_r$, then M is a free R-module by Proposition 3.5. Here we define R-homomorphism $\phi : M \to M$ by $\phi(s_i) = \sum_{j=1}^n a_{ij} s_j$. Then, by Theorem 1.8, $\det(A)$ is regular and

$$\left[\frac{Rs_1 + \cdots + Rs_r}{Rs'_1 + \cdots + Rs'_r}\right] = [M/\phi(M)] = [R/\det(A)R].$$

Moreover,

$$\begin{aligned} h_\sigma(s'_i \otimes 1, s'_j \otimes 1) &= h_\sigma\left(\sum_k a_{ik} s_k \otimes 1, \sum_l a_{jl} s_l \otimes 1\right) \\ &= \sum_{k,l} \sigma(a_{ik}) h_\sigma(s_k \otimes 1, s_l \otimes 1)\overline{\sigma(a_{jl})}, \end{aligned}$$

that is,

$$(h_\sigma(s'_i \otimes 1, s'_j \otimes 1)) = \sigma(A)(h_\sigma(s_i \otimes 1, s_j \otimes 1))^t\overline{\sigma(A)}$$

as matrices. Therefore, we have

$$-\log\det(h_\sigma(s'_i \otimes 1, s'_j \otimes 1)) = -\log\det(h_\sigma(s_i \otimes 1, s_j \otimes 1)) - \log|\sigma(\det(A))|^2.$$

Thus, gathering the above two formulae, we obtain

$$z(s'_1, \ldots, s'_r) = z(s_1, \ldots, s_r) + \widehat{(\det(A))}.$$

Next let us consider a general case. Since $H/(Rs_1 + \cdots + Rs_r)$ is a torsion module as a $\mathbb{Z}$-module, there is a non-zero integer a such that $as'_1, \ldots, as'_r \in Rs_1 + \cdots + Rs_r$, so that, using the previous case,

$$z(as'_1, \ldots, as'_r) - z(s_1, \ldots, s_r) \in \{\widehat{(x)} \mid x \in K \setminus \{0\}\}.$$

In the same way,

$$z(as'_1, \ldots, as'_r) - z(s'_1, \ldots, s'_r) \in \{\widehat{(x)} \mid x \in K \setminus \{0\}\}.$$

Hence the proposition follows. □

The arithmetic degree $\widehat{\deg}(\widehat{c}_1(H,h))$ of $\widehat{c}_1(H,h)$ is called the *arithmetic degree* of (H,h) and it is denoted by $\widehat{\deg}(H,h)$. Moreover, by using the equation (3.1), $\widehat{\deg}(H,h)$ can be described as follows Let $s_1, \ldots, s_r \in H$ such that $H/Rs_1 + \cdots + Rs_r$ is a torsion module as a $\mathbb{Z}$-module. Then

$$\begin{aligned} \widehat{\deg}(H,h) &= \log \#\left(H/(Rs_1 + \cdots + Rs_r)\right) \\ &\quad - \frac{1}{2}\sum_\sigma \log\det\left(h_\sigma(s_i \otimes^\sigma 1, s_j \otimes^\sigma 1)\right)_{\substack{1\le i\le r \\ 1\le j\le r}}. \end{aligned}$$

In the case where $r = 0$, $\widehat{\deg}(H,h) = \log\#(H)$.

We assume that R is an integral domain. In this case, K is the quotient field of R. Let (L,h) be a hermitian R-module such that L is flat over R and the rank of L is 1. Let s be a non-zero element of $L \otimes_R K$. For each $P \in \text{m-Spec}(R)$, let l_P be a local basis of L around P. Then we can find $f_P \in K$ with $s = f_P l_P$. Then $\text{ord}_P(f_P)$ does not depend on the choice of the local basis l_P, so that it is denoted by $\text{ord}_P(s)$. Here we define $\text{div}(s)$ to be

$$\text{div}(s) := \sum_{P \in \text{m-Spec}(R)} \text{ord}_P(s)[P].$$

Then we have the following.

PROPOSITION 3.11.

$$\widehat{\deg}\left(\text{div}(s), \sum\nolimits_{\sigma \in K(\mathbb{C})} -\log h_\sigma(s \otimes^\sigma 1, s \otimes^\sigma 1)[\sigma]\right) = \widehat{\deg}(L,h).$$

PROOF. We choose a non-zero $a \in R$ such that $as \in L$. Then

$$[L/Ras] = \sum_P \operatorname{ord}_P(as)[P] = \sum_P \operatorname{ord}_P(a)[P] + \sum_P \operatorname{ord}_P(s)[P].$$

On the other hand,

$$\log h_\sigma(as \otimes^\sigma 1, as \otimes^\sigma 1) = \log |\sigma(a)|^2 + \log h_\sigma(s \otimes^\sigma 1, s \otimes^\sigma 1).$$

Thus

$$\left([L/Ras], \sum_\sigma -\log h_\sigma(as \otimes^\sigma 1, as \otimes^\sigma 1)[\sigma]\right) = \widehat{(a)} + \left(\operatorname{div}(s), -\sum_\sigma \log h_\sigma(s \otimes^\sigma 1, s \otimes^\sigma 1)[\sigma]\right).$$

Hence we get the proposition. □

Finally we discuss a relation between $\widehat{\deg}$ and $\widehat{\chi}$ in Section 2.5. Let (H, h) be a hermitian $\mathbb{Z}$-module of real type, that is, $h(x \otimes 1, y \otimes 1) \in \mathbb{R}$ $(\forall x, y \in H)$. Let $\|\cdot\|_h$ be a norm of $H_\mathbb{R} := H \otimes \mathbb{R}$ given by $\|x\|_h = \sqrt{h(x,x)}$. Then we have the following.

PROPOSITION 3.12. *$\widehat{\chi}(H, \|\cdot\|_h) = \widehat{\deg}(H, h) + \log V(\operatorname{rk} H)$, where $V(n)$ is the volume of the unit ball in $\mathbb{R}^n$, that is,*

$$V(n) = \operatorname{vol}_n\left(\{(x_1, \ldots, x_n) \in \mathbb{R}^n \mid x_1^2 + \cdots + x_n^2 \leq 1\}\right).$$

PROOF. Let H_{tor} be the torsion part of H. Then

$$\begin{cases} \widehat{\chi}(H, \|\cdot\|_h) = \widehat{\chi}(H/H_{tor}, \|\cdot\|_h) + \log \#(H_{tor}), \\ \widehat{\deg}(H, h) = \widehat{\deg}(H/H_{tor}, h) + \log \#(H_{tor}). \end{cases}$$

Thus we may assume that H is a free $\mathbb{Z}$-module. As h is an inner product of $H_\mathbb{R}$, we can take an orthonormal basis $x_1, \ldots, x_n$ of $H_\mathbb{R}$ with respect to h. Here we fix a Haar measure vol such that

$$\operatorname{vol}(\{\lambda_1 x_1 + \cdots + \lambda_n x_n \mid 0 \leq \lambda_i \leq 1 \ (\forall i)\}) = 1.$$

Then $\operatorname{vol}(\{x \in H_\mathbb{R} \mid \|x\| \leq 1\}) = V(\operatorname{rk} H)$. On the other hand, let $e_1, \ldots, e_n$ be a free basis of H. Then, as we will see in the next section, $\operatorname{vol}(H_\mathbb{R}/H) = \sqrt{\det(h(e_i, e_j))}$. Therefore,

$$\widehat{\chi}(H, \|\cdot\|_h) = \log V(\operatorname{rk} H) - \log \sqrt{\det(h(e_i, e_j))},$$

which proves the proposition. □

3.4. Arithmetic Riemann-Roch formula on arithmetic curves

Let M be a finitely generated $\mathbb{Z}$-module and let $\langle\ ,\ \rangle : M \times M \to \mathbb{C}$ be a positive definite hermitian form on M. Note that $\langle\ ,\ \rangle$ yields a hermitian inner product on $M \otimes_\mathbb{Z} \mathbb{C}$. Thus $(M, \langle\ ,\ \rangle)$ is a hermitian $\mathbb{Z}$-module. We define $\operatorname{vol}(M, \langle,\rangle)$ to be

$$\operatorname{vol}(M, \langle,\rangle) := \exp(-\widehat{\deg}(M, \langle,\rangle))$$

and it is called the *volume of M with respect to $\langle\ ,\ \rangle$*. If M is free and $e_1, \ldots, e_r$ form a free basis of M, then it is easy to see that

$$\operatorname{vol}(M, \langle,\rangle) = \sqrt{\det(\langle e_i, e_j\rangle)}.$$

Thus, if $\langle\ ,\ \rangle$ gives rise to an inner product on $M \otimes_{\mathbb{Z}} \mathbb{R}$ (i.e. $(M, \langle,\rangle)$ is of real type), then $\mathrm{vol}(M, \langle,\rangle)$ is the volume of the fundamental area of the lattice M in $M \otimes_{\mathbb{Z}} \mathbb{R}$.

Let R be a reduced order and let K be the total quotient ring of R. Let $\{\omega_1, \ldots, \omega_n\}$ be a free basis of R as a $\mathbb{Z}$-module. Then we define the absolute discriminant D_R of R to be

$$D_R := \det\left(\mathrm{Tr}_{K/\mathbb{Q}}(\omega_i \cdot \omega_j)\right).$$

In the case where K is an algebraic number field and $R = O_K$, D_R is the usual discriminant $D_{K/\mathbb{Q}}$ of K over $\mathbb{Q}$.

PROPOSITION 3.13. $\mathrm{vol}(R, \langle,\rangle_{h^{can}}) = \sqrt{|D_R|}$.

PROOF. We set $K(\mathbb{C}) = \{\sigma_1, \ldots, \sigma_n\}$. Let us see that

$$\mathrm{Tr}_{K/\mathbb{Q}}(\omega_i \cdot \omega_j) = \sum_{k=1}^{n} \sigma_k(\omega_i) \cdot \sigma_k(\omega_j). \tag{3.2}$$

This is well known in the case where K is an algebraic number field. In general, since K is the product of algebraic number fields, (3.2) follows. On the other hand, as

$$\langle \omega_i, \omega_j \rangle_{h^{can}} = \sum_{k=1}^{n} \sigma_k(\omega_i) \cdot \bar{\sigma}_k(\omega_j),$$

if we set $\Delta = (\sigma_i(\omega_j))$, then

$$(\mathrm{Tr}_{K/\mathbb{Q}}(\omega_i \cdot \omega_j)) = {}^t\Delta \cdot \Delta \quad \text{and} \quad (\langle \omega_i, \omega_j \rangle_{h^{can}}) = {}^t\Delta \cdot \overline{\Delta}.$$

Thus we have $\mathrm{vol}(R, \langle,\rangle_{h^{can}}) = \sqrt{|D_R|}$. □

For a hermitian R-module (H, h), we define $\chi_{Ar}(H, h)$ to be

$$\chi_{Ar}(H, h) := -\log(\mathrm{vol}(H, \langle,\rangle_h)) = \widehat{\deg}(H/\mathbb{Z}, \langle,\rangle_h),$$

where $H/\mathbb{Z}$ means that H is viewed as a $\mathbb{Z}$-module. For example, by Proposition 3.13,

$$\chi_{Ar}(R, h^{can}) = -\frac{1}{2}\log|D_R|.$$

If h is of real type, then, by Proposition 3.12,

$$\widehat{\chi}(H/\mathbb{Z}, \langle,\rangle_h^{1/2}) = \log(V(\mathrm{rk}_{\mathbb{Z}}(H))) + \chi_{Ar}(H, h), \tag{3.3}$$

where $\widehat{\chi}$ is defined in Section 2.5.

THEOREM 3.14 (Riemann-Roch formula over a reduced order). *For a hermitian R-module (H, h) of pure rank r,*

$$\chi_{Ar}(H, h) = \widehat{\deg}(H, h) + r\chi_{Ar}(R, h^{can}).$$

PROOF. In the case where $r = 0$, the assertion of the theorem is obvious, so that we may assume $r > 0$. Let $\omega_1, \ldots, \omega_n$ be a free basis of R as a $\mathbb{Z}$-module. Moreover, let us choose $s_1, \ldots, s_r \in H$ such that $H/Rs_1 + \cdots + Rs_r$ is a torsion module as a $\mathbb{Z}$-module. Here we consider a matrix $A = (\langle \omega_i s_j, \omega_k s_l \rangle_h)$ with respect to the following order:

$$\omega_1 s_1, \ldots, \omega_n s_1, \omega_1 s_2, \ldots, \omega_n s_2, \ldots, \omega_1 s_r, \ldots, \omega_n s_r.$$

Note that A is an $(rn \times rn)$-matrix and

$$\mathbb{Z}\omega_1 s_1 + \cdots + \mathbb{Z}\omega_n s_1 + \cdots + \mathbb{Z}\omega_1 s_r + \cdots + \mathbb{Z}\omega_n s_r = Rs_1 + \cdots + Rs_r.$$

Then

$$\chi_{Ar}(H,h) = \widehat{\deg}(H/\mathbb{Z}, \langle , \rangle_h) = \log \# \left(\frac{H}{Rs_1 + \cdots + Rs_r} \right) - \frac{1}{2} \log(\det A).$$

We divide A into $n \times n$-matrices. Let $A(j,l)$ be the (j,l)-block in the division. We set $K(\mathbb{C}) = \{\sigma_1, \ldots, \sigma_n\}$. Since

$$\begin{aligned} \langle \omega_i s_j, \omega_k s_l \rangle_h &= \sum_{m=1}^{n} h_{\sigma_m}(\omega_i s_j \otimes 1, \omega_k s_l \otimes 1) \\ &= \sum_{m=1}^{n} \sigma_m(\omega_i) h_{\sigma_m}(s_j \otimes 1, s_l \otimes 1) \overline{\sigma_m(\omega_k)}, \end{aligned}$$

if we set $\Delta = (\sigma_i(\omega_j))$ and

$$B(j,l) = \begin{pmatrix} h_{\sigma_1}(s_j \otimes 1, s_l \otimes 1) & & 0 \\ & \ddots & \\ 0 & & h_{\sigma_n}(s_j \otimes 1, s_l \otimes 1) \end{pmatrix},$$

then $A(j,l) = {}^t\Delta B(j,l)\overline{\Delta}$. Therefore,

$$A = \begin{pmatrix} {}^t\Delta & & 0 \\ & \ddots & \\ 0 & & {}^t\Delta \end{pmatrix} B \begin{pmatrix} \overline{\Delta} & & 0 \\ & \ddots & \\ 0 & & \overline{\Delta} \end{pmatrix},$$

where B is an $(rn \times rn)$-matrix whose (j,l)-th block is $B(j,l)$ for every j,l. Therefore, as $|\det(\Delta)| = \operatorname{vol}(R, \langle , \rangle_{h^{can}})$ by the proof of Proposition 3.13, we obtain

$$\det(A) = \operatorname{vol}(R, \langle , \rangle_{h^{can}})^{2r} \det(B),$$

that is,

$$\chi_{Ar}(H,h) = \log \# \left(\frac{H}{Rs_1 + \cdots + Rs_r} \right) - \frac{1}{2} \log(\det B) + r\chi_{Ar}(R, h^{can}).$$

Next we consider $\det(B)$. We set

$$V = \bigoplus_{m=1}^{n} H \otimes_R^{\sigma_m} \mathbb{C}.$$

Let h_V be a hermitian inner product of V given by $h_V = h_{\sigma_1} \oplus^{\perp} \cdots \oplus^{\perp} h_{\sigma_n}$. Let B' be the matrix $(h_V(s_i \otimes^{\sigma_k} 1, s_j \otimes^{\sigma_l} 1))$ with respect to the following basis:

$$s_1 \otimes^{\sigma_1} 1, \ldots, s_r \otimes^{\sigma_1} 1, s_1 \otimes^{\sigma_2} 1, \ldots, s_r \otimes^{\sigma_2} 1, \ldots, s_1 \otimes^{\sigma_n} 1, \ldots, s_r \otimes^{\sigma_n} 1.$$

Then

$$\det(B') = \prod_{m=1}^{n} \det\left(h_{\sigma_m}(s_i \otimes 1, s_j \otimes 1)\right).$$

On the other hand, B is the matrix $(h_V(s_i \otimes^{\sigma_k} 1, s_j \otimes^{\sigma_l} 1))$ with respect to the following basis:

$$s_1 \otimes^{\sigma_1} 1, \ldots, s_1 \otimes^{\sigma_n} 1, s_2 \otimes^{\sigma_1} 1, \ldots, s_2 \otimes^{\sigma_n} 1, \ldots, s_r \otimes^{\sigma_1} 1, \ldots, s_r \otimes^{\sigma_n} 1.$$

Thus we have $\det(B) = \det(B')$, that is,

$$\det(B) = \prod_{m=1}^{n} \det\left(h_{\sigma_m}(s_i \otimes 1, s_j \otimes 1)\right).$$

Therefore,

$$\log \# \left(\frac{H}{Rs_1 + \cdots + Rs_r} \right) - \frac{1}{2} \log(\det B) = \widehat{\deg}(H, h),$$

which yields the theorem. □

3.5. Effective estimate of the number of small sections

Let R be an integral order and K the quotient field. Let $\overline{E} = (E, h)$ be a hermitian R-module. By Proposition 3.8, the natural $\mathbb{C}$-linear map

$$\phi : E \otimes_{\mathbb{Z}} \mathbb{C} \to \bigoplus_{\sigma \in K(\mathbb{C})} E \otimes_R^\sigma \mathbb{C}$$

is an isomorphism. For each $\sigma \in K(\mathbb{C})$, the natural homomorphism $E \otimes_{\mathbb{Z}} \mathbb{C} \to E \otimes_R^\sigma \mathbb{C}$ is denoted by ϕ_σ. As $E \otimes_{\mathbb{Z}} \mathbb{R} \hookrightarrow E \otimes_{\mathbb{Z}} \mathbb{C}$, for $s \in E \otimes_{\mathbb{Z}} \mathbb{R}$, we define $\|s\|_{\sup}^{\overline{E}}$ by

$$\|s\|_{\sup}^{\overline{E}} = \max_{\sigma \in K(\mathbb{C})} \left\{ h_\sigma(\phi_\sigma(s), \phi_\sigma(s))^{1/2} \right\}.$$

Then $(E, \|\cdot\|_{\sup}^{\overline{E}})$ is a normed finitely generated $\mathbb{Z}$-module.

We assume that E is torsion free and h is of real type. Let r_1 and r_2 be the numbers of real embeddings of K into $\mathbb{R}$ and complex embeddings of K into $\mathbb{C}$, respectively. Here, to count the number of complex embeddings, we identify σ with $\bar{\sigma}$. In this section, we consider the exact estimate of $\hat{h}^0(E, \|\cdot\|_{\sup})$ and $\hat{h}^0_{<1}(E, \|\cdot\|_{\sup})$ (for the definition of $\hat{h}^0_{<1}(E, \|\cdot\|_{\sup})$, see Section 2.5). Namely we have the following theorem:

THEOREM 3.15. *Let r be the rank of E over R. Then we have the following:*

(1) $\hat{\chi}(E, \|\cdot\|_{\sup}) = \widehat{\deg}(\overline{E}) + \log\left(\dfrac{V(r)^{r_1}(2^r V(2r))^{r_2}}{|D_R|^{r/2}} \right)$, *where $V(n)$ is the volume of the unit ball in $\mathbb{R}^n$.*

(2) $\hat{h}^0(E, \|\cdot\|_{\sup}) > \widehat{\deg}(\overline{E}) + \log\left(\dfrac{V(r)^{r_1} V(2r)^{r_2}}{2^{r(r_1+r_2)}|D_R|^{r/2}} \right)$. *In particular, if*

$$\widehat{\deg}(\overline{E}) \geq \log\left(\frac{2^{r(r_1+r_2)}|D_R|^{r/2}}{V(r)^{r_1} V(2r)^{r_2}} \right),$$

then there is $s \in E \setminus \{0\}$ such that $\|s\|_{\sup} \leq 1$.

(3) $\hat{h}^0_{<1}(E, \|\cdot\|_{\sup}) \geq \widehat{\deg}(\overline{E}) + \log\left(\dfrac{V(r)^{r_1} V(2r)^{r_2}}{2^{r(r_1+r_2)}|D_R|^{r/2}} \right)$. *In particular, if*

$$\widehat{\deg}(\overline{E}) > \log\left(\frac{2^{r(r_1+r_2)}|D_R|^{r/2}}{V(r)^{r_1} V(2r)^{r_2}} \right),$$

then there is $s \in E \setminus \{0\}$ such that $\|s\|_{\sup} < 1$.

PROOF. (1) Let $\sigma_1, \ldots, \sigma_{r_1}$ be all real embeddings of K into $\mathbb{R}$ and let $\tau_1, \bar{\tau}_1, \ldots, \tau_{r_2}, \bar{\tau}_{r_2}$ be all complex embeddings of K into $\mathbb{C}$. For each $\sigma \in K(\mathbb{C})$, we choose an orthonormal basis $e_{\sigma,1}, \ldots, e_{\sigma,r}$ of E_σ with respect to h_σ. We may assume that

$\iota_{\sigma,\bar{\sigma}}(e_{\sigma,k}) = e_{\bar{\sigma},k}$. Then

$$
\begin{aligned}
&e_{\sigma_1,1},\ldots,e_{\sigma_1,r},\ldots,e_{\sigma_{r_1},1},\ldots,e_{\sigma_{r_1},r},\\
&\quad\frac{e_{\tau_1,1}+e_{\bar{\tau}_1,1}}{\sqrt{2}},\frac{\sqrt{-1}(e_{\tau_1,1}-e_{\bar{\tau}_1,1})}{\sqrt{2}},\ldots,\frac{e_{\tau_1,r}+e_{\bar{\tau}_1,r}}{\sqrt{2}},\frac{\sqrt{-1}(e_{\tau_1,r}-e_{\bar{\tau}_1,r})}{\sqrt{2}},\ldots,\\
&\quad\frac{e_{\tau_{r_2},1}+e_{\bar{\tau}_{r_2},1}}{\sqrt{2}},\frac{\sqrt{-1}(e_{\tau_{r_2},1}-e_{\bar{\tau}_{r_2},1})}{\sqrt{2}},\ldots,\frac{e_{\tau_{r_2},r}+e_{\bar{\tau}_{r_2},r}}{\sqrt{2}},\frac{\sqrt{-1}(e_{\tau_{r_2},r}-e_{\bar{\tau}_{r_2},r})}{\sqrt{2}}
\end{aligned}
$$

form a real orthonormal basis of $\phi(E\otimes_{\mathbb{Z}}\mathbb{R})$ with respect to $\bigoplus_\sigma^\perp h_\sigma$. Let vol be the Haar measure of $\phi(E\otimes_{\mathbb{Z}}\mathbb{R})$ arising from $\bigoplus_\sigma^\perp h_\sigma$. Here we claim the following:

CLAIM 1. $\operatorname{vol}\left(\{(x_\sigma)\in\phi(E\otimes_{\mathbb{Z}}\mathbb{R}) \mid h_\sigma(x_\sigma,x_\sigma)\le 1\ \forall\sigma\}\right) = V(r)^{r_1}(2^rV(2r))^{r_2}$.

PROOF. For $(x_\sigma)\in\phi(E\otimes_{\mathbb{Z}}\mathbb{R})$, if we set

$$
(x_\sigma)=\sum_{i=1}^{r_1}\sum_{k=1}^{r}a_{i,k}e_{\sigma_i,k}+\sum_{j=1}^{r_2}\sum_{l=1}^{r}\left(b_{j,l}\frac{e_{\tau_j,l}+e_{\bar{\tau}_j,l}}{\sqrt{2}}+c_{j,l}\frac{\sqrt{-1}(e_{\tau_j,l}-e_{\bar{\tau}_j,l})}{\sqrt{2}}\right),
$$

then

$$
(x_\sigma)=\sum_{i=1}^{r_1}\sum_{k=1}^{r}a_{i,k}e_{\sigma_i,k}+\sum_{j=1}^{r_2}\sum_{l=1}^{r}\left(\frac{b_{j,l}+\sqrt{-1}c_{j,l}}{\sqrt{2}}e_{\tau_j,l}+\frac{b_{j,l}-\sqrt{-1}c_{j,l}}{\sqrt{2}}e_{\bar{\tau}_j,l}\right).
$$

Thus $h_\sigma(x_\sigma,x_\sigma)\le 1\ \forall\sigma$ if and only if

$$
\begin{cases}
\sum_{k=1}^{r}a_{i,k}^2\le 1 & \text{for all } i=1,\ldots,r_1,\\
\sum_{l=1}^{r}\left(b_{j,l}^2+c_{j,l}^2\right)\le 2 & \text{for all } j=1,\ldots,r_2.
\end{cases}
$$

Therefore, the claim follows. □

Note that $-\log(\operatorname{vol}(\phi(E\otimes_{\mathbb{Z}}\mathbb{R})/\phi(E))) = \chi_{Ar}(\overline{E})$, and hence, by the above claim, $\widehat{\chi}(E,\|\cdot\|_{\sup}) = \log(V(r)^{r_1}(2^rV(2r))^{r_2}) + \chi_{Ar}(\overline{E})$. Therefore, by the arithmetic Riemann-Roch formula on arithmetic curves (cf. Theorem 3.14),

$$
\widehat{\chi}(E,\|\cdot\|_{\sup}) = \widehat{\deg}(\overline{E}) + \log\left(\frac{V(r)^{r_1}(2^rV(2r))^{r_2}}{|D_R|^{r/2}}\right).
$$

(2) By Corollary 2.2,

$$
\hat{h}^0(E,\|\cdot\|_{\sup}) > \widehat{\chi}(E,\|\cdot\|_{\sup}) - \log(2^{rn}) = \widehat{\deg}(\overline{E}) + \log\left(\frac{V(r)^{r_1}V(2r)^{r_2}}{2^{r(r_1+r_2)}|D_R|^{r/2}}\right),
$$

as required.

(3) Note that

$$
\begin{aligned}
&\operatorname{vol}\left(\{(x_\sigma)\in\phi(H\otimes_{\mathbb{Z}}\mathbb{R}) \mid h_\sigma(x_\sigma,x_\sigma)< 1\ \forall\sigma\}\right)\\
&\qquad=\operatorname{vol}\left(\{(x_\sigma)\in\phi(H\otimes_{\mathbb{Z}}\mathbb{R}) \mid h_\sigma(x_\sigma,x_\sigma)\le 1\ \forall\sigma\}\right) = V(r)^{r_1}(2^rV(2r))^{r_2}
\end{aligned}
$$

and

$$
\hat{H}^0_{<1}(E,\|\cdot\|_{\sup}) = \{s\in E \mid h_\sigma(s\otimes^\sigma 1, s\otimes^\sigma 1)<1\ (\forall\sigma)\}.
$$

Thus (3) follows from Corollary 2.2. □

From now on, we assume that R is normal. Let $D = \sum_P n_P[P]$ and $D' = \sum_P n'_P[P]$ be cycles of codimension one on $\operatorname{Spec}(R)$. If $n_P \geq n'_P$ for all P, then it is denoted by $D \geq D'$. Moreover, for arithmetic cycles

$$\overline{D} = \left(\sum_P n_P[P], \sum_\sigma \xi_\sigma[\sigma]\right) \quad \text{and} \quad \overline{D}' = \left(\sum_P n'_P[P], \sum_\sigma \xi'_\sigma[\sigma]\right)$$

of codimension one on $\operatorname{Spec}(R)$, if $n_P \geq n'_P$ for all P and $\xi_\sigma \geq \xi'_\sigma$ for all σ, then we denote it by $\overline{D} \geq \overline{D}'$.

COROLLARY 3.16. *Let $\overline{D} = (D, \xi)$ be an arithmetic cycle of codimension one on* $\operatorname{Spec}(R)$ *such that $\xi_{\bar{\sigma}} = \xi_\sigma$ for all $\sigma \in K(\mathbb{C})$, where $\xi = \sum_\sigma \xi_\sigma[\sigma]$. If $\widehat{\deg}(\overline{D}) \geq \log((\pi/2)^{r_2}\sqrt{|D_R|})$, then there is $x \in K^\times$ such that $\overline{D} + \widehat{(x)} \geq (0,0)$.*

PROOF. We set $\mathscr{O}(D) = \{x \in K^\times \mid D + (x) \geq 0\} \cup \{0\}$. We choose a non-zero integer a such that $D \leq (a)$. Then

$$\mathscr{O}(D) \subseteq \mathscr{O}((a)) = R\frac{1}{a}.$$

Thus $\mathscr{O}(D)$ is a finitely generated torsion free R-module of rank 1. Note that $\mathscr{O}(D)_\sigma = K \otimes_K^\sigma \mathbb{C}$ is naturally isomorphic to $\mathbb{C}$, so that we define h_σ on $\mathscr{O}(D)_\sigma$ to be

$$h_\sigma(z, w) = z\bar{w}\exp(-\xi_\sigma) \quad (z, w \in \mathbb{C}).$$

Then $\mathscr{O}(\overline{D}) = (\mathscr{O}(D), \{h_\sigma\})$ forms a hermitian R-module of real type. We set $D = \sum_P n_P[P]$. Note that 1 is a rational section of $\mathscr{O}(D)$, that is,

$$1 \in \mathscr{O}(D) \otimes_R K = K.$$

Let t_P be a local parameter of R_P. Since $t_P^{-n_P}$ forms a local basis of $\mathscr{O}(D)_P$, we can see $\operatorname{div}(1) = D$. Moreover, $-\log h_\sigma(1,1) = \xi_\sigma$. Thus, by Proposition 3.11, $\widehat{\deg}(\mathscr{O}(\overline{D})) = \widehat{\deg}(\overline{D})$, and hence, by Theorem 3.15, if $\widehat{\deg}(\overline{D}) \geq \log((\pi/2)^{r_2}\sqrt{|D_R|})$, then there is a non-zero x such that $x \in \mathscr{O}(D)$ and $h_\sigma(x, x) \leq 1$ for all σ. Note that

$$h_\sigma(x,x) \leq 1 \iff |\sigma(x)|^2 \exp(-\xi_\sigma) \leq 1 \iff -\log|\sigma(x)|^2 + \xi_\sigma \geq 0,$$

as desired. □

COROLLARY 3.17. *The group $\bigoplus_P \mathbb{Z}[P]$ modulo $\{(x) \mid x \in K^\times\}$ is finite.*

PROOF. For $D = \sum_P n_P[P] \in \bigoplus_P \mathbb{Z}[P]$, we define $\deg(D)$ to be

$$\deg(D) := \sum_P n_P \log \#(R/P).$$

We set $C_R = \log((\pi/2)^{r_2}\sqrt{|D_R|})$ and

$$\Theta = \left\{E \in \bigoplus\nolimits_P \mathbb{Z}[P] \;\middle|\; E \geq 0 \text{ and } \deg(E) \leq C_R\right\}.$$

It is easy to see that Θ is a finite set. For $D \in \bigoplus_P \mathbb{Z}[P]$, if we set

$$\overline{D} = \left(D, \sum_\sigma \frac{2(C_R - \deg(D))}{[K:\mathbb{Q}]}[\sigma]\right),$$

then, as $\widehat{\deg}(\overline{D}) = C_R$, by Corollary 3.16, there is $x \in K^{\times}$ such that $\overline{D} + \widehat{(x)} \geq (0,0)$. By using the product formula (cf. Theorem 3.3),

$$\deg(D + (x)) = \deg(D) + \sum_{\sigma} \log|\sigma(x)| \leq \deg(D) + \sum_{\sigma} \frac{C_R - \deg(D)}{[K:\mathbb{Q}]} = C_R,$$

and hence $D + (x) \in \Theta$. Thus the corollary follows. □

3.6. Several formulae on arithmetic degree

Let R be a reduced order. In this section, we consider several formulae on the arithmetic degree.

LEMMA 3.18. *Let (L,h) and (E,k) be hermitian R-modules of pure rank* 1 *and r, respectively. We assume that L is flat over R. If we set $(L,h) \otimes (E,k) := (L \otimes_R E, h \otimes k)$, then*

$$\widehat{c}_1((L,h) \otimes (E,k)) = r\widehat{c}_1(L,h) + \widehat{c}_1(E,k).$$

In particular,

$$\widehat{\deg}((L,h) \otimes (E,k)) = r\widehat{\deg}(L,h) + \widehat{\deg}(E,k).$$

PROOF. Let us choose $s \in L$ and $s_1, \ldots, s_r \in E$ such that L/Rs and $E/(Rs_1 + \cdots + Rs_r)$ are torsion modules as $\mathbb{Z}$-modules. Then, since $s \otimes s_1, \ldots, s \otimes s_r$ yields a basis of $(L \otimes E)_Q = L_Q \otimes E_Q$ for all minimal primes Q of R, $(L \otimes E)/(R(s \otimes s_1) + \cdots + R(s \otimes s_r))$ is a torsion module as a $\mathbb{Z}$-module. First let us see the following:

$$[(L \otimes E)/(R(s \otimes s_1) + \cdots + R(s \otimes s_r))] = r[L/Rs] + [E/(Rs_1 + \cdots + Rs_r)].$$

It is sufficient to see

$$\begin{aligned} \text{(3.4)} \quad & \operatorname{length}_{R_P}((L_P \otimes E_P)/(R_P(s \otimes s_1) + \cdots + R_P(s \otimes s_r))) \\ & \qquad = r \operatorname{length}_{R_P}(L_P/R_P s) + \operatorname{length}_{R_P}(E_P/(R_P s_1 + \cdots + R_P s_r)) \end{aligned}$$

for all $P \in \text{m-Spec}(R)$. Let ω be a local basis of L_P around P and $s = a\omega$. By Proposition 3.5, $Rs_1 + \cdots + Rs_r$ is a free R-module. Thus, if we set

$$A = \operatorname{length}_{R_P}\left(\frac{L_P \otimes E_P}{R_P(s \otimes s_1) + \cdots + R_P(s \otimes s_r)}\right),$$

then

$$\begin{aligned} A &= \operatorname{length}_{R_P}\left(\frac{E_P}{a(R_P s_1 + \cdots + R_P s_r)}\right) \\ &= \operatorname{length}_{R_P}\left(\frac{E_P}{R_P s_1 + \cdots + R_P s_r}\right) + \operatorname{length}_{R_P}\left(\frac{R_P s_1 + \cdots + R_P s_r}{a(R_P s_1 + \cdots + R_P s_r)}\right) \\ &= \operatorname{length}_{R_P}\left(\frac{E_P}{R_P s_1 + \cdots + R_P s_r}\right) + r \operatorname{length}_{R_P}(R_P/aR_P) \\ &= \operatorname{length}_{R_P}\left(\frac{E_P}{R_P s_1 + \cdots + R_P s_r}\right) + r \operatorname{length}_{R_P}(L_P/R_P s), \end{aligned}$$

Hence we obtain (3.4).

On the other hand, we have

$$\det((h_\sigma \otimes k_\sigma)(s \otimes s_i, s \otimes s_j)) = \det(h_\sigma(s,s) k_\sigma(s_i, s_j)) = h_\sigma(s,s)^r \det(k_\sigma(s_i, s_j)).$$

Therefore, we get

$$\widehat{c}_1((L,h)\otimes(E,k)) = r\widehat{c}_1(L,h)+\widehat{c}_1(E,k). \qquad \square$$

Let R and R' be reduced orders with $R \subseteq R'$. Let K and K' be the total quotient rings of R and R', respectively. For $\sigma \in K(\mathbb{C})$, we set

$$K'(\mathbb{C})_\sigma = \{\sigma' \in K'(\mathbb{C}) \mid \sigma'|_K = \sigma\}.$$

For a moment, we assume that R' is flat over R. For example, if R and R' are normal, then R' is flat over R. Let $\overline{L}' = (L', h_{L'})$ be a hermitian R'-module of pure rank 1 such that L' is flat over R', that is, L' yields an invertible sheaf on $\mathrm{Spec}(R')$. According to Section 1.5, $N_{R'/R}(L')$ is a flat R-module of rank 1. We give a hermitian metric to $N_{R'/R}(L')$ as follows: For $s' \in L' \otimes_{R'} K'$ and $\sigma \in K(\mathbb{C})$, we put

$$H(s',\sigma) = \prod_{\sigma' \in K'(\mathbb{C})_\sigma} h_{L',\sigma'}(s' \otimes^{\sigma'} 1, s' \otimes^{\sigma'} 1).$$

Then, for $x' \in K'$, we have

$$H(x's',\sigma) = |\sigma(N_{K'/K}(x'))|^2 H(s',\sigma). \tag{3.5}$$

Indeed, by Lemma 3.6,

$$\begin{aligned} H(x's',\sigma) &= \prod_{\sigma' \in K'(\mathbb{C})_\sigma} h_{L',\sigma'}(x's' \otimes^{\sigma'} 1, x's' \otimes^{\sigma'} 1) \\ &= \prod_{\sigma' \in K'(\mathbb{C})_\sigma} |\sigma'(x')|^2 h_{L',\sigma'}(s' \otimes^{\sigma'} 1, s' \otimes^{\sigma'} 1) \\ &= |\sigma(N_{K'/K}(x'))|^2 H(s',\sigma). \end{aligned}$$

Let us choose $s' \in L' \otimes_{R'} K'$ such that $N_{K'/K}(s') \otimes^\sigma 1$ gives a basis of $N_{R'/R}(L') \otimes^\sigma \mathbb{C}$. Using this s', we define $h_{N(L'),\sigma}$ by the following formula:

$$h_{N(L'),\sigma}(N_{K'/K}(s') \otimes^\sigma \lambda, N_{K'/K}(s') \otimes^\sigma \mu) = \lambda\bar{\mu} H(s',\sigma).$$

Note that (3.5) guarantees that this definition does not depend on the choice of s'. Here $(N_{R'/R}(L'), h_{N(L')})$ is denoted by $N_{R'/R}(\overline{L}')$. Then we have the following.

LEMMA 3.19. $\widehat{\deg}(\overline{L}') = \widehat{\deg}(N_{R'/R}(\overline{L}'))$.

PROOF. Let $s' \in L'$ such that $L'/R's'$ is a torsion module as a $\mathbb{Z}$-module. By Theorem 1.8,

$$\log \#(L'/R's') = \log \#(N_{R'/R}(L')/R \cdot N_{R'/R}(s')).$$

On the other hand,

$$\begin{aligned} \sum_{\sigma' \in K'(\mathbb{C})} \log(h_{L',\sigma'}(s' \otimes^{\sigma'} 1, s' \otimes^{\sigma'} 1)) &= \sum_{\sigma \in K(\mathbb{C})} \log(H(s',\sigma)) \\ &= \sum_{\sigma \in K(\mathbb{C})} \log(h_{N(L),\sigma}(N(s') \otimes^\sigma 1, N(s') \otimes^\sigma 1)). \end{aligned}$$

Thus $\widehat{\deg}(\overline{L}') = \widehat{\deg}(N_{R'/R}(\overline{L}'))$. $\square$

From now on, R' is not necessarily flat over R. Let (L, h_L) be a hermitian R-module of pure rank 1 such that L is flat over R. As there is a natural isomorphism

$$(L \otimes_R R') \otimes_{R'}^{\sigma'} \mathbb{C} \simeq L \otimes_R^{\sigma} \mathbb{C}$$

for any $\sigma' \in K'(\mathbb{C})_\sigma$, we can give a metric $h_{L\otimes_R R',\sigma'}$ to $(L \otimes_R R') \otimes_{R'}^{\sigma'} \mathbb{C}$ by setting $h_{L\otimes_R R',\sigma'} = h_{L,\sigma}$. In this way, we obtain a hermitian R'-module $(L \otimes_R R', h_{L\otimes_R R'})$. It is denoted by $(L, h_L) \otimes_R R'$.

LEMMA 3.20. *(1) If $K' = K$, then $\widehat{\deg}((L,h) \otimes_R R') = \widehat{\deg}(L,h)$.*
(2) If K is an algebraic number field, then

$$\widehat{\deg}((L,h) \otimes_R R') = \dim_K(K')\widehat{\deg}(L,h).$$

PROOF. (1) We set $L' = L \otimes_R R'$. Let us choose $s \in L$ such that L/Rs is a torsion module as a $\mathbb{Z}$-module. We set $s' = s \otimes 1$. Since $L'/R's' = (L/Rs) \otimes_R R'$, $L'/R's'$ is also a torsion module as a $\mathbb{Z}$-module. Therefore, it is sufficient to show that $\#(L/sR) = \#(L'/s'R')$. Since L' and sR are finitely generated free $\mathbb{Z}$-modules and $L' \otimes_{\mathbb{Z}} \mathbb{Q} = sR \otimes_{\mathbb{Z}} \mathbb{Q}$, L'/sR is a finite group. On the other hand, considering two sequences:

$$sR \subseteq L \subseteq L' \quad \text{and} \quad sR \subseteq s'R' \subseteq L',$$

we have

$$\#(L/sR) = \frac{\#(L'/sR)}{\#(L'/L)}, \qquad \#(L'/s'R') = \frac{\#(L'/sR)}{\#(s'R'/sR)}.$$

Thus we need to show $\#(L'/L) = \#(s'R'/sR)$. For $P \in \text{m-Spec}(R)$, let ω_P be a free basis of L_P as a R_P-module and let $s = a_P\omega_P$. Then $\omega_P \otimes 1$ is a free basis of L'_P as a R'_P-module and $s' = a_P(\omega_P \otimes 1)$. Thus, by Lemma 1.6,

$$L'/L = \bigoplus_{P\in \text{m-Spec}(R)} L'_P/L_P \simeq \bigoplus_{P\in \text{m-Spec}(R)} R'_P/R_P,$$

$$s'R'/sR = \bigoplus_{P\in \text{m-Spec}(R)} s'R'_P/sR_P \simeq \bigoplus_{P\in \text{m-Spec}(R)} a_PR'_P/a_PR_P.$$

Since R_P/a_PR_P is a torsion module as a $\mathbb{Z}$-module, there is a non-zero integer n and $b_P \in R_P$ with $a_Pb_P = n$. In particular, a_P is an R'_P-regular element, that is, the multiplication map $R'_P \xrightarrow{a_P\times} R'_P$ by a_P is injective. Therefore, it is easy to see that the multiplication map $R'_P/R_P \to a_PR'_P/a_PR_P$ by a_P is an isomorphism. Thus $\#(R'_P/R_P) = \#(a_PR'_P/a_PR_P)$, and hence $\#(L'/L) = \#(sR'/sR)$.

(2) First we consider the case where R' is flat over R. Then, by our construction, it is easy to see that

$$N_{R'/R}((L,h) \otimes_R R') = (L,h)^{\otimes \dim_K(K')}.$$

Thus the assertion follows from Lemma 3.19 in this case.

Let us consider a general case. Let $\tilde{R}$ and $\tilde{R}'$ be the normalizations of R and R', respectively. Then, since $\tilde{R}'$ is flat over $\tilde{R}$, we have

$$\widehat{\deg}((L,h) \otimes_R \tilde{R}') = \dim_K(K')\widehat{\deg}((L,h) \otimes_R \tilde{R}).$$

On the other hand, by (1),

$$\begin{cases} \widehat{\deg}(L,h) = \widehat{\deg}((L,h) \otimes_R \tilde{R}), \\ \widehat{\deg}((L,h) \otimes_R R') = \widehat{\deg}((L,h) \otimes_R \tilde{R}'). \end{cases}$$

Thus we obtain (2). □

3.7. Volume exactness

Let $\mathbb{K}$ be either $\mathbb{R}$ or $\mathbb{C}$. Let V be a finite-dimensional vector space over $\mathbb{K}$ and let h be a $\mathbb{K}$-inner product of V. Then we can give a natural $\mathbb{K}$-inner product $\det(h)$ to $\det(V)$ as follows: Let $\Lambda = (x_1, \ldots, x_n)$ be a sequence consisting of a basis of V. Then we define $\det(h)$ by the following formula:

$$\det(h)(\lambda \det(\Lambda), \mu \det(\Lambda)) = \lambda\bar{\mu} \det(h(x_i, x_j)) \quad (\lambda, \mu \in \mathbb{K}),$$

where $\det(\Lambda) = x_1 \wedge \cdots \wedge x_n \in \det(V)$ (for details, see Section 1.10). It is easy to see that the above definition does not depend on the choice of Λ.

Let

$$E : 0 \to V_0 \to V_1 \to \cdots \to V_{n-1} \to V_n \to 0$$

be an exact sequence of finite-dimensional vector spaces over $\mathbb{K}$ and let h_i be a $\mathbb{K}$-inner product of V_i for each $i = 0, \ldots, n$. These data $(E, h_0, \ldots, h_n)$ are denoted by $\overline{E}$. Since $\bigotimes_{i=0}^n (\det V_i)^{\otimes(-1)^i}$ is canonically isomorphic to $\mathbb{K}$ as described in Section 1.10, $\bigotimes_{i=0}^n (\det h_i)^{\otimes(-1)^i}$ yields a $\mathbb{K}$-inner product of $\mathbb{K}$. The *volume difference* of $\overline{E}$ is defined to be

$$\gamma(\overline{E}) = -\frac{1}{2} \log \left(\left(\bigotimes_{i=0}^n (\det h_i)^{\otimes(-1)^i} \right) (1, 1) \right).$$

We say that $\overline{E}$ is *volume exact* if $\gamma(\overline{E}) = 0$. For example, if $n = 2$ and h_0 and h_2 are the subinner product and the quotient inner product induced by h_1, respectively, then $\overline{E}$ is volume exact by Lemma 1.1.

Let R be a reduced order and let K be the total quotient ring. Let

$$E : 0 \to H_0 \xrightarrow{\alpha_0} H_1 \xrightarrow{\alpha_1} \cdots \to H_{n-1} \xrightarrow{\alpha_{n-1}} H_n \to 0$$

be an exact sequence of finitely generated R-modules of pure rank. Let $Q_1, \ldots, Q_l$ be all minimal prime ideals of R. Then $K = \bigoplus_{i=1}^l R_{Q_i}$ and R_{Q_i}'s are algebraic number fields. For $\sigma \in K(\mathbb{C})$, there is $\tau \in R_{Q_i}(\mathbb{C})$ with $\sigma(x_1, \ldots, x_l) = \tau(x_i)$. Thus, $\sigma|_R : R \to R_{Q_i} \xrightarrow{\tau} \mathbb{C}$ and $\sigma : K \to R_{Q_i} \xrightarrow{\tau} \mathbb{C}$ are flat. Therefore,

$$E_\sigma : 0 \to (H_0)_\sigma \to \cdots \to (H_n)_\sigma \to 0$$

is exact for all $\sigma \in K(\mathbb{C})$. Then we have the following.

PROPOSITION 3.21. *We assume that each H_i has a hermitian metric h_i. For each $\sigma \in K(\mathbb{C})$, we set $\overline{E}_\sigma = (E_\sigma, h_{0,\sigma}, \ldots, h_{n,\sigma})$. Then*

$$\sum_{i=0}^n (-1)^i \widehat{\deg}(H_i, h_i) = \sum_{\sigma \in K(\mathbb{C})} \gamma(\overline{E}_\sigma).$$

PROOF. In this book, we use this proposition in the case where R is integral domain, so that we only give a proof for this case. A proof for a general case is left to the reader.

Since $0 \to H_0 \otimes_R K \to \cdots \to H_n \otimes_R K \to 0$ is an exact sequence over K, we can find a system of bases $\{\Lambda_i\}_{i=0}^n$ of the above exact sequence (cf. Section 1.10). By multiplying a non-zero integer to entries of $\Lambda_0, \ldots, \Lambda_n$, we may assume that entries of Λ_i $(i = 0, \ldots, n)$ are elements of H_i. Let us decompose $\Lambda_i = \Lambda'_i \amalg \Lambda''_i$ as in the beginning of Section 1.10. By multiplying a non-zero integer to entries of $\Lambda_0, \ldots, \Lambda_n$, we may assume that $\alpha_i(\Lambda''_i) = \Lambda'_{i+1}$. We denote the sub-$R$-module generated by entries of Λ_i by $R\Lambda_i$. In the same way, we consider $R\Lambda'_i$ and $R\Lambda''_i$.

Then $\alpha_i(H_i) \cap R\Lambda''_{i+1} = \{0\}$. Indeed, if $x \in \alpha_i(H_i) \cap R\Lambda''_{i+1}$, then $\alpha_{i+1}(x) = 0$. On the other hand, it is easy to see that $\alpha_{i+1}|_{R\Lambda''_{i+1}}$ is injective. Thus $x = 0$.

Therefore, $0 \to H_0/R\Lambda_0 \to \cdots \to H_n/R\Lambda_n \to 0$ is exact, so that

$$\sum_{i=0}^{n} (-1)^i \log \#(H_i/R\Lambda_i) = 0.$$

We denote $\det((h_i)_\sigma(s_k \otimes^\sigma 1, s_l \otimes^\sigma 1))$ by $\det((h_i)_\sigma(\Lambda_i))$, where $\Lambda_i = (s_1, \ldots, s_a)$. Then

$$\widehat{\deg}(H_i, h_i) = \log \#(H_i/R\Lambda_i) - \frac{1}{2} \sum_{\sigma \in K(\mathbb{C})} \log \det((h_i)_\sigma(\Lambda_i)).$$

Thus, noting

$$\gamma(\overline{E}_\sigma) = -\frac{1}{2} \sum_{i=0}^{n} (-1)^i \log \det((h_i)_\sigma(\Lambda_i)),$$

we obtain the proposition. □

3.8. Ample invertible sheaves on arithmetic curves

Let R be an integral order and K the quotient field. Let us begin with the following proposition.

PROPOSITION 3.22. *Let $\overline{L} = (L, h)$ and $\overline{E} = (E, k)$ be hermitian R-modules of real type such that L and E are flat over R and the ranks of L and E are 1 and r, respectively. Then we have the following.*

(1) $\widehat{\chi}(L^{\otimes n} \otimes E, \|\cdot\|_{\sup}) = nr\widehat{\deg}(\overline{L}) + O(1)$.
(2) $\hat{h}^0_{<1}(L^{\otimes n} \otimes E, \|\cdot\|_{\sup}) \geq nr\widehat{\deg}(\overline{L}) + O(1)$.
(3) *If $\widehat{\deg}(\overline{L}) > 0$, then, for a sufficiently large n, $L^{\otimes n} \otimes E$ is generated by elements of $\{s \in L^{\otimes n} \otimes E \mid \|s\|_{\sup} < 1\}$.*

PROOF. (1) and (2) are consequences of Theorem 3.15.

Finally, we consider (3). By (2), there are a positive integer n_0 and $s \in L^{\otimes n_0} \setminus \{0\}$ with $\|s\|_{\sup} < 1$. Let c be a positive number with $\|s\|_{\sup} = \exp(-c)$. Let $m_{i,1}, \ldots, m_{i,r[K:\mathbb{Q}]}$ be a free basis of $L^{\otimes i} \otimes E$ as a $\mathbb{Z}$-module. We set

$$d = \log\left(\max_{0 \leq i \leq n_0 - 1} \max\{\|m_{i,1}\|_{\sup}, \ldots, \|m_{i,r[K:\mathbb{Q}]}\|_{\sup}\}\right).$$

Then, since

$$L^{\otimes i} \otimes E \xrightarrow{s^{\otimes n} \otimes} L^{\otimes (nn_0 + i)} \otimes E$$

is injective,

$$s^{\otimes n} \otimes m_{i,1}, \ldots, s^{\otimes n} \otimes m_{i,r[K:\mathbb{Q}]}$$

yields a basis of $(L^{\otimes(nn_0+i)} \otimes E)_{\mathbb{Q}}$. Moreover,

$$\|s^{\otimes n} \otimes m_{i,j}\|_{\sup} \leq \|s\|^n_{\sup} \|m_{i,j}\|_{\sup} \leq \exp(-cn + d) \leq \exp(-c'(nn_0 + i) + d'),$$

where $c' = c/n_0$, $d' = c + d$. Thus, if m is sufficiently large, this means that

$$\lambda_{\mathbb{Q}}(L^{\otimes m} \otimes E, \|\cdot\|_{\sup}) \leq \exp(-c'm + d').$$

Therefore, by Lemma 2.18,

$$\lambda_{\mathbb{Z}}(L^{\otimes m} \otimes E, \|\cdot\|_{\sup}) \leq r[K : \mathbb{Q}] \exp(-c'm + d').$$

Thus we obtain (3). □

Let $\overline{L} = (L, h)$ be a hermitian flat R-module of real type such that the rank of L is 1. We say that $\overline{L}$ is *ample* if $L^{\otimes n}$ is generated by elements with $\|s\|_{\sup} < 1$ for a sufficiently large n. Then, as a corollary of the above proposition, we have the following.

COROLLARY 3.23. *The following are equivalent.*

(1) $\widehat{\deg}(\overline{L}) > 0$.
(2) *Let* $\overline{E} = (E, k)$ *be a hermitian flat* R*-module of real type. Then, for a sufficiently large* n, $L^{\otimes n} \otimes E$ *is generated by elements with* $\|s\|_{\sup} < 1$.
(3) $\overline{L}$ *is ample.*
(4) *For a sufficiently large* n, *there is* $s \in L^{\otimes n} \setminus \{0\}$ *with* $\|s\|_{\sup} < 1$.

PROOF. (1) $\Longrightarrow$ (2) is a consequence of Proposition 3.22. (2) $\Longrightarrow$ (3) is obvious. (3) $\Longrightarrow$ (4) is also obvious. (4) $\Longrightarrow$ (1) follows from the definition of $\widehat{\deg}$. □

REMARK 3.2. Finally we would like to give remarks through easy examples. Let m be a square free integer. We set $R = \mathbb{Z}[\sqrt{m}]$ and $K = \mathbb{Q}(\sqrt{m})$. Then $K(\mathbb{C}) = \{\sigma_1, \sigma_2\}$ $(\sigma_1(\sqrt{m}) = \sqrt{m}, \sigma_2(\sqrt{m}) = -\sqrt{m})$. We put

$$\overline{L} = (O_K, \{\exp(-\lambda_i)|\sigma_i(\cdot)|\}_{i=1,2}) \quad (\lambda_1, \lambda_2 \in \mathbb{R}).$$

Then $\overline{L}^{\otimes n} = (O_K, \{\exp(-n\lambda_i)|\sigma_i(\cdot)|\}_{i=1,2})$ and $\widehat{\deg}(\overline{L}) = \lambda_1 + \lambda_2$.

(1) In order to guarantee Proposition 3.22, the condition that $\overline{L}$ is of real type is necessary. For example, we consider the case where $m = -1$. Let us choose $\lambda_1, \lambda_2 \in \mathbb{R}$ with $\lambda_1 < 0$ and $\lambda_1 + \lambda_2 > 0$. Then $\overline{L}$ is not of real type. For $s = x + \sqrt{-1}y \in R$ $(x, y \in \mathbb{Z})$, $\|s\|_{\sup}^{\overline{L}^{\otimes n}} \leq 1$ is equivalent to $x^2 + y^2 \leq \exp(2n\lambda_1)$. Thus, if $n \geq 1$ and $\|s\|_{\sup}^{\overline{L}^{\otimes n}} \leq 1$, then $s = 0$. This means that (3) in Proposition 3.22 does not hold for $\overline{L}$.

(2) Next we consider the case where $m > 0$. In this case, for any λ_1, λ_2, $\overline{L}$ is of real type. Let λ, ϵ be a positive real number. We set $\lambda_1 = \lambda + \epsilon$ and $\lambda_2 = -\lambda$. Then $\widehat{\deg}(\overline{L}) = \epsilon > 0$. Thus, by (3) in Proposition 3.22, if $n \gg 1$, then there is non-zero $s = x + \sqrt{m}y \in R$ $(x, y \in \mathbb{Z})$ with $\|s\|_{\sup}^{\overline{L}^{\otimes n}} \leq 1$, that is, $|x + \sqrt{m}y| \leq \exp(n(\lambda + \epsilon))$ and $|x - \sqrt{m}y| \leq \exp(-n\lambda)$ hold. Thus we have

$$\left|\frac{x}{y} - \sqrt{m}\right| \leq \frac{1}{|y|\exp(n\lambda)}, \quad |y| \leq \frac{\exp(n(\lambda + \epsilon))}{\sqrt{m}}.$$

This is weaker than the classical result of Diophantine approximation (Dirichlet's theorem). We can, however, get the flavor of Proposition 3.22.

CHAPTER 4

Arakelov Geometry on Arithmetic Surfaces

In this section, we give detailed expositions of the basic theory of Arakelov geometry on arithmetic surfaces. Several applications will be treated on a general arithmetic variety in later sections. On the other hand, the basic theory on a general arithmetic variety (e.g. the arithmetic Riemann-Roch formula on an arithmetic variety) will be introduced without proofs, so that the expositions in this section might be useful in later sections. The intersection theory on a general arithmetic variety will be established in the standard way, so that we use Deligne's pairing to construct the intersection theory on arithmetic surfaces. The most important and difficult theorem in this section is the arithmetic Riemann-Roch formula. Actually we have two formulations of the arithmetic Riemann-Roch formula on arithmetic surfaces. Here we follow the classical way due to Faltings. For this method, the most complicated part is the existence of Faltings' metrics. In later sections, we do not use the context of this section. So, the reader does not necessarily have to read this section completely. It will be a good warm-up for fully general theory.

4.1. Deligne's pairing

Let S be a connected Dedekind scheme, that is, a connected noetherian normal scheme of dimension less than or equal to 1. Let K be the function field of S. Let $\pi : X \to S$ be a projective and flat morphism of relative dimension 1. We assume that X is regular and π is smooth over K. Throughout this section, we fix $\pi : X \to S$ unless otherwise noted. Let X_P denote the fiber $\pi^{-1}(P)$ for a point P of S, and let X_U denote $\pi^{-1}(U)$ for an open set U of S. Moreover, the generic fiber of $\pi : X \to S$ is denoted by X_K. Let $\operatorname{Div}(X)$ be the group of divisors on X. We set

$$\Upsilon = \{(D, E) \in \operatorname{Div}(X) \times \operatorname{Div}(X) \mid D \text{ and } E \text{ have no common component}\}.$$

Notice that, for $(D, E), (D', E), (D, E') \in \Upsilon$,

$$(E, D), (D + D', E), (D, E + E') \in \Upsilon.$$

Let us fix $(D, E) \in \Upsilon$ such that D and E are effective, that is, all coefficients are non-negative in their expressions as the sum of prime divisors. Let d_x and e_x be local equations of D and E at a closed point $x \in X$, respectively. We define the intersection number $I_x(D, E)$ of D and E at x to be

$$I_x(D, E) = \operatorname{length}_{\mathscr{O}_{X,x}}(\mathscr{O}_{X,x}/(d_x\mathscr{O}_{X,x} + e_x\mathscr{O}_{X,x})).$$

In the case where $\dim S = 0$ (i.e. $\dim X = 1$), $I_x(D, E) = 0$. Let us see the following proposition.

PROPOSITION 4.1. *Let D, D', E, E' be effective divisors on X with*

$$(D, E), (D', E), (D, E') \in \Upsilon.$$

Then we have the following.

(1) $I_x(D,E) = I_x(E,D)$.

(2) $\begin{cases} I_x(D+D',E) = I_x(D,E) + I_x(D',E), \\ I_x(D,E+E') = I_x(D,E) + I_x(D,E'). \end{cases}$

PROOF. (1) is obvious by the definition. For (2), let d_x, d'_x, e_x be local equations of D, D', E at x, respectively. It is sufficient to see that

$$\operatorname{length}_{\mathscr{O}_{X,x}}\left(\frac{\mathscr{O}_{X,x}}{(d_x d'_x, e_x)}\right) = \operatorname{length}_{\mathscr{O}_{X,x}}\left(\frac{\mathscr{O}_{X,x}}{(d_x, e_x)}\right) + \operatorname{length}_{\mathscr{O}_{X,x}}\left(\frac{\mathscr{O}_{X,x}}{(d'_x, e_x)}\right).$$

We write

$$R = \mathscr{O}_{X,x}/(e_x), \quad \bar{d}_x = \text{the class of } d_x \text{ in } R \quad \text{and} \quad \bar{d}'_x = \text{the class of } d'_x \text{ in } R.$$

By (3) in Lemma 1.7,

$$\operatorname{length}_{\mathscr{O}_{X,x}}\left(R/\bar{d}_x\bar{d}'_x R\right) = \operatorname{length}_{\mathscr{O}_{X,x}}\left(R/\bar{d}_x R\right) + \operatorname{length}_{\mathscr{O}_{X,x}}\left(R/\bar{d}'_x R\right),$$

so that we have our desired equality. □

For a divisor D, there is a unique decomposition $D = D_+ - D_-$ such that D_+ and D_- are effective and have no common component. Let $E = E_+ - E_-$ be the same decomposition of a divisor E. If $(D,E) \in \Upsilon$, then we define $I_x(D,E)$ to be

$$I_x(D,E) = I_x(D_+,E_+) - I_x(D_+,E_-) - I_x(D_-,E_+) + I_x(D_-,E_-).$$

Then, by the previous proposition, we can see that

$$\begin{cases} I_x(D,E) = I_x(E,D), \\ I_x(D+D',E) = I_x(D,E) + I_x(D',E), \\ I_x(D,E+E') = I_x(D,E) + I_x(D,E') \end{cases}$$

for $(D,E), (D',E), (D,E') \in \Upsilon$.

Here we define the 0-cycle $I(D,E)$ on X to be

$$I(D,E) := \sum_{x\,:\,\text{closed point of } X} I_x(D,E)x. \tag{4.1}$$

Note that if $\dim S = 0$, then $I(D,E) = 0$. Moreover, for $(D,E), (D',E), (D,E') \in \Upsilon$,

$$\begin{cases} I(D,E) = I(E,D), \\ I(D+D',E) = I(D,E) + I(D',E), \\ I(D,E+E') = I(D,E) + I(D,E'). \end{cases}$$

For a 0-cycle $z = \sum_x n_x x$ on X, the push-forward $\pi_*(z)$ is defined by

$$\pi_*(z) := \sum_x n_x[\kappa(x) : \kappa(\pi(x))]\pi(x),$$

where $\pi(x)$ is the point of S given by the image of x by π. If $\dim S = 1$, then $\pi_*(z)$ is a divisor on S. Using the push-forward π_*, we define $\langle D,E\rangle$ to be

$$\langle D,E\rangle := \pi_*(I(D,E)).$$

Thus, if $\dim S = 1$, then we have a map

$$\langle\ ,\ \rangle : \Upsilon \to \operatorname{Div}(S).$$

In the case where $\dim S = 0$, we understand that $\langle D,E\rangle = 0$ for all $(D,E) \in \Upsilon$.

If we need to emphasize that $\langle D, E\rangle$ is considered over S, then $\langle D, E\rangle$ is often denoted by $\langle D, E\rangle_S$. By the properties of $I(D, E)$,

$$\begin{cases} \langle D, E\rangle = \langle E, D\rangle, \\ \langle D + D', E\rangle = \langle D, E\rangle + \langle D', E\rangle, \\ \langle D, E + E'\rangle = \langle D, E\rangle + \langle D, E'\rangle \end{cases}$$

hold for $(D, E), (D', E), (D, E') \in \Upsilon$.

LEMMA 4.2. *We assume* $\dim S = 1$. *Let* Γ *be an irreducible component of the fiber* X_P *over a closed point* $P \in S$. *Then*

$$\langle D, \Gamma\rangle = \deg(D|_\Gamma)P.$$

PROOF. Each side of the above equation is linear with respect to D. Thus we may assume that D is effective. Let x be a closed point of Γ and $\bar{d}_x$ the class of a local equation d_x of D at x in $\mathscr{O}_{\Gamma,x}$. Then, by the definitions,

$$I(D, \Gamma) = \sum_{x\in\Gamma} \operatorname{length}_{\mathscr{O}_{\Gamma,x}}(\mathscr{O}_{\Gamma,x}/\bar{d}_x\mathscr{O}_{\Gamma,x})x.$$

Thus

$$\langle D, \Gamma\rangle = \left(\sum_{x\in\Gamma} \operatorname{length}_{\mathscr{O}_{\Gamma,x}}(\mathscr{O}_{\Gamma,x}/\bar{d}_x\mathscr{O}_{\Gamma,x})[\kappa(x) : \kappa(P)]\right) P.$$

Note that $\sum_{x\in\Gamma} \operatorname{length}_{\mathscr{O}_{\Gamma,x}}(\mathscr{O}_{\Gamma,x}/\bar{d}_x\mathscr{O}_{\Gamma,x})[\kappa(x) : \kappa(P)] = \deg(D|_\Gamma)$, and hence the lemma follows. $\square$

Let Γ be a prime divisor of X and ϕ a non-zero rational function on X such that Γ is not a component of (ϕ). Here (ϕ) is the *principal divisor* of ϕ given by

$$(\phi) = \sum_{E \text{ is a prime divisor}} \operatorname{ord}_E(\phi)E.$$

We define $N_\Gamma(\phi)$ to be

$$N_\Gamma(\phi) := \begin{cases} N_{\Gamma/S}(\phi|_\Gamma) & \Gamma \text{ is horizontal}, \\ 1 & \Gamma \text{ is vertical}, \end{cases}$$

where $N_{\Gamma/S}$ is the usual norm map from the function field of $\mathscr{O}_\Gamma$ to the function field of $\mathscr{O}_S$. Note that $N_\Gamma(\phi) \in K$. More generally, if D is a divisor on X such that D and (ϕ) have no common component, then we define $N_D(\phi)$ by

$$N_D(\phi) = \prod_i N_{\Gamma_i}(\phi)^{n_i},$$

where $D = \sum_i n_i\Gamma_i$. By the definitions, if D and D' are divisors on X and ϕ and ϕ' are non-zero rational functions on X with $(D, (\phi)), (D', (\phi)), (D, (\phi')) \in \Upsilon$, then we can see that

$$\begin{cases} N_{D+D'}(\phi) = N_D(\phi)N_{D'}(\phi), \\ N_D(\phi\phi') = N_D(\phi)N_D(\phi'). \end{cases}$$

If we need to emphasize that $N_D(\phi)$ is considered over S, then $N_D(\phi)$ is often denoted by $N_{D/S}(\phi)$.

PROPOSITION 4.3. *We assume* $\dim S = 1$. *Let* D *be a divisor on* X *and* ϕ *a non-zero rational function on* X *such that* D *and* (ϕ) *have no common component. Then*

$$\langle D, (\phi)\rangle = (N_D(\phi)).$$

PROOF. Since both sides are linear with respect to D, we may assume that D is a prime divisor. If D is vertical, the proposition follows from Proposition 1.18 and Lemma 4.2. If D is horizontal, then the assertion is local with respect to S. Thus it follows from Lemma 1.12. □

LEMMA 4.4. *Let* D *be a divisor on* X *and* ϕ *a non-zero rational function on* X *such that* D *and* (ϕ) *have no common component. For a non-empty open set* U *of* S, *we have*

$$N_{D\cap X_U/U}(\phi|_{X_U}) = N_{D/S}(\phi).$$

PROOF. Let $D = D' + D''$ be the decomposition such that D' is horizontal and D'' is vertical. Then $N_{D/S}(\phi) = N_{D'/S}(\phi)$. In the same way, $N_{D\cap X_U/U}(\phi|_{X_U}) = N_{D'\cap X_U/U}(\phi|_{X_U})$. On the other hand, by the definition, we can see

$$N_{D'/S}(\phi) = N_{D'\cap X_U/U}(\phi|_{X_U}).$$

Thus the lemma follows. □

PROPOSITION 4.5 (Weil's reciprocity law on arithmetic surfaces). *Let* ϕ *and* ψ *be non-zero rational functions on* X *such that* (ϕ) *and* (ψ) *have no common component. Then*

$$N_{(\phi)}(\psi) = N_{(\psi)}(\phi).$$

$N_{(\phi)}(\psi)$ *is often denoted by* $\langle \phi, \psi\rangle$. *In this notation,* $\langle \phi, \psi\rangle = \langle \psi, \phi\rangle$. *Moreover, by Proposition* 4.3, $\langle (\phi), (\psi)\rangle = (\langle \phi, \psi\rangle)$.

PROOF. By Lemma 4.4, we may replace S by a non-empty open set of S, so that it is sufficient to see it on the generic fiber X_K. Thus it follows from Proposition 1.19 (Weil's reciprocity law). □

LEMMA 4.6. *We assume that* $S = \operatorname{Spec}(R)$ *for a discrete valuation ring* R. *Let* o *be the closed point of* S. *Let* L *and* M *be invertible sheaves on* X. *Then there are non-zero rational sections* a *and* b *of* L *and* M, *respectively, such that* $\operatorname{div}(a)$ *and* $\operatorname{div}(b)$ *are horizontal and have no common point on the fiber* X_o *over* o. *In particular,* $\langle \operatorname{div}(a), \operatorname{div}(b)\rangle = 0$.

PROOF. Let $X_o = \Gamma_1 \cup \cdots \cup \Gamma_s$ be the irreducible decomposition of X_o. Let us choose a closed point x_i of Γ_i for each $i = 1, \ldots, s$. By Proposition 1.22, there is a non-zero rational section a with $\operatorname{Supp}(\operatorname{div}(a)) \cap \{x_1, \ldots, x_s\} = \emptyset$. In particular, $\operatorname{Supp}(\operatorname{div}(a))$ does not contain any irreducible component of X_o, so that we set $X_o \cap \operatorname{Supp}(\operatorname{div}(a)) = \{y_1, \ldots, y_r\}$. By Proposition 1.22 again, there is a non-zero rational section b of M with $\operatorname{Supp}(\operatorname{div}(b)) \cap \{x_1, \ldots, x_s, y_1, \ldots, y_r\} = \emptyset$. Thus the lemma follows. □

We have finished every preparation for the definition of Deligne's pairing, that is, we have the following theorem. The advantage of Deligne's pairing is the existence of natural rational sections $\langle l, m\rangle$.

THEOREM 4.7. *We can assign a unique invertible sheaf* $\langle L, M\rangle$ *on* S *to any invertible sheaves* L *and* M *on* X *in such a way that the following properties hold:*

(1) *Let L' and M' be invertible sheaves on X. If $L \simeq L'$ and $M \simeq M'$, then $\langle L, M\rangle \simeq \langle L', M'\rangle$.*

(2) *Let L' and M' be invertible sheaves on X. Then we have the following natural isomorphism.*

$$\begin{cases} \langle L, M\rangle \simeq \langle M, L\rangle, \\ \langle L \otimes L', M\rangle \simeq \langle L, M\rangle \otimes \langle L', M\rangle, \\ \langle L, M \otimes M'\rangle \simeq \langle L, M\rangle \otimes \langle L, M'\rangle. \end{cases}$$

(3) *Let l and m be non-zero rational sections of L and M, respectively, such that $\operatorname{div}(l)$ and $\operatorname{div}(m)$ have no common component. Then there is a non-zero rational section $\langle l, m\rangle$ of $\langle L, M\rangle$ with the following properties.*

(3.1) *Let f and g be rational functions on X with*

$$(\operatorname{div}(fl), \operatorname{div}(m)), (\operatorname{div}(l), \operatorname{div}(gm)) \in \Upsilon.$$

Then

$$\begin{cases} \langle fl, m\rangle = N_{\operatorname{div}(m)}(f)\langle l, m\rangle, \\ \langle l, gm\rangle = N_{\operatorname{div}(l)}(g)\langle l, m\rangle. \end{cases}$$

(3.2) *We assume $\dim S = 1$. There is an isomorphism*

$$\langle L, M\rangle \overset{\sim}{\longrightarrow} \mathscr{C}_S(\langle \operatorname{div}(l), \operatorname{div}(m)\rangle)$$

which sends $\langle l, m\rangle$ to $1_{\langle \operatorname{div}(l), \operatorname{div}(m)\rangle}$. In particular,

$$\operatorname{div}(\langle l, m\rangle) = \langle \operatorname{div}(l), \operatorname{div}(m)\rangle.$$

(4) *Let D be a horizontal divisor on X. Then, for an invertible sheaf L on X, there is a natural isomorphism*

$$\langle L, \mathscr{O}_X(D)\rangle \overset{\sim}{\longrightarrow} N_D(L)$$

which sends $\langle l, 1_D\rangle$ to $N_D(l)$, where if $D = \sum_i n_i\Gamma_i$ is the decomposition of D, then $N_D(L)$ and $N_D(l)$ are given by

$$\begin{cases} N_D(L) = \bigotimes_i N_{\Gamma_i/S}(L|_{\Gamma_i})^{\otimes n_i}, \\ N_D(l) = \bigotimes_i N_{\Gamma_i/S}(l|_{\Gamma_i})^{\otimes n_i} \end{cases}$$

(see Section 1.5 for the definitions of $N_{\Gamma_i/S}(L|_{\Gamma_i})$ and $N_{\Gamma_i/S}(l|_{\Gamma_i})$).

(5) *Let $\rho : S' \to S$ be a flat morphism of connected Dedekind schemes, $X' = X \times_S S'$ and let*

$$\begin{array}{ccc} X & \xleftarrow{\ \mu\ } & X' \\ {\scriptstyle \pi}\big\downarrow & & \big\downarrow{\scriptstyle \pi'} \\ S & \xleftarrow{\ \rho\ } & S' \end{array}$$

be the natural commutative diagram. Then there is a natural isomorphism

$$\rho^*(\langle L, M\rangle) \overset{\sim}{\longrightarrow} \langle \mu^*(L), \mu^*(M)\rangle.$$

PROOF. Let f and g be non-zero rational functions on X such that (f) and (g) have no common component. We denote $N_{(f)}(g)$ by $\langle f, g\rangle$ according to Proposition 4.5. By Weil's reciprocity law, $\langle f, g\rangle = \langle g, f\rangle$. If we need to show that it is considered over S, $\langle f, g\rangle$ is often denoted by $\langle f, g\rangle_S$.

Let K be the function field of S and X_K the generic fiber of $\pi : X \to S$. Here we consider the following sets Υ_K and Σ_K:

$$\Upsilon_K = \{(D, E) \in \mathrm{Div}(X) \times \mathrm{Div}(X) \mid D \text{ and } E \text{ have no common component on } X_K\},$$

$$\Sigma_K = \left\{ (l,m) \;\middle|\; \begin{array}{l} l \text{ and } m \text{ are non-zero rational sections of } L \text{ and } M \\ \text{with } (\mathrm{div}(l), \mathrm{div}(m)) \in \Upsilon_K \end{array} \right\}.$$

We set

$$A = \bigoplus_{(l,m)\in\Sigma_K} K(l,m),$$

$$B = \bigoplus_{\substack{f\in \mathrm{Rat}(X)\setminus\{0\} \\ (l,m),(fl,m)\in\Sigma_K}} K\left((fl,m) - N_{\mathrm{div}(m)}(f)(l,m)\right),$$

$$C = \bigoplus_{\substack{g\in \mathrm{Rat}(X)\setminus\{0\} \\ (l,m),(l,gm)\in\Sigma_K}} K\left((l,gm) - N_{\mathrm{div}(l)}(g)(l,m)\right),$$

where $\mathrm{Rat}(X)$ is the function field of X. We define $\langle L, M\rangle_K$ to be

$$\langle L, M\rangle_K = A/(B+C).$$

The class of (l,m) in $\langle L, M\rangle_K$ is denoted by $\langle l, m\rangle_K$. Then we have

$$\begin{cases} \langle fl, m\rangle_K = N_{\mathrm{div}(m)}(f)\langle l, m\rangle_K, \\ \langle l, gm\rangle_K = N_{\mathrm{div}(l)}(g)\langle l, m\rangle_K. \end{cases} \tag{4.2}$$

First, we have the following claim.

CLAIM 1. *$\langle L, M\rangle_K$ is a 1-dimensional vector space over K.*

PROOF. Let us fix $(l_0, m_0) \in \Sigma_K$. Then, for any $(l,m) \in \Sigma_K$, there are non-zero rational functions f_0 and g_0 on X with $l = f_0 l_0$ and $m = g_0 m_0$. It is easy to see that

$$((f_0),(g_0)),\; ((f_0), \mathrm{div}(m_0)),\; (\mathrm{div}(l_0), (g_0)) \in \Upsilon_K.$$

Here

$$\varphi : \bigoplus_{(l,m)\in\Sigma_K} K(l,m) \to K$$

is defined by

$$\varphi((l,m)) = \langle f_0, g_0\rangle_K N_{\mathrm{div}(m_0)_K/K}(f_0) N_{\mathrm{div}(l_0)_K/K}(g_0).$$

Clearly $\varphi((l_0, m_0)) = 1$. Let us see

$$\begin{cases} \varphi((fl,m)) = N_{\mathrm{div}(m)_K/K}(f)\varphi((l,m)), \\ \varphi((l,gm)) = N_{\mathrm{div}(l)_K/K}(g)\varphi((l,m)). \end{cases}$$

Indeed, by Weil's reciprocity law (cf. Proposition 4.5),

$$\begin{aligned} \varphi((fl,m)) &= \langle ff_0, g_0\rangle_K N_{\mathrm{div}(m_0)_K/K}(ff_0) N_{\mathrm{div}(l_0)_K/K}(g_0) \\ &= \langle f, g_0\rangle_K \langle f_0, g_0\rangle_K N_{\mathrm{div}(m_0)_K/K}(f) N_{\mathrm{div}(m_0)_K/K}(f_0) N_{\mathrm{div}(l_0)_K/K}(g_0) \\ &= \langle f, g_0\rangle_K N_{\mathrm{div}(m_0)_K/K}(f)\varphi((l,m)) \\ &= \langle g_0, f\rangle_K N_{\mathrm{div}(m_0)_K/K}(f)\varphi((l,m)) \\ &= N_{\mathrm{div}(m)_K/K}(f)\varphi((l,m)). \end{aligned}$$

In the same way, we can see $\varphi((l, gm)) = N_{\mathrm{div}(l)_K/K}(g)\varphi((l,m))$. Thus φ yields a surjective homomorphism $\langle L, M\rangle_K \to K$. On the other hand, by the relations (4.2),

$$\langle l, m\rangle_K = \langle f_0, g_0\rangle_K N_{\mathrm{div}(m_0)_K/K}(f_0) N_{\mathrm{div}(l_0)_K/K}(g_0)\langle l_0, m_0\rangle_K.$$

Thus $\langle L, M\rangle_K$ is generated by $\langle l_0, m_0\rangle_K$. Therefore, $\dim_K \langle L, M\rangle_K = 1$. □

For a non-empty open set U of S, we set

$$\Sigma_U = \left\{ (l,m) \;\middle|\; \begin{array}{l} l \text{ and } m \text{ are non-zero rational sections of } L \text{ and } M \text{ such} \\ \text{that } \mathrm{div}(l) \text{ and } \mathrm{div}(m) \text{ have no common component on } X_U \\ \text{and that } \langle \mathrm{div}(l)|_{X_U}, \mathrm{div}(m)|_{X_U}\rangle_U \text{ is effective on } U \end{array} \right\}$$

and

$$\mathscr{A}(U) = \bigoplus_{(l,m)\in\Sigma_U} \mathscr{O}_U(l,m).$$

The correspondence $U \mapsto \mathscr{A}(U)$ yields a sheaf $\mathscr{A}$ on S. Let $\phi : \mathscr{A} \to \langle L, M\rangle_K$ be the natural homomorphism given by $\phi((l,m)) = \langle l, m\rangle_K$. We define $\langle L, M\rangle$ to be

$$\langle L, M\rangle := \mathscr{A}/\operatorname{Ker}(\phi).$$

Moreover, for $(l,m) \in \Sigma_U$, the class of (l,m) in $\langle L, M\rangle(U)$ is denoted by $\langle l, m\rangle_U$. Then we have the following.

CLAIM 2. *Let $(l,m) \in \Sigma_U$ such that $\langle \mathrm{div}(l)|_{X_U}, \mathrm{div}(m)|_{X_U}\rangle_U$ is zero on U. Then, for any $(l', m') \in \Sigma_U$, there is $a \in \mathscr{O}_S(U)$ with $\langle l', m'\rangle_U = a\langle l, m\rangle_U$.*

PROOF. Let us choose non-zero rational functions ϕ and ψ with $l' = \phi l$ and $m' = \psi m$. Then, by Proposition 4.3,

$$\begin{aligned} \langle \mathrm{div}(l'), \mathrm{div}(m')\rangle_U &= \langle \mathrm{div}(l), \mathrm{div}(m)\rangle_U + \langle (\phi), (\psi)\rangle_U \\ &\quad + \langle (\phi), \mathrm{div}(m)\rangle_U + \langle \mathrm{div}(l), (\psi)\rangle_U \\ &= (\langle \phi, \psi\rangle_U)_U + (N_{\mathrm{div}(m)/U}(\phi))_U + (N_{\mathrm{div}(l)/U}(\psi))_U \\ &= (\langle \phi, \psi\rangle_U N_{\mathrm{div}(m)/U}(\phi) N_{\mathrm{div}(l)/U}(\psi))_U, \end{aligned}$$

where l, m, l', m', ϕ, ψ are considered over X_U. Since $\langle \mathrm{div}(l'), \mathrm{div}(m')\rangle_U$ is effective, we have

$$\langle \phi, \psi\rangle_U N_{\mathrm{div}(m)/U}(\phi) N_{\mathrm{div}(l)/U}(\psi) \in \mathscr{O}_S(U).$$

On the other hand, by the relations (4.2),

$$\langle l', m'\rangle_K = \langle \phi, \psi\rangle_K N_{\mathrm{div}(m)_K/K}(\phi) N_{\mathrm{div}(l)_K/K}(\psi)\langle l, m\rangle_K.$$

Therefore, $\langle l', m'\rangle_U = \langle \phi, \psi\rangle_U N_{\mathrm{div}(m)/U}(\phi) N_{\mathrm{div}(l)/U}(\psi)\langle l, m\rangle_U$ holds. Thus the claim follows. □

By Claim 1, Claim 2 and Lemma 4.6, we can see that $\langle L, M\rangle$ is invertible.

In the following, we consider the properties (1)–(5). (1) is obvious. (2) follows from the correspondences

$$\begin{cases} \langle l, m\rangle_K \longleftrightarrow \langle m, l\rangle_K, \\ \langle l \otimes l', m\rangle_K \longleftrightarrow \langle l, m\rangle_K \otimes \langle l', m\rangle_K, \\ \langle l, m \otimes m'\rangle_K \longleftrightarrow \langle l, m\rangle_K \otimes \langle l, m'\rangle_K. \end{cases}$$

Let us consider the property (3). For $(l,m)\in\Sigma_K$, $\langle l,m\rangle_K$ is a non-zero element of $\langle L,M\rangle_K$, which means that $\langle l,m\rangle_K$ is a non-zero rational section of $\langle L,M\rangle$. Moreover, if U is a non-empty open open set of S such that $\operatorname{div}(l)$ and $\operatorname{div}(m)$ have no common component on X_U and that $\langle \operatorname{div}(l)|_{X_U}, \operatorname{div}(m)|_{X_U}\rangle_U$ is zero on U, then $\langle l,m\rangle_K$ is an element of $\langle L,M\rangle(U)$. By abuse of notation, $\langle l,m\rangle_K$ is denoted by $\langle l,m\rangle$. By the relations (4.2), we can check the property (3.1) in the theorem.

Next we show the property (3.2). We assume that $(a,b)\in\Sigma_U$. We set $a=\phi_a l$, $b=\psi_b m$ and define $\Phi(a,b)$ by

$$\Phi(a,b)=\langle\phi_a,\psi_b\rangle N_{\operatorname{div}(m)}(\phi_a)N_{\operatorname{div}(l)}(\psi_b).$$

Then

$$\begin{aligned}
\Phi(fa,b)&=\langle f\phi_a,\psi_b\rangle N_{\operatorname{div}(m)}(f\phi_a)N_{\operatorname{div}(l)}(\psi_b)\\
&=\langle\phi_a,\psi_b\rangle\langle\psi_b,f\rangle N_{\operatorname{div}(m)}(f)N_{\operatorname{div}(m)}(\phi_a)N_{\operatorname{div}(l)}(\psi_b)\\
&=N_{\operatorname{div}(\psi_b m)}(f)\langle\phi_a,\psi_b\rangle N_{\operatorname{div}(m)}(\phi_a)N_{\operatorname{div}(l)}(\psi_b)\\
&=N_{\operatorname{div}(b)}(f)\Phi(a,b).
\end{aligned}$$

In the same way, we can see $\Phi(a,gb)=N_{\operatorname{div}(a)}(g)\Phi(a,b)$. Moreover, it is easy to show

$$\langle\operatorname{div}(l),\operatorname{div}(m)\rangle+(\Phi(a,b))=\langle\operatorname{div}(a),\operatorname{div}(b)\rangle.$$

Thus we can define

$$\alpha:\langle L,M\rangle\to\mathscr{O}_S(\langle\operatorname{div}(l),\operatorname{div}(m)\rangle)$$

by $\alpha(\langle a,b\rangle)=\Phi(a,b)$. Since $\Phi(l,m)=1$, we have $\alpha(\langle l,m\rangle)=1$. Hence it is sufficient to show that α is an isomorphism. We fix $s\in S$. By Lemma 4.6, there are non-zero rational sections a and b of L and M, respectively, such that $\operatorname{div}(a)$ and $\operatorname{div}(b)$ have no intersection on the fiber X_s over s. Then $\langle\operatorname{div}(a),\operatorname{div}(b)\rangle=0$ around s. This means that $\langle a,b\rangle$ forms a basis of $\langle L,M\rangle_s$. On the other hand, around s,

$$0=\langle\operatorname{div}(a),\operatorname{div}(b)\rangle=\langle\operatorname{div}(l),\operatorname{div}(m)\rangle+(\Phi(a,b)).$$

Thus $\Phi(a,b)$ gives rise to a basis of $\mathscr{O}_S(\langle\operatorname{div}(l),\operatorname{div}(m)\rangle)_s$. Thus α is an isomorphism around s.

Next let us consider (4). Let $\Sigma(D)$ denote the set of all non-zero rational sections l of L such that $\operatorname{div}(l)$ and D have no common component. For $l\in\Sigma(D)$, $\langle l,1_D\rangle_K$ gives a basis of $\langle L,\mathscr{O}_X(D)\rangle_K$. Thus $\beta:\langle L,\mathscr{O}_X(D)\rangle_K\to N_D(L)_K$ is defined by $\beta(\langle l,1_D\rangle_K)=N_D(l)$. Then $\beta(\langle l',1_D\rangle_K)=N_D(l')$ holds for $l'\in\Sigma(D)$. Indeed, if we choose a non-zero rational function f with $l'=fl$, then

$$\langle l',1_D\rangle_K=N_D(f)\langle l,1_D\rangle_K\quad\text{and}\quad N_D(fl)=N_D(f)N_D(l).$$

For each $l\in\Sigma(D)$, let U_l be the maximal open set U of S such that

$$X_U\cap\operatorname{Supp}(\operatorname{div}(l))\cap\operatorname{Supp}(D)=\emptyset.$$

Since D is horizontal, by Lemma 4.6, we have $S=\bigcup_{l\in\Sigma(D)}U_l$. Let $D=\sum n_i\Gamma_i$ be the decomposition of D. Then, for each i, $l|_{\Gamma_i\cap X_{U_l}}$ is a basis of $L|_{\Gamma_i\cap X_{U_l}}$. Thus $N_{\Gamma_i/S}(l|_{\Gamma_i})$ gives a basis of $N_{\Gamma_i/S}(L|_{\Gamma_i})$ on U_l. Therefore, $N_D(l)$ is a basis of $N_D(L)$ on U_l. On the other hand, $\langle\operatorname{div}(l),D\rangle$ is zero over U_l. Thus, by Claim 2, $\langle l,1_D\rangle_{U_l}$ is a basis of $\langle L,\mathscr{O}_X(D)\rangle$ on U_l. Namely, β yields an isomorphism $\langle L,\mathscr{O}_X(D)\rangle\xrightarrow{\sim}N_D(L)$.

Finally we consider the property (5). Let K' be the function field of S'. Let $X'_{K'}$ be the generic fiber of $X' \to S'$. Then $X'_{K'} = X_K \times_{\mathrm{Spec}(K)} \mathrm{Spec}(K')$. Let $\rho^* : K \to K'$ be the natural homomorphism. We define a K'-homomorphism

$$\tau : \left(\bigoplus_{(l,m)\in\Sigma_K} K(l,m) \right) \otimes_K K' \to \langle \mu^*(L), \mu^*(M) \rangle_{K'}$$

by $\tau((l,m) \otimes f') = f' \langle \mu^*(l), \mu^*(m) \rangle_{K'}$. Then, by Lemma 1.14, we have

$$\begin{aligned}
\tau(((fl,m) - N_{\mathrm{div}(m)}(f)(l,m)) \otimes 1) \\
= \langle \mu^*(fl), \mu^*(m) \rangle_{K'} - \rho^*(N_{\mathrm{div}(m)}(f)) \langle \mu^*(l), \mu^*(m) \rangle_{K'} \\
= N_{\mu^*(\mathrm{div}(m))}(\mu^*(f)) \langle \mu^*(l), \mu^*(m) \rangle_{K'} \\
- N_{\mu^*(\mathrm{div}(m))}(\mu^*(f)) \langle \mu^*(l), \mu^*(m) \rangle_{K'} = 0.
\end{aligned}$$

In the same way,

$$\tau(((l,gm) - N_{\mathrm{div}(l)}(g)(l,m)) \otimes 1) = 0.$$

Thus τ yields a surjective homomorphism

$$\tilde{\tau} : \langle L, M \rangle_K \otimes_K K' \to \langle \mu^*(L), \mu^*(M) \rangle_{K'}.$$

Comparing the dimensions of both sides, we can conclude that $\tilde{\tau}$ is an isomorphism. For $(l,m) \in \Sigma_K$, let $U_{(l,m)}$ be the maximal open set U of S with $X_U \cap \mathrm{Supp}(\mathrm{div}(l)) \cap \mathrm{Supp}(\mathrm{div}(m)) = \emptyset$. Then, by Claim 2, $\langle l,m \rangle_{U_{(l,m)}}$ is a basis of $\langle L, M \rangle$ over $U_{(l,m)}$. On the other hand, since $\mathrm{div}(\mu^*(l))$ and $\mathrm{div}(\mu^*(m))$ have no intersection over $X'_{\rho^{-1}(U_{(l,m)})}$, by Claim 2 again, $\langle \mu^*(l), \mu^*(m) \rangle_{\rho^{-1}(U_{(l,m)})}$ is a basis of $\langle \mu^*(L), \mu^*(M) \rangle$ over $\rho^{-1}(U_{(l,m)})$. This means that $\tilde{\tau} : \langle L, M \rangle_K \otimes_K K' \to \langle \mu^*(L), \mu^*(M) \rangle_{K'}$ yields an isomorphism $\rho^*(\langle L, M \rangle) \xrightarrow{\sim} \langle \mu^*(L), \mu^*(M) \rangle$. □

REMARK 4.1. In the case where $L = M = \mathcal{O}_X$, non-zero rational functions f and g on X give rise to non-zero rational sections of $\mathcal{O}_X$. If (f) and (g) have no common component, then $\langle f, g \rangle'$ is a non-zero rational section of $\langle \mathcal{O}_X, \mathcal{O}_X \rangle$. Here we use the symbol $\langle f, g \rangle'$ in order to distinguish it from $\langle f, g \rangle$ in Proposition 4.5. By the relation (3.1) of Theorem 4.7, we can see $\langle f, g \rangle' = \langle f, g \rangle \langle 1, 1 \rangle'$. Moreover, by (3.2) of Theorem 4.7, the correspondence $\langle 1, 1 \rangle' \longleftrightarrow 1$ yields an isomorphism $\langle \mathcal{O}_X, \mathcal{O}_X \rangle \xrightarrow{\sim} \mathcal{O}_S$. This means that $\langle f, g \rangle'$ is essentially the same as $\langle f, g \rangle$ in Proposition 4.5.

4.2. Green functions on Riemann surfaces

Let X be a connected compact Riemann surface, that is, $S = \mathrm{Spec}(\mathbb{C})$. Let g be a C^∞-function on a dense Zariski open set of X (a set obtained by excluding finite points from X). We say that g is a *Green function* on X if, for any point $x \in X$, there are a real number a and a C^∞-function u around x such that

$$g = a \log |z|^2 + u$$

around x, where z is a local coordinate of X around x with $z(x) = 0$. The number a does not depend on the choice of the local coordinate z. Indeed, let w be another local coordinate of X around x with $w(x) = 0$. Then there is a holomorphic function f around the origin of $\mathbb{C}$ such that $z = wf(w)$ and $f(0) \neq 0$. Thus

$$g = a \log |z|^2 + u = a \log |w|^2 + \log |f(w)|^2 + u$$

and $\log|f(w)|^2 + u$ is C^∞ around x. We denote $-a$ by $\operatorname{ord}_x^G(g)$, that is, g can be locally written by

$$g = -\operatorname{ord}_x^G(g)\log|z|^2 + (C^\infty\text{-function}).$$

A point x with $\operatorname{ord}_x^G(g) \neq 0$ is called a *singular point* of g. By the definition, we have only finitely many singular points of g. Note that g is not singular at x if and only if g is C^∞ around x.

The set of all Green functions on X is denoted by $G(X)$. The space $G(X)$ forms a real vector space. For $g \in G(X)$, we define $\operatorname{div}^G(g)$ by

$$\operatorname{div}^G(g) = \sum_{x \in X} \operatorname{ord}_x^G(g)x.$$

Then div^G yields a homomorphism $G(X) \to \operatorname{Div}(X)_{\mathbb{R}} (:= \operatorname{Div}(X) \otimes_{\mathbb{Z}} \mathbb{R})$.

LEMMA 4.8. *Let (L, h) be a C^∞-hermitian invertible sheaf on X and s a non-zero rational section of L. Then $-\log(h(s,s))$ is a Green function on X and*

$$\operatorname{div}^G(-\log(h(s,s))) = \operatorname{div}(s).$$

In particular, div^G is surjective. Moreover, if g is a Green function on X with $\operatorname{div}^G(g) \in \operatorname{Div}(X)$, then there are a C^∞-hermitian invertible sheaf (L, h) on X and a non-zero rational section s of L with $-\log(h(s,s)) = g$.

PROOF. Let z be a local coordinate of X around x with $z(x) = 0$. Let l be a local basis of L around x. Then there are an integer a and a holomorphic function $f(z)$ around the origin of $\mathbb{C}$ such that $s = z^a f(z) l$ and $f(0) \neq 0$. Thus $h(s,s) = |z|^{2a} \cdot u$, where u is a C^∞-function around x and $u(x) \neq 0$. Therefore, $-\log(h(s,s)) = -a\log|z|^2 - \log u$ is a Green function around x and $\operatorname{ord}_x^G(-\log(h(s,s))) = \operatorname{ord}_x(\operatorname{div}(s))$.

Let g be a Green function on X with $D = \operatorname{div}^G(g) \in \operatorname{Div}(X)$. We set $L = \mathscr{O}_X(D)$. Notice that L is a subsheaf of $\operatorname{Rat}(X)$ and 1 ($\in \operatorname{Rat}(X)$) is a non-zero rational section of L with $\operatorname{div}(1) = D$. Let h_1 be a C^∞-hermitian metric of L. We set $f = g + \log(h_1(1,1))$. Then $\operatorname{div}^G(f) = 0$. Thus f is a C^∞-function on X, so that $h = \exp(-f)h_1$ is a C^∞-hermitian metric of L and $-\log(h(1,1)) = g$. □

Let $f : X \to \mathbb{R}$ be a function on X, which gives rise to a linear map

$$\tilde{f} : \operatorname{Div}(X)_{\mathbb{R}} \longrightarrow \mathbb{R}$$

defined by $\tilde{f}(\sum_x n_x x) = \sum_x n_x f(x)$. For example, if 1 is a constant function with the value 1, then $\tilde{1} = \deg$.

We define the differential operator d^c by

$$d^c = \frac{\partial - \bar{\partial}}{4\pi\sqrt{-1}}.$$

Then $dd^c = \dfrac{\sqrt{-1}}{2\pi}\partial\bar{\partial}$. Since $dd^c(\log|z|^2) = 0$, $dd^c(g)$ is a C^∞ $(1,1)$-form for a Green function g on X. On the other hand, g is locally integrable. This means that g can be viewed as a distribution function on X. In order to distinguish the function from the distribution, the distribution function induced by g is denoted by $[g]$. Then we have the following Poincaré-Lelong formula for Green functions. This is slightly more general than Theorem 1.39 in the 1-dimensional case. However, the proof is almost same.

THEOREM 4.9 (Poincaré-Lelong formula). *Let g be a Green function on X and $D = \operatorname{div}^G(g)$. Then*

$$dd^c([g]) + \delta_D = [dd^c(g)],$$

that is, for a C^∞-function f on X,

$$\int_X g\,dd^c(f) + \tilde{f}(D) = \int_X f\,dd^c(g).$$

In particular, $\displaystyle\int_X dd^c(g) = \deg(\operatorname{div}^G(g))$.

Let $ds^2 = u(z)dz \otimes d\bar{z} = u(z)(dx^2 + dy^2)$ $(z = x + \sqrt{-1}y)$ be a Kähler metric of X. The associated Käher form Ω with the Kähler metric ds^2 is defined by

$$\Omega = \frac{\sqrt{-1}}{2} u(z)dz \wedge d\bar{z}.$$

Let $\Delta_{\bar\partial}$ be the $\bar\partial$-Laplacian with respect to Ω, that is, $\Delta_{\bar\partial} = \bar\partial^* \circ \bar\partial$. $\Delta_{\bar\partial}$ is locally given by

$$\frac{-1}{2u}\left(\frac{\partial^2}{\partial x^2} + \frac{\partial^2}{\partial y^2}\right).$$

Moreover, it is easy to see that

$$dd^c(f) = \frac{-1}{2\pi}(\Delta_{\bar\partial}(f))\Omega \tag{4.3}$$

(see Proposition 1.37). In particular, for a Green function g on X, $\Delta_{\bar\partial}(g)$ is a C^∞-function. Here we consider the following two spaces.

$$G^\Omega(X) = \{g \in G(X) \mid \Delta_{\bar\partial}(g) \text{ is a constant}\},$$
$$G_0^\Omega(X) = \{g \in G(X) \mid \Delta_{\bar\partial}(g) \text{ is a constant and } \int_X g\Omega = 0\}.$$

THEOREM 4.10. *A homomorphism* $\operatorname{div}^G\big|_{G_0^\Omega(X)} : G_0^\Omega(X) \longrightarrow \operatorname{Div}(X)_{\mathbb{R}}$ *is an isomorphism. Moreover, for $g \in G^\Omega(X)$, there are unique $g_0 \in G_0^\Omega(X)$ and $c \in \mathbb{R}$ with $g = g_0 + c$.*

PROOF. For $g \in G_0^\Omega(X)$, we assume $\operatorname{div}^G(g) = 0$. Then g is a C^∞-function. By (4.3), there is a constant a with $dd^c(g) = a\Omega$. By Theorem 4.9, $\int_X dd^c(g) = \deg(\operatorname{div}^G(g)) = 0$. Therefore, $a = 0$, which means $\Delta_{\bar\partial}(g) = 0$. Thus g is a constant. On the other hand, as $\int_x g\Omega = 0$, we have $g = 0$. Hence, $\operatorname{div}^G\big|_{G_0^\Omega(X)}$ is injective.

We choose $D \in \operatorname{Div}(X)_{\mathbb{R}}$. By Lemma 4.8, there is a Green function g_1 on X with $\operatorname{div}^G(g_1) = D$. We set

$$f = \Delta_{\bar\partial}(g_1) - \frac{1}{\operatorname{vol}(X)}\int_X \Delta_{\bar\partial}(g_1)\Omega,$$

where $\operatorname{vol}(X) = \displaystyle\int_X \Omega$. Then $\displaystyle\int_X f\Omega = 0$. Thus, by Corollary 1.34, there is a C^∞-function u with $\Delta_{\bar\partial}(u) = f$. If we write $g_2 = g_1 - u$, then

$$\Delta_{\bar\partial}(g_2) = \Delta_{\bar\partial}(g_1) - f = \frac{1}{\operatorname{vol}(X)}\int_X \Delta_{\bar\partial}(g_1)\Omega.$$

Therefore, $g_2 \in G^{\Omega}(X)$, so that

$$g = g_2 - \frac{1}{\mathrm{vol}(X)} \int_X g_2 \Omega \in G_0^{\Omega}(X).$$

The last assertion follows from the fact: If $g = g_0 + c \quad (g \in G^{\Omega}(X), g_0 \in G_0^{\Omega}(X))$, then

$$\int_X g\Omega = \int_X g_0 \Omega + \int_X c\Omega = c \,\mathrm{vol}(X). \qquad \square$$

Let L be an invertible sheaf on X and h a C^∞-hermitian metric of L. We say that h is *admissible* with respect to Ω if there is a constant a with $c_1(L,h) = a\Omega$.

COROLLARY 4.11. *Every invertible sheaf on X has an admissible C^∞-hermitian metric with respect to Ω. Moreover, an admissible C^∞-hermitian metric with respect to Ω is uniquely determined up to positive constant multiplication.*

PROOF. Let h_1 be a C^∞-hermitian metric of L and s a non-zero rational section of L. Then, by Lemma 4.8, $-\log(h_1(s,s))$ is a Green function with

$$\mathrm{div}^G(-\log(h_1(s,s))) = \mathrm{div}(s).$$

By Theorem 4.10, there is a Green function g such that $\mathrm{div}^G(g) = \mathrm{div}(s)$ and $\Delta_{\bar\partial}(g)$ is a constant. We set $f = g + \log(h_1(s,s))$. Since $\mathrm{div}^G(f) = 0$, f is a C^∞-function. Therefore, if we write $h = \exp(-f)h_1$, then

$$\Delta_{\bar\partial}(-\log(h(s,s))) = \Delta_{\bar\partial}(g)$$

is a constant, which means that h is admissible with respect to Ω.

Let h_1 and h_2 be two admissible C^∞-hermitian metrics of L. If we set

$$f = \log(h_1(s,s)) - \log(h_2(s,s)),$$

then f is a C^∞-function on X. Moreover, since h_1 and h_2 are admissible, there is a constant a with $dd^c(f) = a\Omega$. On the other hand,

$$\int_X c_1(L,h_1) = \int_X c_1(L,h_2) = \deg(L).$$

Thus $a = 0$, that is, $\Delta_{\bar\partial}(f) = 0$. Therefore, f is a constant and $h_2 = \exp(-f)h_1$. $\square$

By Theorem 4.10, the homomorphism

$$\mathrm{div}^G\Big|_{G_0^{\Omega}(X)} : G_0^{\Omega}(X) \longrightarrow \mathrm{Div}(X)_{\mathbb{R}}$$

is an isomorphism, so that its inverse map is denoted by g^{Ω}. Let D be a divisor on X. Then $\mathscr{O}_X(D)$ is a subsheaf of $\mathrm{Rat}(X)$ and 1 is a non-zero rational section of $\mathscr{O}_X(D)$ with $\mathrm{div}(1) = D$. By Corollary 4.11, $\mathscr{O}_X(D)$ has a unique admissible C^∞-hermitian metric h_D^{Ω} with respect to Ω such that $-\log(h_D^{\Omega}(1,1)) = g^{\Omega}(D)$. The metric h_D^{Ω} is called the *canonical metric* of $\mathscr{O}_X(D)$ with respect to Ω. Moreover, we have the natural isometry

$$(\mathscr{O}_X(D_1 + D_2), h_{D_1+D_2}^{\Omega}) \simeq (\mathscr{O}_X(D_1), h_{D_1}^{\Omega}) \otimes (\mathscr{O}_X(D_2), h_{D_2}^{\Omega}).$$

Let g_1 and g_2 be Green functions on X such that $\mathrm{div}^G(g_1)$ and $\mathrm{div}^G(g_2)$ have no common component. Then the *product* $\langle g_1, g_2 \rangle$ of g_1 and g_2 is defined to be

$$\langle g_1, g_2 \rangle = \widetilde{g_1}(\mathrm{div}^G(g_2)) + \int_X dd^c(g_1) g_2.$$

Note that $\widetilde{g_1}(\operatorname{div}^G(g_2))$ makes sense because g_1 and g_2 do not have a common singular point. In the literature, $\langle g_1, g_2\rangle$ is denoted by $g_1 * g_2$ and it is called the $*$-product.

PROPOSITION 4.12. *$\langle g_1, g_2\rangle = \langle g_2, g_1\rangle$. In particular, $g^\Omega(p)(q) = g^\Omega(q)(p)$ holds for $p, q \in X$ with $p \neq q$. In Theorem 5.6, this proposition will be generalized.*

PROOF. Let $\{x_1, \ldots, x_s\}$ and $\{y_1, \ldots, y_r\}$ be the sets of singular points of g_1 and g_2, respectively. Moreover, let $D_{x_i}(\epsilon)$ (resp. $D_{y_j}(\epsilon)$) be a ϵ-disk at x_i (resp. at y_j). Then, for a sufficiently small positive number ϵ, we can construct C^∞-functions ρ_1^ϵ and ρ_2^ϵ with the following properties.

(a) $0 \le \rho_1^\epsilon \le 1$ and $0 \le \rho_2^\epsilon \le 1$.
(b) $\rho_1^\epsilon = 1$ on $X \setminus \bigcup_i D_{x_i}(\epsilon)$ and $\rho_2^\epsilon = 1$ on $X \setminus \bigcup_j D_{y_j}(\epsilon)$.
(c) $\rho_1^\epsilon = 0$ on $\bigcup_i \overline{D_{x_i}(\epsilon/2)}$ and $\rho_2^\epsilon = 0$ on $\bigcup_j \overline{D_{y_j}(\epsilon/2)}$.

We set $g_1^\epsilon = \rho_1^\epsilon g_1$ and $g_2^\epsilon = \rho_2^\epsilon g_2$. Note that g_1^ϵ and g_2^ϵ are C^∞-functions. Moreover, we write $\omega_1^\epsilon = dd^c(g_1 - g_1^\epsilon)$ and $\omega_2^\epsilon = dd^c(g_2 - g_2^\epsilon)$. Let us see the following.

CLAIM 3. *For any C^∞-function f on X,*

$$\lim_{\epsilon\downarrow 0} \int_X \omega_1^\epsilon f = \tilde{f}(D_1) \quad \text{and} \quad \lim_{\epsilon\downarrow 0} \int_X \omega_2^\epsilon f = \tilde{f}(D_2),$$

where $D_1 = \operatorname{div}^G(g_1)$ and $D_2 = \operatorname{div}^G(g_2)$.

PROOF. First, we have

$$\int_X \omega_1^\epsilon f = \int_X dd^c(g_1)f - \int_X dd^c(g_1^\epsilon)f = \int_X dd^c(g_1)f - \int_X g_1^\epsilon dd^c(f).$$

Thus, by Theorem 4.9 (the Poincaré-Lelong formula),

$$\lim_{\epsilon\downarrow 0} \int_X \omega_1^\epsilon f = \int_X dd^c(g_1)f - \int_X g_1 dd^c(f) = \tilde{f}(D_1).$$

Therefore, the claim follows. □

We set

$$a(\epsilon) = \int_X g_1^\epsilon dd^c(g_2^\epsilon) \left(= \int_X g_2^\epsilon dd^c(g_1^\epsilon)\right).$$

Let us evaluate $\lim_{\epsilon\downarrow 0} a(\epsilon)$. First let us check

$$\lim_{\epsilon\downarrow 0} a(\epsilon) = \int_X g_1\omega_2 - \widetilde{g_1}(D_2)$$

using an expression $a(\epsilon) = \int_X g_1^\epsilon dd^c(g_2^\epsilon)$. Indeed, we write $\omega_1 = dd^c(g_1)$ and $\omega_2 = dd^c(g_2)$. Then

$$a(\epsilon) = \int_X g_1^\epsilon(\omega_2 - \omega_2^\epsilon).$$

Since $\operatorname{Supp}(\omega_2^\epsilon) \subset \bigcup_i \overline{D_{y_i}(\epsilon)}$, there is a positive number ϵ_0 such that

$$a(\epsilon) = \int_X (g_1^\epsilon\omega_2 - g_1^{\epsilon_0}\omega_2^\epsilon)$$

for a sufficiently small positive number ϵ. Therefore, by Claim 3, we have

$$\lim_{\epsilon\downarrow 0} a(\epsilon) = \int_X g_1\omega_2 - \widetilde{g_1}(D_2).$$

In the same way, using an expression $a(\epsilon) = \int_X g_2^\epsilon dd^c(g_1^\epsilon)$, we have

$$\lim_{\epsilon \downarrow 0} a(\epsilon) = \int_X g_2\omega_1 - \widetilde{g_2}(D_1).$$

Thus the proposition follows. □

Let us consider the following two lemmas which will be used later.

PROPOSITION 4.13. *For a non-zero rational function ϕ and a Green function g,*

$$\langle \log|\phi|, g\rangle = \log|N_{\mathrm{div}^G(g)}(\phi)|.$$

PROOF. Note that $dd^c(\log|\phi|) = 0$. Thus, if we set $D = \mathrm{div}^G(g)$, then

$$\langle \log|\phi|, g\rangle = \widetilde{\log|\phi|}(D).$$

Thus, if we set $D = \sum_i n_i P_i$, then

$$\widetilde{\log|\phi|}(D) = \sum_i n_i \log|\phi(P_i)| = \log\left|\prod_i \phi(P_i)^{n_i}\right| = \log|N_D(\phi)|.$$ □

PROPOSITION 4.14. *Let $\mu : Y \to X$ be a surjective morphism of connected compact Riemann surfaces. If g is a Green function on X, then $\mu^*(g)$ is also a Green function on Y and $\mathrm{div}^G(\mu^*(g)) = \mu^*(\mathrm{div}^G(g))$. Moreover, if g_1 and g_2 are Green functions on X such that $\mathrm{div}^G(g_1)$ and $\mathrm{div}^G(g_2)$ have no common component, then the following holds:*

$$\langle \mu^*(g_1), \mu^*(g_2)\rangle = \deg(\mu)\langle g_1, g_2\rangle.$$

PROOF. The first half is almost obvious, so that we consider the last half of the proposition. First,

$$\int_Y dd^c(\mu^*(g_1))\mu^*(g_2) = \int_Y \mu^*(dd^c(g_1)g_2) = \deg(\mu)\int_X dd^c(g_1)g_2.$$

If we set $\mathrm{div}^G(\mu^*(g_2)) = \sum_i a_i P_i$, then

$$\widetilde{\mu^*(g_1)}(\mathrm{div}^G(\mu^*(g_2))) = \sum_i a_i g_1(\mu(P_i)) = \widetilde{g_1}(\mu_*(\mathrm{div}^G(\mu^*(g_2)))).$$

By Lemma 1.12, for a divisor D on X, $\mu_*(\mu^*(D)) = \deg(\mu)D$. Therefore,

$$\widetilde{\mu^*(g_1)}(\mathrm{div}^G(\mu^*(g_2))) = \widetilde{g_1}(\mu_*(\mu^*(\mathrm{div}^G(g_2)))) = \deg(\mu)\widetilde{g_1}(\mathrm{div}^G(g_2))$$

because $\mathrm{div}^G(\mu^*(g_2)) = \mu^*(\mathrm{div}^G(g_2))$. Thus we have the last half of the proposition. □

Finally, we consider the metric of Deligne's pairing. Let L and M be an invertible sheaf on X. Then the Deligne's pairing $\langle L, M\rangle$ of L and M is a 1-dimensional vector space over $\mathbb{C}$. We assume that L and M have metrics h_L and h_M, respectively. We introduce a natural metric $h_{\langle L,M\rangle}$ of $\langle L, M\rangle$ arising from h_L and h_M. Let l and m be non-zero rational sections of L and M such that $\mathrm{div}(l)$ and $\mathrm{div}(m)$ have no common component. Then $\langle l, m\rangle$ is a basis of $\langle L, M\rangle$, so that we would like to define $h_{\langle L,M\rangle}(\langle l, m\rangle, \langle l, m\rangle)$ by

$$h_{\langle L,M\rangle}(\langle l, m\rangle, \langle l, m\rangle) =: \exp\left(-\langle \log(h_L(l,l)), \log(h_M(m,m))\rangle\right), \tag{4.4}$$

that is,

$$-\log\left(h_{\langle L,M\rangle}(\langle l, m\rangle, \langle l, m\rangle)\right) = \langle -\log(h_L(l,l)), -\log(h_M(m,m))\rangle.$$

For this purpose, we need to show that (4.4) does not depend on the choice of l and m. Indeed, let l' and m' be another non-zero rational section. We can write $l' = fl$ and $m' = gm$ for some non-zero rational functions f and g. Then

$$\begin{cases} \operatorname{div}^G(-\log(h_L(l,l))) = \operatorname{div}(l), \\ \operatorname{div}^G(-\log(h_M(m,m))) = \operatorname{div}(m), \\ \operatorname{div}^G(-\log|f|^2) = (f), \\ \operatorname{div}^G(-\log|g|^2) = (g). \end{cases}$$

Thus, by Proposition 4.13,

$$\begin{aligned} &\langle -\log(h_L(l',l')), -\log(h_M(m',m'))\rangle \\ &\quad = \langle -\log(h_L(l,l)) - \log|f|^2, -\log(h_M(m,m)) - \log|g|^2\rangle \\ &\qquad = \langle -\log(h_L(l,l)), -\log(h_M(m,m))\rangle \\ &\qquad\qquad - \log\left(|\langle f,g\rangle \cdot N_{\operatorname{div}(m)}(f)\cdot N_{\operatorname{div}(l)}(g)|^2\right). \end{aligned}$$

On the other hand, we have

$$\langle l',m'\rangle = \langle f,g\rangle N_{\operatorname{div}(m)}(f)N_{\operatorname{div}(m)}(g)\langle l,m\rangle.$$

Thus the metric does not depend on the choice of l and m.

4.3. Arithmetic Chow groups on arithmetic surfaces

Let K be an algebraic number field and O_K the ring of integers in K. Let $\pi : X \to \operatorname{Spec}(O_K)$ be a regular projective arithmetic surface, that is, X is regular, π is flat and projective, $\dim \pi = 1$ and X_K is geometrically irreducible over K. For $\sigma \in K(\mathbb{C})$, that is, an embedding $\sigma : K \to \mathbb{C}$, we set $X_\sigma = X \times_{\operatorname{Spec}(O_K)} \operatorname{Spec}(\mathbb{C})$, where $\operatorname{Spec}(\mathbb{C})$ is a scheme over $\operatorname{Spec}(O_K)$ via $\sigma : O_K \hookrightarrow \mathbb{C}$. Let us introduce the arithmetic Chow group of X. Let $\operatorname{Div}(X)$ be the group of divisors on X. The *group of arithmetic divisors* $\widehat{\operatorname{Div}}(X)$ is defined by the subgroup of

$$\operatorname{Div}(X) \times \left(\bigoplus_{\sigma \in K(\mathbb{C})} G(X_\sigma)\sigma\right)$$

consisting of $(D, \sum_\sigma g_\sigma \sigma)$ such that $\operatorname{div}^G(g_\sigma) = D_\sigma$ for all $\sigma \in K(\mathbb{C})$. Namely,

$$\widehat{\operatorname{Div}}(X) := \left\{ \left(D, \sum\nolimits_\sigma g_\sigma\sigma\right) \;\middle|\; \begin{array}{l} D \in \operatorname{Div}(X) \text{ and, for all } \sigma \in K(\mathbb{C}), \\ g_\sigma \in G(X_\sigma) \text{ and } \operatorname{div}^G(g_\sigma) = D_\sigma \end{array} \right\}.$$

Let f be a non-zero rational function on X. Then the *arithmetic principal divisor* $\widehat{(f)}$ is defined by

$$\widehat{(f)} := \left((f), \sum\nolimits_{\sigma\in K(\mathbb{C})} -\log(|f\otimes^\sigma 1|^2)\sigma\right).$$

Clearly, $\widehat{(f)} \in \widehat{\operatorname{Div}}(X)$, so that $\widehat{\operatorname{CH}}^1(X)$ is defined by

$$\widehat{\operatorname{CH}}^1(X) := \frac{\widehat{\operatorname{Div}}(X)}{\left\{\widehat{(f)} \mid f \text{ is a non-zero rational function on } X\right\}}.$$

Let us fix a Kähler form Ω_σ on X_σ for each $\sigma \in K(\mathbb{C})$, and we set $\Omega = \{\Omega_\sigma\}_{\sigma \in K(\mathbb{C})}$. Then the subgroup $\widehat{\mathrm{CH}}^1_{Ar}(X, \Omega)$ of $\widehat{\mathrm{CH}}^1(X)$ is defined as follows. First we set

$$\mathrm{Div}_{Ar}(X, \Omega) := \left\{ \left(D, \sum\nolimits_\sigma g_\sigma \sigma\right) \,\middle|\, \begin{array}{l} D \in \mathrm{Div}(X) \text{ and, for all } \sigma \in K(\mathbb{C}), \\ g_\sigma \in G^{\Omega_\sigma}(X_\sigma) \text{ and } \mathrm{div}^G(g_\sigma) = D_\sigma \end{array} \right\}.$$

An element of $\mathrm{Div}_{Ar}(X, \Omega)$ is called an *Arakelov divisor* with respect to Ω. Note that $\widehat{(f)} \in \mathrm{Div}_{Ar}(X, \Omega)$. Thus we define $\mathrm{CH}^1_{Ar}(X, \Omega)$ by

$$\mathrm{CH}^1_{Ar}(X, \Omega) := \frac{\mathrm{Div}_{Ar}(X)}{\left\{ \widehat{(f)} \mid f \text{ is a non-zero rational function on } X \right\}}.$$

The natural homomorphisms

$$z : \widehat{\mathrm{Div}}(X) \to \mathrm{Div}(X) \quad \text{and} \quad \omega : \widehat{\mathrm{Div}}(X) \to \bigoplus_\sigma A^{1,1}(X_\sigma)$$

are defined to be

$$\begin{cases} z(D, \sum_\sigma g_\sigma \sigma) = D, \\ \omega(D, \sum_\sigma g_\sigma \sigma) = \sum_\sigma dd^c(g_\sigma), \end{cases}$$

where $A^{1,1}(X_\sigma)$ is the space of C^∞ $(1,1)$-forms on X_σ. It is easy to see that z and ω induce the homomorphisms

$$z : \widehat{\mathrm{CH}}^1(X) \to \mathrm{CH}^1(X) \quad \text{and} \quad \omega : \widehat{\mathrm{CH}}^1(X) \to \bigoplus_\sigma A^{1,1}(X_\sigma).$$

Let L be an invertible sheaf on X. For each $\sigma \in K(\mathbb{C})$, we set $L_\sigma = L \otimes_{O_K} \mathbb{C}$, where the tensor product is taken by viewing $\mathbb{C}$ as an O_K-module via $\sigma : O_K \hookrightarrow \mathbb{C}$. Then L_σ is an invertible sheaf on X_σ. For each $\sigma \in K(\mathbb{C})$, let h_σ be a C^∞-hermitian metric of L_σ. Then a pair $(L, \{h_\sigma\}_{\sigma \in K(\mathbb{C})})$ of L and $\{h_\sigma\}_{\sigma \in K(\mathbb{C})}$ is called a C^∞-*hermitian invertible sheaf* on X, and it is denoted by (L, h) or $\overline{L}$ for short. If h_σ is admissible for every $\sigma \in K(\mathbb{C})$, then (L, h) is called an *admissible* C^∞*-hermitian invertible sheaf* on X. Here we define the arithmetic first Chern class of (L, h) as follows: Let s be a non-zero rational section of L. Then the class

$$\left(\mathrm{div}(s), -\sum\nolimits_\sigma \log h_\sigma(s \otimes^\sigma 1, s \otimes^\sigma 1)\sigma\right)$$

in $\widehat{\mathrm{CH}}^1(X)$ does not depend on the choice of s. Indeed, let s' be another non-zero rational section of L. Then there is a rational function f on X with $s' = fs$. We denote

$$\left(\mathrm{div}(s), -\sum\nolimits_\sigma \log h_\sigma(s \otimes^\sigma 1, s \otimes^\sigma 1)\sigma\right)$$

and

$$\left(\mathrm{div}(s'), -\sum\nolimits_\sigma \log h_\sigma(s' \otimes^\sigma 1, s' \otimes^\sigma 1)\sigma\right)$$

by $z(s)$ and $z(s')$, respectively. Then

$$\begin{aligned} z(s') &= \left(\operatorname{div}(fs), -\sum\nolimits_{\sigma} \log h_{\sigma}(fs \otimes^{\sigma} 1, fs \otimes^{\sigma} 1)\sigma\right) \\ &= \left((f) + \operatorname{div}(s), -\sum\nolimits_{\sigma} \log h_{\sigma}(s \otimes^{\sigma} \sigma(f), s \otimes^{\sigma} \sigma(f))\sigma\right) \\ &= \left((f) + \operatorname{div}(s), -\sum\nolimits_{\sigma} \log |1 \otimes^{\sigma} \sigma(f)|^2 h_{\sigma}(s \otimes^{\sigma} 1, s \otimes^{\sigma} 1)\sigma\right) \\ &= \left((f) + \operatorname{div}(s), -\sum\nolimits_{\sigma} \log |f \otimes^{\sigma} 1|^2 h_{\sigma}(s \otimes^{\sigma} 1, s \otimes^{\sigma} 1)\sigma\right) \\ &= \widehat{(f)} + z(s). \end{aligned}$$

The class $z(s)$ in $\widehat{\mathrm{CH}}^1(X)$ is denoted by $\widehat{c}_1(L,h)$ and it is called the *arithmetic first Chern class* of (L,h). Note that $\omega(\widehat{c}_1(L,h)) = \sum_{\sigma} c_1(L_{\sigma}, h_{\sigma})$. If h is admissible with respect to Ω, then $\widehat{c}_1(L,h) \in \mathrm{CH}^1_{Ar}(X,\Omega)$.

Since $\operatorname{div}^G : G_0^{\Omega_{\sigma}}(X_{\sigma}) \to \operatorname{Div}(X_{\sigma})$ is an isomorphism, its inverse map is denoted by $g^{\Omega_{\sigma}}$. Then, for $D \in \operatorname{Div}(X)$, $\overline{D} \in \operatorname{Div}_{Ar}(X,\Omega)$ is defined by

$$\overline{D} = \left(D, \sum\nolimits_{\sigma \in K(\mathbb{C})} g^{\Omega_{\sigma}}(D_{\sigma})\sigma\right),$$

which is called the *compactification* of D with respect to Ω. Moreover, $(0, 2\sigma)$ is denoted by F_{σ}. More precisely, if we set

$$\delta_{\sigma\sigma'} = \begin{cases} 1 & \text{if } \sigma = \sigma', \\ 0 & \text{if } \sigma \neq \sigma', \end{cases}$$

then $F_{\sigma} = (0, 2\sum_{\sigma' \in K(\mathbb{C})} \delta_{\sigma\sigma'}\sigma')$. This is called the *infinite fiber* at σ and it is sometimes denoted by X_{σ} by abuse of notation.

LEMMA 4.15. *Any element of* $\operatorname{Div}_{Ar}(X,\Omega)$ *can be written uniquely by a form*

$$\overline{D} + \sum\nolimits_{\sigma \in K(\mathbb{C})} a_{\sigma} F_{\sigma},$$

where $D \in \operatorname{Div}(X)$ *and* $a_{\sigma} \in \mathbb{R}$. *In other words, a homomorphism*

$$\varphi^{\Omega} : \operatorname{Div}(X) \oplus \bigoplus_{\sigma \in K(\mathbb{C})} \mathbb{R}F_{\sigma} \longrightarrow \operatorname{Div}_{Ar}(X,\Omega)$$

given by

$$\varphi^{\Omega}\left(D + \sum\nolimits_{\sigma \in K(\mathbb{C})} a_{\sigma} F_{\sigma}\right) = \overline{D} + \sum\nolimits_{\sigma \in K(\mathbb{C})} a_{\sigma} F_{\sigma}$$

is an isomorphism.

PROOF. First let us see the uniqueness. An element x of $\operatorname{Div}_{Ar}(X,\Omega)$ has two expressions

$$x = \overline{D} + \sum\nolimits_{\sigma} a_{\sigma} F_{\sigma} = \overline{D'} + \sum\nolimits_{\sigma} a'_{\sigma} F_{\sigma}.$$

Then, considering $z(x)$, we have $D = D'$. Thus $\overline{D} = \overline{D'}$. Therefore, $a_{\sigma} = a'_{\sigma}$ holds for every $\sigma \in K(\mathbb{C})$.

Let us choose an element $x = (D, \sum_{\sigma} g_{\sigma}\sigma)$ of $\operatorname{Div}_{Ar}(X,\Omega)$. Since $\overline{D}$ is given by $\left(D, \sum_{\sigma} g^{\Omega_{\sigma}}(D_{\sigma})\sigma\right)$,

$$x - \overline{D} = \left(0, \sum\nolimits_{\sigma} (g_{\sigma} - g^{\Omega_{\sigma}}(D_{\sigma}))\sigma\right).$$

Note that $g_\sigma \in G^{\Omega_\sigma}(X_\sigma)$ and $g^{\Omega_\sigma}(D_\sigma) \in G_0^{\Omega_\sigma}(X_\sigma)$. Thus $g_\sigma - g^{\Omega_\sigma}(D_\sigma)$ is a constant. If we set

$$a_\sigma = \frac{g_\sigma - g^{\Omega_\sigma}(D_\sigma)}{2},$$

then $x = \overline{D} + \sum_\sigma a_\sigma F_\sigma$. □

For an Arakelov divisor $\overline{D} + \sum_{\sigma \in K(\mathbb{C})} a_\sigma F_\sigma$, we can give an admissible C^∞-hermitian invertible sheaf whose first Chern class is equal to the class of $\overline{D} + \sum_{\sigma \in K(\mathbb{C})} a_\sigma F_\sigma$. Indeed, there is a unique C^∞-hermitian metric $h^{\Omega_\sigma}_{D_\sigma}$ of $\mathscr{O}_{X_\sigma}(D_\sigma)$ with $-\log h^{\Omega_\sigma}_{D_\sigma}(1,1) = g^{\Omega_\sigma}(D_\sigma)$. The C^∞-hermitian invertible sheaf associated with $\overline{D} + \sum_{\sigma \in K(\mathbb{C})} a_\sigma F_\sigma$ is defined by

$$\left(\mathscr{O}_X(D), \left\{\exp(-2a_\sigma) h^{\Omega_\sigma}_{D_\sigma}\right\}_{\sigma \in K(\mathbb{C})}\right).$$

It is easy to see that $\widehat{c}_1\left(\mathscr{O}_X(D), \left\{\exp(-2a_\sigma) h^{\Omega_\sigma}_{D_\sigma}\right\}_{\sigma \in K(\mathbb{C})}\right)$ coincides with the class of $\overline{D} + \sum_{\sigma \in K(\mathbb{C})} a_\sigma F_\sigma$.

LEMMA 4.16. *Let (L, h) be an admissible C^∞-hermitian invertible sheaf on X. Then there is an Arakelov divisor $\overline{D} + \sum_{\sigma \in K(\mathbb{C})} a_\sigma F_\sigma$ such that (L, h) is isometric to*

$$\left(\mathscr{O}_X(D), \left\{\exp(-2a_\sigma) h^{\Omega_\sigma}_{D_\sigma}\right\}_{\sigma \in K(\mathbb{C})}\right).$$

PROOF. Let s be a non-zero rational section of L. We set $D = \operatorname{div}(s)$. Then there is an isomorphism $\phi : L \to \mathscr{O}_X(D)$ with $\phi(s) = 1$. Here we consider an arithmetic divisor

$$\left(\operatorname{div}(s), -\sum\nolimits_{\sigma \in K(\mathbb{C})} h_\sigma(s \otimes 1, s \otimes 1)\sigma\right).$$

Then, by Lemma 4.15, the above divisor has an expression

$$\overline{D} + \sum\nolimits_{\sigma \in K(\mathbb{C})} a_\sigma F_\sigma.$$

Then (L, h) is isometric to $\left(\mathscr{O}_X(D), \left\{\exp(-2a_\sigma) h^{\Omega_\sigma}_{D_\sigma}\right\}_{\sigma \in K(\mathbb{C})}\right)$. □

4.4. Intersection theory on arithmetic surfaces

As in the previous section, let $\pi : X \to \operatorname{Spec}(O_K)$ be a regular arithmetic surface. We set

$$\widehat{\Upsilon} = \left\{ ((D, g), (D', g')) \in \widehat{\operatorname{Div}}(X) \times \widehat{\operatorname{Div}}(X) \;\middle|\; \begin{array}{l} D \text{ and } D' \text{ have} \\ \text{no common component} \end{array} \right\}.$$

For $((D, g), (D', g')) \in \widehat{\Upsilon}$, an element $\langle (D, g), (D', g') \rangle$ of $\widehat{\operatorname{Div}}(O_K)$ is defined by

$$\langle (D, g), (D', g') \rangle := \left(\langle D, D' \rangle, \sum\nolimits_{\sigma \in K(\mathbb{C})} \langle g_\sigma, g'_\sigma \rangle \sigma \right). \tag{4.5}$$

Then

$$\widehat{\deg}\left(\langle (D, g), (D', g') \rangle\right) = \deg(\langle D, D' \rangle) + \frac{1}{2} \sum_{\sigma \in K(\mathbb{C})} \langle g_\sigma, g'_\sigma \rangle,$$

where, for a 0-cycle $w = \sum_P n_P P$ on $\operatorname{Spec}(O_K)$,

$$\deg(w) := \sum_P n_P \log \#\kappa(P).$$

We denote $\widehat{\deg}\left(\langle (D,g),(D',g')\rangle\right)$ by $\widehat{\deg}((D,g)\cdot(D',g'))$, or, $(D,g)\cdot(D',g')$ for short. It is called the *arithmetic intersection number* of (D,g) and (D',g'). For example,

$$\widehat{\deg}\left(\left(D, \sum_{\sigma \in K(\mathbb{C})} g_\sigma \sigma\right)\cdot F_\sigma\right) = \deg(D_K).$$

For a 0-cycle $z = \sum_x n_x x$ on X, $\deg(z)$ is defined by

$$\deg(z) = \sum_x n_x \log \#\kappa(x). \tag{4.6}$$

Then

$$\deg(\pi_*(z)) = \deg(z). \tag{4.7}$$

Indeed,

$$\begin{aligned}\deg(\pi_*(z)) &= \sum_x n_x [\kappa(x) : \kappa(\pi(x))] \log \#\kappa(\pi(x)) \\ &= \sum_x n_x \log \#\kappa(x) = \deg(z).\end{aligned}$$

Let $((D,g),(D',g')) \in \widehat{\Upsilon}$. Since D and D' have no common component, as in Section 4.1, we can consider $I(D,D')$ (see (4.1)). Note that $\langle D, D'\rangle = \pi_*(I(D,D'))$, so that, by (4.7), we have

$$\widehat{\deg}((D,g)\cdot(D',g')) = \deg(I(D,D')) + \frac{1}{2}\sum_{\sigma\in K(\mathbb{C})} \langle g_\sigma, g'_\sigma\rangle. \tag{4.8}$$

We assume that D and D' are effective. Let d_x and d'_x be local equations of D and D' at $x \in X$. Then

$$\log \#(\mathscr{O}_{X,x}/(d_x, d'_x)) = \operatorname{length}_{\mathscr{O}_{X,x}}(\mathscr{O}_{X,x}/(d_x,d'_x)) \log \#\kappa(x),$$

which shows

$$\widehat{\deg}((D,g)\cdot(D',g')) = \sum_{x\in X} \log \#(\mathscr{O}_{X,x}/(d_x,d'_x)) + \frac{1}{2}\sum_{\sigma\in K(\mathbb{C})} \langle g_\sigma, g'_\sigma\rangle. \tag{4.9}$$

Now we have the following proposition.

Proposition 4.17. (1) *Let*

$$((D,g),(D',g')), \quad ((D',g'),(D'',g'')) \quad \text{and} \quad ((D,g),(D'',g''))$$

be elements of $\widehat{\Upsilon}$. *Then we have*

$$\begin{cases}\langle (D,g),(D',g')\rangle = \langle (D',g'),(D,g)\rangle, \\ \langle (D,g)+(D',g'),(D'',g'')\rangle = \langle (D,g),(D'',g'')\rangle + \langle (D',g'),(D'',g'')\rangle, \\ \langle (D,g),(D',g')+(D'',g'')\rangle = \langle (D,g),(D',g')\rangle + \langle (D,g),(D'',g'')\rangle.\end{cases}$$

(2) *Let D be a divisor on X and ϕ a non-zero rational function on X such that D and (ϕ) have no common component. Then*

$$\langle (D,g),\widehat{(\phi)}\rangle = (\widehat{N_D(\phi)}).$$

In particular, $\widehat{\deg}\left(\langle (D,g),\widehat{(\varphi)}\rangle\right) = 0.$

PROOF. (1) is obvious by the definition (for example, see Proposition 4.12). For (2), as $\langle D, (\phi)\rangle = (N_D(\phi))$, it is sufficient to show

$$\langle g_\sigma, -\log|\phi_\sigma|^2\rangle = -\log|\sigma(N_D(\phi))|^2.$$

By Proposition 4.13, $\langle g_\sigma, -\log|\phi_\sigma|^2\rangle = -\log|N_{D_\sigma}(\phi_\sigma)|^2$. On the other hand, by Lemma 1.14, $N_{D_\sigma}(\phi_\sigma) = \sigma(N_D(\phi))$. Thus (2) follows. □

We have all the tools to define the arithmetic intersection number in general. Let (D, g) and (E, h) be an element of $\widehat{\mathrm{Div}}(X)$. By using Proposition 1.22, there is a non-zero rational function f on X such that $D + (f)$ and E have no common component. Then we would like to define $\widehat{\deg}((D, g)\cdot(E, h))$ to be

$$\widehat{\deg}((D,g)\cdot(E,h)) := \widehat{\deg}\left(\left((D,g)+\widehat{(f)}\right)\cdot(E,h)\right).$$

For this purpose, we need to see that the right side of the above equation does not depend on the choice of f. Namely, let f' be another non-zero rational function on X such that $D + (f')$ and E have no common component. Then

$$\widehat{\deg}\left(\left((D,g)+\widehat{(f)}\right)\cdot(E,h)\right) = \widehat{\deg}\left(\left((D,g)+\widehat{(f')}\right)\cdot(E,h)\right).$$

Indeed, since $D + (f)$ and $D + (f')$ do not have the common component with E, $(f/f') = (D + (f)) - (D + (f'))$ and E have no common component. Thus, by (2) in Proposition 4.17, we have

$$\begin{aligned}&\widehat{\deg}\left(\left((D,g)+\widehat{(f)}\right)\cdot(E,h)\right) - \widehat{\deg}\left(\left((D,g)+\widehat{(f')}\right)\cdot(E,h)\right)\\ &\quad= \widehat{\deg}\left(\widehat{(f/f')}\cdot(E,h)\right) = 0.\end{aligned}$$

From now on, let us see the properties of the arithmetic intersection number. First, by Proposition 4.17, one can easily check the following proposition.

PROPOSITION 4.18. (1) *$\widehat{\deg}(x\cdot\widehat{(f)}) = \widehat{\deg}(\widehat{(f)}\cdot x) = 0$ for all $x \in \widehat{\mathrm{Div}}(X)$ and all non-zero rational functions f on X.*

(2) *$\widehat{\deg}((x+y)\cdot z) = \widehat{\deg}(x\cdot z) + \widehat{\deg}(y\cdot z)$ and $\widehat{\deg}(x\cdot(y+z)) = \widehat{\deg}(x\cdot y) + \widehat{\deg}(x\cdot z)$ for all $x, y, z \in \widehat{\mathrm{Div}}(X)$.*

(3) *$\widehat{\deg}(x\cdot y) = \widehat{\deg}(y\cdot x)$ for all $x, y \in \widehat{\mathrm{Div}}(X)$.*

Next we would like to extend Deligne's pairing to the pairing of C^∞-hermitian invertible sheaves. Let $\overline{L} = (L, h_L)$ and $\overline{M} = (M, h_M)$ be C^∞-hermitian invertible sheaves on X. We define the hermitian O_K-module $\langle(L, h_L), (M, h_M)\rangle$ on $\mathrm{Spec}(O_K)$ to be

$$\langle(L,h_L),(M,h_M)\rangle = \left(\langle L, M\rangle, h_{\langle L,M\rangle}\right), \tag{4.10}$$

where the metric $h_{\langle L,M\rangle}$ is given in the following way. By (5) in Theorem 4.7, $\langle L, M\rangle_\sigma = \langle L, M\rangle \otimes^\sigma \mathbb{C}$ is canonically isomorphic to $\langle L_\sigma, M_\sigma\rangle$. The metric of $\langle L_\sigma, M_\sigma\rangle$ was discussed in the last part of Section 4.2. Namely,

$$h_{\langle L,M\rangle_\sigma}(\langle l,m\rangle_\sigma, \langle l,m\rangle_\sigma) = \exp\left(-\langle \log(h_{L_\sigma}(l_\sigma,l_\sigma)), \log(h_{M_\sigma}(m_\sigma,m_\sigma))\rangle\right).$$

Then we have the following.

PROPOSITION 4.19.

$$\widehat{\deg}\left(\langle(L,h_L),(M,h_M)\rangle\right) = \widehat{\deg}\left(\widehat{c}_1(L,h_L)\cdot\widehat{c}_1(M,h_M)\right).$$

PROOF. Let l and m be non-zero rational sections of L and M, respectively, such that $\operatorname{div}(l)$ and $\operatorname{div}(m)$ have no common component. Then $\langle l, m\rangle$ is a non-zero rational section of $\langle L, M\rangle$ and, by (3.2) in Theorem 4.7, $\operatorname{div}(\langle l, m\rangle)=\langle\operatorname{div}(l), \operatorname{div}(m)\rangle$. First,

$$\begin{aligned}&\widehat{\deg}\left(\widehat{c}_1(L, h_L)\cdot\widehat{c}_1(M, h_M)\right)\\&\qquad=\deg(\langle\operatorname{div}(l), \operatorname{div}(m)\rangle)+\frac{1}{2}\sum\nolimits_{\sigma}\langle-\log(h_{L_\sigma}(l_\sigma, l_\sigma)), -\log(h_{M_\sigma}(m_\sigma, m_\sigma))\rangle.\end{aligned}$$

On the other hand, by Proposition 3.11,

$$\widehat{\deg}\left(\langle(L, h_L), (M, h_M)\rangle\right)=\deg(\operatorname{div}(\langle l, m\rangle))+\frac{1}{2}\sum\nolimits_{\sigma}-\log\left(h_{\langle L, M\rangle_\sigma}(\langle l, m\rangle_\sigma, \langle l, m\rangle_\sigma)\right).$$

Here

$$-\log\left(h_{\langle L, M\rangle_\sigma}(\langle l, m\rangle_\sigma, \langle l, m\rangle_\sigma)\right)=\langle-\log(h_{L_\sigma}(l_\sigma, l_\sigma)), -\log(h_{M_\sigma}(m_\sigma, m_\sigma))\rangle.$$

Thus the proposition follows. □

PROPOSITION 4.20. *As in the previous section, let us fix a Kähler form* $\Omega=\{\Omega_\sigma\}_{\sigma\in K(\mathbb{C})}$. *Let* (L, h) *be an admissible* C^∞*-hermitian invertible sheaf with respect to* Ω *on* X *and* D *a horizontal prime divisor on* X. *Then we have an isometry*

$$\langle(L, h), (\mathscr{O}_X(D), h_D^\Omega)\rangle\simeq N_{D/\operatorname{Spec}(O_K)}((L, h)|_D)$$

(for the definition of $N_{D/\operatorname{Spec}(O_K)}((L, h)|_D)$*, see Section* 3.6*). In particular, together with Lemma* 3.19*,*

$$\widehat{\deg}\left(\widehat{c}_1(L, h)\cdot\widehat{c}_1(\mathscr{O}_X(D), h_D^\Omega)\right)=\widehat{\deg}((L, h)|_D).$$

PROOF. By (4) in Theorem 4.7, there is a natural isomorphism

$$\beta:\langle L, \mathscr{O}_X(D)\rangle\simeq N_{D/\operatorname{Spec}(O_K)}(L|_D).$$

Thus it is sufficient to see that the above isomorphism extends to an isometry. Let l be a non-zero rational section of L such that $\operatorname{div}(l)$ do not have D as an irreducible component. Then $\beta(\langle l, 1_D\rangle)=N_D(l)$. Since

$$\int_{X_\sigma}\log h_{D_\sigma}^{\Omega_\sigma}(1_{D_\sigma}, 1_{D_\sigma})c_1(\overline{L}_\sigma)=0,$$

we can see

$$\begin{aligned}&h_{\langle L, \mathscr{O}_X(D)\rangle_\sigma}(\langle l, 1_D\rangle\otimes^\sigma 1, \langle l, 1_D\rangle\otimes^\sigma 1)\\&\qquad=\exp(-\langle\log h_{L_\sigma}(l_\sigma, l_\sigma), \log h_{D_\sigma}^{\Omega_\sigma}(1_{D_\sigma}, 1_{D_\sigma})\rangle)\\&\qquad\qquad=\exp\left((\log h_{L_\sigma}(l_\sigma, l_\sigma))^{\sim}(D_\sigma)\right).\end{aligned}$$

On the other hand, letting K' be the function field of D, we set

$$K'(\mathbb{C})_\sigma=\{\sigma'\in K'(\mathbb{C})\mid\sigma'|_K=\sigma\}.$$

For $\sigma'\in K'(\mathbb{C})_\sigma$, let $P_{\sigma'}$ be the corresponding $\mathbb{C}$-value point of X_σ to σ' in terms of the morphism $\operatorname{Spec}(K')\to X$ induced by $D\hookrightarrow X$. Then $D_\sigma=\sum_{\sigma'\in K(\mathbb{C})_\sigma}P_{\sigma'}$ and

$$\exp\left((\log h_{L_\sigma}(l_\sigma, l_\sigma))^{\sim}(D_\sigma)\right)=\prod_{\sigma'\in K'(\mathbb{C})_\sigma}h_{L_\sigma}(l_\sigma, l_\sigma)(P_{\sigma'}),$$

which means that

$$h_{\langle L_\sigma, \mathscr{O}_{X_\sigma}(D_\sigma)\rangle}(\langle l, 1_D\rangle \otimes^\sigma 1, \langle l, 1_D\rangle \otimes^\sigma 1)$$
$$= h_{N_{D/\operatorname{Spec}(O_K)}(L|_D),\sigma}(N_{K'/K}(l|_D) \otimes^\sigma 1, N_{K'/K}(l|_D) \otimes^\sigma 1).$$

Thus β is an isometry. □

Let K' be a finite extension field of K and $\pi' : X' \to \operatorname{Spec}(O_{K'})$ a regular projective arithmetic surface. We assume there is a surjective morphism $\mu : X' \to X$ such that the following diagram is commutative.

$$(4.11)\qquad \begin{array}{ccc} X & \xleftarrow{\ \mu\ } & X' \\ {\scriptstyle\pi}\downarrow & & \downarrow{\scriptstyle\pi'} \\ \operatorname{Spec}(O_K) & \longleftarrow & \operatorname{Spec}(O_{K'}) \end{array}$$

For $\sigma \in K(\mathbb{C})$, we set $K'(\mathbb{C})_\sigma = \{\sigma' \in K'(\mathbb{C}) \mid \sigma'|_K = \sigma\}$. Then, for $\sigma' \in K(\mathbb{C})_\sigma$, we have the natural morphism $\mu_{\sigma'} : X'_{\sigma'} \to X_\sigma$ of Riemann surfaces induced by μ. For $x = \left(D, \sum_{\sigma \in K(\mathbb{C})} g_\sigma\right)$, we define $\mu^*(x)$ by

$$\mu^*(x) = \left(\mu^*(D), \sum_{\sigma \in K(\mathbb{C})} \sum_{\sigma' \in K'(\mathbb{C})_\sigma} \mu^*_{\sigma'}(g_\sigma)\sigma'\right).$$

Note that $\mu^*(x) \in \widehat{\operatorname{Div}}(X')$ by Proposition 4.14. We have the following.

PROPOSITION 4.21.
$$\widehat{\deg}(\mu^*(x) \cdot \mu^*(y)) = \deg(\mu)\widehat{\deg}(x \cdot y)$$
for $x, y \in \widehat{\operatorname{Div}}(X)$.

PROOF. Before starting the proof, let us prepare some notation. Let z' be either a 0-cycle on X', or, a divisor. Then the push-forward $\mu_*(z')$ of z' is defined as follows. If z' is either a closed point, or, a prime divisor, then

$$\mu_*(z') = \begin{cases} [\kappa(z') : \kappa(\mu(z'))]\mu(z') & \text{if } \dim \mu(z') = \dim z', \\ 0 & \text{if } \dim \mu(z') < \dim z'. \end{cases}$$

In general, for $z' = \sum_i a'_i z'_i$, we define $\mu_*(z')$ by $\mu_*(z') = \sum_i a'_i \mu_*(z'_i)$. Let us see the following.

CLAIM 4. *Let D and D' be divisors on X and X', respectively, such that $\mu^*(D)$ and D' has no common component. Thus we can consider $I(\mu^*(D), D')$ and $I(D, \mu_*(D'))$ on X and X', respectively, as in Section* 4.1 (*especially see the equation* (4.1)). *Then $\mu_*(I(\mu^*(D), D')) = I(D, \mu_*(D'))$.*

PROOF. By the linearity of the equation, we may assume that D and D' are prime divisors. We set $\Gamma = \mu(D')$. First we assume $\dim \Gamma = 0$. If $D \cap \Gamma \neq \emptyset$, then $\mu^*(D)$ contains D'. Thus $D \cap \Gamma = \emptyset$. Therefore, $\mu^*(D)$ and D' have no intersection, that is, $I(\mu^*(D), D') = 0$. On the other hand, since $\mu_*(D') = 0$, we have $I(D, \mu_*(D')) = 0$, as required.

Next we consider the case where $\dim \Gamma = 1$. Since the claim is local with respect to X, we may replace X by an open set U and X' by $\mu^{-1}(U)$. Therefore, we may assume that D is defined by an equation $f = 0$. Let $\bar{f}$ and $\bar{\mu}$ denote $f|_\Gamma$ and $\mu|_{D'} : D' \to \Gamma$, respectively. Then

$$I(\mu^*(D), D') = \sum_{x' \in D'} \operatorname{length}_{\mathscr{O}_{D',x'}}(\mathscr{O}_{D',x'}/\bar{\mu}^*(\bar{f})\mathscr{O}_{D',x'})x'.$$

Thus

$$\mu_*(I(\mu^*(D), D')) = \sum_{x \in \Gamma} \left(\sum_{\substack{x' \in D' \\ \mu(x') = x}} \operatorname{length}_{\mathscr{O}_{D',x'}}(\mathscr{O}_{D',x'}/\bar{\mu}^*(\bar{f})\mathscr{O}_{D',x'})[\kappa(x') : \kappa(x)] \right) x.$$

On the other hand, by Lemma 1.12,

$$\begin{aligned} \sum_{\substack{x' \in D' \\ \mu(x') = x}} & \operatorname{length}_{\mathscr{O}_{D',x'}}(\mathscr{O}_{D',x'}/\bar{\mu}^*(\bar{f})\mathscr{O}_{D',x'})[\kappa(x') : \kappa(x)] \\ &= \deg(\bar{\mu}) \operatorname{length}_{\mathscr{O}_{\Gamma,x}}(\mathscr{O}_{\Gamma,x}/\bar{f}\mathscr{O}_{\Gamma,x}), \end{aligned}$$

and hence

$$\begin{aligned} \mu_*(I(\mu^*(D), D')) &= \deg(\bar{\mu}) \sum_{x \in \Gamma} \operatorname{length}_{\mathscr{O}_{\Gamma,x}}(\mathscr{O}_{\Gamma,x}/\bar{f}\mathscr{O}_{\Gamma,x})x \\ &= \deg(\bar{\mu}) I(D, \Gamma) = I(D, \mu_*(D')). \end{aligned}$$ □

We set $x = \left(D, \sum_{\sigma \in K(\mathbb{C})} g_\sigma\right)$ and $y = \left(E, \sum_{\sigma \in K(\mathbb{C})} k_\sigma\right)$. By Proposition 1.22, we may assume that E and D have no common component. Moreover, by Proposition 1.22 again, E and D does not pass through a point at which μ is not flat, so that $\mu^*(D)$ and $\mu^*(E)$ have no common component. First, by the equation (4.8), we have

$$\widehat{\deg}(x \cdot y) = \deg(I(D, E)) + \frac{1}{2} \sum_{\sigma \in K(\mathbb{C})} \langle g_\sigma, k_\sigma \rangle.$$

Next we consider $\widehat{\deg}(\mu^*(x) \cdot \mu^*(y))$. Then

$$\widehat{\deg}(\mu^*(x) \cdot \mu^*(y)) = \deg(I(\mu^*(D), \mu^*(E))) + \frac{1}{2} \sum_{\sigma \in K(\mathbb{C})} \sum_{\sigma' \in K'(\mathbb{C})_\sigma} \langle \mu_{\sigma'}^*(g_\sigma), \mu_{\sigma'}^*(k_\sigma) \rangle.$$

In the same way as in (4.7), for a 0-cycle z' of X', we can see that $\deg(\mu_*(z')) = \deg(z')$. Moreover, by Lemma 1.12, for a divisor E on X, we can check that $\mu_*(\mu^*(E)) = \deg(\mu)E$. Thus, by using the claim,

$$\begin{aligned} \deg(I(\mu^*(D), \mu^*(E))) &= \deg(\mu_*(I(\mu^*(D), \mu^*(E)))) \\ &= \deg(I(D, \mu_*(\mu^*(E)))) \\ &= \deg(\mu) \deg(I(D, E)). \end{aligned}$$

Further, by Proposition 4.14,

$$\langle \mu_{\sigma'}^*(g_\sigma), \mu_{\sigma'}^*(k_\sigma) \rangle = \deg(\mu_{\sigma'}) \langle g_\sigma, k_\sigma \rangle.$$

By using the commutative diagram (4.11), there is a morphism

$$\tilde{\mu} : X' \to X \times_{\operatorname{Spec}(O_K)} \operatorname{Spec}(O_{K'})$$

over $\operatorname{Spec}(O_{K'})$. Clearly $\deg(\tilde{\mu}) = \deg(\mu_{\sigma'})$ and $\deg(\mu) = [K' : K]\deg(\tilde{\mu})$. Thus $\deg(\mu_{\sigma'}) = \deg(\mu)/[K' : K]$. Hence

$$\sum_{\sigma \in K(\mathbb{C})} \sum_{\sigma' \in K'(\mathbb{C})_\sigma} \langle \mu_{\sigma'}^*(g_\sigma), \mu_{\sigma'}^*(k_\sigma) \rangle = \deg(\mu) \sum_{\sigma \in K(\mathbb{C})} \langle g_\sigma, k_\sigma \rangle,$$

as required. □

4.5. Arakelov metric of dualizing sheaf and adjunction formula

Let us begin with the following theorem.

THEOREM 4.22. *Let M be a compact Kähler manifold and Ω a Kähler form of M. Let L be an invertible sheaf on M. There is a unique C^∞-hermitian metric h of L up to a positive constant multiplication such that $c_1(L, h)$ is harmonic with respect to Ω.*

PROOF. Let h_1 be a C^∞-hermitian metric of L . Let η be the harmonic representative of the cohomology class of $c_1(L, h_1)$. Then $c_1(L, h_1) - \eta$ is d-exact. Thus, by $\partial\bar{\partial}$-Lemma (cf. Lemma 1.36), there is a real-valued C^∞-function f such that $dd^c(f) = c_1(L, h_1) - \eta$. Thus, if we set $h = \exp(f)h_1$, then $c_1(L, h) = \eta$.

Let h' be another C^∞-hermitian metric of L with $c_1(L, h') = \eta$. There is a positive C^∞-function u with $h' = uh$. Then $dd^c(\log(u)) = 0$. This means that $\log(u)$ is a constant. □

Let C be a connected Riemann surface of genus $g \geq 1$. Let us fix a Kähler form Ω of C. Let $p : C \times C \to C$ and $q : C \times C \to C$ be the projections to the first factor and the second factor, respectively. We set $\Omega_{C\times C} = p^*(\Omega) + q^*(\Omega)$, which is a Kähler form on $C \times C$. Let Δ be the diagonal of $C \times C$, that is, $\Delta = \{(x, x) \in C \times C \mid x \in C\}$. Let 1 be the canonical section of $\mathscr{O}_{C\times C}(\Delta)$. By Theorem 4.22, there is a unique C^∞-hermitian metric h_Δ^Ω of $\mathscr{O}_{C\times C}(\Delta)$ such that $c_1\left(\mathscr{O}_{C\times C}(\Delta), h_\Delta^\Omega\right)$ is harmonic with respect to $\Omega_{C\times C}$ and $\int_{C\times C} \log(h_\Delta^\Omega(1, 1))\Omega_{C\times C}^2 = 0$. Let $\{\omega_1, \ldots, \omega_g\}$ be a basis of $H^0(C, \Omega_C^1)$ such that

$$\frac{\sqrt{-1}}{2} \int_C \omega_i \wedge \overline{\omega}_j = \delta_{ij}$$

for $i, j = 1, \ldots, g$.

Note that $\Omega, \omega_1, \ldots, \omega_g, \bar{\omega}_1, \ldots, \bar{\omega}_g$ are harmonic on C with respect to Ω. Thus, by Künneth formula (Theorem 1.38) in terms of harmonic forms, the set

$$\{p^*(\Omega), q^*(\Omega)\} \cup \{p^*(\omega_i) \wedge q^*(\bar{\omega}_j)\}_{1\leq i,j\leq g} \cup \{q^*(\omega_i) \wedge p^*(\bar{\omega}_j)\}_{1\leq i,j\leq g}$$

forms a basis of the space $\mathbf{H}^{1,1}(C \times C)$ of harmonic $(1, 1)$-forms on $C \times C$ with respect to $\Omega_{C\times C}$. Let $\iota : C \times C \to C \times C$ be a morphism given by $\iota(x, y) = (y, x)$. Then we have a homomorphism

$$\iota^* : \mathbf{H}^{1,1}(C \times C) \to \mathbf{H}^{1,1}(C \times C).$$

Note that Δ is invariant under ι and $c_1\left(\mathscr{O}_{C\times C}(\Delta), h_\Delta^\Omega\right)$ is a harmonic representative of the class of Δ. Thus $\iota^*\left(c_1\left(\mathscr{O}_{C\times C}(\Delta), h_\Delta^\Omega\right)\right) = c_1\left(\mathscr{O}_{C\times C}(\Delta), h_\Delta^\Omega\right)$. Now we have the following lemma.

LEMMA 4.23. *$c_1\left(\mathscr{O}_{C\times C}(\Delta), h^{\Omega}_{\Delta}\right)$ can be written by*

$$\frac{1}{\operatorname{vol}(C)}p^*(\Omega) + \frac{1}{\operatorname{vol}(C)}q^*(\Omega) - \frac{\sqrt{-1}}{2}\sum_{i=1}^{g}\left(p^*(\omega_i)\wedge q^*(\bar{\omega}_i) + q^*(\omega_i)\wedge p^*(\bar{\omega}_i)\right),$$

where $\operatorname{vol}(C) = \int_C \Omega$.

PROOF. We set

$$\begin{aligned}\eta &= c_1(\mathscr{O}_{C\times C}(\Delta), h^{\Omega}_{\Delta})\\ &= cp^*(\Omega) + dq^*(\Omega) + \sum_{i,j} e_{ij}p^*(\omega_i)\wedge q^*(\bar{\omega}_j) + \sum_{i,j} f_{ij}q^*(\omega_i)\wedge p^*(\bar{\omega}_j).\end{aligned}$$

Since $\iota^*(\eta) = \eta$, we have $c = d$ and $e_{ij} = f_{ij}$ $(\forall i, j)$. By applying the Poincaré-Lelong formula (cf. Theorem 1.39) to the case where $f = p^*(\Omega)$, we have

$$\int_{\Delta} p^*(\Omega) = \int_{C\times C}\eta\wedge p^*(\Omega).$$

Therefore,

$$\operatorname{vol}(C) = \int_{C\times C}\eta\wedge p^*(\Omega) = \int_{C\times C} cq^*(\Omega)\wedge p^*(\Omega) = c\left(\int_C \Omega\right)^2 = c\operatorname{vol}(C)^2.$$

Thus $c = 1/\operatorname{vol}(C)$. By applying the Poincaré-Lelong formula to the case where $f = p^*(\omega_i)\wedge q^*(\bar{\omega}_j)$,

$$\int_{\Delta} p^*(\omega_i)\wedge q^*(\bar{\omega}_j) = \int_{C\times C}\eta\wedge p^*(\omega_i)\wedge q^*(\bar{\omega}_j).$$

Thus

$$\begin{aligned}\frac{2}{\sqrt{-1}}\delta_{ij} &= \int_{C\times C}\eta\wedge p^*(\omega_i)\wedge q^*(\bar{\omega}_j)\\ &= \int_{C\times C}\sum_{k,l} e_{kl}q^*(\omega_k)\wedge p^*(\bar{\omega}_l)\wedge p^*(\omega_i)\wedge q^*(\bar{\omega}_j)\\ &= -\int_{C\times C}\sum_{k,l} e_{kl}q^*(\omega_k\wedge\bar{\omega}_j)\wedge p^*(\omega_i\wedge\bar{\omega}_l)\\ &= -\sum_{k,l} e_{kl}\left(\int_C \omega_k\wedge\bar{\omega}_j\right)\left(\int_C \omega_i\wedge\bar{\omega}_l\right)\\ &= -\sum_{k,l} e_{kl}\frac{2\delta_{kj}}{\sqrt{-1}}\frac{2\delta_{il}}{\sqrt{-1}} = 4e_{ji},\end{aligned}$$

so that $e_{ij} = \dfrac{-\sqrt{-1}}{2}\delta_{ij}$. Thus the lemma follows. □

We set $\tilde{g} = -\log(h^{\Omega}_{\Delta}(1,1))$. Now we see that $\tilde{g}$ is symmetric. Indeed, since $dd^c(\tilde{g}) = c_1\left(\mathscr{O}_{C\times C}(\Delta), h^{\Omega}_{\Delta}\right)$, we have

$$dd^c(\iota^*(\tilde{g})) = \iota^*(dd^c(\tilde{g})) = dd^c(\tilde{g}).$$

Moreover,

$$\begin{aligned}\int_{C\times C}\iota^*(\tilde{g})\Omega^2_{C\times C} &= \int_{C\times C}\iota^*(\iota^*(\tilde{g})\Omega^2_{C\times C})\\ &= \int_{C\times C}\tilde{g}\iota^*(\Omega^2_{C\times C}) = \int_{C\times C}\tilde{g}\Omega^2_{C\times C} = 0.\end{aligned}$$

Therefore, $\tilde{g} = \iota^*(\tilde{g})$, which means that $\tilde{g}$ is symmetric.

Here we set $g_P = \tilde{g}|_{\{P\}\times C}$. Then g_P is a C^∞-function on $C \setminus \{P\}$. Now we have the following.

LEMMA 4.24. *Let g^Ω be the inverse map of the isomorphism*

$$\operatorname{div}^G\Big|_{G_0^\Omega(X)} : G_0^\Omega(X) \to \operatorname{Div}(X) \otimes \mathbb{R}$$

as in Section 4.2. Then $g_P = g^\Omega(P)$ holds. In particular, $h^\Omega_\Delta|_{\{P\}\times C} = h^\Omega_P$.

PROOF. By the construction, g_P is a Green function and $\operatorname{div}^G(g_P) = P$. Since $dd^c(\tilde{g}) = c_1\left(\mathscr{O}_{C\times C}(\Delta), h^\Omega_\Delta\right)$, by Lemma 4.23, $dd^c(g_P) = (1/\operatorname{vol}(C))\Omega$, that is, $g_P \in G^\Omega(X)$. Therefore, in order to show $g_P = g^\Omega(P)$, it is sufficient to see $\int_C g_P\Omega = 0$. By Proposition 4.12, we have $\langle g_P, g_Q\rangle = \langle g_Q, g_P\rangle$, which means that

$$\tilde{g}(P,Q) + \frac{1}{\operatorname{vol}(C)}\int_C g_Q\Omega = \tilde{g}(Q,P) + \frac{1}{\operatorname{vol}(C)}\int_C g_P\Omega.$$

Since $\tilde{g}$ is symmetric, $\displaystyle\int_C g_P\Omega = \int_C g_Q\Omega$ holds for all $P \neq Q$. Therefore, $\displaystyle\int_C g_P\Omega$ is a constant. We denote it by a. Thus

$$\begin{aligned}0 = \int_{C\times C}\tilde{g}\Omega^2_{C\times C} &= \int_{C\times C}\tilde{g}p^*(\Omega)\wedge q^*(\Omega)\\ &= \int_C\left(\int_C g_P\Omega\right)\Omega = \int_C a\Omega = a\operatorname{vol}(C),\end{aligned}$$

which implies $a = 0$; that is, $\int_C g_P\Omega = 0$. □

We assume that the genus g of C is greater than or equal to 1. Note that $\tilde{g}$ is a Green function on $C \times C$ and g is the genus of C. Ω^1_C is naturally isomorphic to $\mathscr{O}_{C\times C}(-\Delta)|_\Delta$. Thus $\left(h^\Omega_\Delta\right)^{-1}$ yields a metric of Ω^1_C. Next we consider the admissibility of $\left(\Omega^1_C, \left(h^\Omega_\Delta\right)^{-1}\Big|_\Delta\right)$. Let $\{\omega_1, \dots, \omega_g\}$ be an orthonormal basis of $H^0(C, \Omega^1_C)$ with respect to the inner product given by

$$\langle\omega, \eta\rangle = \frac{\sqrt{-1}}{2}\int_C \omega\wedge\overline{\eta}.$$

Let us consider the following $(1,1)$-form:

$$\frac{\sqrt{-1}}{2g}(\omega_1\wedge\bar{\omega}_1 + \cdots + \omega_g\wedge\bar{\omega}_g).$$

This is a positive $(1,1)$-form and it does not depend on the choice of the orthonormal basis of $H^0(C, \Omega^1_C)$. We denote it by Ω^{can} and it is called the *canonical Kähler form* of C.

LEMMA 4.25. $\left(\Omega^1_C, \left(h^\Omega_\Delta\right)^{-1}\Big|_\Delta\right)$ *is admissible with respect to Ω if and only if there is a positive constant c with $\Omega = c\,\Omega^{can}$.*

PROOF. Since $c_1\left(\Omega^1_C, \left(h^{\Omega}_{\Delta}\right)^{-1}\Big|_{\Delta}\right) = -\,c_1\left(\mathscr{O}_{C\times C}(\Delta), h^{\Omega}_{\Delta}\right)\big|_{\Delta}$, by Lemma 4.23,

$$c_1\left(\Omega^1_C, \left(h^{\Omega}_{\Delta}\right)^{-1}\Big|_{\Delta}\right) = 2g\Omega^{can} - \frac{2}{\mathrm{vol}(C)}\Omega.$$

Thus the lemma follows. □

The C^∞-hermitian metric

$$\left(h^{\Omega^{can}}_{\Delta}\right)^{-1}\Big|_{\Delta}$$

of Ω^1_C arising from the canonical Kähler form Ω^{can} is called the *Arakelov metric* and it is denoted by h_{Ar}.

Let $\pi : X \to \mathrm{Spec}(O_K)$ be a regular arithmetic surface. We assume that the genus of X_K is greater than or equal to 1. We fix the canonical Kähler form Ω^{can}_σ on X_σ for each $\sigma \in K(\mathbb{C})$. We set $\Omega^{can} = \{\Omega^{can}_\sigma\}_{\sigma\in K(\mathbb{C})}$. Let ω_{X/O_K} be the double dual of Ω^1_{X/O_K}, that is,

$$\omega_{X/O_K} = \mathscr{H}om_{\mathscr{O}_X}(\mathscr{H}om_{\mathscr{O}_X}(\Omega^1_{X/O_K}, \mathscr{O}_X), \mathscr{O}_X).$$

Since X is a 2-dimensional regular scheme, ω_{X/O_K} is invertible. Moreover, if π is smooth at x, then $\omega_{X/O_K,x} = \Omega^1_{X/O_K,x}$. The C^∞-hermitian invertible sheaf

$$(\omega_{X/O_K}, \{(h_{Ar})_\sigma\}_{\sigma\in K(\mathbb{C})})$$

is denoted by $\overline{\omega}_{X/O_K}$. Then we have the following.

LEMMA 4.26 (Adjunction formula). *Let B be a section of $\pi : X \to \mathrm{Spec}(O_K)$. Then*

$$\widehat{\deg}\left((\overline{\omega}_{X/O_K} + \overline{B})\cdot \overline{B}\right) = 0.$$

PROOF. Since $\Omega^1_{B/O_K} = 0$, the natural homomorphism $\mathscr{O}_B(-B) \to \Omega^1_{X/O_K}\big|_B$ is an isomorphism. Here let us see the following.

CLAIM 5. *B does not pass through a singular point of $\pi : X \to \mathrm{Spec}(O_K)$.*

PROOF. Let $x \in B$ and $P = \pi(x)$. Let b a local equation of B at x and t a local parameter of $O_{K,P}$. Since $\mathscr{O}_{X,x}/b\mathscr{O}_{X,x}$ is isomorphic to $O_{K,P}$, it is easy to see that the maximal ideal of $\mathscr{O}_{X,x}$ is generated by b and t. Note that $\mathscr{O}_{X,x}$ is a 2-dimensional regular local ring. Thus b and t are a regular parameter of $\mathscr{O}_{X,x}$. In particular, $\mathscr{O}_{X,x}/t\mathscr{O}_{X,x}$ is regular. This means that π is smooth at x. □

By the above claim, $\Omega^1_{X/O_K}\big|_B \to \omega_{X/O_K}\big|_B$ is also an isomorphism. Moreover, by the construction of h_{Ar}, the above isomorphism extends to an isometry

$$\left(\mathscr{O}_X(-B), h^{\Omega^{can}}_{-B}\right)\Big|_B \xrightarrow{\sim} (\omega_{X/O_K}, h_{Ar})\big|_B$$

by Lemma 4.24. Therefore, we have

$$\widehat{\deg}\left(\left(\mathscr{O}_X(-B), h^{\Omega^{can}}_{-B}\right)\Big|_B\right) = \widehat{\deg}\left((\omega_{X/O_K}, h_{Ar})\big|_B\right).$$

Note that the metrics of $\mathscr{O}_X(\overline{B})$ and $\overline{\omega}_{X/O_K}$ are admissible. Thus, by Proposition 4.20, we obtain

$$-\widehat{\deg}\left(\overline{B}\cdot\overline{B}\right) = \widehat{\deg}\left(\widehat{c}_1(\overline{\omega}_{X/O_K})\cdot\overline{B}\right).$$ □

4.6. Determinant bundles for curves

In this section, we will discuss determinant bundles for curves. For general materials of determinant bundles, see Section 1.10.

Let C be a smooth, projective and geometrically irreducible curve over a field K and let L be an invertible sheaf on C. For non-zero effective divisors D and E, we set

$$\begin{cases} A(L,D) = \det H^0(L-D) \otimes \det H^1(L-D)^{-1} \otimes \det H^0(L|_D), \\ \tilde{A}(L,D,E) = \det H^0(L-D-E) \otimes \det H^1(L-D-E)^{-1} \\ \qquad\qquad\qquad\qquad \otimes \det H^0(L-D|_E) \otimes \det H^0(L|_D), \end{cases}$$

where $L-D$ and $L-D-E$ means $L \otimes \mathscr{O}_C(-D)$ and $L \otimes \mathscr{O}_C(-D-E)$, respectively. To take care of the case where either D or E is zero, we set

$$\begin{cases} A(L,0) = \det H^0(L) \otimes \det H^1(L)^{-1}, \\ \tilde{A}(L,0,E) = A(L,E), \ \tilde{A}(L,D,0) = A(L,D), \ \tilde{A}(L,0,0) = A(L,0). \end{cases}$$

If D is a non-zero effective divisor, the isomorphism $A(L,0) \xrightarrow{\sim} A(L,D)$ derived from an exact sequence

$$0 \to H^0(L-D) \to H^0(L) \to H^0(L|_D) \to H^1(L-D) \to H^1(L) \to 0$$

is denoted by $\epsilon(L,D)$. For convenience, if $D=0$, then $\epsilon(L,D)$ is considered as the identity map. Moreover, in the case where D and E are non-zero effective divisors, the isomorphism

$$\det H^0(L|_{D+E}) \xrightarrow{\sim} \det H^0(L-D|_E) \otimes \det H^0(L|_D)$$

derived from an exact sequence

$$0 \to H^0(L-D|_E) \to H^0(L|_{D+E}) \to H^0(L|_D) \to 0$$

is denoted by $\gamma(L,D,E)$. If $D=0$ (resp. $E=0$), then it is considered as the identity map of $\det H^0(L|_E) \to \det H^0(L|_E)$ (resp. $\det H^0(L|_D) \to \det H^0(L|_D)$). Further, if $D=E=0$, then it is considered as the identity map $K \to K$. Then we have the following.

LEMMA 4.27. *For effective divisors D and E, the following diagram*

$$\begin{array}{ccc} A(L,0) & \xrightarrow{\epsilon(L,D)} & A(L,D) \\ {\scriptstyle \epsilon(L,D+E)}\downarrow & & \downarrow{\scriptstyle \epsilon(L-D,E)\otimes \mathrm{id}} \\ A(L,D+E) & \xrightarrow{\mathrm{id}\otimes\gamma(L,D,E)} & \tilde{A}(L,D,E) \end{array}$$

is commutative up to sign, that is,

$$(\epsilon(L-D,E) \otimes \mathrm{id}) \circ \epsilon(L,D) = \pm(\mathrm{id} \otimes \gamma(L,D,E)) \circ \epsilon(L,D+E).$$

PROOF. If either $D = 0$ or $E = 0$, the assertion is obvious, so that we assume that $D \neq 0$ and $E \neq 0$. Let us consider the following commutative diagram:

$$\begin{array}{ccccccccccccc}
 & & & & & & 0 & & & & & & \\
 & & & & & & \downarrow & & & & & & \\
0 \to & H^0(L-D-E) & \xrightarrow{c_1} & H^0(L-D) & \xrightarrow{c_2} & H^0(L-D|_E) & \xrightarrow{c_3} & H^1(L-D-E) & \xrightarrow{c_4} & H^1(L-D) & \to 0 \\
 & \| & & \downarrow{\scriptstyle a_1} & & \downarrow{\scriptstyle d_1} & & \| & & \downarrow{\scriptstyle a_4} & \\
0 \to & H^0(L-D-E) & \xrightarrow{b_1} & H^0(L) & \xrightarrow{b_2} & H^0(L|_{D+E}) & \xrightarrow{b_3} & H^1(L-D-E) & \xrightarrow{b_4} & H^1(L) & \to 0 \\
 & \downarrow{\scriptstyle c_1} & & \| & & \downarrow{\scriptstyle d_2} & & \downarrow{\scriptstyle c_4} & & \| & \\
0 \to & H^0(L-D) & \xrightarrow{a_1} & H^0(L) & \xrightarrow{a_2} & H^0(L|_D) & \xrightarrow{a_3} & H^1(L-D) & \xrightarrow{a_4} & H^1(L) & \to 0 \\
 & & & & & \downarrow & & & & & \\
 & & & & & 0 & & & & &
\end{array}$$

If we can choose bases of nine vector spaces in the above diagram such that they form systems of bases of the three horizontal long exact sequences and the vertical short exact sequence in the center after changing the orders (see Section 3.7 for the definition of a system of bases of the exact sequence), then the lemma follows. The important point is that the basis of a vector space in the diagram must be the same up to order even if the vector space appears in different exact sequences. The proof is complicated, but not so difficult, so that it is left to the reader (the order of basis of $H^0(L|_{D+E})$ must be changed). $\square$

Let C be a smooth and projective irreducible curve over an algebraically closed field k and S a non-singular algebraic variety over k. Let $C_S = C \times S$ and $\pi : C_S \to S$ the projection to the second factor. Let D be a non-zero effective divisor on C_S such that D is flat over S. Let D' be another non-zero effective divisor on C_S such that D' is flat over S. Then, by using a natural exact sequence

$$0 \to \mathscr{O}_C(-D)|_{D'} \to \mathscr{O}_{D+D'} \to \mathscr{O}_D \to 0,$$

we can see that $D + D'$ is also flat over S. Here let us consider the base change of determinant bundles by a morphism which is not necessarily flat.

THEOREM 4.28. *Let L be an invertible sheaf on C_S and $\nu : S' \to S$ a morphism of non-singular algebraic varieties over k. Here we consider the following fiber product:*

$$\begin{array}{ccc}
C_{S'} = C \times S' & \xrightarrow{\nu'} & C_S \\
{\scriptstyle \pi'}\downarrow & & \downarrow{\scriptstyle \pi} \\
S' & \xrightarrow{\nu} & S.
\end{array}$$

If we set $L' = \nu'^*(L)$, then we have the following.

(1) *There is a natural isomorphism*

$$\beta_L : \nu^*(\det \mathbf{R}\pi_*(L)) \xrightarrow{\sim} \det \mathbf{R}\pi'_*(L')$$

up to sign.

(2) *Let D be a non-zero effective divisor on C_S such that D is flat over S. We set $D' = \nu'^*(D)$. Then the following diagram*

$$\begin{array}{ccc} \nu^*(\det \mathbf{R}\pi_* (L)) & \xrightarrow{\nu^*(\epsilon(\pi,L,D))} & \nu^*(\det \mathbf{R}\pi_* (L-D) \otimes \det(\pi_*(L|_D))) \\ {\scriptstyle \beta_L}\downarrow & & \downarrow{\scriptstyle \beta_{L-D}\otimes\iota(\pi,L,D)} \\ \det \mathbf{R}\pi'_* (L') & \xrightarrow{\epsilon(\pi',L',D')} & \det \mathbf{R}\pi'_* (L'-D') \otimes \det(\pi'_*(L'|_{D'})) \end{array}$$

is commutative up to sign, that is,

$$(\beta_{L-D} \otimes \iota(\pi, L, D)) \circ \nu^*(\epsilon(\pi, L, D)) = \pm\epsilon(\pi', L', D') \circ \beta_L,$$

where $\epsilon(\pi, L, D)$ and $\epsilon(\pi', L', D')$ are the natural isomorphisms derived from the exact sequences

$$0 \to \pi_*(L-D) \to \pi_*(L) \to \pi_*(L|_D) \to R^1\pi_*(L-D) \to R^1\pi_*(L) \to 0,$$
$$0 \to \pi'_*(L'-D') \to \pi'_*(L') \to \pi'_*(L'|_{D'}) \to R^1\pi'_*(L'-D') \to R^1\pi'_*(L') \to 0,$$

respectively. Moreover, $\iota(\pi, L, D)$ is the natural isomorphism arising from an isomorphism

$$\nu^*(\pi_*(L|_D)) \xrightarrow{\sim} \pi'_*(L'|_{D'}).$$

PROOF. Here we introduce similar notation as in Lemma 4.27. Let L be an invertible sheaf on C_S and let D and E be non-zero effective divisors on C_S such that D and E are flat over S. We set

$$\begin{cases} A(\pi, L, D) = \det \mathbf{R}\pi_* (L-D) \otimes \det \pi_*(L|_D), \\ \tilde{A}(\pi, L, D, E) = \det \mathbf{R}\pi_* (L-D-E) \otimes \det \pi_*((L-D)|_E) \otimes \det \pi_*(L|_D). \end{cases}$$

In the case where D or E is zero, we set

$$\begin{cases} A(\pi, L, 0) = \det \mathbf{R}\pi_* (L), \ \ \tilde{A}(\pi, L, 0, E) = A(\pi, L, E), \\ \tilde{A}(\pi, L, D, 0) = A(\pi, L, D), \ \ \tilde{A}(\pi, L, 0, 0) = A(\pi, L, 0). \end{cases}$$

If D is non-zero, then the isomorphism

$$A(\pi, L, 0) \xrightarrow{\sim} A(\pi, L, D)$$

derived from an exact sequence

$$0 \to \pi_*(L-D) \to \pi_*(L) \to \pi_*(L|_D) \to R^1\pi_*(L-D) \to R^1\pi_*(L) \to 0$$

is denoted by $\epsilon(\pi, L, D)$. If $D = 0$, $\epsilon(\pi, L, D)$ is considered as the identity map. Let E be a non-zero effective divisor on C_S such that E is flat over S. The isomorphism

$$\det(\pi_*(L|_{D+E})) \xrightarrow{\sim} \det(\pi_*(L-D|_E)) \otimes \det(\pi_*(L|_D))$$

arising from an exact sequence

$$0 \to \pi_*(L-D|_E) \to \pi_*(L|_{D+E}) \to \pi_*(L|_D) \to 0$$

is denoted by $\gamma(\pi, L, D, E)$. If either $D = 0$ or $E = 0$, then it is considered as the identity map on $\det(\pi_*(L|_E))$ or $\det(\pi_*(L|_D))$, respectively. In particular, if $D = E = 0$, then it is the identity map $\mathscr{O}_S \to \mathscr{O}_S$. Here let us see the following claim:

CLAIM 6. *Let D and E be non-zero effective divisors on C_S such that D and E are flat over S. Then the following diagram*

$$\begin{array}{ccc} A(\pi, L, 0) & \xrightarrow{\epsilon(\pi,L,D)} & A(\pi, L, D) \\ {\scriptstyle \epsilon(\pi,L,D+E)}\downarrow & & \downarrow{\scriptstyle \epsilon(\pi,L-D,E)\otimes \mathrm{id}} \\ A(\pi, L, D+E) & \xrightarrow{\mathrm{id}\otimes\gamma(\pi,L,D,E)} & \tilde{A}(\pi, L, D, E) \end{array}$$

is commutative up to sign.

PROOF. By Lemma 4.27, the above diagram is commutative up to sign at the generic point. Thus the claim follows. □

(1) First we consider the case where $H^0(L_s) = 0$ for all $s \in S$. Then $h^1(L_s)$ is a constant with respect to s, so that $R^1\pi_*(L)$ is locally free and the natural homomorphism $R^1\pi_*(L) \otimes \kappa(s) \to H^1(L_s)$ is an isomorphism for all $s \in S$. Let us consider a natural homomorphism

$$\nu^*(R^1\pi_*(L)) \to R^1\pi'_*(L').$$

If it is an isomorphism, then we have the natural isomorphism

$$\beta_L : \nu^*(\det \mathbf{R}\pi_*(L)) \xrightarrow{\sim} \det \mathbf{R}\pi'_*(L').$$

For this, we need to see that $\nu^*(R^1\pi_*(L)) \otimes \kappa(s') \to R^1\pi'_*(L') \otimes \kappa(s')$ is an isomorphism for all $s' \in S'$. If we set $s = \pi(s')$, then we have the following commutative diagram:

$$\begin{array}{ccc} \nu^*(R^1\pi_*(L)) \otimes \kappa(s') & \longrightarrow & R^1\pi'_*(L') \otimes \kappa(s') \\ \downarrow & & \downarrow \\ H^1(L_s) \otimes \kappa(s') & \longrightarrow & H^1(L'_{s'}). \end{array}$$

The left vertical homomorphism is an isomorphism for all $s' \in S'$. The bottom horizontal homomorphism is an isomorphism by the base change theorem in terms of a flat morphism. Moreover, the right vertical homomorphism is also an isomorphism because $H^0(L'_{s'}) = 0$ $(\forall s' \in S')$. Consequently, the upper horizontal homomorphism is an isomorphism.

Let D be a non-zero effective divisor on C_S such that D is flat over S and $H^0(L_s - D_s) = 0$ for all $s \in S$. Then since $\pi|_D : D \to T$ is finite, there is a natural isomorphism

$$\nu^*(\pi_*(L|_D)) \xrightarrow{\sim} \pi'_*(L'|_{D'}),$$

where $D' = \nu'^*(D)$. Let $\iota(\pi, L, D)$ denote the isomorphism

$$\nu^*(\det(\pi_*(L|_D))) \xrightarrow{\sim} \det(\pi'_*(L'|_{D'}))$$

derived from the above isomorphism. Here

$$\beta_{L,D} : \nu^*(\det \mathbf{R}\pi_*(L)) \to \det \mathbf{R}\pi'_*(L')$$

is defined by a homomorphism such that the following diagram is commutative:

$$\begin{array}{ccc} \nu^*(\det \mathbf{R}\pi_*(L)) & \xrightarrow{\nu^*(\epsilon(\pi,L,D))} & \nu^*(A(\pi, L, D)) \\ {\scriptstyle \beta_{L,D}}\downarrow & & \downarrow{\scriptstyle \beta_{L-D}\otimes\iota(\pi,L,D)} \\ \det \mathbf{R}\pi'_*(L') & \xrightarrow{\epsilon(\pi',L',D')} & A(\pi', L', D'). \end{array}$$

It is sufficient to show that, up to sign, $\beta_{L,D}$ does not depend on the choice of D. Let E be another non-zero effective divisor on C_S such that E is flat over S and $H^0(L_s - E_s) = 0$ for all $s \in S$. If we can show $\beta_{L,D} = \pm\beta_{L,D+E}$ and $\beta_{L,E} = \pm\beta_{L,D+E}$, then $\beta_{L,D} = \pm\beta_{L,E}$, so without loss of generality, it is sufficient to see $\beta_{L,D} = \pm\beta_{L,D+E}$. Here we consider the following diagram in a shape of a rectangular parallelepiped:

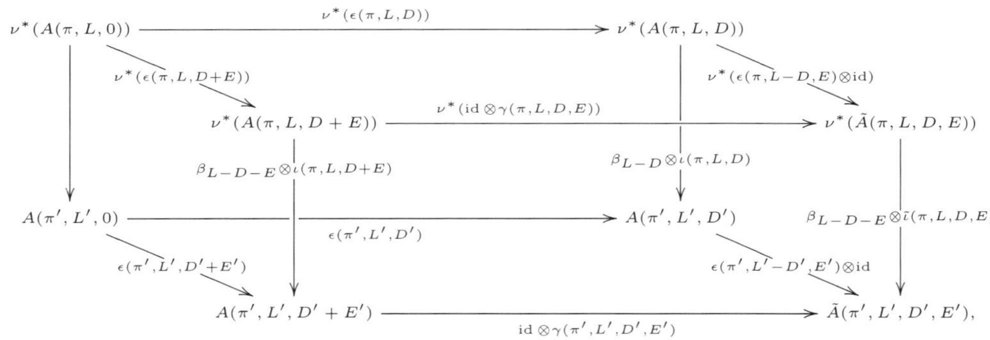

where $\tilde{\iota}(\pi, L, D, E)$ is the natural isomorphism

$$\nu^*(\det(\pi_*(L - D|_E)) \otimes \det(\pi_*(L|_D))) \xrightarrow{\sim} \det(\pi'_*(L' - D'|_{E'})) \otimes \det(\pi'_*(L'|_{D'})).$$

The upper face and the lower face are commutative up to sign by Claim 6. The upper homomorphism of the right face is derived from an exact sequence

$$0 \to \pi_*(L - D|_E) \to R^1\pi_*(L - D) \to R^1\pi_*(L - D - E) \to 0.$$

Moreover, the exact sequence obtained by taking ν^* of the above exact sequence is nothing more than

$$0 \to \pi'_*(L' - D'|_{E'}) \to R^1\pi'_*(L' - D') \to R^1\pi'_*(L' - D' - E') \to 0.$$

These observations show that the right face is commutative. The upper (resp. lower) homomorphism of the front face is derived from an exact sequence

$$0 \to \pi_*(L - D|_E) \to \pi_*(L|_{D+E}) \to \pi_*(L|_D) \to 0$$

(resp. an exact sequence obtained by taking ν^* of the above exact sequence). Thus the front face is commutative. Therefore, it is easy to see $\beta_{L,D} = \pm\beta_{L,D+E}$.

(2) Let E be a non-zero effective divisor on C_S such that E is flat and

$$H^0(L_s - E_s) = 0 \qquad (\forall s \in S).$$

Let us consider the rectangular parallelepiped in (1). The left homomorphism of the back face is β_L. We need to show that the back face is commutative up to sign. Since $H^0(L_s - D_s - E_s) = 0$ holds for all $s \in S$, we choose β_L and β_{L-D} such that the right and left faces are commutative up to sign. We can see that the front face is commutative in the same way as in (1). The upper and lower faces are commutative up to sign by Claim 6. Thus the back face is also commutative up to sign. $\square$

4.7. Faltings' Riemann-Roch theorem on arithmetic surfaces

Let g be an integer with $g \geq 1$ Let C be a connected compact Riemann surface of genus g. Note that $H^0(C, \Omega^1_C)$ has the natural hermitian inner product $\langle\ ,\ \rangle$ defined by

$$\langle \omega, \eta \rangle = \frac{\sqrt{-1}}{2} \int_C \omega \wedge \bar{\eta}.$$

Let $\{\omega_1, \ldots, \omega_g\}$ be an orthonormal basis of $H^0(C, \Omega^1_C)$ with respect to the above inner product. The canonical Kähler form Ω^{can} is defined by

$$\Omega^{can} = \frac{\sqrt{-1}}{2g} \sum_{i=1}^{g} \omega_i \wedge \bar{\omega}_i,$$

which does not depend on the choice of the orthonormal basis $\{\omega_1, \ldots, \omega_g\}$.

For an invertible sheaf L on C, the determinant bundle $\det \mathbf{R}\Gamma\,(C, L)$ of L is defined to be

$$\det \mathbf{R}\Gamma\,(C, L) := \det H^0(C, L) \otimes \left(\det H^1(C, L)\right)^{-1}.$$

This is the determinant bundle with respect to the structure morphism $\pi : C \to \operatorname{Spec}(\mathbb{C})$ viewed C as an algebraic curve. The following theorem due to Faltings [**18**] is the most important part for Faltings' Riemann-Roch formula on arithmetic surfaces.

THEOREM 4.29. *We can assign a hermitian metric* $h^{Fal}_{(L,h)}$ *of* $\det \mathbf{R}\Gamma\,(C, L)$ *to an admissible* C^∞*-hermitian invertible sheaf* (L, h) *with respect to* Ω^{can} *with the following properties. This metric* $h^{Fal}_{(L,h)}$ *is called the* Faltings metric *of* $\det \mathbf{R}\Gamma\,(C, L)$.

(F1) *The determinant bundle* $\det \mathbf{R}\Gamma\,(C, \mathscr{O}_C)$ *of* $\mathscr{O}_C$ *coincides naturally with*

$$\det H^0(C, \mathscr{O}_C) \otimes \det H^0(C, \Omega^1_C).$$

Thus the metric of $h^{Fal}_{(\mathscr{O}_C, h^{can})}$ *is given by the following way: We give the usual absolute value to* $\det H^0(C, \mathscr{O}_C) = \mathbb{C}$. *The hermitian inner product of* $H^0(C, \Omega^1_C)$ *given by*

$$\langle \omega, \eta \rangle = \frac{\sqrt{-1}}{2} \int_C \omega \wedge \bar{\eta} \qquad (\omega, \eta \in H^0(C, \Omega^1_C))$$

yields the metric of $\det H^0(C, \Omega^1_C)$. *Thus we can give the natural metric* $h^{Fal}_{(\mathscr{O}_C, h^{can})}$ *to* $\det \mathbf{R}\Gamma\,(C, \mathscr{O}_C)$

(F2) *An isometry* $(L_1, h_1) \xrightarrow{\sim} (L_2, h_2)$ *gives rise to an isometry*

$$(\det \mathbf{R}\Gamma\,(C, L_1)\,, h^{Fal}_{(L_1 h_1)}) \xrightarrow{\sim} (\det \mathbf{R}\Gamma\,(C, L_2)\,, h^{Fal}_{(L_2 h_2)}).$$

More precisely, an isomorphism $L_1 \xrightarrow{\sim} L_2$ *yields isomorphisms*

$$H^0(C, L_1) \xrightarrow{\sim} H^0(C, L_2) \quad \text{and} \quad H^1(C, L_1) \xrightarrow{\sim} H^1(C, L_2).$$

Thus we obtain a natural isomorphism

$$\det \mathbf{R}\Gamma\,(C, L_1) \xrightarrow{\sim} \det \mathbf{R}\Gamma\,(C, L_2)\,,$$

which gives rise to an isometry

$$(\det \mathbf{R}\Gamma\,(C, L_1)\,, h^{Fal}_{(L_1 h_1)}) \xrightarrow{\sim} (\det \mathbf{R}\Gamma\,(C, L_2)\,, h^{Fal}_{(L_2 h_2)}).$$

(F3) *For* (L,h) *and* $P \in C$, *an exact sequence*

$$0 \to H^0(C, L \otimes \mathscr{O}_C(-P)) \to H^0(C,L) \to L|_P \to H^1(C, L \otimes \mathscr{O}_C(-P)) \to H^1(C,L) \to 0$$

yields an isomerty

$$(\det \mathbf{R}\Gamma\,(C,L)\,, h^{Fal}_{(L,h)}) \simeq \left(\det \mathbf{R}\Gamma\,(C, L \otimes \mathscr{O}_C(-P))\,, h^{Fal}_{(L\otimes \mathscr{O}_C(-P), h\otimes h^{\Omega}_{-P})}\right) \otimes ((L,h)|_P),$$

where h^{Ω}_{-P} *is the canonical admissible metric of* $\mathscr{O}_C(-P)$ *with respect to the canonical Kähler form* Ω^{can}.

(F4) $h^{Fal}_{(L,ch)} = c^{\chi(C,L)} h^{Fal}_{(L,h)}$ *holds for a positive number* c.

The proof of Theorem 4.29 will be done in Section 4.9. Here we assume it and proceed with arguments.

In order to simplify technical aspects, let us introduce the concept of semistable curves. Let k be a field and C a 1-dimensional algebraic scheme over k. First we consider the case where k is an algebraically closed. We say that C is a *semistable curve* if C is reduced and, for any closed point x of C, the completion of the local ring at x is isomorphic to either $k[[x]]$ or $k[[x,y]]/(x \cdot y)$ over k. For a general field k, if $C \times_{\mathrm{Spec}(k)} \mathrm{Spec}(\bar{k})$ is a semistable curve, then the original C is said to be semistable, where $\bar{k}$ is the algebraic closure of k. Let C be a semistable curve over k and $\mu : C' \to C$ the normalization of C. A singular point x of C is said to be *split over* k if all of points in $\mu^{-1}(x)$ are k-rational. It is well known that there is a separable finite extension k' over k such that all singular points of $C \times_{\mathrm{Spec}(k)} \mathrm{Spec}(k')$ are split over k' (for example, see [**43**, Proposition 10.3.7 and Definition 10.3.8]).

More generally, let $\pi : X \to S$ be a morphism of noetherian schemes such that π is flat and of finite type. We say that $\pi : X \to S$ is a *semistable curve* if, for any $s \in S$, the fiber X_s over s is a semistable curve over $\kappa(s)$. If $X \to S$ is a semistable curve and $S' \to S$ is a morphism of noetherian schemes, then $X \times_S S' \to S'$ is also a semistable curve. Moreover, if S is a Dedekind scheme and the generic fiber $X \to S$ is normal, then X is normal (for example, see [**43**, Proposition 10.3.15]).

Let K be an algebraic number field and $\pi : X \to \mathrm{Spec}(O_K)$ a regular projective arithmetic surface. We assume that the genus g of the generic fiber is greater than or equal to 1. Let us fix the canonical Kähler form $\Omega^{can} = \{\Omega^{can}_\sigma\}$ of each X_σ. Let $\overline{L} = (L,h)$ be an admissible C^∞-hermitian invertible sheaf on X. For each $\sigma \in K(\mathbb{C})$, let $h^{Fal}_{\overline{L}_\sigma}$ be the Faltings' metric of $\det \mathbf{R}\Gamma\,(X_\sigma, L_\sigma)$. Let $h^{Fal}_{\overline{L}}$ denote the collection

$$\left\{ h^{Fal}_{\overline{L}_\sigma} \right\}_{\sigma \in K(\mathbb{C})}.$$

Then $\left(\det \mathbf{R}\pi_*\,(L)\,, h^{Fal}_{\overline{L}}\right)$ is a hermitian O_K-module. Here we define $\widehat{\chi}^{Fal}(\overline{L})$ by

$$\widehat{\chi}^{Fal}(\overline{L}) = \widehat{\deg}\left(\det \mathbf{R}\pi_*\,(L)\,, h^{Fal}_{\overline{L}}\right).$$

The following theorem is Faltings' Riemann-Roch theorem [**18**].

THEOREM 4.30 (Faltings' Riemann-Roch theorem). *We assume that* $X \to \mathrm{Spec}(O_K)$ *is a semistable curve. Then*

$$\widehat{\chi}^{Fal}(\overline{L}) = \frac{1}{2}\widehat{\deg}\left(\widehat{c}_1(\overline{L}) \cdot (\widehat{c}_1(\overline{L}) - \widehat{c}_1(\overline{\omega}_{X/O_K}))\right) + \widehat{\chi}^{Fal}(\mathscr{O}_X, h^{can})$$

for any admissible C^∞*-hermitian invertible sheaf* $\overline{L}$.

PROOF. We need several notations. For a divisor D on X and $c \in \mathbb{R}^{K(\mathbb{C})}$, the Arakelov divisor $\overline{D}(c)$ is defined by

$$\overline{D}(c) = \overline{D} + \sum_{\sigma \in K(\mathbb{C})} c(\sigma) F_\sigma$$

and let $\mathscr{O}_X(\overline{D}(c))$ denote the C^∞-hermitian invertible sheaf

$$\left(\mathscr{O}_X(D), \left\{ \exp(-2c(\sigma)) h^{\Omega_\sigma}_{D_\sigma} \right\}_{\sigma \in K(\mathbb{C})} \right)$$

associated with the Arakelov divisor $\overline{D}(c)$. Then

$$\widehat{c}_1(\mathscr{O}_X(\overline{D}(c))) = \overline{D}(c).$$

For details, see Section 4.3. Now we have the following lemma.

LEMMA 4.31. *We assume that D is effective. Then*

$$\widehat{\chi}^{Fal}(\overline{L}) - \widehat{\chi}^{Fal}(\overline{L} \otimes \mathscr{O}_X(-\overline{D}(c))) = \widehat{c}_1(\overline{L}) \cdot \overline{D}(c) - \frac{1}{2}\left((\overline{D}(c) + \widehat{c}_1(\overline{\omega}_{X/O_K})) \cdot \overline{D}(c) \right)$$

for any admissible C^∞-hermitian invertible sheaf $\overline{L}$.

PROOF. We denote the equation with respect to $\overline{L}$ and $\overline{D}(c)$ in the lemma by $E(\overline{L}, \overline{D}(c))$. Let us start the following claim.

CLAIM 7. *We assume that $E(\overline{L}, \overline{D}(c))$ holds. Moreover, $E(\overline{L} \otimes \mathscr{O}_X(-\overline{D}(c)), \overline{D'}(c'))$ holds for another effective divisor D' on X and $c' \in \mathbb{R}^{K(\mathbb{C})}$. Then $E(\overline{L}, \overline{D}(c) + \overline{D'}(c'))$ holds.*

PROOF. By an easy calculation, we can see that

$$\begin{aligned}
\widehat{c}_1(\overline{L}) \cdot \overline{D}(c) &- \frac{1}{2}\left((\overline{D}(c) + \widehat{c}_1(\overline{\omega}_{X/O_K})) \cdot \overline{D}(c) \right) \\
&+ (\widehat{c}_1(\overline{L}) - \overline{D}(c)) \cdot \overline{D'}(c') - \frac{1}{2}\left((\overline{D'}(c') + \widehat{c}_1(\overline{\omega}_{X/O_K})) \cdot \overline{D'}(c') \right) \\
&= \widehat{c}_1(\overline{L}) \cdot (\overline{D}(c) + \overline{D'}(c')) \\
&\qquad - \frac{1}{2}\left((\overline{D}(c) + \overline{D'}(c') + \widehat{c}_1(\overline{\omega}_{X/O_K})) \cdot (\overline{D}(c) + \overline{D'}(c')) \right).
\end{aligned}$$

Thus the claim follows. □

The above claim says that $E(\overline{L}, \overline{D}(c))$ holds in general if we can show the following special cases.

(i) $E(\overline{L}, aF_\sigma)$ holds for a general $\overline{L}$.
(ii) $E(\overline{L}, \overline{D})$ holds for a general $\overline{L}$ and a prime divisor D.

For a horizontal D in the case (ii), we would like to reduce to a simpler case. By Lemma 3.7, there is a finite extension field $K' \supseteq K$ with the following properties.

(a) K' contains the splitting field of the function field of D.
(b) Every singular point of any singular fiber $\pi' : X' = X \times_{\mathrm{Spec}(O_K)} \mathrm{Spec}(O_{K'}) \to \mathrm{Spec}(O_{K'})$ are split over the residue field at the point.

Since every singular point of any singular fiber $\pi' : X' = X \times_{\mathrm{Spec}(O_K)} \mathrm{Spec}(O_{K'}) \to \mathrm{Spec}(O_{K'})$ are split, it is easy to construct a resolution $\tau : X'' \to X'$ of singularities of X' and any exceptional set of τ is the chain of non-singular rational curves (for example, see [**43**, Lemma 10.3.21]). In particular, $R^1\tau_*(\mathscr{O}_{X''}) = 0$. Moreover, since X' is normal, $\tau_*(\mathscr{O}_{X''}) = \mathscr{O}_{X'}$ also holds. We denote morphisms $X' \to X$, $X'' \to X' \to X$ and $X'' \to \operatorname{Spec}(O_{K'})$ by ν, μ and π'', respectively, that is,

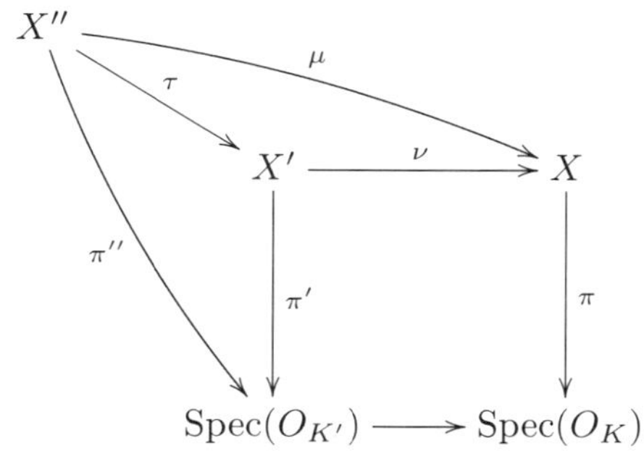

By the property (a), every horizontal component of $\mu^*(D)$ is a section. Moreover, $\pi''_*(\mu^*(L)) = \pi'_*(\nu^*(L))$ and $R^1\pi''_*(\mu^*(L)) = R^1\pi'_*(\nu^*(L))$. Thus $\det \mathbf{R}\pi'_*\,(\nu^*(L)) = \det \mathbf{R}\pi''_*\,(\mu^*(L))$ holds. On the other hand, by (1) in Proposition 1.32, we have $\det \mathbf{R}\pi'_*\,(\nu^*(L)) = \det \mathbf{R}\pi_*\,(L) \otimes_{O_K} O_{K'}$. Therefore, we obtain

$$\det \mathbf{R}\pi''_*\,(\mu^*(L)) = \det \mathbf{R}\pi_*\,(L) \otimes_{O_K} O_{K'},$$

which shows

$$\widehat{\chi}^{Fal}(\mu^*(\overline{L})) = [K' : K]\widehat{\chi}^{Fal}(\overline{L}).$$

In the same way, we can see

$$\widehat{\chi}^{Fal}(\mu^*(\overline{L} \otimes \mathscr{O}_X(-\overline{D}))) = [K' : K]\widehat{\chi}^{Fal}(\overline{L} \otimes \mathscr{O}_X(-\overline{D})).$$

Thus we get

$$\widehat{\chi}^{Fal}(\mu^*(\overline{L})) - \widehat{\chi}^{Fal}(\mu^*(\overline{L} \otimes \mathscr{O}_X(-\overline{D}))) = [K' : K]\left(\widehat{\chi}^{Fal}(\overline{L}) - \widehat{\chi}^{Fal}(\overline{L} \otimes \mathscr{O}_X(-\overline{D}))\right).$$

Moreover, by the construction, it is easy to see that

$$\overline{\omega}_{X''/O_{K'}} = \tau^*(\overline{\omega}_{X'/O_{K'}}) = \tau^*(\nu^*(\overline{\omega}_{X/O_K})) = \mu^*(\overline{\omega}_{X/O_K}),$$

and hence we have

$$\widehat{c}_1(\overline{\omega}_{X''/O_{K'}}) \cdot \mu^*(\overline{D}) = \widehat{c}_1(\mu^*(\overline{\omega}_{X/O_K})) \cdot \mu^*(\overline{D}).$$

Therefore, by Proposition 4.21,

$$\begin{aligned}\widehat{c}_1(\mu^*(\overline{L})) \cdot \mu^*(\overline{D}) - \frac{1}{2}\left(\mu^*(\overline{D})^2 + \widehat{c}_1(\overline{\omega}_{X''/O_{K'}}) \cdot \mu^*(\overline{D})\right) \\ = [K' : K]\left[\widehat{c}_1(\overline{L}) \cdot \overline{D} - \frac{1}{2}\left(\overline{D}^2 + \widehat{c}_1(\overline{\omega}_{X/O_K}) \cdot \overline{D}\right)\right].\end{aligned}$$

This means that we may assume that every horizontal component is a section. Therefore, using Claim 7, we may assume that D is a section.

By the above consideration, in order to show the lemma, it is sufficient to show the following cases for a general $\overline{L}$.

(1) $E(\overline{L}, aF_\sigma)$.
(2) $E(\overline{L}, \overline{D})$ for a vertical prime divisor D.
(3) $E(\overline{L}, \overline{D})$ for a section D.

(1) Let us consider $E(\overline{L}, aF_\sigma)$. In this case, the metric of $\overline{L} \otimes \mathscr{O}_X(-aF_\sigma)$ at σ is the multiplication of the metric of $\overline{L}$ at σ by $\exp(2a)$. Thus the Faltings metric of $\overline{L} \otimes \mathscr{O}_X(-aF_\sigma)$ at σ is the multiplication of the Faltings metric of $\overline{L}$ at σ by $\exp(2a)^{\chi(L_\sigma)}$. Therefore, we have

$$\widehat{\chi}^{Fal}(\overline{L}) - \widehat{\chi}^{Fal}(\overline{L} \otimes \mathscr{O}_X(-aF_\sigma)) = a\chi(L_\sigma).$$

On the other hand,

$$\overline{L} \cdot (aF_\sigma) - \frac{1}{2}\left((aF_\sigma)^2 + \overline{\omega}_{X/O_K} \cdot (aF_\sigma)\right) = \deg(L_\sigma)a - \frac{1}{2}\deg(\omega_{X_\sigma})a = a\chi(L_\sigma),$$

as required.

(2) Let us consider $E(\overline{L}, \overline{D})$ in the case where D is a vertical prime divisor. We set $\pi(D) = P \in \operatorname{Spec}(O_K)$. Note that $\overline{D} = (D, 0)$. Considering the exact sequence

$$\begin{aligned} 0 \to \pi_*(L \otimes \mathscr{O}_X(-D)) \to \pi_*(L) \to H^0(D, L|_D) \\ \to R^1\pi_*(L \otimes \mathscr{O}_X(-D)) \to R^1\pi_*(L) \to H^1(D, L|_D) \to 0, \end{aligned}$$

we obtain

$$\det \mathbf{R}\pi_*(L)) \otimes \det \mathbf{R}\pi_*(L \otimes \mathscr{O}(-D))^{-1} = \det(H^0(D, L|_D)) \otimes \det(H^1(D, L|_D))^{-1}.$$

Since the Faltings metric of $\det \mathbf{R}\pi_*(L)$ coincides with the Faltings metric of

$$\det \mathbf{R}\pi_*(L \otimes \mathscr{O}(-D)),$$

we have

$$\begin{aligned} \widehat{\chi}^{Fal}(\overline{L}) - \widehat{\chi}^{Fal}(\overline{L} \otimes \mathscr{O}_X(-\overline{D})) = \widehat{\deg}\left(\det(H^0(D, L|_D)), 0\right) \\ - \widehat{\deg}\left(\det(H^1(D, L|_D)), 0\right). \end{aligned}$$

Therefore, by (3.1) in Section 3.3,

$$\begin{aligned} \widehat{\chi}^{Fal}(\overline{L}) - \widehat{\chi}^{Fal}(\overline{L} \otimes \mathscr{O}_X(-\overline{D})) &= \log \#\left(H^0(D, L|_D)\right) - \log \#\left(H^1(D, L|_D)\right) \\ &= \log \#(O_K/P) \cdot \chi(D, L|_D). \end{aligned}$$

By using Lemma 4.2 and the geometric adjunction formula

$$(D + \omega_{X/O_K})\big|_D \simeq \omega_D$$

for D, we have

$$\left(\overline{D} + \overline{\omega}_{X/O_K} \cdot \overline{D}\right) = \log \#(O_K/P) \deg\left((D + \omega_{X/O_K})\big|_D\right) = \log \#(O_K/P) \deg(\omega_D).$$

Thus we get

$$-\frac{1}{2}\left(\overline{D} + \overline{\omega}_{X/O_K} \cdot \overline{D}\right) = \log \#(O_K/P) \cdot \chi(D, \mathscr{O}_D).$$

In the same way, $\overline{L} \cdot \overline{D} = \log \#(O_K/P) \deg(L|_D)$. Hence we obtain

$$\begin{aligned} \overline{L} \cdot \overline{D} - \frac{1}{2}\left(\overline{D} + \overline{\omega}_{X/O_K} \cdot \overline{D}\right) &= \log \#(O_K/P) \cdot (\deg(L|_D) + \chi(D, \mathscr{O}_D)) \\ &= \log \#(O_K/P) \cdot \chi(D, L|_D). \end{aligned}$$

Thus this case follows.

(3) Finally we consider $E(\overline{L},\overline{D})$ in the case where D is a section. The natural isomorphism $\det \mathbf{R}\pi_* (L) \xrightarrow{\sim} \det \mathbf{R}\pi_* (L \otimes \mathscr{O}_X(-D)) \otimes \pi_*(L|_D)$ derived from the exact sequence

$$0 \to \pi_*(L\otimes\mathscr{O}_X(-D)) \to \pi_*(L) \to \pi_*(L|_D) \to R^1\pi_*(L\otimes\mathscr{O}_X(-D)) \to R^1\pi_*(L) \to 0$$

yields the isometry

$$\left(\det \mathbf{R}\pi_* (L), h^{Fal}_{(L,h)}\right) \xrightarrow{\sim} \left(\det \mathbf{R}\pi_* (L \otimes \mathscr{O}_X(-D)), h^{Fal}_{(L\otimes\mathscr{O}_X(-D),h\otimes h^{\Omega}_{-D})}\right) \otimes \overline{L}\big|_D$$

by Theorem 4.29. Thus

$$\widehat{\chi}^{Fal}(\overline{L}) - \widehat{\chi}^{Fal}(\overline{L} \otimes \mathscr{O}_X(-\overline{D})) = \widehat{\deg}(\overline{L}\big|_D) = \overline{L}\cdot\overline{D}$$

holds. On the other hand, by the adjunction formula, we have

$$\left(\overline{D} + \widehat{c}_1(\overline{\omega}_{X/O_K})\right)\cdot\overline{D} = 0.$$

Thus case (3) follows. □

Let us go back to the proof of Theorem 4.30. For an admissible C^∞-hermitian invertible sheaf $\overline{L}$, there are effective divisors D_1 and D_2 on X and $c_1, c_2 \in \mathbb{R}^{K(\mathbb{C})}$ such that $\overline{L}$ is isometric to $\mathscr{O}_X\left(\overline{D}_1(c_1) - \overline{D}_2(c_2)\right)$. Using $E(\mathscr{O}_X(\overline{D}_1(c_1)), \overline{D}_1(c_1))$ and $E(\mathscr{O}_X(\overline{D}_1(c_1)), \overline{D}_2(c_2))$, we obtain

$$\begin{aligned}\widehat{\chi}^{Fal}(\mathscr{O}_X(\overline{D}_1(c_1))) &- \widehat{\chi}^{Fal}(\mathscr{O}_X, h^{can})\\ &= \overline{D}_1(c_1)\cdot\overline{D}_1(c_1) - \frac{1}{2}\left((\overline{D}_1(c_1) + \widehat{c}_1(\overline{\omega}_{X/O_K}))\cdot\overline{D}_1(c_1)\right)\end{aligned}$$

and

$$\begin{aligned}\widehat{\chi}^{Fal}(\mathscr{O}_X(\overline{D}_1(c_1))) &- \widehat{\chi}^{Fal}(\mathscr{O}_X(\overline{D}_1(c_1) - \overline{D}_2(c_2)))\\ &= \overline{D}_1(c_1)\cdot\overline{D}_2(c_2) - \frac{1}{2}\left((\overline{D}_2(c_2) + \widehat{c}_1(\overline{\omega}_{X/O_K}))\cdot\overline{D}_2(c_2)\right).\end{aligned}$$

Therefore,

$$\begin{aligned}\widehat{\chi}^{Fal}(\mathscr{O}_X(\overline{D}_1(c_1) - \overline{D}_2(c_2))) &- \widehat{\chi}^{Fal}(\mathscr{O}_X, h^{can})\\ &= \frac{1}{2}\left[\left(\overline{D}_1(c_1) - \overline{D}_2(c_2)\right)\cdot\left(\overline{D}_1(c_1) - \overline{D}_2(c_2) - \widehat{c}_1(\overline{\omega}_{X/O_K})\right)\right],\end{aligned}$$

as required. □

4.8. Determinant bundle and theta divisor

Let C be a connected compact Riemann surface of genus $g \geq 1$. First we recall the theta divisor on the Jacobian variety of C. Let $\mathrm{Pic}^n(C)$ denote the group of isomorphisms classes of invertible sheaves of degree n on C. In particular, $\mathrm{Pic}^0(C)$ is called the *Jacobian variety* of C and it is often denoted by J. Note that J is a g-dimensional abelian variety. We fix $\theta \in \mathrm{Pic}^1(C)$ with $\theta^{\otimes(2g-2)} \simeq \omega_C$. By using an isomorphism

$$\lambda_\theta^n : J \to \mathrm{Pic}^n(C) \quad (x \mapsto x\otimes\theta^n = x + n\theta)$$

by θ, we can identify $\operatorname{Pic}^n(C)$ with J. Let Θ be the locus in $\operatorname{Pic}^{g-1}(C)$ consisting of the classes of invertible sheaves with non-zero global sections, that is, Θ is the image of the following morphism:

$$\overbrace{C \times \cdots \times C}^{g-1 \text{ times}} \to \operatorname{Pic}^{g-1}(C) \quad ((P_1, \ldots, P_{g-1}) \mapsto \mathscr{O}_C(P_1 + \cdots + P_{g-1})).$$

In particular, Θ is an irreducible divisor on $\operatorname{Pic}^{g-1}(C)$. It is called the *theta divisor*. By using $\lambda_\theta^{g-1} : J \to \operatorname{Pic}^{g-1}(C)$, we define Θ_J by $\Theta_J = (\lambda_\theta^{g-1})^*(\Theta)$. For $a \in \operatorname{Pic}^1(C)$, we define the morphism $\varphi_a : C \to J$ to be $\varphi_a(x) = a - x$. By Abel's theorem, it is a closed embedding. We denote φ_θ by φ especially. There is an invertible sheaf L_n on $C \times \operatorname{Pic}^n(C)$, which is called a *Poincaré bundle*, such that $L_n|_{C\times\{c\}}$ is isomorphic to the invertible sheaf corresponding to c for any $c \in \operatorname{Pic}^n(C)$. Let us begin with the following proposition.

PROPOSITION 4.32. (1) *$\varphi_a^*(\mathscr{O}_J(\Theta_J)) \simeq \theta^{\otimes(g-1)} \otimes a$ for any $a \in \operatorname{Pic}^1(C)$.*

(2) *Let $q_i : J \times J \to J$ be the projection to the i-th factor. By abuse of notation, let $q_i : C \times C \to C$ be the same as above. Then*

$$\begin{aligned}(\varphi \times \varphi)^*(\mathscr{O}_{J\times J}((q_1 + q_2)^*(\Theta_J) - q_1^*(\Theta_J) - q_2^*(\Theta_J)))& \\ \simeq q_1^*(\theta) \otimes q_2^*(\theta) \otimes \mathscr{O}_C(-\Delta)&,\end{aligned}$$

where Δ is the diagonal of $C \times C$.

PROOF. In the following proof, we identify invertible sheaves with divisors, and the tensor product of invertible sheaves are denoted by the additive notation.

(1) First we consider the case where a is a general point. As

$$C^g \to \operatorname{Pic}^g(C) \quad ((P_1, \ldots, P_g) \mapsto \mathscr{O}_C(P_1 + \cdots + P_g))$$

is surjective, if a is a general point, then $h^0(a + (g-1)\theta) = 1$ and $|a + (g-1)\theta|$ consists of the sum of points of C, that is, there are unique points $P_1, \ldots, P_g$ up to order with $a + (g-1)\theta \sim P_1 + \cdots + P_g$. Thus

$$\begin{aligned} a - x \in \Theta_J &\iff a - x + (g-1)\theta \in \Theta \\ &\iff a - x + (g-1)\theta \sim Q_1 + \cdots + Q_{g-1} \quad (\exists Q_i \in C) \\ &\iff a + (g-1)\theta \sim x + Q_1 + \cdots + Q_{g-1} \quad (\exists Q_i \in C) \\ &\iff x \in \{P_1, \ldots, P_g\}. \end{aligned}$$

Therefore, $\varphi_a^{-1}(\Theta_J) = \{P_1, \ldots, P_g\}$, that is,

$$\varphi_a^*(\Theta_J) \sim P_1 + \cdots + P_g \sim a + (g-1)\theta.$$

Next we consider an arbitrary point $a \in \operatorname{Pic}^1(C)$. Let $\pi_0 : C \times \operatorname{Pic}^1(C) \to C$ be the projection to C and let L_1 be a Poincaré bundle on $C \times \operatorname{Pic}^1(C)$. Let $\Phi : C \times \operatorname{Pic}^1(C) \to J$ be the morphism given by $(x, a) \mapsto a - x$ and $M = \Phi^*(\mathscr{O}_J(\Theta_J)) - L_1 - (g-1)\pi_0^*(\theta)$. Then

$$M|_{C\times\{a\}} = \varphi_a^*(\mathscr{O}_J(\Theta_J)) - a - (g-1)\theta.$$

If a is a general point, then $h^0\left(M|_{C\times\{a\}}\right) \neq 0$. Thus, by the upper semicontinuity of the dimension of cohomology groups, for any $a \in \operatorname{Pic}^1(C)$, $h^0\left(M|_{C\times\{a\}}\right) \neq 0$. Note that $\deg\left(M|_{C\times\{a\}}\right) = 0$. Therefore, $M|_{C\times\{a\}} \simeq \mathscr{O}_C$, as required.

(2) Let

$P = \mathscr{O}_{J\times J}((q_1+q_2)^*(\Theta_J)-q_1^*(\Theta_J)-q_2^*(\Theta_J))$ and $N = q_1^*(\theta)+q_2^*(\theta)-\Delta-(\varphi\times\varphi)^*(P)$.

In order to see that $N \simeq \mathscr{O}_{C\times C}$, by the seesaw theorem (for example, [**55**]), it is sufficient to show that $N|_{\{x_0\}\times \mathscr{O}_C} \simeq \mathscr{O}_C$ and $N|_{C\times\{y_0\}} \simeq \mathscr{O}_C$ for all $x_0, y_0 \in C$.

Let us fix an arbitrary x_0. As $\varphi(x_0)+\varphi(x) = (2\theta - x_0) - x$, the composition of

$$C \xrightarrow{(\varphi\times\varphi)|_{\{x_0\}\times C}} J\times J \xrightarrow{q_1+q_2} J$$

is $\varphi_{2\theta - x_0}$. Thus, by (1),

$$(\varphi\times\varphi)^*(P)|_{\{x_0\}\times C} \sim \varphi_{2\theta-x_0}^*(\mathscr{O}_J(\Theta_J)) - \varphi^*(\mathscr{O}_J(\Theta_J)) \sim \theta - x_0.$$

On the other hand, $(q_1^*(\theta)+q_2^*(\theta)-\Delta)|_{\{x_0\}\times C} = \theta - x_0$. Therefore, $N|_{\{x_0\}\times C} \simeq \mathscr{O}_C$. In the same way, we have $N|_{C\times\{y_0\}} \simeq \mathscr{O}_C$. □

Let $\pi_0 : C\times \mathrm{Pic}^{g-1}(C) \to C$ be the projection to C and $\pi : C\times\mathrm{Pic}^{g-1}(C) \to \mathrm{Pic}^{g-1}(C)$ the projection to $\mathrm{Pic}^{g-1}(C)$. Let L be a Poincaré bundle on $C\times \mathrm{Pic}^{g-1}(C)$. We set $U = \mathrm{Pic}^{g-1}(C)\setminus\Theta$. As $H^0\left(L|_{C\times\{a\}}\right) = H^1\left(L|_{C\times\{a\}}\right) = 0$ for $a\in U$, we have

$$\pi_*(L)|_U = R^1\pi_*(L)\big|_U = 0.$$

Thus there is the natural isomorphism

$$\tau : \mathscr{O}_U \to \det \mathbf{R}\pi_*(L)|_U.$$

Then we have the following.

THEOREM 4.33. *There is an injective homomorphism*

$$\alpha : \det\mathbf{R}\pi_*(L) \to \mathscr{O}_{\mathrm{Pic}^{g-1}(C)}$$

with the following properties.

(1) Image $\left(\alpha : \det\mathbf{R}\pi_*(L) \to \mathscr{O}_{\mathrm{Pic}^{g-1}(C)}\right) = \mathscr{O}_{\mathrm{Pic}^{g-1}(C)}(-\Theta)$.

(2) *The composition $\alpha\circ\tau$ of homomorphisms*

$$\mathscr{O}_U \xrightarrow{\tau} \det\mathbf{R}\pi_*(L)|_U \xrightarrow{\alpha} \mathscr{O}_U$$

is the identity map.

PROOF. First, note that $\det\mathbf{R}\pi_*(L)$ does not depend on the choice of the Poincaré bundle. Indeed, if L' is another Poincaré bundle, then there is an invertible sheaf M on $\mathrm{Pic}^{g-1}(C)$ such that $L' \simeq L\otimes\pi^*(M)$. As $\chi(L_t) = 0$ for $t\in\mathrm{Pic}^{g-1}(C)$, by (2) in Proposition 1.32, we have $\det\mathbf{R}\pi_*(L') \simeq \det\mathbf{R}\pi_*(L)$.

First let us see the following: if there is an injective homomorphism

$$\alpha' : \det\mathbf{R}\pi_*(L) \to \mathscr{O}_{\mathrm{Pic}^{g-1}(C)}$$

satisfying the property (1), then we have the homomorphism satisfying (1) and (2). Let s be the rational section of $\det\mathbf{R}\pi_*(L)$ given by $\tau(1)$, and s' the rational section given by $(\alpha'|_U)^{-1}(1)$. Then there is a non-zero rational function f on $\mathrm{Pic}^{g-1}(C)$ with $s' = fs$. Since s and s' have no zero and no pole on U, f has also no zero and no pole on U. Thus there is an integer l with $(f) = l\Theta$ because Θ is irreducible. On the other hand, by (1) in Proposition 4.32, we have $l = 0$. Thus f is a constant. Therefore, if we set $\alpha = f\alpha'$, then α satisfies (1) and (2).

From now on, let us consider the construction of an injection homomorphism satisfying (1). Let us choose distinct g points $P_1, \ldots, P_g \in C$ with

$$h^0(\mathscr{O}_C(P_1 + \cdots + P_g)) = 1.$$

We set $E = P_1 + \cdots + P_g$ and $D = \pi_0^*(E)$. Then, $H^0(C, L_t \otimes \mathscr{O}_C(-E)) = 0$ for all $t \in \operatorname{Pic}^{g-1}(C)$. Moreover, as $\pi_*(L)$ is torsion free and zero at the generic point, we have $\pi_*(L) = 0$. Thus the exact sequence

$$0 \to \pi_*(L|_D) \to R^1\pi_*(L \otimes \mathscr{O}(-D)) \to R^1\pi_*(L) \to 0$$

yields the isomorphism

$$\det \mathbf{R}\pi_*(L) \xrightarrow{\sim} \det \pi_*(L|_D) \otimes \left(\det R^1\pi_*(L \otimes \mathscr{O}(-D))\right)^{-1}.$$

Let us denote the homomorphism $\pi_*(L|_D) \to R^1\pi_*(L \otimes \mathscr{O}(-D))$ by T. As $R^i\pi_*(L) = 0$ $(i \geq 2)$, by the base change theorem of cohomology groups (for example, see [**29**, Theorem 12.11 in Chapter III]), for all $t \in \operatorname{Pic}^{g-1}(C)$,

$$R^1\pi_*(L) \otimes \kappa(t) \to H^1(C, L_t)$$

is an isomorphism, so that, by the definition of the theta divisor Θ, T_t is an isomorphism if and only if $t \notin \Theta$. As Θ is irreducible, there is a non-negative integer a such that

$$\det(\pi_*(L|_D)) \otimes \mathscr{O}_{\operatorname{Pic}^{g-1}(C)}(a\Theta) \simeq \det(R^1\pi_*(L \otimes \mathscr{O}(-D))).$$

Therefore, it is enough to show $a = 1$.

Let $\phi_E : C \to \operatorname{Pic}^{g-1}(C)$ be the morphism given by $\phi_E(P) = \mathscr{O}_C(E - P)$. As $\phi_E = \lambda_\theta^{g-1} \circ \varphi_{E-(g-1)\theta}$, ϕ_E is an embedding and, by Proposition 4.32,

$$\deg(\phi_E^*(\Theta)) = \deg(\varphi_{E-(g-1)\theta}^*((\lambda_\theta^{g-1})^*(\Theta))) = \deg(\varphi_{E-(g-1)\theta}^*(\Theta_J)) = g.$$

By abuse of notation, let $\pi_0 : C \times C \to C$ and $\pi : C \times C \to C$ be the morphisms given by $\pi_0(x, y) = x$ and $\pi(x, y) = y$. Let M be the invertible sheaf on $C \times C$ given by $M = \mathscr{O}_{C \times C}(\pi_0^*(E) - \Delta)$, where Δ is the diagonal of $C \times C$. Let us consider the following commutative diagram:

$$\begin{array}{ccc} C \times C & \xrightarrow{\operatorname{id} \times \phi_E} & C \times \operatorname{Pic}^{g-1}(C) \\ {\scriptstyle\pi}\big\downarrow & & \big\downarrow{\scriptstyle\pi} \\ C & \xrightarrow[\phi_E]{} & \operatorname{Pic}^{g-1}(C). \end{array}$$

Note that this is the fiber product. As $M|_{C \times \{P\}} = \phi_E(P)$ and

$$(\operatorname{id} \times \phi_E)^*(L)|_{C \times \{P\}} = L|_{C \times \{\phi_E(P)\}},$$

there is an invertible sheaf Q on C such that $(\operatorname{id} \times \varphi_d)^*(L) \simeq M \otimes \pi^*(Q)$. Thus $\phi_E^*(\det \mathbf{R}\pi_*(L)) \simeq \det \mathbf{R}\pi_*(M)$. In order to see $a = 1$, it is sufficient to show $\deg(\det \mathbf{R}\pi_*(M)) = -g$.

By abuse of notation, we set $D = \pi_0^*(E)$ on $C \times C$. By the exact sequence

$$0 \to \pi_*(M \otimes \mathscr{O}(-D)) \to \pi_*(M) \to \pi_*(M|_D) \to R^1\pi_*(L \otimes \mathscr{O}(-D)) \to R^1\pi_*(M) \to 0,$$

we have

$$\deg(\det \mathbf{R}\pi_*(M)) = \deg(\det \mathbf{R}\pi_*(M \otimes \mathscr{O}(-D))) + \deg(\pi_*(M|_D)).$$

Here since $M|_D \simeq \mathscr{O}(-\Delta)|_D$, we have $\deg(\pi_*(M|_D)) = -g$. On the other hand, as $M \otimes \mathscr{O}(-D) = \mathscr{O}(-\Delta)$, the exact sequence

$$0 \to \pi_*(M \otimes \mathscr{O}(-D)) \to \pi_*(\mathscr{O}) \to \pi_*(\mathscr{O}_\Delta) \to R^1\pi_*(M \otimes \mathscr{O}(-D)) \to R^1\pi_*(\mathscr{O}) \to 0$$

yields $\deg(\det \mathbf{R}\pi_* (\mathscr{O})) = \deg(\det \mathbf{R}\pi_* (M \otimes \mathscr{O}(-D))) + \deg(\pi_*(\mathscr{O}_\Delta))$. Note that

$$\deg(\det \mathbf{R}\pi_* (\mathscr{O})) = \deg(\pi_*(\mathscr{O}_\Delta)) = 0,$$

and hence,

$$\deg(\det \mathbf{R}\pi_* (M \otimes \mathscr{O}(-D))) = 0.$$

Therefore, $\deg(\det \mathbf{R}\pi_* (M)) = -g$. □

Let us consider a generalization of Theorem 4.33. We would like to generalize ϕ_E in the last half of the proof of the above theorem. Let E be a divisor on C such that $\deg(E) \geq g$ and $h^0(\mathscr{O}_C(E)) = \deg(E) + 1 - g$. We write $r = \deg(E) + 1 - g$. Then $r \geq 1$. We set $C_0 = C_1 = \cdots = C_r = C$ and consider $C_0 \times C_1 \times \cdots \times C_r$. Let $\pi_{i_1\cdots i_s} : C_0 \times \cdots \times C_r \to C_{i_1} \times \cdots \times C_{i_s}$ be the projection to $C_{i_1} \times \cdots \times C_{i_s}$. For simplicity, $\pi_{1\cdots r}$ is denoted by π. We set

$$\tilde{D}_E = \pi_0^*(E) - \sum_{i=1}^{r} \pi_{0i}^* \text{ (the diagonal of } C_0 \times C_i).$$

For $P = (P_1, \ldots, P_r) \in C_1 \times \cdots C_r$, if $\tilde{D}_E\big|_{\pi^{-1}(P_1,\ldots,P_r)}$ is denoted by $\tilde{D}_{E,P}$, then $\tilde{D}_{E,P} = E - P_1 - \cdots - P_r$. Let

$$\phi_E : C_1 \times \cdots \times C_r \to \operatorname{Pic}^{g-1}(C)$$

be the morphism given by $\phi_E(P_1, \ldots, P_r) = E - P_1 - \cdots - P_r$. By the choice of E, $\phi_E(C_1 \times \cdots \times C_r) \not\subseteq \Theta$. We write

$$U_E = C_1 \times \cdots \times C_r \setminus \phi_E^{-1}(\Theta).$$

As $H^0(\mathscr{O}_C(\tilde{D}_{E,P})) = H^1(\mathscr{O}_C(\tilde{D}_{E,P})) = 0$ for $P \in U_E$,

$$\pi_*(\mathscr{O}(\tilde{D}_E))\Big|_{U_E} = R^1\pi_*(\mathscr{O}(\tilde{D}_E))\Big|_{U_E} = 0.$$

Thus we have the natural homomorphism

$$\tau_E : \mathscr{O}_{U_E} \to \det \mathbf{R}\pi_* \left(\mathscr{O}(\tilde{D}_E)\right)\Big|_{U_E}.$$

Then we have the following.

COROLLARY 4.34. *There is an injective homomorphism*

$$\alpha_E : \det \mathbf{R}\pi_* \left(\mathscr{O}(\tilde{D}_E)\right) \to \mathscr{O}_{C_1 \times \cdots \times C_r}$$

with the following properties.

(1) $\operatorname{Image}\left(\alpha_E : \det \mathbf{R}\pi_* \left(\mathscr{O}(\tilde{D}_E)\right) \to \mathscr{O}_{C_1 \times \cdots \times C_r}\right) = \mathscr{O}_{C_1 \times \cdots \times C_r}(-\phi_E^*(\Theta))$.

(2) *The composition* $\alpha_E \circ \tau_E$ *of* $\mathscr{O}_{U_E} \xrightarrow{\tau_E} \det \mathbf{R}\pi_* \left(\mathscr{O}(\tilde{D}_E)\right)\Big|_{U_E} \xrightarrow{\alpha_E} \mathscr{O}_{U_E}$ *is the identity map.*

PROOF. The following commutative diagram

$$\begin{array}{ccc} C_0 \times C_1 \times \cdots \times C_r & \xrightarrow{\mathrm{id}_C \times \phi_E} & C \times \mathrm{Pic}^{g-1}(C) \\ {\scriptstyle\pi}\downarrow & & \downarrow{\scriptstyle\pi} \\ C_1 \times \cdots \times C_r & \xrightarrow[\phi_E]{} & \mathrm{Pic}^{g-1}(C) \end{array}$$

is the fiber product. Thus there is an invertible sheaf N on $C_1 \times \cdots \times C_r$ such that $\mathscr{O}(\tilde{D}_E) \simeq (\mathrm{id}_C \times \phi_E)^*(L) \otimes \pi^*(N)$. Since

$$\pi_*((\mathrm{id}_C \times \phi_E)^*(L))|_{U_E} = R^1\pi_*((\mathrm{id}_C \times \phi_E)^*(L))\big|_{U_E} = 0,$$

there is the natural homomorphism $\tau'_E : \mathscr{O}_{U_E} \to \det \mathbf{R}\pi_* \left((\mathrm{id}_C \times \phi_E)^*(L)\right)|_{U_E}$. By (2) in Proposition 1.32, there is an isomorphism

$$\beta : \det \mathbf{R}\pi_* \left(\mathscr{O}(\tilde{D}_E)\right) \to \det \mathbf{R}\pi_* \left((\mathrm{id}_C \times \phi_E)^*(L)\right)$$

with $\beta(\tau_E(1)) = \tau'_E(1)$. On the other hand, by (1) in Theorem 4.28, there is an isomorphism

$$\gamma : \det \mathbf{R}\pi_* \left((\mathrm{id}_C \times \phi_E)^*(L)\right) \to \phi_E^*(\det \mathbf{R}\pi_* (L))$$

with $\gamma(\tau'_E(1)) = \pm\phi_E^*(\tau(1))$. Replacing γ by $-\gamma$ if necessary, we may assume $\gamma(\tau'_E(1)) = \phi_E^*(\tau(1))$. Thus, if we set $\delta = \gamma \circ \beta$, then

$$\delta : \det \mathbf{R}\pi_* \left(\mathscr{O}(\tilde{D}_E)\right) \xrightarrow{\sim} \phi_E^*(\det \mathbf{R}\pi_* (L))$$

and $\delta(\tau_E(1)) = \phi_E^*(\tau(1))$. Let $\alpha : \det \mathbf{R}\pi_* (L) \to \mathscr{O}_{\mathrm{Pic}^{g-1}(C)}$ be the homomorphism given in Theorem 4.33 and let α_E be the composition of the above δ and the pull-back $\phi_E^*(\alpha)$ of α, that is,

$$\alpha_E : \det \mathbf{R}\pi_* \left(\mathscr{O}(\tilde{D}_E)\right) \xrightarrow{\delta} \phi_E^*(\det \mathbf{R}\pi_* (L)) \xrightarrow{\phi_E^*(\alpha)} \mathscr{O}_{C_1 \times \cdots \times C_r}.$$

Then the corollary follows. □

4.9. Existence of Faltings' metric

In this section, we will give the proof of Theorem 4.29. Let C be a connected compact Riemann surface of genus $g \geq 1$ and Ω^{can} the canonical Kähler form of C. In the following, Ω^{can} is denoted by Ω for short. For an admissible C^∞-hermitian invertible sheaf (L, h) with respect to Ω, we would like to construct the unique Faltings metric $h^{Fal}_{(L,h)}$ of

$$\det \mathbf{R}\Gamma\,(C, L) = \det(H^0(C, L)) \otimes \det(H^1(C, L))^{-1}$$

satisfying (F1)–(F4) in Theorem 4.29.

Roughly speaking, the Faltings metric of an admissible C^∞-hermitian invertible sheaf can be constructed using (F1), (F3) and (F4). In particular, (F3) gives an inductive step for the construction.

Let (L, h) be an admissible C^∞-hermitian invertible sheaf on C. We assume that we have already constructed the Faltings metric $h^{Fal}_{(L,h)}$ of (L, h). Then

$$h^{Fal}_{(L\otimes\mathscr{O}_C(P), h\otimes h^{\Omega}_P)} \quad\text{and}\quad h^{Fal}_{(L\otimes\mathscr{O}_C(-P), h\otimes h^{\Omega}_{-P})}$$

can be constructed by using the following rules (4.12) and (4.13) according to (F3):

$$h^{Fal}_{(L\otimes\mathscr{O}_C(P),h\otimes h^{\Omega}_P)} = h^{Fal}_{(L,h)} \otimes (h\otimes h^{\Omega}_P)|_P, \tag{4.12}$$

$$h^{Fal}_{(L\otimes\mathscr{O}_C(-P),h\otimes h^{\Omega}_{-P})} = h^{Fal}_{(L,h)} \otimes (h|_P)^{-1}. \tag{4.13}$$

First, we need to show that the metric constructed by using the above rules does not depend on the order of chosen points. In order to treat it rigorously, let us consider a finite sequence of signed points such as $(P, -Q, -R)$. Let (L, h) be an admissible C^{∞}-hermitian invertible sheaf and $(Q_1, \ldots, Q_n)$ a finite sequence of signed points. We assume that we have already constructed the Faltings metric $h^{Fal}_{(L,h)}$ of (L, h). By using rules (4.12) and (4.13) in the order $Q_1 \to Q_2 \to \cdots \to Q_n$, we can construct the Faltings metric of $(L\otimes\mathscr{O}_C(Q_1+\cdots+Q_n), h\otimes h^{\Omega}_{Q_1+\cdots+Q_n})$. This is denoted by $h^{Fal}(Q_1,\ldots,Q_n)$. Clearly,

$$h^{Fal}(Q_1,\ldots,Q_i)(Q_{i+1},\ldots,Q_n) = h^{Fal}(Q_1,\ldots,Q_n).$$

Then we have the following.

LEMMA 4.35. *Let $(Q_1,\ldots,Q_n)$ and $(Q'_1,\cdots,Q'_{n'})$ be two finite sequences of signed points. If $Q_1+\cdots+Q_n = Q'_1+\cdots+Q'_{n'}$ as a divisor on C, then*

$$h^{Fal}(Q_1,\ldots,Q_n) = h^{Fal}(Q'_1,\cdots,Q'_{n'}).$$

PROOF. **Step 1.** First let us consider $h^{Fal}(Q_1,Q_2) = h^{Fal}(Q_2,Q_1)$. We need to take care of four cases. (I) $(P,-P)$, $(-P,P)$, (II) (P,Q), (Q,P), (III) $(-P,-Q)$, $(-Q,-P)$, (IV) $(P,-Q)$,$(-Q,P)$, where P and Q are distinct points.

(I) Let us consider $h^{Fal}(P,-P) = h^{Fal}(-P,P) = h^{Fal}_{(L,h)}$. Indeed, by the definition, we have

$$\begin{aligned} h^{Fal}(P,-P) &= h^{Fal}_{(L\otimes\mathscr{O}_C(P),h\otimes h^{\Omega}_P)} \otimes (h\otimes h^{\Omega}_P|_P)^{-1} \\ &= h^{Fal}_{(L,h)} \otimes (h\otimes h^{\Omega}_P|_P)\otimes(h\otimes h^{\Omega}_P|_P)^{-1} = h^{Fal}_{(L,h)} \end{aligned}$$

and

$$\begin{aligned} h^{Fal}(-P,P) &= h^{Fal}_{(L\otimes\mathscr{O}_C(-P),h\otimes h^{\Omega}_{-P})} \otimes (h\otimes h^{\Omega}_{-P}\otimes h^{\Omega}_P|_P) \\ &= h^{Fal}_{(L,h)} \otimes (h|_P)^{-1}\otimes(h|_P) = h^{Fal}_{(L,h)}. \end{aligned}$$

Thus

$$h^{Fal}(P,-P) = h^{Fal}(-P,P) = h^{Fal}_{(L,h)}.$$

(II) By using our rules, we obtain

$$\begin{aligned} h^{Fal}(P,Q) &= \left(h^{Fal}_{(L,h)} \otimes (h\otimes h^{\Omega}_P)|_P\right) \otimes (h\otimes h^{\Omega}_P\otimes h^{\Omega}_Q)|_Q \\ &= h^{Fal}_{(L,h)} \otimes h|_P\otimes h|_Q\otimes h^{\Omega}_P|_P\otimes h^{\Omega}_Q|_Q\otimes h^{\Omega}_P|_Q. \end{aligned}$$

Let h_{Δ} be the natural metric of $\mathscr{O}_{C\times C}(\Delta)$ as in Section 4.5. Then, since $h_{\Delta}|_{\{P\}\times C} = h^{\Omega}_P$, we have $h^{\Omega}_P|_Q = h_{\Delta}|_{(P,Q)}$. Therefore,

$$h^{Fal}(P,Q) = h^{Fal}_{(L,h)} \otimes h|_P\otimes h|_Q\otimes h^{\Omega}_P|_P\otimes h^{\Omega}_Q|_Q\otimes h_{\Delta}|_{(P,Q)}.$$

In the same way,

$$h^{Fal}(Q,P) = h^{Fal}_{(L,h)} \otimes h|_P\otimes h|_Q\otimes h^{\Omega}_P|_P\otimes h^{\Omega}_Q|_Q\otimes h_{\Delta}|_{(Q,P)}.$$

Since h_{Δ} is symmetric, $h_{\Delta}|_{(P,Q)} = h_{\Delta}|_{(Q,P)}$. Thus we get $h^{Fal}(P,Q) = h^{Fal}(Q,P)$.

(III) In order to see $h^{Fal}(-P,-Q) = h^{Fal}(-Q,-P)$, it is sufficient to see

$$h^{Fal}(-P,-Q,Q,P) = h^{Fal}(-Q,-P,Q,P).$$

Indeed, using (I) and (II),

$$\begin{cases} h^{Fal}(-P,-Q,Q,P) = h^{Fal}(-P,P) = h^{Fal}_{(L,h)}, \\ h^{Fal}(-Q,-P,Q,P) = h^{Fal}(-Q,-P,P,Q) = h^{Fal}_{(L,h)}. \end{cases}$$

(IV) It is sufficient to see $h^{Fal}(P,-Q,-P,Q) = h^{Fal}(-Q,P,-P,Q)$. By using (I) and (III), we have

$$\begin{cases} h^{Fal}(P,-Q,-P,Q) = h^{Fal}(P,-P,-Q,Q) = h^{Fal}_{(L,h)}, \\ h^{Fal}(-Q,P,-P,Q) = h^{Fal}_{(L,h)}. \end{cases}$$

Step 2. Next we consider the case where $n = n'$ and $(Q'_1, \ldots, Q'_n)$ is obtained by changing the order of $(Q_1, \ldots, Q_n)$. By Step 1,

$$h^{Fal}(Q_1, \ldots, Q_i, Q_{i+1}, \ldots, Q_n) = h^{Fal}(Q_1, \ldots, Q_{i+1}, Q_i, \ldots, Q_n),$$

so that Step 2 follows from Step 1.

Step 3. We consider a general case. We set

$$Q_1 + \cdots + Q_n = P_1 + \cdots + P_s - (P'_1 + \cdots + P'_t),$$

where P_i and P'_j are points of C and $\{P_1, \ldots, P_s\} \cap \{P'_1, \ldots, P'_t\} = \emptyset$. Then there are $R_1, \ldots, R_k \in C$ such that $(Q_1, \ldots, Q_n)$ can be rearranged as follows:

$$(P_1, \ldots, P_s, -P'_1, \ldots, -P'_t, R_1, -R_1, R_2, -R_2, \ldots R_k, -R_k),$$

that is,

$$(Q_{\tau(1)}, \ldots, Q_{\tau(n)}) = (P_1, \ldots, P_s, -P'_1, \ldots, -P'_t, R_1, -R_1, R_2, -R_2, \ldots R_k, -R_k)$$

for some permutation τ of $\{1, \ldots, n\}$. Therefore, using Step 1 and Step 2,

$$h^{Fal}(Q_1, \ldots, Q_n) = h^{Fal}(Q_{\tau(1)}, \ldots, Q_{\tau(n)}) = h^{Fal}(P_1, \ldots, P_s, -P'_1, \ldots, -P'_t).$$

In the same way,

$$h^{Fal}(Q'_1, \ldots, Q'_{n'}) = h^{Fal}(P_1, \ldots, P_s, -P'_1, \ldots, -P'_t).$$

Thus the lemma follows. □

By Lemma 4.35, we can define the Faltings metric $h^{Fal}_{(\mathscr{O}_C(D), h^{\Omega}_D)}$ of $(\mathscr{O}_C(D), h^{\Omega}_D)$ for a divisor D on C. Moreover, if $(\mathscr{O}_X(D), h)$ is admissible, then there is a positive constant c with $h = ch^{\Omega}_D$, so that we define $h^{Fal}_{(\mathscr{O}_X(D),h)}$ to be

$$h^{Fal}_{(\mathscr{O}_X(D),h)} := c^{\chi(\mathscr{O}_X(D))} h^{Fal}_{(\mathscr{O}_X(D), h^{\Omega}_D)}.$$

The most difficult part of the existence theorem of the Faltings metrics is to check the property (F2). In the meantime, we see that $h^{Fal}_{(\mathscr{O}_X(D),h)}$ satisfies (F3) and (F4).

PROPOSITION 4.36. *$h^{Fal}_{(\mathscr{O}_X(D),h)}$ satisfies* (F3) *and* (F4).

PROOF. Let h be an admissible C^∞-hermitian metric $\mathscr{O}_C(D)$. Then there is a positive constant a with $h = ah^\Omega_D$. First we consider (F3). By our construction,

$$\begin{aligned}&\left(\det \mathbf{R}\Gamma\left(\mathscr{O}_C(D)\right), h^{Fal}_{(\mathscr{O}_C(D),h^\Omega_D)}\right)\\ &\qquad\simeq \left(\det \mathbf{R}\Gamma\left(\mathscr{O}_C(D-P)\right), h^{Fal}_{(\mathscr{O}_C(D-P),h^\Omega_{D-P})}\right) \otimes \left(\left.(\mathscr{O}_C(D), h^\Omega_D)\right|_P\right).\end{aligned}$$

Therefore, we obtain

$$\begin{aligned}&\left(\det \mathbf{R}\Gamma\left(\mathscr{O}_C(D)\right), a^{\chi(\mathscr{O}_C(D))}h^{Fal}_{(\mathscr{O}_C(D),h^\Omega_D)}\right)\\ &\qquad\simeq \left(\det \mathbf{R}\Gamma\left(\mathscr{O}_C(D-P)\right), a^{\chi(\mathscr{O}_C(D))-1}h^{Fal}_{(\mathscr{O}_C(D-P),h^\Omega_{D-P})}\right)\\ &\qquad\qquad\otimes \left(\left.(\mathscr{O}_C(D), ah^\Omega_D)\right|_P\right),\end{aligned}$$

which means (F3) for $(\mathscr{O}_C(D), h)$.

Next we consider (F4). Since $(\mathscr{O}_C(D), ch) = (\mathscr{O}_C(D), cah^\Omega_D)$, we have

$$\left(\det \mathbf{R}\Gamma(L), h^{Fal}_{(L,h)}\right) \simeq \left(\det \mathbf{R}\Gamma\left(\mathscr{O}_C(D)\right), a^{\chi(L)}h^{Fal}_{(\mathscr{O}_C(D),h^\Omega_D)}\right)$$

and

$$\left(\det \mathbf{R}\Gamma(L), h^{Fal}_{(L,ch)}\right) \simeq \left(\det \mathbf{R}\Gamma\left(\mathscr{O}_C(D)\right), (ac)^{\chi(L)}h^{Fal}_{(\mathscr{O}_C(D),h^\Omega_D)}\right).$$

Thus (F4) follows. □

Let E be a divisor on C such that $\deg(E) \geq g$ and $h^0(\mathscr{O}_C(E)) = \deg(E)+1-g$. We set $r = \deg(E)+1-g$. As in the previous section, we set $C_0 = C_1 = \cdots = C_r = C$ and we consider $C_0 \times C_1 \times \cdots \times C_r$. Let $\pi_{i_1\cdots i_s} : C_0 \times \cdots \times C_r \to C_{i_1} \times \cdots \times C_{i_s}$ be the projection and $\pi = \pi_{1\cdots r}$. Moreover, we set

$$\tilde{D}_E = \pi_0^*(E) - \sum_{i=1}^r \pi_{0i}^*(\text{the diagonal of } C_0 \times C_i).$$

Let $\phi_E : C_1 \times \cdots \times C_r \to \operatorname{Pic}^{g-1}(C)$ be the morphism given by

$$\phi_E(P_1, \ldots, P_r) = E - P_1 - \cdots - P_r.$$

By Theorem 4.28, the fiber of $\det \mathbf{R}\pi_*\left(\mathscr{O}(\tilde{D}_E)\right)$ at $P = (P_1, \ldots, P_r)$ is naturally isomorphic to

$$\det \mathbf{R}\Gamma\left(\mathscr{O}_C(E - P_1 - \cdots - P_r)\right)$$

up to sign. Thus we can assign the Faltings metric

$$h^{Fal}_{(\mathscr{O}_C(E-P_1-\cdots-P_r),h^\Omega_{E-P_1-\cdots-P_r})}$$

to the fiber $\det \mathbf{R}\pi_*\left(\mathscr{O}(\tilde{D}_E)\right)$ at P. Note that the difference of sign does not matter to give the metric. In this way, we have a hermitian metric h^{Fal} of $\det \mathbf{R}\pi_*\left(\mathscr{O}(\tilde{D}_E)\right)$.

On the other hand, by Proposition 4.34, we have the isomorphism

$$\alpha_E : \det \mathbf{R}\pi_*\left(\mathscr{O}(\tilde{D}_E)\right) \xrightarrow{\sim} (\phi_E)^*(\mathscr{O}_{\operatorname{Pic}^{g-1}(C)}(-\Theta))$$

with the following properties. We set $U_E = C_1 \times \cdots \times C_r \setminus \phi_E^{-1}(\Theta)$. Note that

$$\left.\pi_*(\mathscr{O}(\tilde{D}_E))\right|_{U_E} = \left.R^1\pi_*(\mathscr{O}(\tilde{D}_E))\right|_{U_E} = 0.$$

Thus we have the canonical isomorphism $\tau_E : \mathscr{O}_{U_E} \to \det \mathbf{R}\pi_* \left(\mathscr{O}(\tilde{D}_E)\right)\Big|_{U_E}$. Then

$$\mathscr{O}_{U_E} \xrightarrow{\alpha_E \circ \tau_E} \mathscr{O}_{U_E}$$

is the identity map. The main theorem of this section is the following.

THEOREM 4.37. *There is a C^∞-hermitian metric h of $\mathscr{O}_{\mathrm{Pic}^{g-1}(C)}(-\Theta)$ such that α_E extends to the isometry*

$$(\det \mathbf{R}\pi_* \left(\mathscr{O}(\tilde{D}_E)\right), h^{Fal}) \xrightarrow{\sim} (\phi_E)^*(\mathscr{O}_{\mathrm{Pic}^{g-1}(C)}(-\Theta), h).$$

Moreover, h does not depend on E.

Assuming Theorem 4.37, we proceed with our argument.

PROPOSITION 4.38. *An isometry $(\mathscr{O}_C(D_1), h_1) \xrightarrow{\sim} (\mathscr{O}_C(D_2), h_2)$ induces*

$$(\det \mathbf{R}\Gamma\left(\mathscr{O}_C(D_1)\right), h^{Fal}_{(\mathscr{O}_C(D_1),h_1)}) \xrightarrow{\sim} (\det \mathbf{R}\Gamma\left(\mathscr{O}_C(D_2)\right), h^{Fal}_{(\mathscr{O}_C(D_2),h_2)}).$$

PROOF. Let us begin with the following.

CLAIM 8. *We assume that an isometry $(\mathscr{O}_C(D_1), h_1) \xrightarrow{\sim} (\mathscr{O}_C(D_2), h_2)$ induces an isometry*

$$(\det \mathbf{R}\Gamma\left(\mathscr{O}_C(D_1)\right), h^{Fal}_{(\mathscr{O}_C(D_1),h_1)}) \xrightarrow{\sim} (\det \mathbf{R}\Gamma\left(\mathscr{O}_C(D_2)\right), h^{Fal}_{(\mathscr{O}_C(D_2),h_2)}).$$

Then the isometries

$$\begin{aligned}(\det \mathbf{R}\Gamma&\left(\mathscr{O}_C(D_1 - P)\right), h^{Fal}_{(\mathscr{O}_C(D_1-P),h_1\otimes h^{\Omega}_{-P})}) \\ &\xrightarrow{\sim} (\det \mathbf{R}\Gamma\left(\mathscr{O}_C(D_2 - P)\right), h^{Fal}_{(\mathscr{O}_C(D_2-P),h_2\otimes h^{\Omega}_{-P})})\end{aligned}$$

and

$$\begin{aligned}(\det \mathbf{R}\Gamma&\left(\mathscr{O}_C(D_1 + P)\right), h^{Fal}_{(\mathscr{O}_C(D_1+P),h_1\otimes h^{\Omega}_{P})}) \\ &\xrightarrow{\sim} (\det \mathbf{R}\Gamma\left(\mathscr{O}_C(D_2 + P)\right), h^{Fal}_{(\mathscr{O}_C(D_2+P),h_2\otimes h^{\Omega}_{P})})\end{aligned}$$

are induced.

PROOF. We consider a commutative diagram:

$$\begin{array}{ccccccccccc} 0 \to & H^0(\mathscr{O}_C(D_1 - P)) & \to & H^0(\mathscr{O}_C(D_1)) & \to & \mathscr{O}_C(D_1)|_P & \to & H^1(\mathscr{O}_C(D_1 - P)) & \to & H^1(\mathscr{O}_C(D_1)) & \to 0 \\ & \downarrow\wr & & \downarrow\wr & & \downarrow\wr & & \downarrow\wr & & \downarrow\wr & \\ 0 \to & H^0(\mathscr{O}_C(D_2 - P)) & \to & H^0(\mathscr{O}_C(D_2)) & \to & \mathscr{O}_C(D_2)|_P & \to & H^1(\mathscr{O}_C(D_2 - P)) & \to & H^1(\mathscr{O}_C(D_2)) & \to 0 \end{array}$$

Then we have the natural commutative diagram:

$$\begin{array}{ccc} \det \mathbf{R}\Gamma\left(\mathscr{O}_C(D_1)\right) & \xrightarrow{\sim} & \det \mathbf{R}\Gamma\left(\mathscr{O}_C(D_1 - P)\right) \otimes \mathscr{O}_C(D_1)|_P \\ \downarrow\wr & & \downarrow\wr \\ \det \mathbf{R}\Gamma\left(\mathscr{O}_C(D_2)\right) & \xrightarrow{\sim} & \det \mathbf{R}\Gamma\left(\mathscr{O}_C(D_2 - P)\right) \otimes \mathscr{O}_C(D_2)|_P . \end{array}$$

Thus the first assertion follows from the above commutative diagram and (F3). The second assertion can be also shown in the same way. $\square$

Next we claim the following.

CLAIM 9. *Let D_1 and D_2 be divisors on C such that $\mathscr{O}_C(D_1) \simeq \mathscr{O}_C(D_2)$, $\deg(D_1) = \deg(D_2) = g-1$ and $h^0(\mathscr{O}_C(D_1)) = h^0(\mathscr{O}_C(D_2)) = 0$. Let h_1 and h_2 be admissible C^∞-hermitian metric of $\mathscr{O}_C(D_1)$ and $\mathscr{O}_C(D_2)$, respectively. Then the isomorphism $\det \mathbf{R}\Gamma\,(C, \mathscr{O}_C(D_1)) \overset{\sim}{\longrightarrow} \det \mathbf{R}\Gamma\,(C, \mathscr{O}_C(D_2))$ induced by the isomorphism $\mathscr{O}_C(D_1) \overset{\sim}{\longrightarrow} \mathscr{O}_C(D_2)$ extends an isometry*

$$(\det \mathbf{R}\Gamma\,(\mathscr{O}_C(D_1))\,, h^{Fal}_{(\mathscr{O}_C(D_1),h_1)}) \overset{\sim}{\longrightarrow} (\det \mathbf{R}\Gamma\,(\mathscr{O}_C(D_2))\,, h^{Fal}_{(\mathscr{O}_C(D_2),h_2)}).$$

PROOF. Note that

$$h^0(\mathscr{O}_C(D_1)) = h^0(\mathscr{O}_C(D_2)) = h^1(\mathscr{O}_C(D_1)) = h^1(\mathscr{O}_C(D_2)) = 0.$$

Thus $\det \mathbf{R}\Gamma\,(C, \mathscr{O}_C(D_1))$ and $\det \mathbf{R}\Gamma\,(C, \mathscr{O}_C(D_2))$ is naturally isomorphic to $\mathbb{C}$ and the isomorphism

$$\det \mathbf{R}\Gamma\,(C, \mathscr{O}_C(D_1)) \overset{\sim}{\longrightarrow} \det \mathbf{R}\Gamma\,(C, \mathscr{O}_C(D_2))$$

is the identity map from $\mathbb{C}$ to $\mathbb{C}$ independent from an isomorphism $\mathscr{O}_C(D_1) \overset{\sim}{\longrightarrow} \mathscr{O}_C(D_2)$.

We choose effective divisors D_1^+ and D_1^- with $D_1 = D_1^+ - D_1^-$. In the same way, we set $D_2 = D_2^+ - D_2^-$. Let P be a point of C. We set $E = D_1^+ + D_2^+ + nP$. Then $D_1 = E - (D_1^- + D_2^+ + nP)$ and $D_2 = E - (D_2^- + D_1^+ + nP)$. Moreover, if n is sufficiently large, then $\deg(E) \geq 2g-1$. Then $h^0(\mathscr{O}_C(E)) = \deg(E) + 1 - g$. Therefore, by Theorem 4.37, we have the isometry

$$(\det \mathbf{R}\Gamma\,(\mathscr{O}_C(D_1))\,, h^{Fal}_{(\mathscr{O}_C(D_1),h^{\Omega}_{D_1})}) \simeq (\det \mathbf{R}\Gamma\,(\mathscr{O}_C(D_2))\,, h^{Fal}_{(\mathscr{O}_C(D_2),h^{\Omega}_{D_2})}).$$

We can set $h_1 = c_1 h^{\Omega}_{D_1}$ and $h_2 = c_2 h^{\Omega}_{D_2}$ for some positive real numbers c_1 and c_2. By (F3),

$$\begin{cases} h^{Fal}_{(\mathscr{O}_C(D_1),h_1)} = h^{Fal}_{(\mathscr{O}_C(D_1),h^{\Omega}_{D_1})}, \\ h^{Fal}_{(\mathscr{O}_C(D_2),h_2)} = h^{Fal}_{(\mathscr{O}_C(D_2),h^{\Omega}_{D_2})}. \end{cases}$$

Thus the claim follows. □

Let us consider the proof of Proposition 4.38. We choose a suitable divisor E such that $\deg(D_1 + E) = g - 1$ and $h^0(\mathscr{O}_C(D_1 + E)) = 0$. Since

$$(\mathscr{O}_C(D_1 + E), h_1 \otimes h^{\Omega}_E) \overset{\sim}{\longrightarrow} (\mathscr{O}_C(D_2 + E), h_2 \otimes h^{\Omega}_E),$$

by using Claim 9, we have

$$\begin{aligned} &(\det \mathbf{R}\Gamma\,(\mathscr{O}_C(D_1 + E))\,, h^{Fal}_{(\mathscr{O}_C(D_1+E),h_1\otimes h^{\Omega}_E)}) \\ &\qquad \overset{\sim}{\longrightarrow} (\det \mathbf{R}\Gamma\,(\mathscr{O}_C(D_2 + E))\,, h^{Fal}_{(\mathscr{O}_C(D_2+E),h_2\otimes h^{\Omega}_E)}). \end{aligned}$$

Therefore, using Claim 8 inductively, we can see our conclusion. □

Let L be an invertible sheaf and h an admissible C^∞-hermitian metric of L. Let us define the Faltings metric $h^{Fal}_{(L,h)}$ of $\det \mathbf{R}\Gamma\,(C, L)$. We choose a divisor D and an admissible C^∞-hermitian metric k of $\mathscr{O}_C(D)$ such that there is an isometry $(L, h) \overset{\sim}{\longrightarrow} (\mathscr{O}_C(D), k)$. Let $\det \mathbf{R}\Gamma\,(C, L) \overset{\sim}{\longrightarrow} \det \mathbf{R}\Gamma\,(C, \mathscr{O}_C(D))$ be the isomorphism induced by the isomorphism $L \overset{\sim}{\longrightarrow} \mathscr{O}_C(D)$. Then $h^{Fal}_{(L,h)}$ is defined in such a way that the above isomorphism $\det \mathbf{R}\Gamma\,(C, L) \overset{\sim}{\longrightarrow} \det \mathbf{R}\Gamma\,(C, \mathscr{O}_C(D))$ extends to the isometry

$$(\det \mathbf{R}\Gamma\,(C, L)\,, h^{Fal}_{(L,h)}) \overset{\sim}{\longrightarrow} (\det \mathbf{R}\Gamma\,(C, \mathscr{O}_C(D))\,, h^{Fal}_{(\mathscr{O}_C(D),k)}).$$

We need to show $h^{Fal}_{(L,h)}$ does not depend on the choice of D and k. Indeed, let D' and k' be another divisor and metric such that $(L,h) \xrightarrow{\sim} (\mathscr{O}_C(D'), k')$. Then there is an isometry $(\mathscr{O}_C(D), k) \xrightarrow{\sim} (\mathscr{O}_C(D'), k')$ such that the following diagram is commutative:

$$\begin{array}{ccc} (\mathscr{O}_C(D), k) & \xrightarrow{\quad\sim\quad} & (\mathscr{O}_C(D'), k') \\ & \nwarrow^{\wr} \quad\quad {}^{\wr}\nearrow & \\ & (L, h) & \end{array}$$

Then, by Proposition 4.38, it induces the isometry

$$(\det \mathbf{R}\Gamma\left(\mathscr{O}_C(D)\right), h^{Fal}_{(\mathscr{O}_C(D),k)}) \simeq (\det \mathbf{R}\Gamma\left(\mathscr{O}_C(D')\right), h^{Fal}_{(\mathscr{O}_C(D'),k')}).$$

Moreover,

$$\begin{array}{ccc} \det \mathbf{R}\Gamma\left(C, \mathscr{O}_C(D)\right) & \xrightarrow{\quad\sim\quad} & \det \mathbf{R}\Gamma\left(C, \mathscr{O}_C(D')\right) \\ & \nwarrow^{\wr} \quad\quad {}^{\wr}\nearrow & \\ & \det \mathbf{R}\Gamma\left(C, L\right) & \end{array}$$

is commutative. Thus the metric of $\det \mathbf{R}\Gamma\left(C, L\right)$ induced by $h^{Fal}_{(\mathscr{O}_C(D'),k')}$ is the same as the metric induced by $h^{Fal}_{(\mathscr{O}_C(D),k)}$.

Finally, we need to see that $h^{Fal}_{(L,h)}$ satisfies the properties (F1)–(F4). This follows from Proposition 4.36 and Proposition 4.38.

Now we begin the proof of Theorem 4.37. For this purpose, several preparations are needed. We use the same notation on the Jacobi variety and the theta divisor of a connected compact Riemann surface as in the beginning of Section 4.8. Let us consider $\varphi_E : C_1 \times \cdots \times C_r \to J$ such that the following diagram is commutative:

$$\begin{array}{ccc} C_1 \times \cdots \times C_r & \xrightarrow{\varphi_E} & J \\ & \searrow^{\phi_E} & \downarrow{\scriptstyle \lambda_\theta^{g-1}} \\ & & \operatorname{Pic}^{g-1}(C) \end{array}$$

Namely,

$$\varphi_E(P_1, \ldots, P_r) = \varphi(P_1) + \cdots + \varphi(P_r) + (E - (r+g-1)\theta).$$

Let $\{\omega_1, \ldots, \omega_g\}$ be a basis of $H^0(C, \Omega^1_C)$ with

$$\frac{\sqrt{-1}}{2} \int_C \omega_i \wedge \bar{\omega}_j = \delta_{ij}.$$

Since $\varphi^* : H^0(J, \Omega^1_J) \xrightarrow{\sim} H^0(C, \Omega^1_C)$, if we set $\omega^J_i = (\varphi^*)^{-1}(\omega_i)$, then $\{\omega^J_1, \ldots, \omega^J_g\}$ forms a basis of $H^0(J, \Omega^1_J)$. Here we consider a Kähler form Ω_J on J given by

$$\Omega_J = \frac{\sqrt{-1}}{2g} \sum_{k=1}^{g} \omega^J_k \wedge \bar{\omega}^J_k.$$

We denote the projection from $C_1 \times \cdots \times C_r$ to $C_{i_1} \times \cdots \times C_{i_e}$ by $q_{i_1 \cdots i_e}$. Note that $p_{i_1 \cdots i_s}$ is the projection from $C_0 \times C_1 \times \cdots \times C_r$ to $C_{i_1} \times \cdots \times C_{i_s}$. Then we have the following.

LEMMA 4.39.

$$\varphi_E^*(\Omega_J) = \frac{\sqrt{-1}}{2g} \sum_{k=1}^{g} \left(\sum_{i=1}^{r} q_i^*(\omega_k) \right) \wedge \left(\sum_{j=1}^{r} q_j^*(\bar{\omega}_k) \right).$$

PROOF. First let us recall the following fact.

SUBLEMMA 4.40. *Let $\alpha, \beta : A \to B$ be a homomorphism of complex abelian varieties. Then, for any holomorphic 1-form ω on B, $(\alpha+\beta)^*(\omega) = \alpha^*(\omega) + \beta^*(\omega)$.*

PROOF. The proof is easy. □

Let us go back to the proof of the lemma. Note that ϕ_E is the composition of the following morphisms:

$$C^r \xrightarrow{\varphi \times \cdots \times \varphi} J^r \xrightarrow{q_1 + \cdots + q_r} J \xrightarrow{\text{translation}} J,$$

where $q_i : J^r \to J$ is the projection to the i-th factor by abuse of notation. Since a holomorphic 1-form is invariant by a translation, we have

$$\varphi_E^*(\omega_k^J) = q_1^*(\omega_k) + \cdots + q_r^*(\omega_k),$$

so that the lemma follows. □

Moreover, let us see the following lemma.

LEMMA 4.41. *There is a unique (up to a positive constant multiplication) C^∞-hermitian metric h'_J of $\mathscr{O}_J(-\Theta_J)$ with $c_1(\mathscr{O}_J(-\Theta_J), h'_J) = -g\Omega_J$.*

PROOF. Let η be the harmonic representative of the cohomology class of Θ. By Theorem 4.22, it is sufficient to see $\eta = g\Omega_J$. Note that $\{\omega_i^J \wedge \bar{\omega}_j^J\}_{i,j}$ is a basis of the space $\mathbf{H}^{1,1}(J)$ of harmonic $(1,1)$-forms with respect to Ω_J. Thus we can set

$$\eta = \sum_{i,j} a_{ij} \omega_i^J \wedge \bar{\omega}_j^J.$$

Let $r_i : J \times J \to J$ be the projection to the i-th factor. Here we consider the form

$$-(r_1 + r_2)^*(\eta) + r_1^*(\eta) + r_2^*(\eta)$$

on $J \times J$. Then, by Sublemma 4.40, we have

$$\begin{aligned}
&-(r_1 + r_2)^*(\eta) + r_1^*(\eta) + r_2^*(\eta) \\
&\qquad = -\sum_{i,j} a_{ij} \left(r_1^*(\omega_i^J) + r_2^*(\omega_i^J) \right) \wedge \left(r_1^*(\bar{\omega}_j^J) + r_2^*(\bar{\omega}_j^J) \right) \\
&\qquad\qquad + \sum_{i,j} a_{ij} r_1^*(\omega_i^J) \wedge r_1^*(\bar{\omega}_j^J) + \sum_{i,j} a_{ij} r_2^*(\omega_i^J) \wedge r_2^*(\bar{\omega}_j^J) \\
&\qquad\qquad = -\sum_{i,j} a_{ij} \left(r_1^*(\omega_i^J) \wedge r_2^*(\bar{\omega}_j^J) + r_2^*(\omega_i^J) \wedge r_1^*(\bar{\omega}_j^J) \right).
\end{aligned}$$

Therefore,

$$(\varphi \times \varphi)^*(-(r_1 + r_2)^*(\eta) + r_1^*(\eta) + r_2^*(\eta)) \\ = -\sum_{i,j} a_{ij} \left(r_1^*(\omega_i) \wedge r_2^*(\bar{\omega}_j) + r_2^*(\omega_i) \wedge r_1^*(\bar{\omega}_j)\right),$$

where the projection $C \times C \to C$ to the i-th factor is denoted by r_i by abuse of notation. On the other hand, by Proposition 4.32,

$$(\varphi \times \varphi)^*(-(r_1 + r_2)^*(\Theta_J) + r_1^*(\Theta_J) + r_2^*(\Theta_J))$$

is linearly equivalent to

$$\Delta - r_1^*(\theta) - r_2^*(\theta).$$

Thus, by Lemma 4.23, we can see $a_{ij} = \frac{\sqrt{-1}}{2}\delta_{ij}$, and hence $\eta = g\Omega_J$. □

Next let us calculate $c_1\left(\det \mathbf{R}\pi_*\left(\mathscr{O}(\tilde{D}_E)\right), h^{Fal}\right)$. First, we need the following lemma.

LEMMA 4.42. *There is an isometry*

$$\left(\det \mathbf{R}\pi_*\left(\mathscr{O}(\tilde{D}_E)\right), h^{Fal}\right) \otimes \bigotimes_{i=1}^{r} q_i^*(\mathscr{O}_C(E), h_E^{\Omega}) \otimes \bigotimes_{1\le i<j\le r} q_{ji}^*(\mathscr{O}(\Delta), h_\Delta)^{-1} \\ \simeq \left(\det \mathbf{R}\Gamma\left(\mathscr{O}_C(E)\right), h^{Fal}_{(\mathscr{O}_C(E),h_E^{\Omega})}\right) \otimes \mathscr{O}_{C_1\times\cdots\times C_r},$$

where h_Δ is the canonical metric of $\mathscr{O}_{C\times C}(\Delta)$ induced by Ω.

PROOF. For simplicity, we set $\Delta_{ij} = \pi_{ij}^*(\Delta)$ and $h_{\Delta_{ij}} = \pi_{ij}^*(h_\Delta)$. Here we have the following claim.

CLAIM 10. (1) $\pi_*(\mathscr{O}(\pi_0^*(E))|_{\Delta_{0i}}) = q_i^*(\mathscr{O}(E))$ *for* $i \ge 1$.
(2) $\pi_*(\mathscr{O}(\Delta_{0i})|_{\Delta_{0j}}) = q_{ji}^*(\mathscr{O}(\Delta))$ *for* $1 \le i \ne j \le r$.

PROOF. (1) As $\pi_0 = \pi_i$ on Δ_{0i} and $q_i \circ \pi = \pi_i$, we have

$$\pi_*(\mathscr{O}(\pi_0^*(E))|_{\Delta_{0i}}) = \pi_*(\pi_i^*(\mathscr{O}(E))|_{\Delta_{0i}}) = \pi_*(\pi^*(q_i^*(\mathscr{O}(E)))|_{\Delta_{0i}}) = q_i^*(\mathscr{O}(E)).$$

(2) Note that $\pi_{0i} = \pi_{ji}$ on Δ_{0j} and $q_{ji} \circ \pi = \pi_{ji}$. Thus

$$\pi_*(\mathscr{O}(\Delta_{0i})|_{\Delta_{0j}}) = \pi_*(\pi_{ji}^*(\mathscr{O}(\Delta))\big|_{\Delta_{0j}}) = \pi_*(\pi^*(q_{ji}^*\mathscr{O}(\Delta))\big|_{\Delta_{0j}}) = q_{ji}^*(\mathscr{O}(\Delta)).$$

□

In general, we will see that

$$\left(\det \mathbf{R}\pi_*\left(\mathscr{O}(\pi_0^*(E) - \pi_{01}^*(\Delta) - \cdots - \pi_{0s}^*(\Delta))\right), h^{Fal}\right) \otimes \\ \bigotimes_{i=1}^{s} q_i^*(\mathscr{O}_C(E), h_E^{\Omega}) \otimes \bigotimes_{1\le i<j\le s} q_{ji}^*(\mathscr{O}(\Delta), h_\Delta)^{-1} \\ \simeq \left(\det \mathbf{R}\Gamma\left(\mathscr{O}_C(E)\right), h^{Fal}_{(\mathscr{O}_C(E),h_E^{\Omega})}\right) \otimes \mathscr{O}_{C_1\times\cdots\times C_r}$$

holds for $1 \le s \le r$. We prove this by induction on s.

First we consider the case where $s = 1$. By an exact sequence

$$0 \to \pi_*(\mathscr{O}(\pi_0^*(E) - \Delta_{01})) \to H^0(\mathscr{O}_C(E)) \otimes \mathscr{O}_{C_1\times\cdots\times C_r} \to \pi_*(\mathscr{O}(\pi_0^*(E))|_{\Delta_{01}}) \\ \to R^1\pi_*(\mathscr{O}(\pi_0^*(E) - \Delta_{01})) \to H^1(\mathscr{O}_C(E)) \otimes \mathscr{O}_{C_1\times\cdots\times C_r} \to 0,$$

we have the natural isomorphism

$$\epsilon : \det \mathbf{R}\Gamma\left(\mathscr{O}_C(E)\right)\otimes\mathscr{O}_{C_1\times\cdots\times C_r} \overset{\sim}{\longrightarrow} \det \mathbf{R}\pi_*\left(\mathscr{O}(\pi_0^*(E)-\Delta_{01})\right)\otimes\pi_*(\mathscr{O}(\pi_0^*(E))|_{\Delta_{01}}).$$

Therefore, by Claim 10, ϵ is

$$\det \mathbf{R}\Gamma\left(\mathscr{O}_C(E)\right)\otimes\mathscr{O}_{C_1\times\cdots\times C_r} \overset{\sim}{\longrightarrow} \det \mathbf{R}\pi_*\left(\mathscr{O}(\pi_0^*(E)-\Delta_{01})\right)\otimes q_1^*(\mathscr{O}(E)).$$

For $P=(P_1,\ldots,P_r)\in C_1\times\cdots\times C_r$, by (2) in Theorem 4.28, there is the natural isomorphism

$$\beta_P : \det \mathbf{R}\pi_*\left(\mathscr{O}(\pi_0^*(E)-\Delta_{01})\right)\otimes\kappa(P) \overset{\sim}{\longrightarrow} \det \mathbf{R}\Gamma\left(\mathscr{O}_C(E-P_1)\right)$$

up to sign such that a diagram

$$\begin{array}{ccc}
\det \mathbf{R}\Gamma\left(\mathscr{O}_C(E)\right)\otimes\mathscr{O}_{C_1\times\cdots\times C_r}\otimes\kappa(P) & \xrightarrow{\epsilon\otimes\mathrm{id}} & \det \mathbf{R}\pi_*\left(\mathscr{O}(\pi_0^*(E)-\Delta_{01})\right)\otimes q_1^*(\mathscr{O}(E))\otimes\kappa(P) \\
\Big\| & & \Big\downarrow{\scriptstyle \beta_P\otimes\mathrm{id}} \\
\det \mathbf{R}\Gamma\left(\mathscr{O}_C(E)\right) & \xrightarrow{\epsilon'} & \det \mathbf{R}\Gamma\left(\mathscr{O}_C(E-P_1)\right)\otimes\mathscr{O}_C(E)|_{P_1}
\end{array}$$

is commutative up to sign, where ϵ' is the isomorphism induced by the following exact sequence.

$$\begin{aligned}
0\to H^0(\mathscr{O}_C(E-P_1))\to H^0(\mathscr{O}_C(E))\to \mathscr{O}_C(E)|_{P_1}& \\
\to H^1(\mathscr{O}_C(E-P_1))\to H^1(\mathscr{O}_C(E))\to 0.&
\end{aligned}$$

Thus, by (F3), the above isomorphism extends to the isometry

$$\begin{aligned}
&\left(\det \mathbf{R}\Gamma\left(\mathscr{O}_C(E)\right), h^{Fal}_{(\mathscr{O}_C(E),h)}\right)\otimes\mathscr{O}_{C_1\times\cdots\times C_r} \\
&\qquad\overset{\sim}{\longrightarrow}\left(\det \mathbf{R}\pi_*\left(\mathscr{O}(\pi_0^*(E)-\Delta_{01})\right), h^{Fal}\right)\otimes q_1^*(\mathscr{O}(E),h_E^{\Omega}).
\end{aligned}$$

Next we assume $s>1$. In the same way as the case $s=1$, using the exact sequence

$$\begin{aligned}
0\to \pi_*(\mathscr{O}(\pi_0^*(E)-\Delta_{01}-\cdots-\Delta_{0s}))\to \pi_*(\mathscr{O}(\pi_0^*(E)-\Delta_{01}-\cdots-\Delta_{0s-1}))& \\
\to \pi_*(\mathscr{O}(\pi_0^*(E)-\Delta_{01}-\cdots-\Delta_{0s-1})|_{\Delta_{0s}})& \\
\to R^1\pi_*(\mathscr{O}(\pi_0^*(E)-\Delta_{01}-\cdots-\Delta_{0s}))& \\
\to R^1\pi_*(\mathscr{O}(\pi_0^*(E)-\Delta_{01}-\cdots-\Delta_{0s-1}))\to 0&
\end{aligned}$$

together with Claim 10, Theorem 4.28 and (F3), we obtain the isometry

$$\begin{aligned}
&\left(\det \mathbf{R}\pi_*\left(\mathscr{O}(\pi_0^*(E)-\Delta_{01}-\cdots-\Delta_{0s-1})\right), h^{Fal}\right) \\
&\qquad\overset{\sim}{\longrightarrow}\left(\det \mathbf{R}\pi_*\left(\mathscr{O}(\pi_0^*(E)-\Delta_{01}-\cdots-\Delta_{0s})\right), h^{Fal}\right) \\
&\qquad\qquad\otimes q_s^*(\mathscr{O}_C(E),h_E^{\Omega})\otimes\bigotimes_{i=1}^{s-1} q_{si}^*(\mathscr{O}(\Delta),h_\Delta)^{-1}.
\end{aligned}$$

Thus, combining it with the hypothesis of induction, the lemma follows. □

LEMMA 4.43.

$$c_1\left(\det \mathbf{R}\pi_*\left(\mathscr{O}(\tilde{D}_E)\right), h^{Fal}\right)=\frac{-\sqrt{-1}}{2}\sum_{k=1}^{g}\left(\sum_{i=1}^{r}q_i^*(\omega_k)\right)\wedge\left(\sum_{j=1}^{r}q_j^*(\bar{\omega}_k)\right).$$

PROOF. Note that $c_1(\mathcal{O}_C(E), h^{\Omega}_E) = (r+g-1)\Omega$ and

$$c_1(\mathcal{O}_{C\times C}(\Delta), h^{\Omega}_{\Delta}) = r_1^*(\Omega) + r_2^*(\Omega) - \frac{\sqrt{-1}}{2}\sum_{k=1}^{g}\left(r_1^*(\omega_k)\wedge r_2^*(\bar{\omega}_k) + r_2^*(\omega_k)\wedge r_1^*(\bar{\omega}_k)\right),$$

where $r_i : C \times C \to C$ is the projection to the i-th factor. Thus, by Lemma 4.42, h^{Fal} is C^∞ and

$$\begin{aligned}
c_1&\left(\det \mathbf{R}\pi_*\left(\mathcal{O}(\tilde{D}_E)\right), h^{Fal}\right)\\
&= -\sum_{i=1}^{r} q_i^*(c_1(\mathcal{O}_C(E), h^{\Omega}_E)) + \sum_{1\le i<j\le r} q_{ji}^*(c_1(\mathcal{O}_{C\times C}(\Delta), h^{\Omega}_{\Delta}))\\
&= -(r+g-1)\sum_{i=1}^{r} q_i^*(\Omega) + \sum_{i<j}\Bigg[q_i^*(\Omega) + q_j^*(\Omega)\\
&\qquad - \frac{\sqrt{-1}}{2}\sum_{k=1}^{g}\left(q_i^*(\omega_k)\wedge q_j^*(\bar{\omega}_k) + q_j^*(\omega_k)\wedge q_i^*(\bar{\omega}_k)\right)\Bigg]\\
&= -g\sum_{i=1}^{r} q_i^*(\Omega) - \frac{\sqrt{-1}}{2}\sum_{i\neq j}\sum_{k=1}^{g} q_i^*(\omega_k)\wedge q_j^*(\bar{\omega}_k)\\
&= \frac{-\sqrt{-1}}{2}\left[\sum_{i=1}^{r}\sum_{k=1}^{g} q_i^*(\omega_k)\wedge q_i^*(\bar{\omega}_k) + \sum_{i\neq j}\sum_{k=1}^{g} q_i^*(\omega_k)\wedge q_j^*(\bar{\omega}_k)\right]\\
&= \frac{-\sqrt{-1}}{2}\sum_{k=1}^{g}\left(\sum_{i=1}^{r} q_i^*(\omega_k)\right)\wedge\left(\sum_{j=1}^{r} q_j^*(\bar{\omega}_k)\right).
\end{aligned}$$

□

PROOF OF THEOREM 4.37. Let us go back to the proof of Theorem 4.37. By Lemma 4.39, Lemma 4.41 and Lemma 4.43, we have

$$c_1\left(\det \mathbf{R}\pi_*\left(\mathcal{O}(\tilde{D}_E)\right), h^{Fal}\right) = \varphi_E^*(c_1(\mathcal{O}_J(-\Theta_J), h'_J)).$$

Therefore, if we set $h_J = ch'_J$ for a suitable positive constant c, then

$$\left(\det \mathbf{R}\pi_*\left(\mathcal{O}(\tilde{D}_E)\right), h^{Fal}\right) \simeq \varphi_E^*(\mathcal{O}_J(-\Theta_J), h_J).$$

Thus, if we choose a C^∞-hermitian metric h of $\mathcal{O}_{\mathrm{Pic}^{g-1}(C)}(-\Theta)$ with

$$(\lambda_\theta^{g-1})^*(\mathcal{O}_{\mathrm{Pic}^{g-1}(C)}(-\Theta), h) \simeq (\mathcal{O}_J(-\Theta_J), h_J),$$

then

$$\left(\det \mathbf{R}\pi_*\left(\mathcal{O}(\tilde{D}_E)\right), h^{Fal}\right) \simeq \phi_E^*(\mathcal{O}_{\mathrm{Pic}^{g-1}(C)}(-\Theta), h).$$

Finally we need to show that h does not depend on E. Let E' be another divisor. Then there are effective divisors D_1 and D_2 with $E + D_1 = E' + D_2$. Thus, it is sufficient to see that h does not depend on E and $E + P$ for a point P of C. Note that $h^0(\mathcal{O}_C(E+P)) = \deg(E+P) + 1 - g$. Indeed,

$$\begin{aligned}
\deg(E+P) + 1 - g &= \chi(\mathcal{O}_C(E+P)) \le h^0(\mathcal{O}_C(E+P))\\
&\le h^0(\mathcal{O}_C(E)) + 1 = (\deg(E) + 1 - g) + 1 = \deg(E+P) + 1 - g.
\end{aligned}$$

Let k be a C^∞-hermitian metric of $\mathscr{O}_{\mathrm{Pic}^{g-1}(C)}(-\Theta)$ by $E+P$. By our construction, there is a positive constant a with $k = ah$. Let $(P_1, \dots, P_r)$ be a point of $C_1 \times \cdots \times C_r$ with $\phi_E(P_1, \dots, P_r) \notin \Theta$. Then, as $\phi_{E+P}(P, P_1, \dots, P_r) = \phi_E(P_1, \dots, P_r)$, we have

$$\begin{aligned} h|_{\phi_E(P_1,\dots,P_r)}(1,1) &= h^{Fal}_{(\mathscr{O}_C(E-P_1-\cdots-P_r), h^{\Omega}_{E-P_1-\cdots-P_r})}(1,1) \\ &= h^{Fal}_{(\mathscr{O}_C((E+P)-P-P_1-\cdots-P_r), h^{\Omega}_{(E+P)-P-P_1-\cdots-P_r})}(1,1) \\ &= k|_{\phi_{E+P}(P,P_1,\dots,P_r)}(1,1) = a\, h|_{\phi_{E+P}(P,P_1,\dots,P_r)}(1,1) \\ &= a\, h|_{\phi_E(P_1,\dots,P_r)}(1,1), \end{aligned}$$

and hence $a = 1$. □

CHAPTER 5

Arakelov Geometry on General Arithmetic Varieties

In this section, we will discuss basic materials of Arakelov geometry including the intersection theory on general arithmetic varieties and the arithmetic Hilbert-Samuel formula. A rigorous treatment of the intersection theory for general intermediate-dimensional arithmetic cycles is algebraically and analytically very hard. Nevertheless, it is very rare to apply it to a concrete problem. In this book, we only consider the intersection cycles of arithmetic Cartier divisors with arithmetic cycles. The arithmetic Hilbert-Samuel formula is usually derived as an application of the arithmetic Riemann-Roch theorem [**26**] due to Gillet-Soulé. However, we only introduce the arithmetic Riemann-Roch theorem. Instead, we give a direct proof of the arithmetic Hilbert-Samuel formula assuming hard technical results (Bismut-Lebeau formula [**7**] and the asymptotic estimate of analytic torsions due to Bismut-Vasserot [**8**]). The arithmetic Hilbert-Samuel formula is sufficient for the purpose of this book. The Faltings Riemann-Roch formula on arithmetic surface, which is slightly different from Gillet-Soulé's version, can be found in Section 4. Moreover, the arithmetic characteristic classes and the Bott-Chern secondary characteristic forms will be discussed for the proof of the arithmetic Bogomolov inequality in Section 8. Besides them, the Gromov inequality and several kinds of the positivity of C^∞-hermitian invertible sheaves (for example, ample, nef and big) and so on will be considered.

5.1. Preliminaries on algebraic geometry and complex geometry

Before beginning a discussion on Arakelov geometry on general arithmetic varieties, let us recall several necessary facts on algebraic geometry and complex geometry.

5.1.1. Generic resolution of singularities. Let X be an integral scheme quasi-projective and flat over $\mathbb{Z}$, which will be called an arithmetic variety later. We say that X is generically smooth if X is smooth over $\mathbb{Q}$. Let $\pi : Y \to X$ be a projective and birational morphism of integral schemes quasi-projective and flat over $\mathbb{Z}$. The morphism $\pi : Y \to X$ is called a *generic resolution of singularities of* X if Y is generically smooth.

THEOREM 5.1. *Let X be an integral scheme quasi-projective and flat over $\mathbb{Z}$. Then there is a generic resolution of singularities $\pi : Y \to X$ of X.*

PROOF. Clearly we may assume that X is projective. Let $X_{\mathbb{Q}}$ be the generic fiber of $X \to \operatorname{Spec}(\mathbb{Z})$. Then, by Hironaka's theorem of resolution of singularities [**30**], there is a birational morphism $\pi_{\mathbb{Q}} : Y_{\mathbb{Q}} \to X_{\mathbb{Q}}$ over $\mathbb{Q}$ such that $Y_{\mathbb{Q}}$ is a smooth and projective variety over $\mathbb{Q}$. We can construct an integral scheme Y_1

projective and flat over $\mathbb{Z}$ such that the generic fiber of $Y_1 \to \operatorname{Spec}(\mathbb{Z})$ is $Y_{\mathbb{Q}}$. Then $\pi_{\mathbb{Q}} : Y_{\mathbb{Q}} \to X_{\mathbb{Q}}$ induces a birational map $\pi_1 : Y_1 \dashrightarrow X$. Therefore, if we take the graph of $\pi_1 : Y_1 \dashrightarrow X$, we obtain a generic resolution of singularities. □

5.1.2. Algebraic schemes over $\mathbb{R}$. Let X be a reduced algebraic scheme over $\mathbb{R}$. We denote by $X(\mathbb{C})$ the set of all $\mathbb{C}$-valued points of X, that is, the set of all scheme morphisms $\operatorname{Spec}(\mathbb{C}) \to X$ over $\operatorname{Spec}(\mathbb{R})$. To give a $\mathbb{C}$-valued point x of X is equivalent to give a scheme point p of X and a local ring homomorphism $\phi : \mathscr{O}_{X,p} \to \mathbb{C}$ over $\mathbb{R}$, so that, for $x \in X(\mathbb{C})$, the above p and ϕ are denoted by p_x and ϕ_x, respectively. Since $\mathscr{O}_{X,p_x} \to \mathbb{C}$ are a local ring homomorphism over $\mathbb{R}$, it induces two injective homomorphisms

$$\mathbb{R} \hookrightarrow \kappa(p_x) \quad \text{and} \quad \tilde{\phi}_x : \kappa(p_x) \hookrightarrow \mathbb{C}.$$

Thus p_x is a closed point of X and $\kappa(p_x)$ is isomorphic to either $\mathbb{R}$ or $\mathbb{C}$. The $\mathbb{C}$-valued point x is called a *real point* (resp. a *complex point*) if $\kappa(p_x)$ is isomorphic to $\mathbb{R}$ (resp. $\mathbb{C}$).

We set $X_{\mathbb{C}} = X \times_{\operatorname{Spec}(\mathbb{R})} \operatorname{Spec}(\mathbb{C})$. We denote by $X_{\mathbb{C}}(\mathbb{C})$ the set of all scheme morphisms $\operatorname{Spec}(\mathbb{C}) \to X_{\mathbb{C}}$ over $\operatorname{Spec}(\mathbb{C})$. Then a natural correspondence $x \mapsto (x, \mathrm{id}_{\operatorname{Spec}(\mathbb{C})})$ yields a bijection $X(\mathbb{C}) \to X_{\mathbb{C}}(\mathbb{C})$, that is, $X(\mathbb{C})$ has the natural complex space structure.

Let $-^a : \operatorname{Spec}(\mathbb{C}) \to \operatorname{Spec}(\mathbb{C})$ be a scheme morphism over $\mathbb{R}$ arising from the complex conjugation $- : \mathbb{C} \to \mathbb{C}$. For $x \in X(\mathbb{C})$, the composition $\operatorname{Spec}(\mathbb{C}) \xrightarrow{-^a} \operatorname{Spec}(\mathbb{C}) \xrightarrow{x} X$ is denoted by $\bar{x}$ and it is called the *complex conjugate point* of x. Note that $p_{\bar{x}} = p_x$ and $\phi_{\bar{x}} : \mathscr{O}_{X,p_{\bar{x}}} \to \mathbb{C}$ is the composition $\mathscr{O}_{X,p_x} \xrightarrow{\phi_x} \mathbb{C} \xrightarrow{-} \mathbb{C}$. The map $X(\mathbb{C}) \to X(\mathbb{C})$ given by $x \mapsto \bar{x}$ is denoted by F_∞, that is, $F_\infty(x) = \bar{x}$. If X is given by

$$X = \operatorname{Spec}(\mathbb{R}[T_1, \ldots, T_n]/(f_1(T_1, \ldots, T_n), \ldots, f_l(T_1, \ldots, T_n)),$$

then

$$X(\mathbb{C}) = \{(x_1, \ldots, x_n) \in \mathbb{C}^n \mid f_1(x_1, \ldots, x_n) = \cdots = f_l(x_1, \ldots, x_l) = 0\}$$

and $F_\infty(x_1, \ldots, x_n) = (\bar{x}_1, \ldots, \bar{x}_n)$. Thus it is easy to see that F_∞ is a C^∞-map in the sense of Section 1.14.

Let F be a coherent sheaf on X. For $x \in X(\mathbb{C})$, we set

$$F(x) = F_{p_x} \otimes_{\mathscr{O}_{X,p_x}} \mathbb{C} = (F_{p_x} \otimes_{\mathscr{O}_{X,p_x}} \kappa(p_x)) \otimes_{\kappa(p_x)} \mathbb{C},$$

where $\mathbb{C}$ is viewed as an $\mathscr{O}_{X,p_x}$-module via the local ring homomorphism $\phi_x : \mathscr{O}_{X,p_x} \to \mathbb{C}$, and it is also viewed as a $\kappa(p_x)$-module via $\tilde{\phi}_x : \kappa(p_x) \to \mathbb{C}$. For $s \in F_{p_x}$ and $\alpha \in \mathbb{C}$, $s \otimes^x \alpha$ means $s \otimes \alpha$ as an element of $F_{p_x} \otimes_{\mathscr{O}_{X,p_x}} \mathbb{C}$. We set $F_{\mathbb{C}} = F \otimes_{\mathbb{R}} \mathbb{C}$. Then $F_{\mathbb{C}}$ is a coherent sheaf on $X_{\mathbb{C}}$, so that it gives rise to a coherent sheaf on $X(\mathbb{C})$. By abuse of notation, it is also denoted by $F_{\mathbb{C}}$. Notice that $F(x)$ coincides with $F_{\mathbb{C},x} \otimes_{\mathscr{O}_{X(\mathbb{C}),x}} \kappa(x)$, where $\kappa(x)$ is the residue field of the local ring $\mathscr{O}_{X(\mathbb{C}),x}$. For $s \in F_{\mathbb{C},x}$, we denote $s \otimes_{\mathscr{O}_{X(\mathbb{C}),x}} 1$ in $F(x)$ by $s(x)$. By the universal property of tensor products, we have a homomorphism $\iota_{x,\bar{x}} : F(x) \to F(\bar{x})$ given by $s \otimes^x \alpha \mapsto s \otimes^{\bar{x}} \bar{\alpha}$. This is an anti-$\mathbb{C}$-linear map.

Since $H^0(X(\mathbb{C}), F_{\mathbb{C}}) = H^0(X, F) \otimes_{\mathbb{R}} \mathbb{C}$, we can define

$$F_\infty^* : H^0(X(\mathbb{C}), F_{\mathbb{C}}) \to H^0(X(\mathbb{C}), F_{\mathbb{C}})$$

by $F_\infty^* = \mathrm{id}_{H^0(X,F)} \otimes_{\mathbb{R}} {}^-$. Then, for $s \in H^0(X(\mathbb{C}), F_{\mathbb{C}})$ and $x \in X(\mathbb{C})$,

$$F_\infty^*(s)(x) = \iota_{\bar{x},x}(s(\bar{x})). \tag{5.1}$$

Indeed, if we set $s = \sum s_i \otimes \alpha_i$ ($s_i \in H^0(X, F)$, $\alpha_i \in \mathbb{C}$), then $F_\infty^*(s) = \sum s_i \otimes \bar{\alpha}_i$, so that

$$\begin{aligned} F_\infty^*(s)(x) &= \sum s_i \otimes^x \bar{\alpha}_i = \sum \iota_{\bar{x},x}(s_i \otimes^{\bar{x}} \alpha_i) \\ &= \iota_{\bar{x},x}\left(\sum s_i \otimes^{\bar{x}} \alpha_i\right) = \iota_{\bar{x},x}(s(\bar{x})). \end{aligned}$$

Let $h = \{h_x\}_{x \in X(\mathbb{C})}$ be a collection of hermitian metrics h_x of $F(x)$ at each point $x \in X(\mathbb{C})$. This collection h is said to be *of real type* if, for any $x \in X(\mathbb{C})$ and any $v, v' \in F(x)$,

$$h_x(v, v') = \overline{h_{\bar{x}}(\iota_{x,\bar{x}}(v), \iota_{x,\bar{x}}(v'))},$$

that is, for $s, s' \in F_{p_x}$,

$$h_x(s \otimes^x 1, s' \otimes^x 1) = \overline{h_{\bar{x}}(s \otimes^{\bar{x}} 1, s' \otimes^{\bar{x}} 1)}.$$

We define $F_\infty^*(h)$ to be

$$F_\infty^*(h)_x(s, s') = h_{\bar{x}}(\iota_{x,\bar{x}}(s), \iota_{x,\bar{x}}(s')).$$

Then h is of real type if and only if $h = \overline{F_\infty^*(h)}$ holds. Note that $F_\infty^*(\overline{F_\infty^*(h)}) = \bar{h}$. Thus

$$\frac{1}{2}(h + \overline{F_\infty^*(h)})$$

is of real type. Moreover, if h is of real type, then, by (5.1),

$$h_x(F_\infty^*(s)(x), F_\infty^*(s')(x)) = \overline{h_{\bar{x}}(s(\bar{x}), s'(\bar{x}))}. \tag{5.2}$$

A similar definition is possible for a collection $|\cdot| = \{|\cdot|_x\}_{x \in X(\mathbb{C})}$ of norms of $F(x)$ at each point $x \in X(\mathbb{C})$, that is, $|\cdot|$ is said to be of real type if $|v|_x = |\iota_{x,\bar{x}}(v)|_{\bar{x}}$ holds for all $v \in F(x)$ and $x \in X(\mathbb{C})$. Therefore, if $|\cdot|$ is of real type, then $|F_\infty^*(s)(x)|_x = |s(\bar{x})|_{\bar{x}}$, and hence

$$\sup_{x \in X(\mathbb{C})} \{|F_\infty^*(s)|_x\} = \sup_{x \in X(\mathbb{C})} \{|s(x)|_x\}. \tag{5.3}$$

Moreover, for $v \in F(x)$ and $x \in X(\mathbb{C})$, if we set

$$|v|'_x = \frac{1}{2}(|v|_x + |\iota_{x,\bar{x}}(v)|_{\bar{x}}),$$

then $|\cdot|' = \{|\cdot|'_x\}_{x \in X(\mathbb{C})}$ is of real type.

We assume that F is locally free and h is C^∞ in the sense of Section 1.14. Then $F_\infty^*(h)$ is also C^∞. Indeed, let s, s' be sections over an open set U of X and let g, g' be functions on U given by

$$\begin{cases} g(x) = h_x(s \otimes^x 1, s' \otimes^x 1), \\ g'(x) = F_\infty^*(h)_x(s \otimes^x 1, s' \otimes^x 1). \end{cases}$$

Then, since

$$g'(x) = h_{\bar{x}}(s \otimes^{\bar{x}} 1, s \otimes^{\bar{x}} 1),$$

we have $g' = g \circ F_\infty$. Note that F_∞ is a bijective C^∞-map. Thus g' is also C^∞.

By the above consideration, if h is C^∞, then

$$\frac{1}{2}(h + \overline{F_\infty^*(h)})$$

is a C^∞-hermitian metric of real type. In particular, $F_{\mathbb{C}}$ has a C^∞-hermitian metric of real type.

We assume that X is smooth over $\mathbb{R}$. Then $X(\mathbb{C})$ is a complex manifold. A (p,p)-form η on $X(\mathbb{C})$ is said to be of real type if $F_\infty^*(\eta) = (-1)^p\eta$ holds. Note that the concept of 'a form of real type' is different from the one of 'a real form' (i.e. a form η with $\eta = \bar{\eta}$).

We further assume that X is proper over $\mathbb{R}$. Then $X(\mathbb{C})$ is a compact complex manifold. Let Φ be a volume form of real type. Note that a volume form is a real form, so that it satisfies $\Phi = \bar{\Phi}$. Moreover, let F be a locally free coherent sheaf on X and $h = \{h_x\}_{x\in X(\mathbb{C})}$ a C^∞-hermitian metric of real type. Then we can see that

$$\int_{X(\mathbb{C})} h_x(s \otimes^x 1, s' \otimes^x 1)\Phi \in \mathbb{R}$$

for $s, s' \in H^0(X, F)$ in the following way: Let n be the dimension of X. Note that F_∞ preserve the orientation of $X(\mathbb{C})$ if n is even, and that F_∞ changes the orientation of $X(\mathbb{C})$ if n is odd. Therefore,

$$\int_{X(\mathbb{C})} h_x(s \otimes^x 1, s' \otimes^x 1)\Phi = (-1)^n \int_{X(\mathbb{C})} F_\infty^*(h_x(s \otimes^x 1, s' \otimes^x 1)\Phi).$$

On the other hand,

$$\begin{aligned}\int_{X(\mathbb{C})} F_\infty^*(h_x(s \otimes^x 1, s' \otimes^x 1)\Phi) &= \int_{X(\mathbb{C})} \overline{h_x(s \otimes^x 1, s' \otimes^x 1)} F_\infty^*(\Phi) \\ &= (-1)^n \int_{X(\mathbb{C})} \overline{h_x(s \otimes^x 1, s' \otimes^x 1)}\Phi \\ &= (-1)^n \int_{X(\mathbb{C})} \overline{h_x(s \otimes^x 1, s' \otimes^x 1)\Phi}.\end{aligned}$$

Thus

$$\int_{X(\mathbb{C})} h_x(s \otimes^x 1, s' \otimes^x 1)\Phi = \overline{\int_{X(\mathbb{C})} h_x(s \otimes^x 1, s' \otimes^x 1)\Phi},$$

which means that $\displaystyle\int_{X(\mathbb{C})} h_x(s \otimes^x 1, s' \otimes^x 1)\Phi \in \mathbb{R}$.

5.1.3. Norms and metrics. Let X be a d-dimensional compact complex manifold. Let $\overline{L} = (L, h)$ be a C^∞-hermitian invertible sheaf on X. The norm $\sqrt{h(\cdot,\cdot)}$ induced by h is denoted by $|\cdot|_{\overline{L}}$ or $|\cdot|$ for simplicity. We define an inner product $\langle\ ,\ \rangle$ of $H^0(X, L)$ induced by $\overline{L}$ and an volume form Φ on X to be

$$\langle s, s'\rangle := \int_X h(s, s')\Phi \qquad (s, s' \in H^0(X, L)).$$

We denote $\sqrt{\langle s, s\rangle}$ by $\|s\|_{L^2}^{\overline{L},\Phi}$, $\|s\|_{L^2}^{\overline{L}}$ or $\|s\|_{L^2}$ depending on the contexts. Moreover,

$$\sup\{|s|_{\overline{L}}(x) \mid x \in X\}$$

is dented by $\|s\|_{\sup}^{\overline{L}}$ or $\|s\|_{\sup}$. If $\overline{L}$ is positive (see Section 1.12), then

$$\frac{c_1(\overline{L})^{\wedge d}}{\displaystyle\int_X c_1(\overline{L})^{\wedge d}}$$

yields a volume form on X. This volume form is denoted by $\Phi(\overline{L})$.

We assume that there is a proper and smooth scheme X' over $\mathbb{R}$ with $X = X'(\mathbb{C})$. Moreover, we assume that $L = L'_{\mathbb{C}}$ for some invertible sheaf L' on X', h is of real type and that Φ is of real type. Then the above $\langle \cdot, \cdot \rangle$ gives rise to an $\mathbb{R}$-inner product of a real vector space $H^0(X', L')$. Further, if $\overline{L}$ is positive, then $\Phi(\overline{L})$ is of real type. See Section 5.1.2 for the definition 'of real type' in details.

5.1.4. Currents and Green currents. Let X be a d-dimensional complex manifold. Let $A^{p,q}(X)$ and $A_c^{p,q}(X)$ be the space of C^∞ (p,q)-forms on X and the space of C^∞ (p,q)-forms with compact support on X, respectively. Here we give Schwartz's topology on $A_c^{d-p,d-q}(X)$. A continuous $\mathbb{C}$-linear map $T : A_c^{d-p,d-q}(X) \to \mathbb{C}$ is called a *current of type* (p,q) on X. In this book, you do not need to care about the topology on $A_c^{d-p,d-q}(X)$. For details, see [**27**] or [**58**]. The space of all currents of type (p,q) is denoted by $D^{p,q}(X)$.

Let us consider several examples. Let ω be a C^∞ (p,q)-form on X. Then a natural map

$$[\omega] : A_c^{d-p,d-q}(X) \to \mathbb{C} \quad \left(\eta \mapsto \int_X \omega \wedge \eta\right)$$

yields a current of type (p,q). In this sense, $A^{p,q}(X)$ is naturally a subspace of $D^{p,q}(X)$. A (p,q)-form with locally integrable functions as coefficients also gives rise to a current of type (p,q).

Let Y be a codimension p reduced analytic subvariety of X. Then a *current of Dirac type* $\delta_Y \in D^{p,p}(X)$ arising from Y is given by

$$\delta_Y : A_c^{d-p,d-p}(X) \to \mathbb{C} \quad \left(\eta \mapsto \int_{Y_{\mathrm{reg}}} \eta\right),$$

where Y_{reg} is a non-singular locus of Y. More generally, for a codimension p cycle $Y = \sum n_i Y_i$, δ_Y is defined by $\delta_Y = \sum n_i \delta_{Y_i}$, that is,

$$\delta_Y(\eta) = \sum n_i \int_{Y_{i,\mathrm{reg}}} \eta.$$

Next we consider several operations on $D^{p,q}(X)$. For $T \in D^{p,q}(X)$, $\partial(T) \in D^{p+1,q}(X)$ and $\bar{\partial}(T) \in D^{p,q+1}(X)$ are defined by

$$\begin{cases} \partial(T)(\eta) = (-1)^{p+q+1} T(\partial(\eta)) & (\eta \in A_c^{d-(p+1),d-q}(X)), \\ \bar{\partial}(T)(\eta) = (-1)^{p+q+1} T(\bar{\partial}(\eta)) & (\eta \in A_c^{d-p,d-(q+1)}(X)). \end{cases}$$

By using Stokes' theorem, $\partial([\omega]) = [\partial(\omega)]$ and $\bar{\partial}([\omega]) = [\bar{\partial}(\omega)]$ hold for $\omega \in A^{p,q}(X)$. Moreover, we set

$$d^c := \frac{1}{4\pi\sqrt{-1}}(\partial - \bar{\partial}).$$

Then $dd^c = \frac{\sqrt{-1}}{2\pi}\partial\bar{\partial}$.

Let $f : X \to Y$ be a proper morphism of complex manifolds. Then, for $T \in D^{p,q}(X)$, we define $f_*(T) \in D^{p-\dim X+\dim Y, q-\dim X+\dim Y}(Y)$ to be

$$f_*(T)(\eta) = T(f^*(\eta)) \quad (\eta \in A_c^{\dim X-p,\dim X-q}(Y)).$$

The above $f_*(T)$ is called the *push-forward* of the current T. It is easy to see that

$$f_* \circ \partial = \partial \circ f_*, \qquad f_* \circ \bar{\partial} = \bar{\partial} \circ f_*. \tag{5.4}$$

Let Z be a codimension p cycle on X. Then $g \in D^{p-1,p-1}(X)$ is called a *Green Current* of Z if there is $\omega \in A^{p,p}(X)$ with

$$dd^c(g) + \delta_Z = [\omega].$$

As ω is uniquely determined by g, we denote ω by $\omega(Z, g)$ or $\omega(g)$.

For example, let X be a non-singular complex quasi-projective variety. Let (L, h) be a C^∞-hermitian invertible sheaf and s a non-zero rational section of L. By the Poincaré-Lelong formula (Theorem 1.39), $-\log h(s, s)$ is a Green current of $\mathrm{div}(s)$. In this case, ω is the first Chern form $c_1(L, h)$ of (L, h).

More generally, by [**23**, Theorem in 1.3.5] or [**64**, Theorem 3 in Chapter 2], a Green current exists for any cycle.

5.2. Intersection theory of Cartier divisors on excellent schemes

Let X be an excellent integral scheme. For the definition of excellent scheme, see [**43**, §8.2] or [**45**, §32]. Let $\mathrm{Rat}(X)$ be the field of rational functions on X, that is, the local ring at the generic point of X. Let $X^{(p)}$ denote the set of codimension p points (i.e. points x with $\dim \mathscr{O}_{X,x} = p$) of X. A free abelian group generated by $\left\{\overline{\{x\}}\right\}_{x \in X^{(p)}}$ is denoted by $Z^p(X)$ and its element is called a *cycle of codimension* p. For a cycle $Z = \sum_{x \in X^{(p)}} n_x \overline{\{x\}}$ of codimension p, we set

$$\mathrm{Supp}(Z) := \bigcup_{\substack{x \in X^{(p)} \\ n_x \neq 0}} \overline{\{x\}}.$$

Similarly, let $X_{(l)}$ denote the set of dimension l points (i.e. points x with $\dim \overline{\{x\}} = l$) of X. A free abelian group generated by $\left\{\overline{\{x\}}\right\}_{x \in X_{(l)}}$ is denoted by $Z_l(X)$ and its element is called a *cycle of dimension* l. If $p + l = \dim X$, then $X^{(p)} = X_{(l)}$ because X is excellent. In particular, $Z^p(X) = Z_l(X)$.

Let L be an invertible sheaf on X and s a non-zero rational section of L. See Section 1.7 for the definitions of $\mathrm{Supp}(L, s)$ and $\mathrm{div}(L, s)$. Let $Z = \sum_{i=1}^r n_i Z_i$ be a cycle of codimension p. We say that (L, s) *meets properly* with Z if

$$\mathrm{Supp}(L, s) \cap \mathrm{Supp}(Z) \cap X^{(p)} = \emptyset.$$

Under this assumption, the intersection cycle $(L, s) \cdot Z$ of (L, s) with Z is defined as follows: Let γ_i be the generic points of Z_i. The rational section s is a unit at γ_i because $\gamma_i \notin \mathrm{Supp}(L, s)$. Thus the class $s|_{Z_i}$ of s in $L_{\gamma_i}/m_{\gamma_i} L_{\gamma_i}$ is a non-zero element, that is, $s|_{Z_i}$ gives rise to a non-zero rational section of $L|_{Z_i}$, so that we defined $(L, s) \cdot Z$ to be

$$(L, s) \cdot Z = \sum_{i=1}^r n_i \,\mathrm{div}(L|_{Z_i}, s|_{Z_i}).$$

This is a cycle of codimension $p + 1$ on X.

PROPOSITION 5.2. *Let L_1 and L_2 be invertible sheaves on X, and let s_1 and s_2 be non-zero rational sections of L_1 and L_2, respectively. Then we have the following.*

(1) *If (L_1, s_1) and (L_2, s_2) meet properly with Z, that is,*

$$\mathrm{Supp}(L_1, s_1) \cap \mathrm{Supp}(Z) \cap X^{(p)} = \mathrm{Supp}(L_2, s_2) \cap \mathrm{Supp}(Z) \cap X^{(p)} = \emptyset,$$

then $(L_1 \otimes L_2, s_1 \otimes s_2) \cdot Z = (L_1, s_1) \cdot Z + (L_2, s_2) \cdot Z$.

(2) (Higher dimensional generalization of Weil's reciprocity law) *If*

$$\operatorname{Supp}(L_1, s_1) \cap \operatorname{Supp}(L_2, s_2) \cap X^{(1)} = \emptyset,$$

then

$$(L_1, s_1) \cdot \operatorname{div}(L_2, s_2) = (L_2, s_2) \cdot \operatorname{div}(L_1, s_1).$$

PROOF. (1) Since the order function is a homomorphism (cf. Section 1.3), we can see that

$$\operatorname{div}(L_1 \otimes L_2, s_1 \otimes s_2) = \operatorname{div}(L_1, s_1) + \operatorname{div}(L_2, s_2)$$

holds in general. Thus, if we set $Z = \sum_i n_i Z_i$, then

$$\begin{aligned}(L_1 \otimes L_2, s_1 \otimes s_2) \cdot Z &= \sum\nolimits_i n_i \operatorname{div}(L_1 \otimes L_2|_{Z_i}, s_1 \otimes s_2|_{Z_i}) \\ &= \sum\nolimits_i n_i \operatorname{div}(L_1|_{Z_i}, s_1|_{Z_i}) + \sum\nolimits_i n_i \operatorname{div}(L_2|_{Z_i}, s_2|_{Z_i}) \\ &= (L_1, s_1) \cdot Z + (L_2, s_2) \cdot Z.\end{aligned}$$

(2) This is a local problem, so that we may assume that $X = \operatorname{Spec}(A)$, $L_1 = L_2 = A$ and that s_1 and s_2 coincide with elements ϕ and ψ of the quotient field K of A, respectively. Moreover, by the definition of the intersection cycle, we may assume that A is a two-dimensional local excellent integral domain. Let m be the maximal ideal of A. The coefficient of $(A, \phi) \cdot \operatorname{div}(A, \psi)$ at m is

$$\sum_{P \in \operatorname{Supp}(\operatorname{div}(\psi)) \cap \operatorname{Spec}(A)^{(1)}} \operatorname{ord}_P(\psi) \operatorname{ord}_{m/P}(\phi|_{V(P)}),$$

where $V(P)$ is the closed set of X defined by P and $\phi|_{V(P)}$ is the class of ϕ in A_P/PA_P. The above coefficient is denoted by $I_m((A, \phi) \cdot \operatorname{div}(A, \psi))$. Similarly we can define $I_m((A, \psi) \cdot \operatorname{div}(A, \phi))$ as above. In order to get the proposition, it is sufficient to see

$$I_m((A, \phi) \cdot \operatorname{div}(A, \psi)) = I_m((A, \psi) \cdot \operatorname{div}(A, \phi))$$

under the assumption that $\operatorname{Supp}(A, \phi) \cap \operatorname{Supp}(A, \psi) \cap \operatorname{Spec}(A)^{(1)} = \emptyset$. First let us see the following claim.

CLAIM 1. *There is $a \in A \setminus \{0\}$ such that $a\phi \in A$ and*

$$\operatorname{Supp}(A, a) \cap \operatorname{Supp}(A, \psi) \cap \operatorname{Spec}(A)^{(1)} = \emptyset.$$

PROOF. We set $I = \{a \in A \mid a\phi \in A\}$. Then $\operatorname{Supp}(A/I) \subseteq \operatorname{Supp}(A, \phi)$. Thus, for any $Q \in \operatorname{Supp}(A, \psi) \cap \operatorname{Spec}(A)^{(1)}$, we have $A_Q = I_Q$. In particular, $I \not\subseteq Q$. Since $\operatorname{Supp}(A, \psi) \cap \operatorname{Spec}(A)^{(1)}$ is a finite set of prime ideals,

$$I \setminus \bigcup_{Q \in \operatorname{Supp}(A,\psi) \cap \operatorname{Spec}(A)^{(1)}} Q \neq \emptyset$$

(for example, see [**44**, (1.B)]). Let a be an element of the above non-empty set. Note that $QA_Q \cap A = Q$. Thus a is a unit of A_Q for all $Q \in \operatorname{Supp}(A, \psi) \cap \operatorname{Spec}(A)^{(1)}$, which shows $\operatorname{Supp}(A, a) \cap \operatorname{Supp}(A, \psi) \cap \operatorname{Spec}(A)^{(1)} = \emptyset$. □

In the case where $\phi, \psi \in A$, by (6) in Lemma 1.7,

$$\begin{cases} I_m((A, \phi) \cdot \operatorname{div}(A, \psi)) = e_A(\phi, A/\psi A), \\ I_m((A, \psi) \cdot \operatorname{div}(A, \phi)) = e_A(\psi, A/\phi A). \end{cases}$$

Thus (2) follows from (5) in Lemma 1.7.

Next we consider the case where either $\phi \in A$ or $\psi \in A$. Without loss of generality, we may assume that $\psi \in A$. By the above claim, there is $a \in A \setminus \{0\}$ such that $a\phi \in A$ and $\mathrm{Supp}(A, a) \cap \mathrm{Supp}(A, \psi) \cap \mathrm{Spec}(A)^{(1)} = \emptyset$. As $\mathrm{Supp}(A, a\phi) \subseteq \mathrm{Supp}(A, a) \cup \mathrm{Supp}(A, \phi)$, we have $\mathrm{Supp}(A, a\phi) \cap \mathrm{Supp}(A, \psi) \cap \mathrm{Spec}(A)^{(1)} = \emptyset$. Therefore,

$$\begin{cases} I_m((A, a) \cdot \mathrm{div}(A, \psi)) = I_m((A, \psi) \cdot \mathrm{div}(A, a)), \\ I_m((A, a\phi) \cdot \mathrm{div}(A, \psi)) = I_m((A, \psi) \cdot \mathrm{div}(A, a\phi)). \end{cases}$$

Moreover,

$$\begin{cases} I_m((A, a\phi) \cdot \mathrm{div}(A, \psi)) = I_m((A, a) \cdot \mathrm{div}(A, \psi)) + I_m((A, \phi) \cdot \mathrm{div}(A, \psi)), \\ I_m((A, \psi) \cdot \mathrm{div}(A, a\phi)) = I_m((A, \psi) \cdot \mathrm{div}(A, a)) + I_m((A, \psi) \cdot \mathrm{div}(A, \phi)). \end{cases}$$

Thus (2) follows in this case.

Finally, we consider a general case. By the above claim, there is $a \in A \setminus \{0\}$ such that $a\phi \in A$ and $\mathrm{Supp}(A, a) \cap \mathrm{Supp}(A, \psi) \cap \mathrm{Spec}(A)^{(1)} = \emptyset$. Therefore, in the same way as before, using the commutativity for a, ψ and $a\phi, \psi$, we have (2). $\square$

Let $f : X \to Y$ be a proper morphism of excellent integral schemes. The *push-forward of cycle* $f_* : Z_l(X) \to Z_l(Y)$ by f is defined as follows: Let Γ be an l-dimensional integral subscheme of X and $f(\Gamma)$ the image of Γ by f, which is an integral subscheme of Y. Then

$$f_*(\Gamma) := \begin{cases} [\mathrm{Rat}(\Gamma) : \mathrm{Rat}(f(\Gamma))] f(\Gamma) & \text{if } \dim \Gamma = \dim f(\Gamma), \\ 0 & \text{if } \dim \Gamma > \dim f(\Gamma). \end{cases}$$

In general, for an l-dimensional cycle $Z = \sum_i n_i Z_i$, we define $f_*(Z)$ to be

$$f_*(Z) := \sum_i n_i f_*(Z_i).$$

PROPOSITION 5.3. *Let $f : X \to Y$ be a proper surjective morphism of excellent integral schemes. Then,*

$$f_* \mathrm{div}(\mathscr{O}_X, \phi) = \begin{cases} \mathrm{div}(\mathscr{O}_Y, N_{\mathrm{Rat}(X)/\mathrm{Rat}(Y)}(\phi)) & \textit{if } \dim X = \dim Y, \\ 0 & \textit{if } \dim X > \dim Y, \end{cases}$$

for $\phi \in \mathrm{Rat}(X)^\times$.

PROOF. First we consider the case where $\dim X = \dim Y$. Let L and K be the function fields of X and Y, respectively. Then L is a finite extension field over K. Since the problem is local with respect to Y, we may assume that Y is the spectrum of a 1-dimensional excellent local integral domain A, that is, $Y = \mathrm{Spec}(A)$.

Let us see that $f : X \to Y$ is a finite morphism. Let $\pi : \widetilde{X} \to X$ and $\mu : \widetilde{Y} \to Y$ be the normalizations of X and Y in L and K, respectively. Then there is a proper morphism $\tilde{f} : \widetilde{X} \to \widetilde{Y}$ such that the following diagram is commutative:

$$\begin{array}{ccc} X & \xleftarrow{\ \pi\ } & \widetilde{X} \\ {\scriptstyle f}\downarrow & & \downarrow{\scriptstyle \tilde{f}} \\ Y & \xleftarrow{\ \mu\ } & \widetilde{Y} \end{array}$$

Since X and Y are excellent, π and μ are finite. Moreover, $\tilde{Y}$ is a Dedekind scheme. Thus $\tilde{f}$ is flat. In particular, $\tilde{f}$ is finite. Therefore, f is also finite.

Hence, there is a 1-dimensional excellent integral domain B such that $X = \operatorname{Spec}(B)$ and B is finitely generated as an A-module. Therefore, the assertion of the proposition follows from (2) of Lemma 1.12.

Next we consider the case where $\dim X > \dim Y$.

We assume that there is an irreducible subscheme Γ of dimension $\dim X - 1$ with $\dim f(\Gamma) = \dim \Gamma$. Then

$$\dim X - 1 = \dim \Gamma = \dim f(\Gamma) \leq \dim Y \leq \dim X - 1,$$

and hence $\dim X = \dim Y + 1$ and $f(\Gamma) = Y$. Let X_η be the generic fiber of $f : X \to Y$. Then $f_* \operatorname{div}(\mathscr{O}_X, \phi) = \deg(\operatorname{div}(\mathscr{O}_{X_\eta}, \phi))Y$. Therefore, the assertion follows from Proposition 1.18.

Finally, we assume that there is no subscheme Γ as above. In this case,

$$f_* \operatorname{div}(\mathscr{O}_X, \phi) = 0$$

by the definition of the push-forward. □

COROLLARY 5.4. *Let $f : X \to Y$ be a proper surjective morphism of excellent integral schemes. Let L be an invertible sheaf on Y and s a non-zero rational rational section of L. Then*

$$f_* \operatorname{div}(f^*(L), f^*s) = \begin{cases} [\operatorname{Rat}(X) : \operatorname{Rat}(Y)] \operatorname{div}(L, s) & \textit{if } \dim X = \dim Y, \\ 0 & \textit{if } \dim X > \dim Y. \end{cases}$$

PROOF. This is a local problem with respect to Y. Thus we may assume that $L = \mathscr{O}_Y$ and s is a non-zero rational function. Therefore, the corollary follows from Proposition 5.3 and the well-known formula on norms, that is,

$$N_{L/K}(x) = x^{[L:K]} \qquad (x \in K),$$

where L/K is a finite field extension. □

PROPOSITION 5.5 (Projection formula). *Let $f : X \to Y$ be a proper morphism of excellent integral schemes. Let L be an invertible sheaf on Y and s a non-zero rational rational section of L. Let η be the generic point of X. We assume $f(\eta) \notin \operatorname{Supp}(L, s)$. Let f^*s be the image of s via the natural homomorphism $L_{f(\eta)} \to f^*(L)_\eta$. Then f^*s is a non-zero rational section of $f^*(L)$. Let Z be a cycle of dimension l on X with*

$$\operatorname{Supp}(f^*(L), f^*s) \cap \operatorname{Supp}(Z) \cap X_{(l)} = \emptyset.$$

Then $\operatorname{Supp}(L, s) \cap \operatorname{Supp}(f_*Z) \cap Y_{(l)} = \emptyset$ and the following projection formula holds:

$$f_*((f^*(L), f^*s) \cdot Z) = (L, s) \cdot f_*Z.$$

PROOF. First let us see $\operatorname{Supp}(f^*(L), f^*s) = f^{-1}(\operatorname{Supp}(L, s))$. As this is a local problem with respect to Y, we may assume there is a local basis ω of L with $s = a\omega$, where $a \in \operatorname{Rat}(Y)^\times$. For $x \in X$, the natural local ring homomorphism $\mathscr{O}_{Y, f(x)} \to \mathscr{O}_{X,x}$ is denoted by f_x^*. Then, since

$$\begin{aligned} x \notin \operatorname{Supp}(f^*(L), f^*s) &\iff f_x^*(a) \text{ is a unit at } x \iff a \text{ is a unit at } f(x) \\ &\iff f(x) \notin \operatorname{Supp}(L, s) \iff x \notin f^{-1}(\operatorname{Supp}(L, s)), \end{aligned}$$

we have $\operatorname{Supp}(f^*(L), f^*s) = f^{-1}(\operatorname{Supp}(L, s))$.

Let $\gamma_1, \ldots, \gamma_r$ be the generic points of components of Z. As

$$\operatorname{Supp}(f^*(L), f^*s) = f^{-1}(\operatorname{Supp}(L, s)),$$

we have $f(\gamma_i) \notin \mathrm{Supp}(L, s)$ for all i. In particular,

$$\mathrm{Supp}(L, s) \cap \mathrm{Supp}(f_* Z) \cap Y_{(l)} = \emptyset.$$

Finally, we consider the projection formula. Each side of the projection formula is linear with respect to Z. Thus we may assume that Z is integral. Let $g = f|_Z : Z \to f(Z)$ and γ the generic point of Z. Then, by the previous fact we have proved, $f(\gamma) \notin \mathrm{Supp}(L, s)$. Therefore, if we set $M = L|_{f(Z)}$, then $t = s|_{f(Z)}$ is a non-zero rational section of M. Thus

$$(f^*(L), f^*s) \cdot Z = \mathrm{div}\left(f^*(L)|_Z, f^*s|_Z \right) = \mathrm{div}(g^*(M), g^*t),$$

so that the projection formula follows from Corollary 5.4. □

5.3. Higher dimensional generalization of Weil's reciprocity law in complex geometry

In this section, we consider a higher dimensional generalization of Weil's reciprocity law in complex geometry, which is a generalization of Proposition 4.12.

Let X be a d-dimensional non-singular complex algebraic variety. Let (L, h) be a C^∞-hermitian invertible sheaf on X and s a non-zero rational section of L. Let D be a divisor on X such that $\mathrm{div}(s)$ and D have no common component. Then a current $[-\log h(s, s)]_D$ of type $(1, 1)$ is defined by

$$[-\log h(s, s)]_D(\eta) := \int_D (-\log h(s, s))\eta \quad (\eta \in A_c^{d-1,d-1}(X)).$$

Moreover, for a $(1, 1)$-form ω on X, a current $[-\log h(s, s)\omega]$ of type $(1, 1)$ is defined by

$$[-\log h(s, s)\omega](\eta) := \int_X -\log h(s, s)\omega \wedge \eta \quad (\eta \in A_c^{d-1,d-1}(X)).$$

Then we have the following.

THEOREM 5.6 (Weil's reciprocity law in complex geometry). *Let (L_1, h_1) and (L_2, h_2) be C^∞-hermitian invertible sheaves on X, and let s_1 and s_2 be non-zero rational sections of L_1 and L_2, respectively. We assume that $\mathrm{div}(s_1)$ and $\mathrm{div}(s_2)$ have no common component. If we write*

$$g_1 = -\log h_1(s_1, s_1) \quad \textit{and} \quad g_2 = -\log h_2(s_2, s_2),$$

then

$$\left([g_1]_{\mathrm{div}(s_2)} + [g_2 c_1(L_1, h_1)]\right) - \left([g_2]_{\mathrm{div}(s_1)} + [g_1 c_1(L_2, h_2)]\right) = \frac{\sqrt{-1}}{2\pi} \left(\bar{\partial}[g_1 \wedge \partial g_2] + \partial[g_2 \wedge \bar{\partial} g_1]\right),$$

where $[g_1 \wedge \partial g_2]$ and $[g_2 \wedge \bar{\partial} g_1]$ are currents of type $(1, 0)$ and of type $(0, 1)$ given by

$$\omega \mapsto \int_X g_1 \wedge \partial g_2 \wedge \omega \quad \textit{and} \quad \omega' \mapsto \int_X g_2 \wedge \bar{\partial} g_1 \wedge \omega',$$

respectively.

First we consider the problem on $\mathbb{C}^d$. Let $(z_1, \ldots, z_d)$ be the coordinate of $\mathbb{C}^d$. For a subset Λ of $\{1, \ldots, d\}$, let $A_{c,\Lambda}^{d-1,d-1}(\mathbb{C}^d)$ be the space of C^∞ $(d-1, d-1)$-forms η on $\mathbb{C}^d$ with compact supports such that, for all $j \notin \Lambda$,

$$\frac{dz_j}{z_j} \wedge \eta, \quad \frac{d\bar{z}_j}{\bar{z}_j} \wedge \eta \quad \text{and} \quad \frac{dz_j \wedge d\bar{z}_j}{|z_j|^2} \wedge \eta$$

are C^∞-forms. By its definition, it is easy to see that $\dfrac{dz_j}{z_j} \wedge \bar{\partial}(\eta)$ and $\dfrac{d\bar{z}_j}{\bar{z}_j} \wedge \partial(\eta)$ are C^∞-forms for $\eta \in A_{c,\Lambda}^{d-1,d-1}(\mathbb{C}^d)$ and $j \notin \Lambda$. Moreover, η is zero on $\{z_j = 0\}$ for $\eta \in A_{c,\Lambda}^{d-1,d-1}(\mathbb{C}^d)$ and $j \notin \Lambda$. For a rational function $f = z_1^{a_1} \cdots z_d^{a_d}$ $(a_1, \ldots, a_d \in \mathbb{Z})$ on $\mathbb{C}^d$, the principal divisor $(f)_\Lambda$ restricted to Λ is defined by

$$(f)_\Lambda = \sum_{i \in \Lambda} a_i (z_i).$$

Then we have the following.

LEMMA 5.7. *We set $f_1 = z_1^{a_1} \cdots z_d^{a_d}$ and $f_2 = z_1^{b_1} \cdots z_d^{b_d}$, where*

$$a_1, \ldots, a_d, b_1, \ldots, b_d \in \mathbb{Z}.$$

We assume that either $a_i = 0$ or $b_i = 0$ for all $i \in \Lambda$, that is, $(f_1)_\Lambda$ and $(f_2)_\Lambda$ have no common component. Moreover, let $g_1 = -\log|f_1|^2 + u_1$ and $g_2 = -\log|f_2|^2 + u_2$, where u_1 and u_2 are C^∞-functions on $\mathbb{C}^d$. Then, for any $\eta \in A_{c,\Lambda}^{d-1,d-1}(\mathbb{C}^d)$,

$$\begin{aligned}\left(2\pi\sqrt{-1}\int_{(f_1)_\Lambda} g_2\eta - \int_{\mathbb{C}^d} g_1 \partial\bar{\partial}(g_2) \wedge \eta\right) \\ - \left(2\pi\sqrt{-1}\int_{(f_2)_\Lambda} g_1\eta - \int_{\mathbb{C}^d} g_2 \partial\bar{\partial}(g_1) \wedge \eta\right) \\ = \int_{\mathbb{C}^d} g_1 \wedge \partial g_2 \wedge \bar{\partial}\eta + \int_{\mathbb{C}^d} g_2 \wedge \bar{\partial} g_1 \wedge \partial\eta.\end{aligned}$$

PROOF. Let us see that

$$\int_{\mathbb{C}^d} g_1 \partial g_2 \wedge \bar{\partial}\eta = \int_{\mathbb{C}^d} \bar{\partial} g_1 \wedge \partial g_2 \wedge \eta - \int_{\mathbb{C}^d} g_1 \partial\bar{\partial}(g_2) \wedge \eta - 2\pi\sqrt{-1}\int_{(f_2)_\Lambda} g_1\eta \tag{5.5}$$

for $\eta \in A_{c,\Lambda}^{d-1,d-1}(\mathbb{C}^d)$. We can easily check that $g_1 \partial g_2 \wedge \bar{\partial}\eta$ and $\bar{\partial} g_1 \wedge \partial g_2 \wedge \eta$ are integrable because $\eta \in A_{c,\Lambda}^{d-1,d-1}(\mathbb{C}^d)$ (the details are left to the reader). Let us begin with the following claim.

CLAIM 2. *If we set $I_k^{i,j} = \lim_{\epsilon \to 0} \int_{|z_k| = \epsilon} (-\log|z_i|^2) \partial(-\log|z_j|^2) \wedge \eta$, then we have the following.*

(a) $I_k^{i,j} = 0$ *if* $k \neq j$.

(b) $I_k^{i,j} = -2\pi\sqrt{-1}\displaystyle\int_{z_j = 0} (-\log|z_i|^2)\eta$ *if* $k = j$ *and* $k \neq i$.

(c) $I_k^{i,j} = 0$ *if* $k = i = j \notin \Lambda$.

PROOF. (a) is a direct calculation. (b) is same as (b) in Section 1.13, Claim 8. (c) follows from (a) in Section 1.13, Claim 8 because $\partial(-\log|z_j|^2) \wedge \eta$ is a C^∞ $(d, d-1)$-form. $\square$

We set $U_\epsilon = \{(z_1, \ldots, z_d) \in \mathbb{C}^d \mid |z_i| \geq \epsilon \ (i = 1, \ldots, d)\}$. Then $\partial(U_\epsilon) = \bigcup_{k=1}^d \{|z_k| = \epsilon\}$. Since

$$d(g_1 \partial g_2 \wedge \eta) = \bar{\partial}(g_1 \partial g_2 \wedge \eta) = \bar{\partial} g_1 \wedge \partial g_2 \wedge \eta - g_1 \partial\bar{\partial}(g_2) \wedge \eta - g_1 \partial g_2 \wedge \bar{\partial}\eta,$$

we have

$$\begin{aligned}\int_{U_\epsilon} g_1 \partial g_2 \wedge \bar{\partial}\eta &= \int_{U_\epsilon} \bar{\partial} g_1 \wedge \partial g_2 \wedge \eta - \int_{U_\epsilon} g_1 \partial\bar{\partial}(g_2) \wedge \eta - \int_{U_\epsilon} d(g_1 \partial g_2 \wedge \eta) \\ &= \int_{U_\epsilon} \bar{\partial} g_1 \wedge \partial g_2 \wedge \eta - \int_{U_\epsilon} g_1 \partial\bar{\partial}(g_2) \wedge \eta + \sum_{k=1}^d \int_{|z_k|=\epsilon} g_1 \partial g_2 \wedge \eta.\end{aligned}$$

Therefore, for (5.5), it is sufficient to see

$$I = \lim_{\epsilon \to 0} \sum_{k=1}^d \int_{|z_k|=\epsilon} g_1 \partial g_2 \wedge \eta = -2\pi\sqrt{-1} \int_{(f_2)_\Lambda} g_1 \eta.$$

We write $h_1 = -\log|f_1|^2$ and $h_2 = -\log|f_2|^2$. Then

$$g_1 \partial g_2 \wedge \eta = h_1 \partial h_2 \wedge \eta + h_1 \partial u_2 \wedge \eta + u_1 \partial h_2 \wedge \eta + u_1 \partial u_2 \wedge \eta.$$

First we can check easily that

$$\lim_{\epsilon \to 0} \int_{|z_k|=\epsilon} u_1 \partial u_2 \wedge \eta = \lim_{\epsilon \to 0} \int_{|z_k|=\epsilon} h_1 \partial u_2 \wedge \eta = 0.$$

Moreover, as η is zero on $\{z_j = 0\}$ $(j \notin \Lambda)$, by (b) in Section 1.13, Claim 8,

$$\begin{aligned}\lim_{\epsilon \to 0} \sum_{k=1}^d \int_{|z_k|=\epsilon} u_1 \partial h_2 \wedge \eta &= \lim_{\epsilon \to 0} \sum_{k,j} b_j \int_{|z_k|=\epsilon} u_1 \partial(-\log|z_j|^2) \wedge \eta \\ &= 2\pi\sqrt{-1} \sum_{j=1}^d b_j \int_{z_j=0} u_1 \eta \\ &= 2\pi\sqrt{-1} \int_{(f_2)_\Lambda} u_1 \eta.\end{aligned}$$

Notice that

$$J = \sum_{k=1}^d \int_{|z_k|=\epsilon} h_1 \partial h_2 \wedge \eta = \sum_{i,j,k} a_i b_j \int_{|z_k|=\epsilon} (-\log|z_i|^2) \partial(-\log|z_j|^2) \wedge \eta,$$

$a_i b_i = 0$ $(i \in \Lambda)$ and η is zero on $\{z_j = 0\}$ $(j \notin \Lambda)$. Thus, by using (a), (b), (c) of the previous claim,

$$\begin{aligned}J &= \lim_{\epsilon \to 0} \sum_{i,j} a_i b_j \int_{|z_j|=\epsilon} (-\log|z_i|^2) \partial(-\log|z_j|^2) \wedge \eta \\ &= \lim_{\epsilon \to 0} \sum_{i \neq j} a_i b_j \int_{|z_j|=\epsilon} (-\log|z_i|^2) \partial(-\log|z_j|^2) \wedge \eta \\ &\quad + \lim_{\epsilon \to 0} \sum_{i \notin \Lambda} a_i b_i \int_{|z_i|=\epsilon} (-\log|z_i|^2) \partial(-\log|z_i|^2) \wedge \eta \\ &= -2\pi\sqrt{-1} \sum_{i \neq j} a_i b_j \int_{z_j=0} (-\log|z_i|^2) \eta\end{aligned}$$

$$= -2\pi\sqrt{-1} \sum_{\substack{i=1,\ldots,d,\\ j\in\Lambda}} a_i b_j \int_{z_j=0} (-\log|z_i|^2)\eta$$

$$= -2\pi\sqrt{-1} \int_{(f_2)_\Lambda} h_1\eta.$$

Therefore, we obtain $I = -2\pi\sqrt{-1}\int_{(f_2)_\Lambda} g_1\eta$, so that (5.5) follows.

Note that (5.5) still holds even if we replace f_1 and u_1 by f_2 and u_2, respectively. Thus we have

$$\int_{\mathbb{C}^d} g_2\partial g_1 \wedge \bar\partial\bar\eta = \int_{\mathbb{C}^d} \bar\partial g_2 \wedge \partial g_1 \wedge \bar\eta - \int_{\mathbb{C}^d} g_2\partial\bar\partial(g_1)\wedge\bar\eta - 2\pi\sqrt{-1}\int_{(f_1)_\Lambda} g_2\bar\eta.$$

Considering the complex conjugation of the above equation, we have

$$\int_{\mathbb{C}^d} g_2\bar\partial g_1 \wedge \partial\eta = \int_{\mathbb{C}^d} \partial g_2 \wedge \bar\partial g_1 \wedge \eta + \int_{\mathbb{C}^d} g_2\partial\bar\partial(g_1)\wedge\eta + 2\pi\sqrt{-1}\int_{(f_1)_\Lambda} g_2\eta. \tag{5.6}$$

Thus the lemma follows from (5.5) and (5.6). □

LEMMA 5.8. *Let X be a d-dimensional non-singular complex algebraic variety. Let*

$$Y = X_N \overset{\mu_{N-1}}{\longrightarrow} X_{N-1} \overset{\mu_{N-2}}{\longrightarrow} \cdots \overset{\mu_0}{\longrightarrow} X_0 = X$$

be a sequence of blowing-ups along an irreducible smooth subvariety of codimension ≥ 2, that is, for each $0 \leq \alpha < N$, there is an irreducible smooth subvariety C_α of codimension ≥ 2 on X_α such that $\mu_\alpha : X_{\alpha+1} \to X_\alpha$ is the blowing-up along C_α. We assume that the exceptional set D of $\mu = \mu_0 \circ \cdots \circ \mu_{N-1} : Y \to X$ is a normal crossing divisor. For $y \in D$, let $(z_1, \ldots, z_d)$ be a local coordinate of Y at y such that $z_1(y) = \cdots = z_d(y) = 0$ and D is given by $z_1 \cdots z_l = 0$ around y. Then, for $\eta \in A^{d-1,d-1}(X)$,

$$\frac{dz_j}{z_j}\wedge\mu^*(\eta), \quad \frac{d\bar z_j}{\bar z_j}\wedge\mu^*(\eta) \quad \text{and} \quad \frac{dz_j\wedge d\bar z_j}{|z_j|^2}\wedge\mu^*(\eta)$$

are C^∞ around y for $j = 1, \ldots, l$.

PROOF. We can find $0 \leq \alpha < N$ such that $\{z_j = 0\}$ is an irreducible component of $\nu_\alpha^{-1}(C_\alpha)$ around y, where $\nu_\alpha = \mu_\alpha \circ \cdots \circ \mu_{N-1} : Y \to X_\alpha$. Let $(w_1, \ldots, w_d)$ be a local coordinate of X_α at $\nu_\alpha(y)$ such that $w_1(\nu_\alpha(y)) = \cdots = w_d(\nu_\alpha(y)) = 0$ and C_α is given by $w_1 = \cdots = w_b = 0$ around $\nu_\alpha(y)$. As the codimension of C_α is greater than or equal to 2, we have $b \geq 2$. We set $\phi_i = \nu_\alpha^*(w_i)$ $(i = 1, \ldots, d)$. Then, there are holomorphic functions $f_1, \ldots, f_b$ around y such that $\phi_1 = z_j f_1, \ldots, \phi_b = z_j f_b$. We write

$$\begin{aligned}(\mu_0 \circ \cdots \circ \mu_{\alpha-1})^*(\eta) \\ = \sum_{s,t} \eta_{st}\left(dw_1 \wedge \cdots \wedge \widehat{dw_s} \wedge \cdots \wedge dw_d\right) \wedge \left(d\bar w_1 \wedge \cdots \wedge \widehat{d\bar w_t} \wedge \cdots \wedge d\bar w_d\right)\end{aligned}$$

around $\nu_\alpha(y)$. Then

$$\mu^*(\eta) = \sum_{s,t} \nu_\alpha^*(\eta_{st})\left(d\phi_1 \wedge \cdots \wedge \widehat{d\phi_s} \wedge \cdots \wedge d\phi_d\right) \wedge \left(d\bar\phi_1 \wedge \cdots \wedge \widehat{d\bar\phi_t} \wedge \cdots \wedge d\bar\phi_d\right).$$

On the other hand, note that

$$\begin{cases} \dfrac{dz_j}{z_j} \wedge d\phi_s = \dfrac{dz_j}{z_j} \wedge ((dz_j)f_s + z_j df_s) = dz_i \wedge df_s, \\ \\ \dfrac{d\bar{z}_j}{\bar{z}_j} \wedge d\bar{\phi}_t = \dfrac{d\bar{z}_j}{\bar{z}_j} \wedge ((d\bar{z}_j)\bar{f}_t + \bar{z}_j d\bar{f}_t) = d\bar{z}_i \wedge d\bar{f}_t, \end{cases}$$

for $1 \leq s, t \leq b$. Thus, since $b \geq 2$,

$$\frac{dz_j}{z_j} \wedge \mu^*(\eta), \quad \frac{d\bar{z}_j}{\bar{z}_j} \wedge \mu^*(\eta) \quad \text{and} \quad \frac{dz_j \wedge d\bar{z}_j}{|z_j|^2} \wedge \mu^*(\eta)$$

are C^∞-forms for $j = 1, \ldots, l$. □

THE PROOF OF THEOREM 5.6. For the proof of the theorem, it is sufficient to show that

$$\begin{aligned} &\left(\int_{\mathrm{div}(s_2)} g_1 \eta + \int_X g_2 c_1(L_1, h_1) \wedge \eta \right) \\ &\qquad - \left(\int_{\mathrm{div}(s_1)} g_2 \eta + \int_X g_1 c_1(L_2, h_2) \wedge \eta \right) \\ &\qquad\qquad = \frac{\sqrt{-1}}{2\pi} \left(\int_X g_1 \wedge \partial g_2 \wedge \bar{\partial}\eta + \int_X g_2 \wedge \bar{\partial} g_1 \wedge \partial\eta \right) \end{aligned}$$

for all $\eta \in A_c^{d-1,d-1}(X)$. For this purpose, we construct a sequence

$$Y = X_N \overset{\mu_{N-1}}{\longrightarrow} X_{N-1} \overset{\mu_{N-2}}{\longrightarrow} \cdots \overset{\mu_0}{\longrightarrow} X_0 = X$$

of blowing-ups along an irreducible smooth subvariety of codimension ≥ 2 as in Lemma 5.8 such that

$$\mathrm{Supp}\left((\text{the exceptional set of } \mu) + \mu^*(\mathrm{div}(s_1)) + \mu^*(\mathrm{div}(s_2))\right)$$

is a normal crossing divisor, where $\mu : Y \to X$ is the birational morphism given by $\mu = \mu_0 \circ \cdots \circ \mu_{N-1}$ as in Lemma 5.8. Let D_1 and D_2 be the strict transforms of $\mathrm{div}(s_1)$ and $\mathrm{div}(s_2)$ by μ, that is, D_1 and D_2 are divisors obtained by subtracting the exception divisors of μ from $\mu^*(\mathrm{div}(s_1))$ and $\mu^*(\mathrm{div}(s_2))$, respectively. To show the previous equation, it is sufficient to see

$$\begin{aligned} (5.7) \quad &\left(\int_{D_2} \mu^*(g_1)\mu^*(\eta) + \int_Y \mu^*(g_2)\mu^*(c_1(L_1, h_1)) \wedge \mu^*(\eta) \right) \\ &\qquad - \left(\int_{D_1} \mu^*(g_2)\mu^*(\eta) + \int_Y \mu^*(g_1)\mu^*(c_1(L_2, h_2)) \wedge \mu^*(\eta) \right) \\ &= \frac{\sqrt{-1}}{2\pi} \left(\int_Y \mu^*(g_1) \wedge \partial\mu^*(g_2) \wedge \bar{\partial}\mu^*(\eta) + \int_Y \mu^*(g_2) \wedge \bar{\partial}\mu^*(g_1) \wedge \partial\mu^*(\eta) \right). \end{aligned}$$

Let ϕ be a non-negative C^∞-function on Y whose support is sufficiently small. Then Lemma 5.7 and Lemma 5.8 show that (5.7) holds if we replace $\mu^*(\eta)$ by $\phi\mu^*(\eta)$. Thus, using a partition of unity, (5.7) follows. □

5.4. Intersection theory on arithmetic varieties

A scheme X is called an *arithmetic variety* if X is an integral scheme flat and quasi-projective over $\mathbb{Z}$. Note that an arithmetic variety is excellent. An arithmetic variety X is said to be *generically smooth* if the generic fiber of the structure morphism $X \to \operatorname{Spec}(\mathbb{Z})$ is smooth over $\mathbb{Q}$. If X is generically smooth, then the set $X(\mathbb{C})$ of all $\mathbb{C}$-valued points is a non-singular complex quasi-projective variety. Note that $X(\mathbb{C})$ is not necessarily connected. Moreover, X is called a *projective arithmetic variety* if $X \to \operatorname{Spec}(\mathbb{Z})$ is projective. We assume that X is a generically smooth and normal projective arithmetic variety. Let $X \to \operatorname{Spec}(O_K) \to \operatorname{Spec}(\mathbb{Z})$ be the Stein factorization of $X \to \operatorname{Spec}(\mathbb{Z})$, where K is a number field and O_K is the ring of integers in K. For $\sigma \in K(\mathbb{C})$, we set $X_\sigma = X \times^{\sigma}_{\operatorname{Spec}(O_K)} \operatorname{Spec}(\mathbb{C})$ viewing $\operatorname{Spec}(\mathbb{C})$ as a scheme over $\operatorname{Spec}(O_K)$ via σ. Then $\{X_\sigma(\mathbb{C})\}_{\sigma\in K(\mathbb{C})}$ forms a connected component of $X(\mathbb{C})$.

Let X be an arithmetic variety. For $x \in X(\mathbb{C})$, the scheme theoretical point of X given by x is denoted by p_x and the local ring homomorphism $\mathscr{O}_{X,p_x} \to \mathbb{C}$ determined by x is denoted by ϕ_x as in Subsection 5.1.2. Let E be a locally free coherent sheaf on X. Then, for $x \in X(\mathbb{C})$, we write $E(x) = E_{p_x} \otimes_{\mathscr{O}_{X,p_x}} \mathbb{C}$, where $\mathbb{C}$ is viewed as an $\mathscr{O}_{X,p_x}$-module via ϕ_x. For each $x \in X(\mathbb{C})$, we give a hermitian inner product h_x of $E(x)$ and the collection $\{h_x\}_{x\in X(\mathbb{C})}$ is denoted by h. If h is C^∞ (resp. continuous) in the sense of Section 1.14, h is called a *C^∞-hermitian metric* (resp. *continuous hermitian metric*) of E. Moreover, a pair $\overline{E} = (E, h)$ consisting of E and a C^∞-hermitian metric (resp. continuous hermitian metric) h of E is called a *C^∞-hermitian locally free coherent sheaf* (resp. *continuous hermitian locally free coherent sheaf*). If E is invertible, then $\overline{E}$ is called a *C^∞-hermitian invertible sheaf* (resp. *continuous hermitian invertible sheaf*). We set $X_{\mathbb{R}} = X \times_{\operatorname{Spec}(\mathbb{Z})} \operatorname{Spec}(\mathbb{R})$. Then $X_{\mathbb{R}}(\mathbb{C}) = X(\mathbb{C})$. We say that h is *of real type* if h is of real type in the sense of Subsection 5.1.2. Namely,

$$h_x(s \otimes^x 1, s' \otimes^x 1) = \overline{h_{\bar{x}}(s \otimes^{\bar{x}} 1, s' \otimes^{\bar{x}} 1)} \qquad (\forall s, s' \in \mathscr{O}_{X,p_x})$$

for any $x \in X(\mathbb{C})$, where $\bar{x}$ is the composition of morphisms $\operatorname{Spec}(\mathbb{C}) \xrightarrow{-a} \operatorname{Spec}(\mathbb{C})$ given by the complex conjugation and $\operatorname{Spec}(\mathbb{C}) \xrightarrow{x} X$, and $\otimes^x$ and $\otimes^{\bar{x}}$ are the tensor products with respect to ϕ_x and $\phi_{\bar{x}}$. In this book, the hermitian metric of a C^∞-hermitian locally free coherent sheaf is usually assumed to be of real type. However, in this section, we do not need to assume that the hermitian metric is of real type.

Let X be a generically smooth arithmetic variety. Let Z be a cycle of codimension p on X and T a current of type $(p-1, p-1)$ on $X(\mathbb{C})$. Note that $T = 0$ if $p = 0$. A pair (Z, T) is called an *arithmetic cycle of codimension p*. Instead of codimension, if Z is a cycle of dimension l ($l = \dim X - p$), then (Z, T) is called an *arithmetic cycle of dimension l*. An arithmetic cycle (Z, T) is said to be *of Green type* if

$$\omega(Z(\mathbb{C}), T) := dd^c(T) + \delta_{Z(\mathbb{C})} \in A^{p,p}(X(\mathbb{C})),$$

that is, T is a Green current of $Z(\mathbb{C})$ (cf. Subsection 5.1.4). The group of all arithmetic cycles of codimension p on X is denoted by $\widehat{Z}^p_D(X)$ and a subgroup consisting of arithmetic cycles of Green type is denoted by $\widehat{Z}^p(X)$. Moreover, the group of all arithmetic cycles of dimension l on X is denoted by $\widehat{Z}_{D,l}(X)$ and a subgroup consisting of arithmetic cycles of Green type is denoted by $\widehat{Z}_l(X)$. By their definitions, if $p + l = \dim X$, then $\widehat{Z}_{D,l}(X) = \widehat{Z}^p_D(X)$ and $\widehat{Z}_l(X) = \widehat{Z}^p(X)$. Let us consider several examples.

EXAMPLE 5.9. Let $\overline{L} = (L, h)$ be a C^∞-hermitian invertible sheaf on X and s a non-zero rational section of L. Then, by the Poincaré-Lelong formula (cf. Theorem 1.39),

$$(\operatorname{div}(s), -\log h(s,s))$$

is a codimension one arithmetic cycle of Green type. This is denoted by $\widehat{\operatorname{div}}(\overline{L}, s)$.

EXAMPLE 5.10. Let Y be a $p-1$ codimensional integral subscheme of X and ϕ a non-zero rational section on Y. Let us consider a current

$$\left[-\log|\phi|^2\right]_{Y(\mathbb{C})} : A_c^{\dim X(\mathbb{C})-(p-1),\dim X(\mathbb{C})-(p-1)}(X(\mathbb{C})) \to \mathbb{C}$$

of type $(p-1, p-1)$ given by

$$\eta \mapsto \int_{Y(\mathbb{C})} (-\log|\phi|^2)\eta.$$

Then

$$\left(\operatorname{div}(\mathscr{O}_Y, \phi), \left[-\log|\phi|^2\right]_{Y(\mathbb{C})}\right)$$

is a p codimensional arithmetic cycle of Green type by the Poincaré-Lelong formula. It is denoted by $\widehat{(\phi/Y)}$ or $\widehat{(\phi)}$ for simplicity. If (ϕ) has a long expression, then $\widehat{(\phi)}$ is sometime denoted by $(\phi)\widehat{\ }$. If Y is not flat over $\operatorname{Spec}(\mathbb{Z})$, then $\widehat{(\phi/Y)} = (\operatorname{div}(\mathscr{O}_Y, \phi), 0)$ because $Y(\mathbb{C}) = \emptyset$.

EXAMPLE 5.11. Let u and v be currents on $X(\mathbb{C})$ of type $(p-2, p-1)$ and $(p-1, p-2)$, respectively. Then $(0, \partial(u) + \bar{\partial}(v))$ is a codimension p arithmetic cycle of Green type.

DEFINITION 5.12. We assume that X is generically smooth. Let $\widehat{\operatorname{Rat}}^p(X)$ be a subgroup of $\widehat{Z}^p(X)$ generated by arithmetic cycles of Example 5.10 and Example 5.11. Then $\widehat{\operatorname{CH}}^p_D(X)$ and $\widehat{\operatorname{CH}}^p(X)$ are defined to be

$$\widehat{\operatorname{CH}}^p_D(X) := \widehat{Z}^p_D(X)/\widehat{\operatorname{Rat}}^p(X) \quad \text{and} \quad \widehat{\operatorname{CH}}^p(X) := \widehat{Z}^p(X)/\widehat{\operatorname{Rat}}^p(X),$$

respectively. They are called an *arithmetic Chow group* of X and an *arithmetic Chow group of Green type* of X, respectively. If $p + l = \dim X$, then $\widehat{\operatorname{Rat}}^p(X)$ is a subgroup of $\widehat{Z}_{D,l}(X)$. We denote $\widehat{\operatorname{Rat}}^p(X)$ by $\widehat{\operatorname{Rat}}_l(X)$ as a subgroup of $\widehat{Z}_{D,l}(X)$. Then we define $\widehat{\operatorname{CH}}_{D,l}(X)$ and $\widehat{\operatorname{CH}}_l(X)$ by

$$\widehat{\operatorname{CH}}_{D,l}(X) := \widehat{Z}_{D,l}(X)/\widehat{\operatorname{Rat}}_l(X) \quad \text{and} \quad \widehat{\operatorname{CH}}_l(X) := \widehat{Z}_l(X)/\widehat{\operatorname{Rat}}_l(X).$$

Thus, if $p + l = \dim X$, then $\widehat{\operatorname{CH}}^p_D(X) = \widehat{\operatorname{CH}}_{D,l}(X)$ and $\widehat{\operatorname{CH}}^p(X) = \widehat{\operatorname{CH}}_l(X)$.

DEFINITION 5.13. Let $f : X \to Y$ be a projective morphism of generically smooth arithmetic varieties. For $(Z, T) \in \widehat{Z}_{D,l}(X)$, the *push-forward of arithmetic cycle* by f is defined by

$$f_*(Z, T) = (f_*Z, f_*T).$$

For the definition of the push-forward f_*T of the current T, see Subsection 5.1.4. In this way, we obtain a homomorphism $f_* : \widehat{Z}_{D,l}(X) \to \widehat{Z}_{D,l}(Y)$. In Example 5.10, if we assume that Y is flat over $\operatorname{Spec}(\mathbb{Z})$ and generically smooth, then $\widehat{(\phi/Y)} = j_* \widehat{\operatorname{div}}(\overline{\mathscr{O}}_Y^{can}, \phi)$, where j is the inclusion map $Y \hookrightarrow X$.

PROPOSITION 5.14. *Let $f : X \to Y$ be a projective morphism of generically smooth arithmetic varieties. Let Z be an integral subscheme of X and $Z' = f(Z)$. Let K and K' be the function fields of Z and Z', respectively. Then, for a non-zero rational function ϕ on Z,*

$$f_*\left(\widehat{(\phi/Z)}\right) = \begin{cases} (N_{K/K'}(\phi)/Z')^{\widehat{}} & \text{if } \dim Z = \dim Z', \\ 0 & \text{if } \dim Z > \dim Z'. \end{cases}$$

In particular, using (5.4) *in Subsection* 5.1.4, *we have*

$$f_*(\widehat{\mathrm{Rat}}_l(X)) \subseteq \widehat{\mathrm{Rat}}_l(Y).$$

Therefore, the homomorphism $f_* : \widehat{Z}_{D,l}(X) \to \widehat{Z}_{D,l}(Y)$ induces a homomorphism

$$f_* : \widehat{\mathrm{CH}}_{D,l}(X) \to \widehat{\mathrm{CH}}_{D,l}(Y).$$

PROOF. It follows from Proposition 5.3 and the following lemma. □

LEMMA 5.15. *Let $f : X \to Y$ be a proper morphism of non-singular complex algebraic varieties. Let Z be a d-dimensional subvariety of X and $Z' = f(Z)$. Let K and K' be the function fields of Z and Z', respectively. Then, for a non-zero rational function ϕ on Z and a C^∞ (d, d)-form η on Y with compact support,*

$$\int_Z -\log|\phi|^2 f^*(\eta) = \begin{cases} \displaystyle\int_{Z'} -\log|N_{K/K'}(\phi)|^2\eta & \text{if } \dim Z = \dim Z', \\ 0 & \text{if } \dim Z > \dim Z'. \end{cases}$$

PROOF. If $\dim Z > \dim Z'$, then η is zero on Z'. Thus the assertion is obvious, so that we assume $\dim Z = \dim Z'$. We set $g = f|_Z$. Since $-\log|N_{K/K'}(\phi)|^2\eta$ and $-\log|\phi|^2 f^*(\eta)$ are locally integrable on Z' and Z, respectively (the integrability of them is considered on desingularizations of Z and Z'), it is sufficient to see that

$$\int_U -\log|\phi|^2 f^*(\eta) = \int_{g^{-1}(U)} -\log|N_{K/K'}(\phi)|^2\eta$$

for a non-empty Zariski open set U of Z'. Here we take a non-empty Zariski open set U such that (1) U is non-singular, (2) g is etale over U, (3) $\mathrm{Supp}(\mathscr{O}_Z, \phi) \cap g^{-1}(U) = \emptyset$. Let us fix y of U. Note that $N_{K/K'}(\phi)$ is the determinant of the multiplication map by ϕ of a free $\mathscr{O}_{Z',y}$-module $g_*(\mathscr{O}_Z)_y$. We set $g^{-1}(y) = \{x_1, \ldots, x_r\}$ and consider $\mathrm{rest}_{x_i} : g_*(\mathscr{O}_Z)_y^{an} \to \mathscr{O}_{Z,x_i}^{an}$. As g is etale over U, $\mathrm{rest}_{x_1}, \ldots, \mathrm{rest}_{x_r}$ yields an isomorphism

$$g_*(\mathscr{O}_Z)_y^{an} \to \bigoplus_{i=1}^r \mathscr{O}_{Z,x_i}^{an} \quad (e \mapsto (\mathrm{rest}_{x_1}(e), \ldots, \mathrm{rest}_{x_r}(e))).$$

Moreover, $g_{x_i}^* : \mathscr{O}_{Z',y}^{an} \to \mathscr{O}_{Z,x_i}^{an}$ is also an isomorphism. Therefore, if we choose $\phi_i \in \mathscr{O}_{Z',y}^{an}$ with $g_{x_i}^*(\phi_i) = \mathrm{rest}_{x_i}(\phi)$, then $N_{K/K'}(\phi) = \phi_1 \cdots \phi_r$. On the other hand, if we take a small neighborhood V_y of y, then there are neighborhoods $V_{x_1}, \ldots, V_{x_r}$ such that V_{x_i} is isomorphic to V_y and $g^{-1}(V_y) = V_{x_1} \amalg \cdots \amalg V_{x_r}$. Further, if V_y is sufficiently small, then $\mathrm{rest}_{x_i}(\phi)$ is a holomorphic function on V_{x_i}. If $\mathrm{Supp}(\eta) \subseteq V_y$,

then

$$\begin{aligned}\int_{g^{-1}(V_y)} -\log|\phi|^2 f^*(\eta) &= \sum_{i=1}^{r} \int_{V_{x_i}} -\log|\operatorname{rest}_{x_i}(\phi)|^2 f^*(\eta) \\ &= \sum_{i=1}^{n} \int_{V_y} -\log|\phi_i|^2 \eta \\ &= \int_{V_y} -\log|N_{K/K'}(\phi)|^2 \eta.\end{aligned}$$

Therefore, using a partition of unity, we obtain

$$\int_{g^{-1}(U)} -\log|\phi|^2 f^*(\eta) = \int_U -\log|N_{K/K'}(\phi)|^2 \eta,$$

as required. □

DEFINITION 5.16 (Arithmetic intersection cycle). Let X be a generically smooth arithmetic variety. Let $\overline{L} = (L, h)$ be a C^∞-hermitian invertible sheaf on X and s a non-zero rational section of L. Let (Z, T) be an arithmetic cycle of codimension p on X such that (L, s) meets properly with Z, that is, $\operatorname{Supp}(L, s) \cap \operatorname{Supp}(Z) \cap X^{(p)} = \emptyset$. Then the arithmetic intersection cycle $(\overline{L}, s) \cdot (Z, T)$ of $(\overline{L}, s)$ with (Z, T) is defined to be

$$(\overline{L}, s) \cdot (Z, T) := \left((L, s) \cdot Z,\ [-\log h(s, s)]_{Z(\mathbb{C})} + c_1(\overline{L}) \wedge T\right),$$

where $[-\log h(s, s)]_{Z(\mathbb{C})}$ is a current of type (p, p) given by

$$\eta \mapsto \int_{Z(\mathbb{C})} (-\log h(s, s))\eta.$$

This is an arithmetic cycle of codimension $p + 1$.

From now on, we discuss several properties of arithmetic intersection cycles. Let us begin with the following proposition.

PROPOSITION 5.17. *Let $\overline{L}_1 = (L_1, h_1)$ and $\overline{L}_2 = (L_2, h_2)$ be C^∞-hermitian invertible sheaves on X and (Z, T) an arithmetic cycle of codimension p on X. Moreover, let s_1 and s_2 be non-zero rational sections of L_1 and L_2, respectively. We assume that*

$$\operatorname{Supp}(L_1, s_1) \cap \operatorname{Supp}(Z) \cap X^{(p)} = \emptyset \quad \textit{and} \quad \operatorname{Supp}(L_2, s_2) \cap \operatorname{Supp}(Z) \cap X^{(p)} = \emptyset.$$

Then

$$(\overline{L}_1 \otimes \overline{L}_2, s_1 \otimes s_2) \cdot (Z, T) = (\overline{L}_1, s_1) \cdot (Z, T) + (\overline{L}_2, s_2) \cdot (Z, T).$$

PROOF. By (1) in Proposition 5.2,

$$(L_1 \otimes L_2, s_1 \otimes s_2) \cdot (Z, T) = (L_1, s_1) \cdot (Z, T) + (L_2, s_2) \cdot (Z, T).$$

Thus it is sufficient to see that

$$\begin{aligned}&[-\log(h_1 \otimes h_2)(s_1 \otimes s_2, s_1 \otimes s_2)]_{Z(\mathbb{C})} + c_1(\overline{L}_1 \otimes \overline{L}_2) \wedge T \\ &\qquad = [-\log h_1(s_1, s_1)]_{Z(\mathbb{C})} + c_1(\overline{L}_1) \wedge T \\ &\qquad\qquad + [-\log h_2(s_2, s_2)]_{Z(\mathbb{C})} + c_1(\overline{L}_2) \wedge T,\end{aligned}$$

which is obvious. □

Next we consider the projection formula.

PROPOSITION 5.18 (Projection formula for arithmetic cycles). *Let $f : X \to Y$ be a projective morphism of generically smooth arithmetic varieties. Let $\overline{L} = (L, h)$ be a C^∞-hermitian invertible sheaf on Y and s a non-zero rational section of L. Let η be the generic point of X and we assume that $f(\eta) \not\in \mathrm{Supp}(L, s)$. We denote the image of s by the natural homomorphism $L_{f(\eta)} \to f^*(L)_\eta$ by f^*s. Let (Z, T) be an arithmetic cycle of dimension l on X such that* $\mathrm{Supp}(f^*(L), f^*s) \cap \mathrm{Supp}(Z) \cap X_{(l)} = \emptyset$. *Then*

$$f_*((f^*(\overline{L}), f^*s) \cdot (Z, T)) = (\overline{L}, s) \cdot f_*(Z, T).$$

PROOF. First we consider a case where $T = 0$. Since each side of the formula is linear with respect to Z, we may assume that Z is integral. First,

$$\begin{cases} (f^*(\overline{L}), f^*s) \cdot (Z, 0) = \left((f^*(L), f^*s) \cdot Z,\ [-f^*(\log h(s, s))]_{Z(\mathbb{C})}\right), \\ (\overline{L}, s) \cdot f_*(Z, 0) = \left((L, s) \cdot f_*Z,\ [-\log h(s, s)]_{f_*(Z(\mathbb{C}))}\right). \end{cases}$$

Since $f_*((f^*(L), f^*s) \cdot Z) = (L, s) \cdot f_*Z$ by Proposition 5.5, it is sufficient to see that

$$f_*([-f^*(\log h(s, s))]_{Z(\mathbb{C})}) = [-\log h(s, s)]_{f_*(Z(\mathbb{C}))}. \tag{5.8}$$

If Z is not flat over $\mathrm{Spec}(\mathbb{Z})$, then (5.8) is obvious because $Z(\mathbb{C}) = \emptyset$. Thus we assume that Z is flat over $\mathrm{Spec}(\mathbb{Z})$. Let ω be a C^∞ form on $Y(\mathbb{C})$. If $\dim Z = \dim f(Z)$, then

$$\begin{aligned} f_*([-f^*(\log h(s, s))]_{Z(\mathbb{C})})(\omega) &= \int_{Z(\mathbb{C})} -f^*(\log h(s, s)\omega) \\ &= [\mathrm{Rat}(Z) : \mathrm{Rat}(f(Z))] \int_{f(Z(\mathbb{C}))} (-\log h(s, s))\omega \\ &= \int_{f_*(Z(\mathbb{C}))} (-\log h(s, s))\omega = [-\log h(s, s)]_{f_*(Z(\mathbb{C}))}(\omega). \end{aligned}$$

Thus (5.8) holds. If $\dim Z > \dim f(Z)$, then ω is zero on $f(Z(\mathbb{C}))$, so that

$$f_*([-f^*(\log h(s, s))]_{Z(\mathbb{C})})(\omega) = \int_{Z(\mathbb{C})} -f^*(\log h(s, s)\omega) = 0,$$

as required.

Next we consider the case where $Z = 0$. In this case,

$$\begin{cases} (f^*(\overline{L}), f^*s) \cdot (0, T) = \left(0, c_1(f^*\overline{L}) \wedge T\right), \\ (\overline{L}, s) \cdot f_*(0, T) = \left(0, c_1(\overline{L}) \wedge f_*T\right). \end{cases}$$

Let ω be a C^∞ form on $Y(\mathbb{C})$. Then

$$\begin{aligned} f_*(c_1(f^*\overline{L}) \wedge T)(\omega) &= T(f^*(c_1(\overline{L}) \wedge \omega)) = f_*(T)(c_1(\overline{L}) \wedge \omega) \\ &= (c_1(\overline{L}) \wedge f_*(T))(\omega). \end{aligned}$$

Therefore, $f_*(c_1(f^*\overline{L}) \wedge T) = c_1(\overline{L}) \wedge f_*T$. Thus we get the case.

In general, we have a decomposition $(Z, T) = (Z, 0) + (0, T)$, so that, using the previous cases, we have the proposition. □

Moreover, the following proposition holds.

PROPOSITION 5.19. *Let X be a generically smooth arithmetic variety, Y an integral subscheme of codimension $p-1$ on X, and $j : Y \hookrightarrow X$ an inclusion map. Let L_1 and L_2 be invertible sheaves on Y, and let s_1 and s_2 be non-zero rational sections of L_1 and L_2 on Y, respectively, such that* $\operatorname{Supp}(L_1, s_1) \cap \operatorname{Supp}(L_2, s_2) \cap Y^{(1)} = \emptyset$. *Then we have the following.*

(1) *If Y is not flat over* $\operatorname{Spec}(\mathbb{Z})$, *then*
$$j_*\left((L_1, s_1) \cdot \operatorname{div}(L_2.s_2)\right) = j_*\left((L_2, s_2) \cdot \operatorname{div}(L_1, s_1)\right).$$

(2) *We assume that Y is flat over* $\operatorname{Spec}(\mathbb{Z})$ *and generically smooth. Let h_1 and h_2 be C^∞-hermitian metrics of L_1 and L_2 on $Y(\mathbb{C})$, respectively, and let $\overline{L}_1 = (L_1, h_1)$ and $\overline{L}_2 = (L_2, h_2)$. Then, there are $A \in D^{(p-1,p)}(X(\mathbb{C}))$ and $B \in D^{(p,p-1)}(X(\mathbb{C}))$ such that*
$$j_*\left((\overline{L}_1, s_1) \cdot \widehat{\operatorname{div}}(\overline{L}_2, s_2)\right) = j_*\left((\overline{L}_2, s_2) \cdot \widehat{\operatorname{div}}(\overline{L}_1, s_1)\right) + (0, \partial(A) + \bar{\partial}(B)).$$

PROOF. (1) follows from (2) in Proposition 5.2, and (2) follows from (2) in Proposition 5.2 and Theorem 5.6. □

Here let us consider the fundamental theorem.

THEOREM 5.20. *Let X be a generically smooth arithmetic variety and $\overline{L}$ a C^∞-hermitian invertible sheaf on X. Then there is a homomorphism*
$$\widehat{c}_1(\overline{L})\cdot : \widehat{\operatorname{CH}}^p_D(X) \to \widehat{\operatorname{CH}}^{p+1}_D(X) \quad (\alpha \mapsto \widehat{c}_1(\overline{L}) \cdot \alpha)$$
constructed as follow: Let $(Z, T) \in \widehat{Z}^p_D(X)$ be a representative of α and let s be a non-zero rational section of L such that $\operatorname{Supp}(L, s) \cap \operatorname{Supp}(Z) \cap X^{(p)} = \emptyset$. *Then $\widehat{c}_1(\overline{L}) \cdot \alpha$ is the class of $(\overline{L}, s) \cdot (Z, T)$. The meaning of this theorem is that this definition does not depend on the choice of (Z, T) and s. Moreover, the homomorphism $\widehat{c}_1(\overline{L})\cdot : \widehat{\operatorname{CH}}^p_D(X) \to \widehat{\operatorname{CH}}^{p+1}_D(X)$ satisfies the following properties.*

(1) *Let $\overline{L}_1$ and $\overline{L}_2$ be C^∞-hermitian invertible sheaves on X. Then, for any $\alpha \in \widehat{\operatorname{CH}}^p_D(X)$,*
$$\widehat{c}_1(\overline{L}_1 \otimes \overline{L}_2) \cdot \alpha = \widehat{c}_1(\overline{L}_1) \cdot \alpha + \widehat{c}_1(\overline{L}_2) \cdot \alpha.$$
Moreover, $\widehat{c}_1(\overline{\mathscr{O}}_X^{can}) \cdot \alpha = 0$.

(2) (Arithmetic projection formula) *Let $f : X \to Y$ be a projective morphism of generically smooth varieties and $\overline{M}$ a C^∞-hermitian invertible sheaf on Y. Then, for any $\alpha \in \widehat{\operatorname{CH}}^p_D(X)$,*
$$f_*(\widehat{c}_1(f^*(\overline{M})) \cdot \alpha) = \widehat{c}_1(\overline{M}) \cdot f_*\alpha.$$

(3) (Commutativity of product) *Let $\overline{L}_1$ and $\overline{L}_2$ be C^∞-hermitian invertible sheaves on X. Then, for any $\alpha \in \widehat{\operatorname{CH}}^p_D(X)$,*
$$\widehat{c}_1(\overline{L}_1) \cdot (\widehat{c}_1(L_2) \cdot \alpha) = \widehat{c}_1(\overline{L}_2) \cdot (\widehat{c}_1(L_1) \cdot \alpha).$$

PROOF. Let (Z, T) be an arithmetic cycle of codimension p on X. By Proposition 1.22, there is a non-zero rational section s of L with
$$\operatorname{Supp}(L, s) \cap \operatorname{Supp}(Z) \cap X^{(p)} = \emptyset.$$
First we would like to show that the class of $(\overline{L}, s) \cdot (Z, T)$ in $\widehat{\operatorname{CH}}^{p+1}_D(X)$ does not depend on the choice of s. Let s' be another non-zero rational section of L with

$\operatorname{Supp}(L, s') \cap \operatorname{Supp}(Z) \cap X^{(p)} = \emptyset$. Then there is a non-zero rational function ϕ with $s' = \phi s$. Note that $\operatorname{Supp}(\mathscr{O}_X, \phi) \cap \operatorname{Supp}(Z) \cap X^{(p)} = \emptyset$ and $(\overline{L}, s') = (\overline{L} \otimes \overline{\mathscr{O}}_X^{can}, s \otimes \phi)$. Therefore, by Proposition 5.17,

$$(\overline{L}, s') \cdot (Z, T) = (\overline{L}, s) \cdot (Z, T) + (\overline{\mathscr{O}}_X^{can}, \phi) \cdot (Z, T).$$

Thus, if we set $Z = \sum_i n_i Z_i$, then

$$(\overline{\mathscr{O}}_X^{can}, \phi) \cdot (Z, T) = \sum_i n_i \widehat{(\phi/Z_i)}$$

because $\partial\bar{\partial}(\log|\phi|) = 0$. Therefore,

$$(\overline{L}, s') \cdot (Z, T) - (\overline{L}, s) \cdot (Z, T) \in \widehat{\operatorname{Rat}}^{p+1}(X),$$

as required.

By the above discussion, we have a homomorphism

$$\gamma(\overline{L}) : \widehat{Z}_D^p(X) \to \widehat{\operatorname{CH}}_D^{p+1}(X)$$

given by $(Z, T) \mapsto$ the class of $(\overline{L}, s) \cdot (Z, T)$. In order to show the first part of the theorem, it is sufficient to show the following:

$$\gamma(\overline{L})(\widehat{\operatorname{Rat}}^p(X)) = 0. \tag{5.9}$$

Since $d(c_1(\overline{L})) = 0$, we have

$$\gamma(\overline{L})((0, \partial(A) + \bar{\partial}(B))) = \text{the class of } (0, \partial(c_1(\overline{L}) \wedge A) + \bar{\partial}(c_1(\overline{L}) \wedge B)) = 0$$

for $A \in D^{p-2,p-1}(X(\mathbb{C}))$ and $B \in D^{p-1,p-2}(X(\mathbb{C}))$. Thus, in order to see (5.9), it is sufficient to show that $\gamma(\overline{L})(\widehat{(\phi/Y)}) = 0$ for a codimension $p-1$ integral subscheme Y on X and $\phi \in \operatorname{Rat}(Y)^\times$. Let $j : Y \hookrightarrow X$ be the inclusion map.

First we consider the case where Y is not flat over $\operatorname{Spec}(\mathbb{Z})$. Let s be a non-zero rational section with $\operatorname{Supp}(L, s) \cap \operatorname{Supp}(\mathscr{O}_Y, \phi) \cap Y^{(1)} = \emptyset$. Then, by Proposition 5.18 and (1) in Proposition 5.19,

$$\begin{aligned}(\overline{L}, s) \cdot \widehat{(\phi/Y)} &= j_*\left(j^*(\overline{L}, s) \cdot (\operatorname{div}(\mathscr{O}_Y, \phi), 0)\right) \\ &= j_*\left(\left(L|_Y, s|_Y\right) \cdot \operatorname{div}(\mathscr{O}_Y, \phi), 0\right) \\ &= j_*\left((\mathscr{O}_Y, \phi) \cdot \operatorname{div}(L|_Y, s|_Y), 0\right).\end{aligned}$$

Therefore, if we set $\operatorname{div}(L|_Y, s|_Y) = \sum_i n_i Z_i$, then we have $(\overline{L}, s) \cdot \widehat{(\phi/Y)} \in \widehat{\operatorname{Rat}}^{p+1}(X)$ because

$$(\mathscr{O}_Y, \phi) \cdot \operatorname{div}(L|_Y, s|_Y) = \sum_i n_i \operatorname{div}(\mathscr{O}_{Z_i}, \phi|_{Z_i}).$$

Thus (5.9) holds in the case where Y is not flat over $\operatorname{Spec}(\mathbb{Z})$.

Next we consider the case where Y is flat over $\operatorname{Spec}(\mathbb{Z})$ and generically smooth. Let s be a non-zero rational section of L with $\operatorname{Supp}(L, s) \cap \operatorname{Supp}(\mathscr{O}_Y, \phi) \cap Y^{(1)} = \emptyset$. Then, by Proposition 5.18 and (2) in Proposition 5.19, there are $A \in D^{(p-1,p)}(X(\mathbb{C}))$ and $B \in D^{(p,p-1)}(X(\mathbb{C}))$ such that

$$\begin{aligned}(\overline{L}, s) \cdot \widehat{(\phi/Y)} &= j_*\left(\left(\overline{L}|_Y, s|_Y\right) \cdot \widehat{\operatorname{div}}(\overline{\mathscr{O}}_Y^{can}, \phi)\right) \\ &= j_*\left((\overline{\mathscr{O}}_Y^{can}, \phi) \cdot \widehat{\operatorname{div}}(\overline{L}|_Y, s|_Y)\right) + (0, \partial(A) + \bar{\partial}(B)).\end{aligned}$$

We set $\operatorname{div}(L|_Y, s|_Y) = \sum_i n_i Z_i$. Then, as $\partial\bar{\partial}(\log|\phi|) = 0$, we have

$$j_*\left((\overline{\mathscr{O}}_Y^{can}, \phi)\cdot \widehat{\operatorname{div}}(\overline{L}\big|_Y, s|_Y)\right) = \sum_i n_i \left((\phi|_{Z_i})/Z_i\right)^{\widehat{}},$$

that is, $(\overline{L}, s)\cdot \widehat{(\phi/Y)} \in \widehat{\operatorname{Rat}}^{p+1}(X)$.

Finally we consider the case where Y is flat over $\operatorname{Spec}(\mathbb{Z})$ and is not necessarily generically smooth. First we consider the following claim.

CLAIM 3. *We can construct a projective birational morphism $\mu : X' \to X$ of generically smooth arithmetic varieties with the following properties.*

(a) *There is a non-empty Zariski opens set U of X such that $\mu^{-1}(U) \to U$ is an isomorphism and $U \cap Y \neq \emptyset$.*
(b) *Let Y' be the strict transform of Y by μ, that is,*

$$Y' = \text{the Zariski closure of } \mu^{-1}(U\cap Y).$$

Then Y' is generically smooth.

Moreover, we set the induced morphisms as follows:

$$\begin{array}{ccc} Y' & \xrightarrow{j'} & X' \\ {\scriptstyle\nu}\downarrow & & \downarrow{\scriptstyle\mu} \\ Y & \xrightarrow{j} & X \end{array}$$

PROOF. This follow from Hironaka's resolution of singularities [**30**]. □

Let s' be a non-zero rational section of $\mu^*(L)$ with

$$\operatorname{Supp}(\mu^*(L), s') \cap \operatorname{Supp}(\mathscr{O}_{Y'}, \nu^*\phi) \cap {Y'}^{(1)} = \emptyset.$$

Let η' and η be the generic points of X' and X. Then, since $L_\eta \to \mu^*(L)_{\eta'}$ is an isomorphism, there is a non-zero rational section s of L with $s' = \mu^* s$. Note that

$$(\mu^*(\overline{L}), \mu^* s)\cdot(\nu^*\phi/Y')^{\widehat{}} \in \widehat{\operatorname{Rat}}^{p+1}(X')$$

by the previous case. Thus, by Proposition 5.14,

$$\mu_*\left((\mu^*(\overline{L}), \mu^* s)\cdot(\nu^*\phi/Y')^{\widehat{}}\right) \in \widehat{\operatorname{Rat}}^{p+1}(X).$$

Since $N_{\operatorname{Rat}(Y')/\operatorname{Rat}(Y)}(\nu^*\phi) = \phi$, by Proposition 5.14, $\mu_*(\nu^*\phi/Y')^{\widehat{}} = \widehat{(\phi/Y)}$. Therefore, by using Proposition 5.18, we obtain

$$\mu_*\left((\mu^*(\overline{L}), \mu^* s)\cdot(\nu^*\phi/Y')^{\widehat{}}\right) = (\overline{L}, s)\cdot \mu_*(\nu^*\phi/Y')^{\widehat{}} = (\overline{L}, s)\cdot\widehat{(\phi/Y)},$$

so that $(\overline{L}, s)\cdot\widehat{(\phi/Y)} \in \widehat{\operatorname{Rat}}^{p+1}(X)$. Thus we get (5.9).

From now on, we consider the properties (1), (2) and (3). The properties (1) and (2) follow from Proposition 5.17 and Proposition 5.18, respectively, so we consider the property (3). Let Y be a codimension p integral subscheme of X. We assume that α is the class of $(Y, 0)$.

First let us begin with the case where either Y is not flat over $\operatorname{Spec}(\mathbb{Z})$, or Y is flat over $\operatorname{Spec}(\mathbb{Z})$ and generically smooth. Let s_1 and s_2 be non-zero rational sections of L_1 and L_2 with $\operatorname{Supp}(L_1, s_1)\cap\operatorname{Supp}(L_2, s_2)\cap Y^{(1)} = \emptyset$. Then, by Proposition 5.19

and Proposition 5.18, there are $A \in D^{(p-1,p)}(X(\mathbb{C}))$ and $B \in D^{(p,p-1)}(X(\mathbb{C}))$ such that

$$(5.10)\quad (\overline{L}_1, s_1) \cdot ((\overline{L}_2, s_2) \cdot (Y, 0)) = (\overline{L}_2, s_2) \cdot ((\overline{L}_1, s_1) \cdot (Y, 0)) + (0, \partial(A) + \bar{\partial}(B)),$$

which shows (3) for the above Y.

Next we consider the case where Y is flat over $\mathrm{Spec}(\mathbb{Z})$ and not generically smooth. Let us consider a projective birational morphism $\mu : X' \to X$ of generically smooth arithmetic varieties as in Claim 3. In the following, we use the same notation as in Claim 3. Let s'_1 and s'_2 be non-zero rational sections of $\mu^*(L_1)$ and $\mu^*(L_2)$, respectively, such that

$$\mathrm{Supp}(\mu^*(L_1), s'_1) \cap \mathrm{Supp}(\mu^*(L_2), s'_2) \cap Y'^{(1)} = \emptyset.$$

Then there are non-zero rational sections s_1 and s_2 of L_1 and L_2 with $s'_1 = \mu^* s_1$ and $s'_2 = \mu^* s_2$. Thus, the formula (5.10) on Y' implies

$$\begin{aligned}(\mu^*\overline{L}_1, \mu^* s_1) &\cdot ((\mu^*\overline{L}_2, \mu^* s_2) \cdot (Y', 0)) \\ &= (\mu^*\overline{L}_2, \mu^* s_2) \cdot ((\mu^*\overline{L}_1, \mu^* s_1) \cdot (Y', 0)) + (0, \partial(A') + \bar{\partial}(B')),\end{aligned}$$

where $A' \in D^{(p-1,p)}(X'(\mathbb{C}))$ and $B' \in D^{(p,p-1)}(X'(\mathbb{C}))$. By applying μ_* to the above equation and using Proposition 5.18, we obtain the same equation as in (5.10) if we set $A = \mu_* A'$ and $B = \mu_* B'$.

Note that each side of the assertion of (3) is linear with respect to α. By the above discussion, if α is the class of $(Y, 0)$, then (3) holds. Therefore, (3) also holds even if α is the class of $(Z, 0)$ for some codimension p cycle, so that it is sufficient to consider the case where α is the class of $(0, T)$. Then, for non-zero rational sections s_1 and s_2 of L_1 and L_2,

$$\begin{cases} (\overline{L}_1, s_1) \cdot ((\overline{L}_2, s_2) \cdot (0, T)) = (0, c_1(\overline{L}_1) \wedge c_1(\overline{L}_2) \wedge T), \\ (\overline{L}_2, s_2) \cdot ((\overline{L}_1, s_1) \cdot (0, T)) = (0, c_1(\overline{L}_2) \wedge c_1(\overline{L}_2) \wedge T). \end{cases}$$

Thus (3) follows. □

DEFINITION 5.21. Let X be a generically smooth arithmetic variety and $\overline{L}$ a C^∞-hermitian invertible sheaf on X. Then the element $\widehat{c}_1(\overline{L}) \cdot (X, 0) \in \widehat{\mathrm{CH}}^1(X)$ is called the *arithmetic first Chern class* of $\overline{L}$. By abuse of notation, it is often denoted by the same symbol $\widehat{c}_1(\overline{L})$. The class of $\widehat{\mathrm{div}}(\overline{L}, s)$ in $\widehat{\mathrm{CH}}^1(X)$ is nothing more than $\widehat{c}_1(\overline{L}) \cdot (X, 0)$.

DEFINITION 5.22. Let X be a d-dimensional generically smooth arithmetic variety and let (Z, T) be an element of $\widehat{Z}^d_D(X) = \widehat{Z}_{D,0}(X)$, that is, Z is a cycle consisting of closed points of X, and T is a current of type $(d-1, d-1)$. If we write $Z = \sum_{x \in X_{(0)}} n_x x$, then $\widehat{\deg}(Z, T)$ is defined to be

$$\widehat{\deg}(Z, T) = \sum_{x \in X_{(0)}} n_x \log \#(\kappa(x)) + \frac{1}{2} T(1).$$

It is called the *arithmetic degree* of (Z, T)

PROPOSITION 5.23. *The arithmetic degree has the following basis properties:*

(1) *Let $f : X \to Y$ be a projective morphism of generically smooth arithmetic varieties. Then*
$$\widehat{\deg}(Z,T) = \widehat{\deg}\ f_*(Z,T)$$
for $(Z,T) \in \widehat{Z}_{D,0}(X)$.

(2) *If $\alpha \in \widehat{\mathrm{Rat}}_0(X)$, then $\widehat{\deg}(\alpha) = 0$. In particular, $\widehat{\deg} : \widehat{Z}_{D,0}(X) \to \mathbb{R}$ induces*
$$\widehat{\deg} : \widehat{\mathrm{CH}}_{D,0}(X) \to \mathbb{R}.$$

(3) *Let $z = \sum_{i=1}^r n_i P_i$ be a cycle consisting of closed point of X such that all P_i's belong to a fiber X_p of $X \to \mathrm{Spec}(\mathbb{Z})$ over a prime p. Then*
$$\widehat{\deg}(z,0) = (\log p)\deg(z; X_p),$$
where $\deg(z;X_p)$ is the degree of the 0-cycle z in the algebraic scheme X_p defined over $\mathbb{F}_p$.

(4) *Let Z be an l-dimensional subscheme of X such that Z is a subscheme of the fiber X_p of $X \to \mathrm{Spec}(\mathbb{Z})$ over a prime p. Then*
$$\widehat{\deg}\left(\widehat{c}_1(\overline{L})^l \cdot (Z,0)\right) = (\log p)\deg(c_1(L)^l \cdot Z; X_p).$$

PROOF. (1) If $x \in X_{(0)}$, then $f_*x = [\kappa(x) : \kappa(f(x))]f(x)$. Thus
$$\widehat{\deg}(f_*x,0) = [\kappa(x):\kappa(f(x))]\log\#(\kappa(f(x))) = \log\#(\kappa(x)) = \widehat{\deg}(x,0).$$
Therefore, $\widehat{\deg}(Z,0) = \widehat{\deg}(f_*Z,0)$ for any $Z \in Z_0(X)$. On the other hand, for $T \in D^{\dim X-1,\dim X-1}(X(\mathbb{C}))$, $f_*T(1) = T(f^*1) = T(1)$. Thus $\widehat{\deg}(0,T) = \widehat{\deg}(0,f_*T)$. Hence (1) follows.

(2) By Proposition 5.14, $f_*(\widehat{\mathrm{Rat}}_0(X)) \subseteq \widehat{\mathrm{Rat}}_0(\mathrm{Spec}(\mathbb{Z}))$. Thus (2) follows from (1) and Theorem 3.3.

(3) Let V be a finite-dimensional vector space over $\mathbb{F}_p$. Then $\log\#(V) = (\log p)\dim_{\mathbb{F}_p} V$. Thus (3) follows.

(4) Since $\widehat{c}_1(\overline{L})^l \cdot (Z,0) = (c_1(L)^l \cdot Z, 0)$, it follows from (3). □

DEFINITION 5.24. Let X be a d-dimensional projective arithmetic variety and let $\overline{L}_1,\ldots,\overline{L}_d$ be C^∞-hermitian invertible sheaves on X. If X is generically smooth, then the number
$$\widehat{\deg}\left(\widehat{c}_1(\overline{L}_1)\cdots\widehat{c}_1(\overline{L}_d)\cdot(X,0)\right)$$
is called the *arithmetic intersection number* of $\overline{L}_1,\ldots,\overline{L}_d$ and it is denoted by
$$\widehat{\deg}\left(\widehat{c}_1(\overline{L}_1)\cdots\widehat{c}_1(\overline{L}_d)\right)$$
for short.

Even if X is not generically smooth, the arithmetic intersection number
$$\widehat{\deg}\left(\widehat{c}_1(\overline{L}_1)\cdots\widehat{c}_1(\overline{L}_d)\right)$$
is defined as follows: Let $\mu : X' \to X$ be a generic resolution of singularities of X. Let us consider the arithmetic intersection number
$$\widehat{\deg}\left(\widehat{c}_1(\mu^*(\overline{L}_1))\cdots\widehat{c}_1(\mu^*(\overline{L}_d))\right)$$
of $\mu^*(\overline{L}_1),\ldots,\mu^*(\overline{L}_d)$ on X'. We can check that this does not depend on the choice of a generic resolution of singularities. Indeed, let $\mu' : X'' \to X$ be another generic resolution of singularities of X. Then there is a generic resolution of singularities

$\nu : Y \to X$ such that $\nu : Y \to X$ factors $Y \xrightarrow{\nu'} X' \xrightarrow{\mu} X$ and $Y \xrightarrow{\nu''} X'' \xrightarrow{\mu'} X$, that is, the following diagram is commutative:

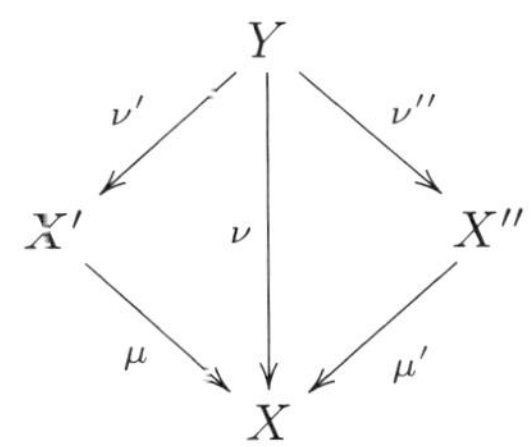

Applying the projection formula (cf. Proposition 5.18) to $\nu' : Y \to X'$ and $\nu'' : Y \to X''$, we have

$$\begin{cases} \widehat{\deg}\left(\widehat{c}_1(\nu^*(\overline{L}_1)) \cdots \widehat{c}_1(\nu^*(\overline{L}_d))\right) = \widehat{\deg}\left(\widehat{c}_1(\mu^*(\overline{L}_1)) \cdots \widehat{c}_1(\mu^*(\overline{L}_d))\right), \\ \widehat{\deg}\left(\widehat{c}_1(\nu^*(\overline{L}_1)) \cdots \widehat{c}_1(\nu^*(\overline{L}_d))\right) = \widehat{\deg}\left(\widehat{c}_1(\mu'^*(\overline{L}_1)) \cdots \widehat{c}_1(\mu'^*(\overline{L}_d))\right), \end{cases}$$

that is, the arithmetic intersection number $\widehat{\deg}\left(\widehat{c}_1(\mu^*(\overline{L}_1)) \cdots \widehat{c}_1(\mu^*(\overline{L}_d))\right)$ does not depend on the choice of the generic resolution of singularities, so that it is called the arithmetic intersection number of $\overline{L}_1, \ldots, \overline{L}_d$ and it is denoted by

$$\widehat{\deg}\left(\widehat{c}_1(\overline{L}_1) \cdots \widehat{c}_1(\overline{L}_d)\right).$$

For $(Y, T_Y) \in \widehat{Z}^p(X)$ and $(Z, T_Z) \in \widehat{Z}^q(X)$, a way to define the intersection cycle $(Y, T_Y) \cdot (Z, T_Z)$ is one of important subjects of Arakelov geometry. Let $\widehat{\mathrm{Pic}}(X)$ be the abelian group consisting of isomorphism classes of C^∞-hermitian invertible sheaves on X. By the arguments as above, we can construct a map

$$\widehat{\mathrm{Pic}}^1(X) \times \widehat{\mathrm{CH}}^p_D(X) \to \widehat{\mathrm{CH}}^{p+1}_D(X) \quad ((\overline{L}, \alpha) \mapsto \widehat{c}_1(\overline{L}) \cdot \alpha).$$

It is known that if X is generically smooth and α is of Green type, then so is $\widehat{c}_1(\overline{L}) \cdot \alpha$. We only recall the intersection theory of general arithmetic cycles. For details, see [**23**, Theorem 4.2.3] or [**64**, Theorem 2 in Chapter 3].

THEOREM 5.25. *We assume that X is regular. Then, for all integers $p, q \geq 0$, there are commutative and associative bilinear maps*

$$\widehat{\mathrm{CH}}^p(X) \times \widehat{\mathrm{CH}}^q(X) \to \widehat{\mathrm{CH}}^{p+q}(X) \otimes_{\mathbb{Z}} \mathbb{Q}.$$

Therefore, the above maps yield a ring structure on $\bigoplus_{p \geq 0} \widehat{\mathrm{CH}}^p(X) \otimes_{\mathbb{Z}} \mathbb{Q}$.

Moreover, they have the following functoriality (for details, see [**64**, Theorem 3 in Chapter 3]).

THEOREM 5.26. *Let $f : X \to Y$ be a morphism of regular arithmetic varieties. Then we have the following.*

(1) *There are the pull-back homomorphisms $f^* : \widehat{\mathrm{CH}}^p(Y) \to \widehat{\mathrm{CH}}^p(X) \otimes \mathbb{Q}$ for all $p \geq 0$ such that they yield the ring homomorphism*

$$\bigoplus_{p \geq 0} \widehat{\mathrm{CH}}^p(Y) \otimes_{\mathbb{Z}} \mathbb{Q} \to \bigoplus_{p \geq 0} \widehat{\mathrm{CH}}^p(X) \otimes_{\mathbb{Z}} \mathbb{Q}.$$

(2) *If f is projective and $f_{\mathbb{Q}} : X_{\mathbb{Q}} \to Y_{\mathbb{Q}}$ is smooth, then the push-forward of Definition 5.13 gives rise to $f_* : \widehat{\mathrm{CH}}_l(X) \to \widehat{\mathrm{CH}}_l(Y)$.*

(3) (Projection formula) *If f is the same as above, then $f_*(f^*(\alpha)\cdot\beta) = \alpha\cdot f_*(\beta)$ for $\alpha \in \widehat{\mathrm{CH}}^p(Y)$ and $\beta \in \widehat{\mathrm{CH}}^q(X)$.*

(4) *If $f : X \to Y$ and $g : Y \to Z$ are morphisms of regular arithmetic varieties, then $(g \circ f)^* = f^* \circ g^*$ holds.*

5.5. Characteristic classes and Bott-Chern secondary characteristic form

Let k be a field and $f(T_1, \ldots, T_r)$ a symmetric k-coefficient polynomial with variables $T_1, \ldots, T_r$. Then we can find a unique k-coefficient polynomial $\boldsymbol{f}(\mathbb{T})$ with variables $r \times r$-matrix $\mathbb{T} = (T_{ij})_{\substack{1\leq i\leq r\\1\leq j\leq r}}$ satisfying the properties (1) and (2) below.

(1) $\boldsymbol{f}\begin{pmatrix} T_1 & & 0 \\ & \ddots & \\ 0 & & T_r \end{pmatrix} = f(T_1, \ldots, T_r)$.

(2) If $\mathbb{T}' = (T'_{ij})$ are another matrix variables, then $\boldsymbol{f}(\mathbb{T}\cdot\mathbb{T}') = \boldsymbol{f}(\mathbb{T}'\cdot\mathbb{T})$.

A proof is left to the reader. By property (2), $\boldsymbol{f}$ is invariant for similar matrices.

Let $\phi \in \mathbb{R}[[T_1, \ldots, T_r]]$ be a symmetric formal power series over $\mathbb{R}$ and ϕ_p the degree p homogeneous part of ϕ. Notice that $\phi = \sum_{p\geq 0}\phi_p$. Let X be a complex manifold and $\overline{E} = (E, h)$ a rank r C^∞-hermitian locally free coherent sheaf on X. We define $\phi(\overline{E})$ to be

$$\phi(\overline{E}) := \sum_{p\geq 0}\boldsymbol{\phi}_p\left(\frac{\sqrt{-1}}{2\pi}\Theta_{\overline{E}}\right) \in \bigoplus_{p\geq 0} A^{p,p}(X)$$

and it is called the *characteristic from* of $\overline{E}$ associated with ϕ. As examples of ϕ, we have the following:

(i) the k-th Chern form : $c_k = \sum_{1\leq i_1<\cdots<i_k\leq r} T_{i_1}\cdots T_{i_k}$,
(ii) the total Chern form : $c = \sum_{k\geq 0} c_k$,
(iii) the Chern characteristic form : $\mathrm{ch}(T_1, \ldots, T_r) = \sum_{i=1}^r \exp(T_i)$,
(iv) the Todd form: $\mathrm{td}(T_1, \ldots, T_r) = \prod_{i=1}^r \frac{T_i}{1-\exp(-T_i)}$.

We define $\phi_0, \phi_1, \ldots$ by the following formula:

$$\phi(T_1 + T, \ldots, T_r + T) = \sum_{i\geq 0}\phi_i(T_1, \ldots, T_r)T^i.$$

Then we can see

$$\phi(\overline{E}\otimes\overline{L}) = \sum_{i\geq 0}\phi_i(\overline{E})c_1(\overline{L})^i \tag{5.11}$$

for a C^∞-hermitian invertible sheaf $\overline{L}$.

Next we discuss the Bott-Chern secondary characteristic form. First, we set

$$\tilde{A}^{p,p}(X) = A^{p,p}(X)/(\mathrm{Image}(\partial) + \mathrm{Image}(\bar{\partial})) \quad \text{and} \quad \tilde{A}(X) = \bigoplus_{p\geq 0}\tilde{A}^{p,p}(X).$$

We fix a symmetric formal power series $\phi \in \mathbb{R}[[T_1, \ldots, T_r]]$. Let

$$\overline{\mathcal{E}} : 0 \to (S, h') \xrightarrow{\alpha} (E, h) \xrightarrow{\beta} (Q, h'') \to 0$$

be an exact sequence of locally free coherent sheaves on X with C^∞-hermitian metrics, that is, forgetting metrics, the sequence $0 \to S \xrightarrow{\alpha} E \xrightarrow{\beta} Q \to 0$ is exact and the metrics h', h, h'' of S, E, Q are independent from the exact sequence, where the rank of E is r. Then it is well known that there is $\eta \in \tilde{A}(X)$ with

$$\phi((S,h') \oplus (Q,h'')) - \phi((E,h)) = \frac{\sqrt{-1}}{2\pi}\partial\bar{\partial}(\eta).$$

For example, if X is a compact Kähler manifold, it follows from $\partial\bar{\partial}$-Lemma (cf. Lemma 1.36). Here we recall the construction of η.

Let $p_1 : X \times \mathbb{P}^1 \to X$ and $p_2 : X \times \mathbb{P}^1 \to \mathbb{P}^1$ be the projections to the first factor and the second factor, respectively. By abuse of notation, $p_1^*(S)$, $p_1^*(E)$, $p_1^*(Q)$ and $p_1^*(\alpha)$, $p_1^*(\beta)$ are also denoted by S, E, Q and α, β, respectively. Moreover, we set $S(1) = S \otimes p_2^*(\mathscr{O}(1))$ on $X \times \mathbb{P}^1$. Let t be a global section of $\mathscr{O}(1)$ such that t is zero at ∞. A homomorphism

$$p_2^*(t) : \mathscr{O}_{X\times\mathbb{P}^1} \to p_2^*(\mathscr{O}(1))$$

induced by $p_2^*(t)$ is also denoted by t by abuse of notation. We define $\widetilde{E}$ by

$$\widetilde{E} = \operatorname{Coker}\left(\alpha \oplus \operatorname{id}_S \otimes t) : S \to E \oplus S(1)\right).$$

Then, as the kernel of the natural homomorphism $\widetilde{E} \to Q$ is $S(1)$, we have an exact sequence

$$0 \to S(1) \to \widetilde{E} \to Q \to 0.$$

Note that $\widetilde{E}\big|_{X\times\{0\}} \simeq E$ and $\widetilde{E}\big|_{X\times\{\infty\}} \simeq S \oplus Q$. Here we take a C^∞-hermitian metric $\widetilde{h}$ of $\widetilde{E}$ satisfying the following conditions.

$$(\widetilde{E},\widetilde{h})|_{X\times\{0\}} \simeq (E,h) \quad\text{and}\quad (\widetilde{E},\widetilde{h})|_{X\times\{\infty\}} \simeq (S,h') \oplus (Q,h''). \tag{5.12}$$

Let z be the inhomogeneous coordinate of $\mathbb{P}^1$. Then $\widetilde{\phi}(\overline{\mathcal{E}})$ is defined by the fiber integral with respect to $p_1 : X \times \mathbb{P}^1 \to X$, that is,

$$\widetilde{\phi}(\overline{\mathcal{E}}) = -\int_{\mathbb{P}^1} \phi(\widetilde{E},\widetilde{h}) \ \log|z|^2,$$

which gives the above η and it does not depend on the choice of $\widetilde{h}$ in $\tilde{A}(X)$. It is called the *Bott-Chern secondary characteristic form* of $\overline{\mathcal{E}}$ associated with ϕ. It is known that it satisfies the following properties.

(1) $\phi(\overline{S} \oplus \overline{Q}) - \phi(\overline{E}) = \dfrac{\sqrt{-1}}{2\pi}\partial\bar{\partial}(\widetilde{\phi}(\overline{\mathcal{E}}))$.

(2) For an analytic map $\pi : Y \to X$ of complex manifolds, $\widetilde{\phi}(\pi^*(\overline{\mathcal{E}})) = \pi^*\widetilde{\phi}(\overline{\mathcal{E}})$.

(3) If $\overline{E} = \overline{S} \oplus \overline{Q}$, then $\widetilde{\phi}(\overline{\mathcal{E}}) = 0$.

For details, see [**24**, Theorem 1.2.2] or [**64**, Theorem 2 in Chapter 4]. Let h and h' be two C^∞-metrics of E. From now on, we denote $\widetilde{\phi}(0 \to (E,h) \to (E,h') \to (0,0) \to 0)$ by $\widetilde{\phi}(E,h,h')$. For example, $\widetilde{c}_1(E,h,h') = \log(\det(h')/\det(h))$.

As before, we define $\phi_0, \phi_1, \ldots$ by

$$\phi(T_1 + T, \ldots, T_r + T) = \sum\nolimits_{i\ge 0} \phi_i(T_1,\ldots,T_r)T^i.$$

Then, by (5.11) and its construction, we can see

$$\widetilde{\phi}(\overline{E} \otimes \overline{L}) = \sum_{i\ge 0} \widetilde{\phi}_i(\overline{E}) c_1(\overline{L})^i \tag{5.13}$$

for a C^∞-hermitian invertible sheaf $\overline{L}$.

Let h_{FS} be the Fubini-Study metric of $\mathscr{O}(1)$. Let $\overline{\mathcal{E}}(\widetilde{h})$ denote the exact sequence

$$0 \to (S(1), h' \otimes h_{FS}) \to (\widetilde{E}, \widetilde{h}) \to (Q, h'') \to 0$$

with hermitian metrics. Finally we show the following proposition on the uniqueness of the secondary characteristic forms.

PROPOSITION 5.27. *Let us consider a certain kind of property* $\mathbf{P}$ *on* $\overline{\mathcal{E}} : 0 \to \overline{S} \to \overline{E} \to \overline{Q} \to 0$ *with the following conditions.*

(a) *Let* $\pi : Y \to X$ *be an analytic map of complex manifolds. If* $\overline{\mathcal{E}}$ *has the property* $\mathbf{P}$*, then* $\pi^*(\overline{\mathcal{E}})$ *has also the property* $\mathbf{P}$*.*

(b) *We assume that* $\overline{\mathcal{E}}$ *has the property* $\mathbf{P}$*. Then there is a hermitian metric* $\widetilde{h}$ *of* $\widetilde{E}$ *on* $X \times \mathbb{P}^1$ *such that* $\widetilde{h}$ *satisfies* (5.12) *and* $\overline{\mathcal{E}}(\widetilde{h})$ *has the property* $\mathbf{P}$*.*

We assume that we can assign $\check{\phi}(\overline{\mathcal{E}}) \in \tilde{A}(X)$ *to each* $\overline{\mathcal{E}}$ *having the property* $\mathbf{P}$ *in such a way that the following conditions are satisfied.*

(1) $\phi(\overline{S} \oplus \overline{Q}) - \phi(\overline{E}) = \dfrac{\sqrt{-1}}{2\pi}\partial\bar{\partial}(\check{\phi}(\overline{\mathcal{E}}))$.

(2) *For an analytic map* $\pi : Y \to X$ *of complex manifolds,* $\check{\phi}(\pi^*(\overline{\mathcal{E}})) = \pi^*\check{\phi}(\overline{\mathcal{E}})$.

(3) *If* $\overline{E} = \overline{S} \oplus \overline{Q}$*, then* $\check{\phi}(\overline{\mathcal{E}}) = 0$.

Then $\check{\phi}(\overline{\mathcal{E}}) = \widetilde{\phi}(\overline{\mathcal{E}})$ *in* $\tilde{A}(X)$.

PROOF. By (b), there is a hermitian metric $\widetilde{h}$ of $\widetilde{E}$ such that

$$\overline{\mathcal{E}}(\widetilde{h}) : 0 \to (S(1), h' \otimes h_{FS}) \to (\widetilde{E}, \widetilde{h}) \to (Q, h'') \to 0$$

has the property $\mathbf{P}$ on $X \times \mathbb{P}^1$. Let us consider the following integral I.

$$I = \int_{\mathbb{P}^1} \frac{\sqrt{-1}}{2\pi}\partial\bar{\partial}(\check{\phi}(\overline{\mathcal{E}}(\widetilde{h}))) \log|z|^2.$$

We try to calculate it in two ways. First, by using (1),

$$I = \int_{\mathbb{P}^1} \phi((S(1), h' \otimes h_{FS}) \oplus (Q, h'')) \log|z|^2 - \int_{\mathbb{P}^1} \phi(\widetilde{E}, \widetilde{h}) \log|z|^2.$$

Let $\iota : X \times \mathbb{P}^1 \to X \times \mathbb{P}^1$ be a map given by $\iota(x, z) = (x, 1/z)$. Then

$$\iota^*(S(1), h' \otimes h_{FS}) = (S(1), h' \otimes h_{FS}), \quad \iota^*(Q, h'') = (Q, h'').$$

Thus

$$\begin{aligned}
\int_{\mathbb{P}^1} & \phi((S(1), h' \otimes h_{FS}) \oplus (Q, h'')) \log|z|^2 \\
&= \int_{\mathbb{P}^1} \iota^*(\phi((S(1), h' \otimes h_{FS}) \oplus (Q, h'')) \log|z|^2) \\
&= \int_{\mathbb{P}^1} \phi(\iota^*(S(1), h' \otimes h_{FS}) \oplus \iota^*(Q, h'')) \log|1/z|^2 \\
&= -\int_{\mathbb{P}^1} \phi((S(1), h' \otimes h_{FS}) \oplus (Q, h'')) \log|z|^2.
\end{aligned}$$

Therefore, $\int_{\mathbb{P}^1} \phi((S(1), h' \otimes h_{FS}) \oplus (Q, h'')) \log|z|^2 = 0$, that is,

$$I = -\int_{\mathbb{P}^1} \phi(\widetilde{E}, \widetilde{h}) \log|z|^2 = \widetilde{\phi}(\overline{\mathcal{E}}). \tag{5.14}$$

Let ∂_x (resp. ∂_z) be the X-direction of ∂ (resp. $\mathbb{P}^1$-direction of ∂). In the same way, we define $\bar{\partial}_x$ and $\bar{\partial}_z$. Then let us see

$$\int_{\mathbb{P}^1} \partial_z\bar{\partial}_z(\eta)\log|z|^2 = -2\pi\sqrt{-1}\left(\eta|_{X\times\{0\}} - \eta|_{X\times\{\infty\}}\right) \tag{5.15}$$

for $\eta \in A^{p,p}(X\times\mathbb{P}^1)$. If η is an element of either $A^{p-1,p}(X\times\mathbb{P}^1)\wedge A^{1,0}(\mathbb{P}^1)$ or $A^{p,p-1}(X\times\mathbb{P}^1)\wedge A^{0,1}(\mathbb{P}^1)$ or $A^{p-1,p-1}(X\times\mathbb{P}^1)\wedge A^{1,1}(\mathbb{P}^1)$, then $\partial_z\bar{\partial}_z(\eta)=0$, so that (5.15) holds. Thus we may assume that η can be written by the form

$$\sum_i a_i\theta_i \quad (a_i \in A^0(X\times\mathbb{P}^1),\ \theta_i\in A^{p,p}(X)).$$

Then

$$\begin{cases} \int_{\mathbb{P}^1} \partial_z\bar{\partial}_z(\eta)\log|z|^2 = \sum_i\left(\int_{\mathbb{P}^1}\partial_z\bar{\partial}_z(a_i)\log|z|^2\right)\theta_i \\ \eta|_{X\times\{0\}} - \eta|_{X\times\{\infty\}} = \sum_i\left(a_i|_{X\times\{0\}} - a_i|_{X\times\{\infty\}}\right)\theta_i. \end{cases}$$

Therefore we can reduce the case where η is an element of $A^0(X\times\mathbb{P}^1)$. In this case, (5.15) follows from the Poincaré-Lelong formula (cf. Theorem 1.39).

Moreover, it is easy to see that

$$\begin{cases} \int_{\mathbb{P}^1}\partial_x(\bar{\partial}_x(\eta)\log|z|^2) = \partial_x\left(\int_{\mathbb{P}^1}\bar{\partial}_x(\eta)\log|z|^2\right), \\ \int_{\mathbb{P}^1}\partial_x(\bar{\partial}_z(\eta)\log|z|^2) = \partial_x\left(\int_{\mathbb{P}^1}\bar{\partial}_z(\eta)\log|z|^2\right), \\ \int_{\mathbb{P}^1}\bar{\partial}_x(\partial_z(\eta)\log|z|^2) = \bar{\partial}_x\left(\int_{\mathbb{P}^1}\partial_z(\eta)\log|z|^2\right), \end{cases}$$

for $\eta\in A^{p,p}(X\times\mathbb{P}^1)$. Thus,

$$\int_{\mathbb{P}^1}\partial_x\bar{\partial}_x(\eta)\log|z|^2,\quad \int_{\mathbb{P}^1}\partial_x\bar{\partial}_z(\eta)\log|z|^2 \quad\text{and}\quad \int_{\mathbb{P}^1}\bar{\partial}_x\partial_z(\eta)\log|z|^2$$

belong to $\mathrm{Image}(\partial)+\mathrm{Image}(\bar{\partial})$. Further, we have

$$\partial\bar{\partial} = (\partial_x+\partial_z)(\bar{\partial}_x+\bar{\partial}_z) = \partial_x\bar{\partial}_x + \partial_x\bar{\partial}_z - \bar{\partial}_x\partial_z + \partial_z\bar{\partial}_z.$$

Therefore,

$$\int_{\mathbb{P}^1}\partial\bar{\partial}(\eta)\log|z|^2 \equiv -2\pi\sqrt{-1}\left(\eta|_{X\times\{0\}} - \eta|_{X\times\{\infty\}}\right) \quad (\mathrm{mod}\ \mathrm{Image}(\partial)+\mathrm{Image}(\bar{\partial})) \tag{5.16}$$

for $\eta\in\bigoplus_{p\geq 0}A^{p,p}(X\times\mathbb{P}^1)$, and hence

$$I = \check{\phi}(\overline{\mathcal{E}}(\widetilde{h}))\Big|_{X\times\{0\}} - \check{\phi}(\overline{\mathcal{E}}(\widetilde{h}))\Big|_{X\times\{\infty\}}$$

in $\tilde{A}(X)$. Thus, using conditions (2) and (3) of $\check{\phi}$, we have

$$I = \check{\phi}(\overline{\mathcal{E}}) - \check{\phi}(\overline{E}\oplus\overline{S}) = \check{\phi}(\overline{\mathcal{E}})$$

in $\tilde{A}(X)$. Therefore, the proposition follows from the above equation and (5.14). $\square$

Remark 5.1. As examples of the property **P** satisfying conditions (a) and (b) in Proposition 5.27, we have the following. It is left to the reader to check conditions (a) and (b) for each example.

Example 1. h' is the submetric of h and h'' is the quotient metric of h.
Example 2. $Q = 0$.

Finally, let us consider the explicit formula of $\widetilde{\mathrm{ch}}_2$ for each example as in the above remark. First, since

$$\mathrm{ch}_2(T_1, \ldots, T_r) = \frac{1}{2}(T_1^2 + \cdots + T_r^2),$$

we have

$$\mathbf{ch}_2(X) = \frac{1}{2}\operatorname{tr}(\mathbb{T}^2).$$

Let

$$\overline{\mathcal{E}} : 0 \to (S, h') \xrightarrow{\alpha} (E, h) \xrightarrow{\beta} (Q, h'') \to 0$$

be an exact sequence of locally free coherent sheaves on X with C^∞-hermitian metrics. We assume that h' is the submetric of h and h'' is the quotient metric of h. Let $E = S \oplus S^\perp$ be the orthogonal decomposition with respect to h in the category of C^∞ vector bundles on X. Then there is $A \in A^{1,0}(\mathscr{H}om(S, S^\perp))$ such that

$$\nabla_{(E,h)} = \begin{pmatrix} \nabla_{(E',h')} & -A^* \\ A & \nabla_{(E'',h'')} \end{pmatrix}.$$

(for example, see [**37**, Proposition 6.4 and Proposition 6.6 in Chapter 1]), where A^* is the adjoint operator of A, that is, A^* is an element of $A^{0,1}(\mathscr{H}om(S^\perp, S))$ with

$$h(A\xi, \xi') = h(\xi, A^*\xi') \quad (\forall \xi \in A^0(S), \forall \xi' \in A^0(S^\perp)).$$

Note that $A = 0$ if and only if $(E, h) \simeq (S, h') \oplus (Q, h'')$.

PROPOSITION 5.28. *In the above situation,*

$$\widetilde{\mathrm{ch}}_2(\overline{\mathcal{E}}) = -\frac{\sqrt{-1}}{2\pi}\operatorname{tr}(A^* \wedge A).$$

In particular, if X is a compact Riemann surface, then $\int_X \widetilde{\mathrm{ch}}_2(\overline{\mathcal{E}}) \geq 0$.

PROOF. By Proposition 5.27 and Remark 5.1, it is sufficient to see that

$$\mathrm{ch}_2((S, h') \oplus (Q, h'')) - \mathrm{ch}_2((E, h)) = \frac{1}{4\pi^2}\partial\bar{\partial}(\operatorname{tr}(A^* \wedge A)).$$

On the other hand, we have

$$\begin{aligned} &\mathrm{ch}_2((S, h') \oplus (Q, h'')) - \mathrm{ch}_2((E, h)) \\ &\quad = \frac{1}{2}\operatorname{tr}\left(\left(\frac{\sqrt{-1}}{2\pi}\Theta_{(S,h')\oplus(Q,h'')}\right)^2\right) - \frac{1}{2}\operatorname{tr}\left(\left(\frac{\sqrt{-1}}{2\pi}\Theta_{(E,h)}\right)^2\right) \\ &\quad = -\frac{1}{8\pi^2}\left(\operatorname{tr}\left(\Theta^2_{(S,h')\oplus(Q,h'')}\right) - \operatorname{tr}\left(\Theta^2_{(E,h)}\right)\right). \end{aligned}$$

Thus the above equation is equivalent to

$$\operatorname{tr}\left(\Theta^2_{(E,h)}\right) - \operatorname{tr}\left(\Theta^2_{(S,h')\oplus(Q,h'')}\right) = 2\partial\bar{\partial}(\operatorname{tr}(A^* \wedge A)).$$

We set

$$\nabla_t = \nabla_{(E,h)} + (e^t - 1)\begin{pmatrix} 0 & 0 \\ A & 0 \end{pmatrix} \quad \text{and} \quad \Theta_t = (\nabla_t)^2.$$

Then, by [**9**, Proposition 3.28 and Lemma 4.7], we obtain

$$\operatorname{tr}(\Theta^2_{(E,h)}) - \operatorname{tr}(\Theta^2_t) = -\partial\bar{\partial}\left(\int_t^0 \left\{\operatorname{tr}\left(\Theta_t \cdot \begin{pmatrix} 1 & 0 \\ 0 & 0 \end{pmatrix}\right) + \operatorname{tr}\left(\begin{pmatrix} 1 & 0 \\ 0 & 0 \end{pmatrix} \cdot \Theta_t\right)\right\} dt\right).$$

Since $\operatorname{tr}(\Theta_{(S,h')})$ is d-closed, by an easy calculation, we can see

$$\operatorname{tr}(\Theta^2_{(E,h)}) - \operatorname{tr}(\Theta^2_t) = 2(1 - e^t)\partial\bar{\partial}(\operatorname{tr}(A^* \wedge A)).$$

Note that

$$\lim_{t \to -\infty} \operatorname{tr}(\Theta^2_t) = \operatorname{tr}\left(\Theta^2_{(S,h') \oplus (Q,h'')}\right).$$

Thus we have the first half of the proposition.

Next we assume that X is a compact Riemann surface. We set $A = \phi dz$ ($\phi \in A^0(\mathscr{H}\!om(S, S^\perp))$) locally. Then $A^* = \phi^* d\bar{z}$, where ϕ^* is the adjoint of ϕ with respect to h. Thus

$$-\sqrt{-1}\operatorname{tr}(A^* \wedge A) = -\sqrt{-1}\operatorname{tr}(\phi^* \cdot \phi) d\bar{z} \wedge dz = \operatorname{tr}(\phi^* \cdot \phi)(\sqrt{-1} dz \wedge d\bar{z}),$$

which shows the last half of the proposition. □

Finally we consider the case where $Q = 0$, that is,

$$\widetilde{\operatorname{ch}}_2(E, h, h') = \widetilde{\operatorname{ch}}_2(0 \to (E,h) \to (E,h') \to (0,0) \to 0).$$

Let $\{h_t\}_{0 \le t \le 1}$ be a C^∞ deformation of metric of E with $h_0 = h'$ and $h_1 = h$. If we set

$$Q_2(E, h, h') = \sqrt{-1} \int_0^1 \operatorname{tr}(h_t^{-1} \partial_t(h_t) \Theta_{(E,h_t)}) dt,$$

the the following holds.

PROPOSITION 5.29. *$\widetilde{\operatorname{ch}}_2(E, h, h') = -(1/2\pi) Q_2(E, h, h')$ in $\widetilde{A}^{1,1}(X)$.*

PROOF. By Proposition 5.27 and Remark 5.1, it is sufficient to see that

$$\operatorname{ch}_2((E,h)) - \operatorname{ch}_2((E,h')) = -\frac{\sqrt{-1}}{4\pi^2}\partial\bar{\partial}(Q_2(E,h,h')),$$

that is,

$$\operatorname{tr}(\Theta^2_{(E,h)}) - \operatorname{tr}(\Theta^2_{(E,h')}) = 2\sqrt{-1}\partial\bar{\partial}(Q_2(E,h,h')),$$

which follows from [**15**, Proposition 6] or [**37**, the proof of Lemma 3.23 in Chapter 6]. □

5.6. Arithmetic characteristic classes

In this section, let us introduce the arithmetic characteristic classes due to Gillet-Soulé.

Let X be a regular arithmetic variety. Let $\widehat{\operatorname{CH}}^p(X)$ be the arithmetic Chow group consisting of p codimensional arithmetic cycles of Green type. Then a and ω are defined to be

$$\begin{cases} a : \tilde{A}^{p-1,p-1}(X) \to \widehat{\operatorname{CH}}^p(X) & (\eta \mapsto (0, \eta)), \\ \omega : \widehat{\operatorname{CH}}^p(X) \to A^{p,p}(X) & ((Z,T) \mapsto \delta_{Z(\mathbb{C})} + dd^c(T)). \end{cases}$$

By [**23**, Theorem 4.29], ω is a ring homomorphism with respect to the ring structure given by Theorem 5.25.

Let $\phi \in \mathbb{Q}[[T_1, \ldots, T_r]]$ be a symmetric formal power series. Then we can uniquely assign

$$\hat{\phi}(\overline{E}) \in \bigoplus_{p \geq 0} \widehat{\mathrm{CH}}^p(X) \otimes \mathbb{Q}$$

to any rank r C^∞-hermitian locally free coherent sheaf $\overline{E} = (E, h)$ on any regular arithmetic variety X (i.e. E is a locally free coherent sheaf of rank r and h is C^∞-hermitian metric of E) with the following properties (for details, see [**24**, Theorem 4.1] or [**64**, Theorem 3 in Chapter 4]).

(1) $f^*(\hat{\phi}(\overline{E})) = \hat{\phi}(f^*(\overline{E}))$ for any morphism $f : Y \to X$ of regular arithmetic varieties.
(2) If there are C^∞-hermitian invertible sheaves $\overline{L}_1, \ldots, \overline{L}_r$ with $\overline{E} = \overline{L}_1 \oplus \cdots \oplus \overline{L}_r$, then $\hat{\phi}(\overline{E}) = \phi(\widehat{c}_1(\overline{L}_1), \ldots, \widehat{c}_1(\overline{L}_r))$.
(3) We define $\phi_0, \phi_1, \ldots$ by the equation

$$\phi(T_1 + T, \ldots, T_r + T) = \sum_{i \geq 0} \phi_i(T_1, \ldots, T_r) T^i.$$

Then, for C^∞-hermitian invertible sheaf $\overline{L}$,

$$\hat{\phi}(\overline{E} \otimes \overline{L}) = \sum_{i \geq 0} \hat{\phi}_i(\overline{E}) \cdot \widehat{c}_1(\overline{L})^i.$$

(4) $\omega(\hat{\phi}(\overline{E})) = \phi(\overline{E}_{\mathbb{C}})$.

We call $\hat{\phi}(\overline{E})$ the *arithmetic characteristic class* associated with ϕ.

Moreover, it is known that if $\overline{\mathcal{E}} : 0 \to \overline{S} \to \overline{E} \to \overline{Q} \to 0$ is an exact sequence of C^∞-hermitian locally free sheaves (metrics are independent from the exact sequence), then

$$\hat{\phi}(\overline{S} \oplus \overline{Q}) - \hat{\phi}(\overline{E}) = a(\tilde{\phi}(\overline{\mathcal{E}}_{\mathbb{C}})) \tag{5.17}$$

(see [**24**, Theorem 4.8]), where $\tilde{\phi}(\overline{\mathcal{E}}_{\mathbb{C}})$ is the Bott-Chern secondary characteristic form in Section 5.5.

Further, let $\psi_1, \ldots, \psi_n \in \mathbb{Q}[[T_1, \ldots, T_r]]$ be symmetric power series,

$$f(X_1, \ldots, X_n) \in \mathbb{Q}[X_1, \ldots, X_n] \quad \text{and} \quad \phi = f(\psi_1, \ldots, \psi_n).$$

Then

$$\hat{\phi}(\overline{E}) = f(\hat{\psi}_1(\overline{E}), \ldots, \hat{\psi}_n(\overline{E})). \tag{5.18}$$

To see (5.18), it is sufficient to see it in cases where either $\phi = a\psi_1 + b\psi_2$ $(a, b \in \mathbb{Q})$ or $\phi = \psi_1\psi_2$. For each case, the assertion follows from the uniqueness of the arithmetic characteristic classes.

Let $c_i(T_1, \ldots, T_r)$ be the i-th fundamental symmetric polynomial, that is,

$$c_i(T_1, \ldots, T_r) = \sum_{1 \leq j_1 < \cdots < j_i \leq r} T_{j_1} \cdots T_{j_i}.$$

Then $\widehat{c}_i(\overline{E})$ is called *arithmetic i-th Chern class*, which can be calculated as follows.

Let $\pi : P = \mathrm{Proj}\left(\bigoplus_{m \geq 0} \mathrm{Sym}^m(E)\right) \to X$ be the projective bundle of E and $\mathscr{O}(1)$ the tautological invertible sheaf of P. Then there is a natural surjective homomorphism $\pi^*(E) \to \mathscr{O}(1)$. We denote its kernel by S. We can give the natural metric $\pi^*(h)$ to $\pi^*(E)$. Let $h_{\mathscr{O}(1)}$ be the quotient metric of $\mathscr{O}(1)$ by $\pi^*(h)$ and h_S

the submetric of S by $\pi^*(h)$. We write $\overline{\mathscr{O}}(1) = (\mathscr{O}(1), h_{\mathscr{O}(1)})$ and $\overline{S} = (S, h_S)$. Then we have an exact sequence

$$\mathcal{E}(\overline{E}) : 0 \to \overline{S} \to \pi^*(\overline{E}) \to \overline{\mathscr{O}}(1) \to 0$$

with hermitian metrics. For $i \geq -r+1$, we define $\hat{s}'_i(\overline{E})$ to be

$$\hat{s}'_i(\overline{E}) = \pi_*(\widehat{c}_1(\overline{\mathscr{O}}(1))^{r-1+i}),$$

where $r = \operatorname{rk} E$. Then, by its definition, we can easily see that $\hat{s}'_i(\overline{E}) = 0$ $(-r+1 \leq i < 0)$ and $\hat{s}'_0(\overline{E}) = [X]$. Moreover, considering $\widetilde{c}_r(\mathcal{E}(\overline{E}) \otimes \overline{\mathscr{O}}(-1))$ with respect to an exact sequence

$$\mathcal{E}(\overline{E}) \otimes \overline{\mathscr{O}}(-1) : 0 \to \overline{S} \otimes \overline{\mathscr{O}}(-1) \to \pi^*(\overline{E}) \otimes \overline{\mathscr{O}}(-1) \to \overline{\mathscr{O}}^{can} \to 0,$$

for $i \geq 1$, we set

$$\theta_i = (-1)^{r-1} \int_{P(\mathbb{C}) \to X(\mathbb{C})} c_1(\overline{\mathscr{O}}(1))^{i-1} \widetilde{c}_r(\mathcal{E}(\overline{E}) \otimes \overline{\mathscr{O}}(-1)).$$

Then the following proposition says that $\widehat{c}_i(\overline{E})$ can be calculated by using $\hat{s}'_i(\overline{E})$ $(i \geq 0)$ and θ_i $(i \geq 1)$.

PROPOSITION 5.30. *Let t be a variable. Then*

$$\left(\sum_{i \geq 0} \widehat{c}_i(\overline{E})(-t)^i\right)\left(\sum_{i \geq 0} \hat{s}'_i(\overline{E}) t^i\right) = [X] + \sum_{i \geq 1} (0, \theta_i)\, t^i,$$

where $\widehat{c}_0(\overline{E}) = [X]$. In particular,

$$\begin{cases} \widehat{c}_1(\overline{E}) = \hat{s}'_1(\overline{E}) - (0, \theta_1), \\ \widehat{c}_2(\overline{E}) = \hat{s}'_1(\overline{E})^2 - \hat{s}'_2(\overline{E}) + (0, \theta_2) - \hat{s}'_1(\overline{E})(0, \theta_1), \end{cases}$$

and

$$\begin{cases} \hat{s}'_1(\overline{E}) = \widehat{c}_1(\overline{E}) + (0, \theta_1), \\ \hat{s}'_2(\overline{E}) = \widehat{c}_1(\overline{E})^2 - \widehat{c}_2(\overline{E}) + (0, \theta_2) + \widehat{c}_1(\overline{E})(0, \theta_1). \end{cases}$$

PROOF. First we consider the following claim.

CLAIM 4.

$$\sum_{i=0}^{r} (-1)^i \pi^*(\widehat{c}_i(\overline{E})) \widehat{c}_1(\overline{\mathscr{O}}(1))^{r-i} = \left(0, (-1)^{r-1} \widetilde{c}_r(\mathcal{E}(\overline{E}) \otimes \overline{\mathscr{O}}(-1))\right).$$

PROOF. Note that

$$\widehat{c}_r\left(\pi^*(\overline{E}) \otimes \overline{\mathscr{O}}(-1)\right) = \widehat{c}_r\left((\overline{S} \otimes \overline{\mathscr{O}}(-1)) \oplus \overline{\mathscr{O}}^{can}\right) - \left(0, \widetilde{c}_r(\mathcal{E}(\overline{E}) \otimes \overline{\mathscr{O}}(-1))\right)$$

by (5.17) and that

$$\widehat{c}_r\left((\overline{S} \otimes \overline{\mathscr{O}}(-1)) \oplus \overline{\mathscr{O}}^{can}\right) = \widehat{c}_{r-1}\left(\overline{S} \otimes \overline{\mathscr{O}}(-1)\right) \cdot \widehat{c}_1(\overline{\mathscr{O}}^{can}) = 0$$

by [**24**, Theorem 4.8, (iii)]. Thus, using properties (1) and (3) of the arithmetic characteristic classes,

$$\widehat{c}_r\left(\pi^*(\overline{E}) \otimes \overline{\mathscr{O}}(-1)\right) = \sum_{i=0}^{r} \pi^*(\widehat{c}_i(\overline{E})) \widehat{c}_1(\overline{\mathscr{O}}(-1))^{r-i}.$$

Therefore the assertion follows. □

We assume $k \geq 1$. Multiplying the equation in the above claim by $\widehat{c}_1(\overline{\mathscr{O}}(1))^{k-1}$, we have

$$\sum_{i=0}^{r}(-1)^i\pi^*(\widehat{c}_i(\overline{E}))\widehat{c}_1(\overline{\mathscr{O}}(1))^{r+k-1-i} = \left(0, (-1)^{r-1}c_1(\overline{\mathscr{O}}(1))^{k-1}\widetilde{c}_r(\mathcal{E}(\overline{E}) \otimes \overline{\mathscr{O}}(-1))\right).$$

Thus, applying π_* to the above equation,

$$\sum_{i=0}^{r}(-1)^i\ \widehat{c}_i(\overline{E})\hat{s}'_{k-i}(\overline{E}) = (0, \theta_k)$$

is obtained. Therefore, the assertion of the proposition follows. □

5.7. Arithmetic Riemann-Roch formula

In this section, we would like to introduce the Arithmetic Riemann-Roch formula. First let us define the analytic torsion and the Quillen metric.

Let M be a compact Kähler manifold and Ω a Kähler form of M. Let $\overline{E} = (E, h)$ be a C^∞-hermitian locally free coherent sheaf on M. Let $\Delta_{\bar{\partial},q}$ be the $\bar{\partial}$-Laplacian on $A^{0,q}(E)$ arising from Ω and h. Let $\{\lambda_{q,1} \leq \lambda_{q_2} \leq \cdots\}$ be positive eigenvalues of $\Delta_{\bar{\partial},q}$ with counting the multiplicity of each eigenvalue. Then we define $\zeta_q(s)$ to be

$$\zeta_q(s) = \sum_{n\geq 1}\lambda_{q,n}^{-s}.$$

It is known that $\zeta_q(s)$ can be extended to the whole $\mathbb{C}$ as a meromorphic function and it is holomorphic at the origin (for example, see [**61**]). The *analytic torsion* $T(\overline{E})$ of $\overline{E}$ is defined by

$$T(\overline{E}) = \sum_{q\geq 0}(-1)^{q+1}q\zeta'_q(0).$$

By using the Kähler form Ω and the hermitian metric h of E, $A^{0,q}(E)$ has a natural inner product $\langle\ ,\ \rangle_q$. Identifying $H^q(M, E)$ with the space $\mathbf{H}^{0,q}(M, E)$ of harmonic forms of $A^{0,q}(E)$, the metric of $\mathbf{H}^{0,q}(M, E)$ induces a metric on $H^q(M, E)$, so that we have a metric of

$$\det \mathbf{R}\Gamma\,(M, E) = \bigotimes_{q\geq 0}\det(H^q(M, E))^{(-1)^q}.$$

This metric is denoted by h_{L^2}. The *Quillen metric* h_Q of $\det \mathbf{R}\Gamma\,(M, E)$ introduced by Quillen [**60**] is defined by

$$h_Q = h_{L^2}\cdot \exp(T(\overline{E})).$$

Let us consider the arithmetic Riemann-Roch formula due to Gillet-Soulé. Let X be a d-dimensional projective regular arithmetic variety. We fix a metric ω of the relative tangent bundle $T_{X/\mathbb{Z}}$ of $\pi : X \to \operatorname{Spec}(\mathbb{Z})$ and we assume that it gives rise to a Kähler matrix of $X(\mathbb{C})$. Let $R(T) \in \mathbb{R}[[T]]$ be a formal power series given by

$$R(T) = \sum_{\substack{m \text{ is even}\\ m\geq 1}}\left(2\zeta'(-m) + \zeta(-m)\left(1 + \frac{1}{2} + \cdots + \frac{1}{m}\right)\right)\frac{T^m}{m!},$$

where $\zeta(s)$ is the Riemann zeta function. This $R(T)$ is called the R-genus due to Gillet-Soulé. Here we define $\widehat{\mathrm{td}}^{R}(T_{X/\mathbb{Z}},\omega)$ to be

$$\widehat{\mathrm{td}}^{R}(T_{X/\mathbb{Z}},\omega)=\widehat{\mathrm{td}}(T_{X/\mathbb{Z}},\omega)(1-a(R_{d-1}(T_{X(\mathbb{C})},\omega))),$$

where $R_{d-1}(T_1,\ldots,T_{d-1})=R(T_1)+\cdots+R(T_{d-1})$. Then the following theorem is the arithmetic Riemann-Roch formula due to Gillet-Soulé. For details, see [**26**], [**64**] and [**20**].

THEOREM 5.31 (Arithmetic Riemann-Roch formula). *Let $\overline{E}$ be a C^∞-hermitian locally free coherent sheaf of real type on X. Then we have the following formula*

$$\widehat{\deg}(\det \mathbf{R}\pi_*(E), h_Q)=\widehat{\deg}\left(\left(\widehat{\mathrm{ch}}(E,h)\widehat{\mathrm{td}}^{R}(T_{X/\mathbb{Z}},\omega)\right)^{(d)}\right),$$

where $\left(\widehat{\mathrm{ch}}(E,h)\,\mathrm{td}^{R}(T_{X/\mathbb{Z}},\omega)\right)^{(d)}$ is the degree d part of

$$\widehat{\mathrm{ch}}(E,h)\,\mathrm{td}^{R}(T_{X/\mathbb{Z}},\omega)\in\bigoplus_{p\geq 0}\widehat{\mathrm{CH}}^{p}(X)\otimes\mathbb{Q}.$$

5.8. Multi-indexed version of Gromov's inequality

For a C^∞-hermitian invertible sheaf $\overline{L}$ on a compact complex manifold, $H^0(L^{\otimes m})$ has various norms. Since $H^0(L^{\otimes m})$ is finite-dimensional, they are equivalent norms. Gromov's inequalities gives information about such an equivalence. The original version can be found in [**26**]. Here we consider the multi-indexed version of Gromov's inequality.

PROPOSITION 5.32 (Multi-indexed version of Gromov's inequality). *Let X be an n-dimensional compact complex manifold, Φ a volume form on X, and $\overline{H}_1,\ldots,\overline{H}_l$ C^∞-hermitian invertible sheaves on X. There is a constant C depending only on $\overline{H}_1,\ldots,\overline{H}_l$, Φ and X such that, for all positive real number p and all non-negative integers $m_1,\ldots,m_l$, the following inequality*

$$\max_{x\in X}\left\{|s|^p_{(m_1,\ldots,m_l)}(x)\right\}\leq C(\lceil p\rceil)^{2n}(m_1+\cdots+m_l+1)^{2n}\left(\int_X |s|^p_{(m_1,\ldots,m_l)}\Phi\right)$$

holds for all $s\in H^0(X,H_1^{\otimes m_1}\otimes\cdots\otimes H_l^{\otimes m_l})$, where $|\cdot|_{(m_1,\ldots,m_l)}$ is the L^2-norm of $\overline{H}_1^{\otimes m_1}\otimes\cdots\otimes\overline{H}_l^{\otimes m_l}$ and $\lceil p\rceil$ is the round-up of p, that is, $\lceil p\rceil=\min\{a\in\mathbb{Z}\mid p\leq a\}$.

PROOF. We choose an open covering $\{U_i\}_{i=1,\ldots,N}$ of X with the following properties.

(1) U_i is isomorphic to $\{z\in\mathbb{C}^n\mid |z|<1\}$ by using a local coordinate $z_i(x)=(z_{i1}(x),\ldots,z_{in}(x))$ of U_i. We set

$$V_i=\{x\in U_i\mid |z_i(x)|<1/2\}\quad\text{and}\quad W_i=\{x\in U_i\mid |z_i(x)|<1/4\}.$$

(2) There are local bases of $\omega_{i1},\ldots,\omega_{il}$ of $H_1,\ldots,H_l$ on U_i, respectively.

(3) $\bigcup_{i=1}^N W_i=X$.

Let us show that the proposition follows from the following local Gromov's inequality, that is, Lemma 5.33, which guarantees a constant C_i for U_i,V_i,W_i. We set $C = \max_{i=1,\dots,N}\{C_i\}$. We choose x_0 and i with $\max_{x\in X}\{|s|^p_{(m_1,\dots,m_l)}(x)\} = |s|^p_{(m_1,\dots,m_l)}(x_0)$ and $x_0 \in W_i$. Then

$$\begin{aligned}\max_{x\in X}\{|s|^p_{(m_1,\dots,m_l)}(x)\} &= \max_{x\in \overline{W}_i}\{|s|^p_{(m_1,\dots,m_l)}(x)\}\\ &\leq C_i(\lceil p\rceil)^{2n}(m_1+\cdots+m_l+1)^{2n}\left(\int_{V_i}|s|^p_{(m_1,\dots,m_l)}\Phi\right)\\ &\leq C(\lceil p\rceil)^{2n}(m_1+\cdots+m_l+1)^{2n}\left(\int_{X}|s|^p_{(m_1,\dots,m_l)}\Phi\right),\end{aligned}$$

as required. □

LEMMA 5.33 (Multi-indexed version of the local Gromov inequality). *We set $U = \{z \in \mathbb{C}^n \mid |z| < a\}$, $V = \{z \in \mathbb{C}^n \mid |z| < b\}$, $W = \{z \in \mathbb{C}^n \mid |z| < c\}$ for real numbers a, b, c with $a > b > c > 0$. Let Φ be a volume form on U. Let $\overline{H}_1, \dots, \overline{H}_l$ be C^∞-hermitian invertible sheaves on U and $\omega_1, \dots, \omega_l$ bases of $H_1, \dots, H_l$ over U, respectively. Then there is a constant C depending only on $\overline{H}_1, \dots, \overline{H}_l$, $\omega_1, \dots, \omega_l$, Φ, a, b, c and n such that, for all positive real number p and non-negative integers $m_1, \dots, m_l$, the following inequality*

$$\max_{x\in\overline{W}}\left\{|s|^p_{(m_1,\dots,m_l)}(x)\right\} \leq C(\lceil p\rceil)^{2n}(m_1+\cdots+m_l+1)^{2n}\left(\int_V |s|^p_{(m_1,\dots,m_l)}\Phi\right)$$

holds for all $s \in H^0(U, H_1^{\otimes m_1}\otimes\cdots\otimes H_l^{\otimes m_l})$.

PROOF. Let $|\cdot|_i$ be the L^2-norm of $\overline{H}_i$. We write $u_i = |\omega_i|_i$ over U. Considering the maximal of partial derivatives of u_i over $\overline{V}$, we can find a constants K_i such that

$$|u_i(x) - u_i(y)| \leq K_i|x-y|'$$

for all $x, y \in \overline{V}$, where $|z|' = |z_1| + \cdots + |z_n|$ for $z = (z_1, \dots, z_n)$. We set

$$D = \max\left\{\max_{x\in\overline{V}}\left\{\frac{K_1}{u_1(x)}\right\}, \dots, \max_{x\in\overline{V}}\left\{\frac{K_l}{u_l(x)}\right\}, \frac{1}{b-c}\right\} \quad\text{and}\quad R = 1/D.$$

Then, for $x_0, x \in \overline{V}$,

$$\begin{aligned}u_i(x) \geq u_i(x_0) - K_i|x-x_0|' &= u_i(x_0)\left(1 - \frac{K_i}{u_i(x_0)}|x-x_0|'\right)\\ &\geq u_i(x_0)(1 - D|x-x_0|').\end{aligned}$$

Thus, if we set $B(x_0, R) = \{x \in \mathbb{C}^n \mid |x-x_0|' \leq R\}$, then

$$1 - D|x-x_0|' \geq 0$$

for all $x \in B(x_0, R)$. Moreover, since

$$|x-x_0| \leq |x-x_0|' \leq R \leq b-c,$$

we have $B(x_0, R) \subseteq \overline{V}$ if $x_0 \in \overline{W}$. Here let us see the following.

CLAIM 5.

$$\int_0^1\cdots\int_0^1 x_1\cdots x_n\left(1 - \frac{1}{n}(x_1+\cdots+x_n)\right)^m dx_1\cdots dx_n \geq \frac{1}{(\lceil m\rceil+1)^n(\lceil m\rceil+2)^n}$$

for non-negative real number m.

PROOF. First we consider the case where m is an integer. If $m = 0$, then the assertion of the claim is obvious, so that we assume that $m \geq 1$. Note that

$$\left(1 - \frac{1}{n}(x_1 + \cdots + x_n)\right)^m = \frac{1}{n^m}\left(\sum_{i=1}^{n}(1 - x_i)\right)^m$$
$$= \frac{1}{n^m} \sum_{\substack{m_1+\cdots+m_n=m \\ m_1\geq 0,\ldots,m_n\geq 0}} \frac{m!}{m_1!\cdots m_n!}(1-x_1)^{m_1}\cdots(1-x_n)^{m_n}$$

and that

$$\int_0^1 x(1-x)^d dx = \frac{1}{(d+1)(d+2)}$$

for a non-negative integer d. Thus the integral I in the claim is equal to

$$\frac{1}{n^m} \sum_{\substack{m_1+\cdots+m_n=m \\ m_1\geq 0,\ldots,m_n\geq 0}} \frac{m!}{m_1!\cdots m_n!} \frac{1}{(m_1+1)(m_1+2)\cdots(m_n+1)(m_n+2)}.$$

Therefore,

$$I \geq \frac{1}{(m+1)^n(m+2)^n n^m} \sum_{\substack{m_1+\cdots+m_n=m \\ m_1\geq 0,\ldots,m_n\geq 0}} \frac{m!}{m_1!\cdots m_n!} = \frac{1}{(m+1)^n(m+2)^n}.$$

If m is not an integer, then

$$\left(1 - \frac{1}{n}(x_1 + \cdots + x_n)\right)^m \geq \left(1 - \frac{1}{n}(x_1 + \cdots + x_n)\right)^{\lceil m \rceil}$$

because $0 \leq 1 - \frac{1}{n}(x_1 + \cdots + x_n) \leq 1$. Thus the claim follows. □

Let us choose a positive constant e with $\Phi \geq e\Phi_{can}$ over $\overline{V}$, where

$$\Phi_{can} = \left(\frac{\sqrt{-1}}{2}\right)^n dz_1 \wedge d\bar{z}_1 \wedge \cdots \wedge dz_n \wedge d\bar{z}_n.$$

Let s be an element of $H^0(U, H_1^{\otimes m_1} \otimes \cdots \otimes H_l^{\otimes m_l})$. Then there is a holomorphic function f over U with $s = f\omega_1^{\otimes m_1} \otimes \cdots \otimes \omega_l^{\otimes m_l}$. We assume that a continuous function $|s|_{(m_1,\ldots,m_l)}$ takes the maximal value at x_0 over $\overline{W}$. Then

$$\int_V |s|^p_{(m_1,\ldots,m_l)}\Phi \geq e\int_{B(x_0,R)} |s|^p_{(m_1,\ldots,m_l)}\Phi_{can} = e\int_{B(x_0,R)} |f|^p u_1^{pm_1}\cdots u_l^{pm_l}\Phi_{can}$$
$$\geq eu_1(x_0)^{pm_1}\cdots u_l(x_0)^{pm_l}\int_{B(x_0,R)} |f|^p(1 - D|x-x_0|')^m\Phi_{can},$$

where $m = p(m_1 + \cdots + m_l)$. Moreover, if we write

$$x - x_0 = (r_1\exp(\sqrt{-1}\theta_1), \ldots, r_n\exp(\sqrt{-1}\theta_n)),$$

then

$$\begin{aligned}\int_{B(x_0,R)} &|f|^p(1-D|x-x_0|')^m\Phi_{can}\\ &= \int_{\substack{r_1+\cdots+r_n\le R\\ r_1\ge 0,\ldots,r_n\ge 0}} \left(\int_0^{2\pi}\cdots\int_0^{2\pi}|f|^p d\theta_1\cdots d\theta_n\right)\\ &\qquad\qquad \times r_1\cdots r_n(1-D(r_1+\cdots+r_n))^m dr_1\cdots dr_n.\end{aligned}$$

Since $|f|^p$ is a pluri-subharmonic function, we have

$$\int_0^{2\pi}\cdots\int_0^{2\pi}|f|^p d\theta_1\cdots d\theta_n \ge (2\pi)^n|f(x_0)|^p.$$

Thus, by Claim 5,

$$\begin{aligned}\int_{B(x_0,R)} &|f|^p(1-D|x-x_0|')^m\Phi_{can} \ge (2\pi)^n|f(x_0)|^p\\ &\times \int_{\substack{r_1+\cdots+r_n\le R\\ r_1\ge 0,\ldots,r_n\ge 0}} r_1\cdots r_n(1-D(r_1+\cdots+r_n))^m dr_1\cdots dr_n\\ &\ge (2\pi)^n|f(x_0)|^p\int_{[0,R/n]^n} r_1\cdots r_n(1-D(r_1+\cdots+r_n))^m dr_1\cdots dr_n\\ &\ge \frac{(2\pi)^n|f(x_0)|^p}{(nD)^{2n}}\frac{1}{(\lceil m\rceil+1)^n(\lceil m\rceil+2)^n}.\end{aligned}$$

Therefore, if we set $C'=e(2\pi)^n/(nD)^{2n}$, then

$$\int_V |s|^p_{(m_1,\ldots,m_l)}\Phi \ge \frac{C'|s(x_0)|^p_{(m_1,\ldots,m_l)}}{(\lceil m\rceil+1)^n(\lceil m\rceil+2)^n}.$$

Moreover, as $\lceil m\rceil \le \lceil p\rceil(m_1+\cdots+m_l)$, we obtain

$$\begin{aligned}(\lceil m\rceil+1)^n(\lceil m\rceil+2)^n &\le (\lceil p\rceil(m_1+\cdots+m_l)+1)^n(\lceil p\rceil(m_1+\cdots+m_l)+2)^n\\ &\le (\lceil p\rceil(m_1+\cdots+m_l+1))^n(2\lceil p\rceil(m_1+\cdots+m_l+1))^n\\ &= 2^n(\lceil p\rceil)^{2n}(m_1+\cdots+m_l+1)^{2n},\end{aligned}$$

and hence the lemma follows. □

COROLLARY 5.34. *Let X be an n-dimensional compact complex manifold and $\overline{H}_1,\ldots,\overline{H}_l$ C^∞-hermitian invertible sheaves on X. Let Y be a closed complex submanifold of X, and Φ_X and Φ_Y volume forms on X and Y, respectively. Then there is a positive constant C such that*

$$C(m_1+\cdots+m_l+1)^{2n}\int_X |s|^2\Phi_X \ge \int_Y |\, s|_Y\,|^2\Phi_Y$$

holds for all non-negative integers $m_1,\ldots,m_l$ and all $s\in H^0(X,H_1^{\otimes m_1}\otimes\cdots\otimes H_l^{\otimes m_l})$.

PROOF. Note that

$$\|s\|^2_{\sup} \ge \|\, s|_Y\,\|^2_{\sup} \ge \frac{\int_Y |\, s|_Y\,|^2\Phi_Y}{\int_Y\Phi_Y}.$$

Thus the corollary follows from Gromov's inequality. □

Let us consider the following analogue of the multi-indexed version of Gromov's inequality. The proof of the following lemma is due to Takuro Mochizuki.

LEMMA 5.35. *Let X be an n-dimensional compact Kähler manifold and Ω a Kähler form of X. Let $\overline{H}_1, \ldots, \overline{H}_l$ be C^∞-hermitian invertible sheaves on X. Let U be an open set of X such that U is not empty on each connected component of X. Then there are positive constants C, C' and D' such that*

$$\sup_{x \in X}\{|s|_{(m_1,\ldots,m_l)}(x)\} \leq C^{m_1+\cdots+m_l} \sup_{x\in U}\{|s|_{(m_1,\ldots,m_l)}(x)\}$$

and

$$\int_X |s|^2_{(m_1,\ldots,m_l)}\Omega^n \leq D' \cdot C'^{m_1+\cdots+m_l} \int_U |s|^2_{(m_1,\ldots,m_l)}\Omega^n$$

for all non-negative integers $m_1, \ldots, m_l$ and all

$$s \in H^0(X, H_1^{\otimes m_1} \otimes \cdots \otimes H_l^{\otimes m_l}),$$

where $|\cdot|_{(m_1,\ldots,m_l)}$ is the L^2-norm of $\overline{H}_1^{\otimes m_1} \otimes \cdots \otimes \overline{H}_l^{\otimes m_l}$.

PROOF. Clearly we may assume that X is connected. Replacing U with a smaller open set, we may assume that U can be identified with $\{x \in \mathbb{C}^n \mid |x| < 1\}$. We set $W = \{x \in \mathbb{C}^n \mid |x| < 1/2\}$. In this proof, the Laplacian Δ is defined by the formula

$$-\frac{\sqrt{-1}}{2\pi}\partial\bar{\partial}(g) \wedge \Omega^{n-1} = \Delta(g)\Omega^n$$

(see Proposition 1.37). We take a C^∞-function a_i with $c_1(\overline{H}_i) \wedge \Omega^{n-1} = a_i\Omega^n$. Let ϕ_i be a C^∞-function on X such that

$$\int_X a_i\Omega^n = \int_X \phi_i\Omega^n$$

and ϕ_i is zero on $X \setminus W$. Then, by Corollary 1.34, there is a C^∞-function F_i on X with $\Delta(F_i) = a_i - \phi_i$. By our construction, $\Delta(F_i) = a_i$ on $X \setminus W$.

We set

$$f = |s|^2_{(m_1,\ldots,m_l)} \exp(-(m_1F_1 + \cdots + m_lF_l)).$$

for $s \in H^0(X, H_1^{\otimes m_1} \otimes \cdots \otimes H_l^{\otimes m_l})$.

CLAIM 6. $\max_{x\in X\setminus W}\{f(x)\} = \max_{x\in\partial(W)}\{f(x)\}$.

PROOF. If f is a constant on $X \setminus W$, the assertion is obvious, so that we may assume that f is not a constant on $X \setminus W$. In particular, $s \neq 0$. Since

$$-\frac{\sqrt{-1}}{2\pi}\partial\bar{\partial}(\log(|s|^2_{(m_1,\ldots,m_l)})) = c_1(m_1\overline{H}_1+\cdots+m_l\overline{H}_l) = m_1c_1(\overline{H}_1)+\cdots+m_lc_1(\overline{H}_l),$$

we have $\Delta(\log(f)) = 0$ on $X\setminus(W\cup\mathrm{Supp}(\mathrm{div}(s)))$. We assume that the C^∞-function f takes the maximal value at x_0 over $X \setminus W$. Then

$$x_0 \in X \setminus (W \cup \mathrm{Supp}(\mathrm{div}(s))).$$

Indeed, if $x_0 \in \mathrm{Supp}(\mathrm{div}(s))$, then $f(x_0) = 0$, which implies that $f \equiv 0$ on $X \setminus W$. This is a contradiction.

Therefore, $\log(f)$ is a non-constant harmonic function on

$$X \setminus (W \cup \mathrm{Supp}(\mathrm{div}(s)))$$

and takes the maximal value at x_0. Thus, by virtue of the maximal principle of harmonic functions, $x_0 \in \partial(W)$. Hence the claim follows. □

Here we write

$$d_i = \min_{x \in X \setminus W} \{\exp(-F_i)\}, \quad D_i = \max_{x \in \partial(W)} \{\exp(-F_i)\} \quad \text{and} \quad C = \max_{i=1,\ldots,l} \{D_i/d_i\}.$$

Then

$$d_1^{m_1} \cdots d_l^{m_l} |s|^2_{(m_1,\ldots,m_l)} \leq f$$

over $X \setminus W$ and

$$f \leq D_1^{m_1} \cdots D_l^{m_l} |s|^2_{(m_1,\ldots,m_l)}$$

over $\partial(W)$. Thus we have

$$\begin{aligned} \max_{x \in X \setminus W} \{|s|^2_{(m_1,\ldots,m_l)}\} &\leq C^{m_1+\cdots+m_l} \max_{x \in \partial(W)} \{|s|^2_{(m_1,\ldots,m_l)}\} \\ &\leq C^{m_1+\cdots+m_l} \max_{x \in \overline{W}} \{|s|^2_{(m_1,\ldots,m_l)}\}, \end{aligned}$$

which implies

$$\max_{x \in X} \{|s|^2_{(m_1,\ldots,m_l)}\} \leq C^{m_1+\cdots+m_l} \max_{x \in \overline{W}} \{|s|^2_{(m_1,\ldots,m_l)}\}.$$

This is the first inequality of the lemma.

Note that $e^x \geq x+1$ $(x \geq 0)$. Thus, by Gromov's inequality, there are constants C_1 and D_1 such that

$$\max_{x \in \overline{W}} \{|s|^2_{(m_1,\ldots,m_l)}\} \leq D_1 \cdot C_1^{m_1+\cdots+m_l} \int_U |s|^2_{(m_1,\ldots,m_l)} \Phi$$

for all non-negative integers $m_1, \ldots, m_l$ and all $s \in H^0(X, H_1^{\otimes m_1} \otimes \cdots \otimes H_l^{\otimes m_l})$. Thus the second inequality follows. □

5.9. Arithmetic Hilbert-Samuel formula

The following arithmetic Hilbert-Samuel formula follows from the arithmetic Riemann-Roch formula due to Gillet-Soulé [**26**] in the previous section and the asymptotic estimate of analytic torsions due to Bismut-Vasserot [**8**]. In this book, we would like to introduce a direct proof due to Abbes-Bouche [**1**].

THEOREM 5.36 (Arithmetic Hilbert-Samuel formula). *Let X be a d-dimensional generically smooth projective arithmetic variety and $\overline{L}$ a C^∞-hermitian invertible sheaf of real type on X. We assume that L is ample on the generic fiber of $X \to$ $\operatorname{Spec}(\mathbb{Z})$ and $\overline{L}$ is positive on $X(\mathbb{C})$. Then, for a sufficiently large integer n,*

$$\widehat{\chi}\left(H^0(X, L^{\otimes n}), \|\cdot\|_{L^2}\right) + \sum_{i>0} (-1)^i \log \#\left(H^i(X, L^{\otimes n})\right) = \frac{\widehat{\deg}\left(\widehat{c}_1(\overline{L})^{\cdot d}\right)}{d!} n^d + o(n^d).$$

Similarly, for a sufficiently large integer n,

$$\widehat{\chi}\left(H^0(X, L^{\otimes n}), \|\cdot\|_{\sup}\right) + \sum_{i>0} (-1)^i \log \#\left(H^i(X, L^{\otimes n})\right) = \frac{\widehat{\deg}\left(\widehat{c}_1(\overline{L})^{\cdot d}\right)}{d!} n^d + o(n^d).$$

PROOF. First, we see that the L^2-version implies the sup-version. By Gromov's inequality (cf. Proposition 5.32), there are constants C_1 and C_2 such that

$$C_1 \|\cdot\|_{L^2} \leq \|\cdot\|_{\sup} \leq C_2 (n+1)^d \|\cdot\|_{L^2}$$

on $H^0(X, L^{\otimes n})$. Therefore, by Lemma 2.16, we have

$$\widehat{\chi}\left(H^0(X, L^{\otimes n}), \|\cdot\|_{\sup}\right) = \widehat{\chi}\left(H^0(X, L^{\otimes n}), \|\cdot\|_{L^2}\right) + o(n^d),$$

as required. □

Before starting the proof of the arithmetic Hilbert-Samuel formula, we need to show the following theorem. The complete proof is very hard and long, so that we gives only ideas of the proof.

THEOREM 5.37. *Let M be an l-dimensional non-singular projective variety. Let $\overline{L}$ be a C^∞-hermitian invertible sheaf on M such that L is ample and $\overline{L}$ is positive. Let $\langle\,,\rangle_n$ be the natural inner product of $H^0(M, L^{\otimes n})$ by using the volume form $\Phi(\overline{L})$ induced by $\overline{L}$ as in Section 5.1.3 and the hermitian metric of $\overline{L}^{\otimes n}$. For a positive integer N, let s be a non-zero element of $H^0(M, L^{\otimes N})$ such that $T = \operatorname{div}(s)$ is non-singular. Then a sequence*

$$0 \to H^0(M, L^{\otimes n-N}) \xrightarrow{\otimes s} H^0(M, L^{\otimes n}) \to H^0(T, L|_T^{\otimes n}) \to 0$$

induced by s is exact if n is sufficiently large. In the same way as above, the natural inner product of $H^0(T, L|_T^{\otimes n})$ induced by the volume form arising from $\overline{L}|_T$ and the hermitian metric of $\overline{L}|_T^{\otimes n}$ is denoted by $\langle\,,\rangle_{n,T}$. Let

$$\begin{aligned}\overline{\mathbf{E}}_n : 0 \to (H^0(M, L^{\otimes n-N}), \langle\,,\rangle_{n-N}) &\to (H^0(M, L^{\otimes n}), \langle\,,\rangle_n) \\ &\to (H^0(T, L|_T^{\otimes n}), \langle\,,\rangle_{n,T}) \to 0\end{aligned}$$

be an exact sequence with metrics. Then the volume difference $\gamma(\overline{\mathbf{E}}_n)$ introduced in Section 3.7 has the following limit formula

$$\lim_{n\to\infty} \frac{\gamma(\overline{\mathbf{E}}_n)}{n^l/l!} = \int_M \log|s| c_1(\overline{L})^l.$$

PROOF. Here we give a proof due to Yoshikawa using the Bismut-Lebeau formula [**7**] and the asymptotic estimate of analytic torsions due to Bismut-Vasserot [**8**], which is different from the original proof [**1**, Section 3] due to Abbes-Bouche.

First, we set

$$\sigma(n) = -2\gamma(\overline{\mathbf{E}}_n) \quad \text{and} \quad \sigma(n)_Q = \sigma(n) + T\left(\overline{L}^{\otimes n-N}\right) - T\left(\overline{L}^{\otimes n}\right) + T\left(\overline{L}|_T^{\otimes n}\right),$$

where $T\left(\overline{L}^{\otimes n-N}\right)$, $T\left(\overline{L}^{\otimes n}\right)$ and $T\left(\overline{L}|_T^{\otimes n}\right)$ are the analytic torsions of $\overline{L}^{\otimes n-N}$, $\overline{L}^{\otimes n}$ and $\overline{L}|_T^{\otimes n}$, respectively. Then, by using the Bismut-Lebeau formula [**7**], we have

$$\sigma(n)_Q = -n^l \int_M \log|s|^2 \frac{c_1(\overline{L})^l}{l!} + O(n^{l-1}).$$

To get the above estimate from the Bismut-Lebeau formula, we need several arguments. The details are left to the reader. On the other hand, by [**8**], there are constants A and B such that, if n is sufficiently large, then

$$T\left(\overline{L}^{\otimes n}\right) = An^l\log(n) + o(n^l) \quad \text{and} \quad T\left(\overline{L}|_T^{\otimes n}\right) = Bn^{l-1}\log(n) + o(n^{l-1}).$$

Therefore,

$$\begin{aligned}T\left(\overline{L}^{\otimes n-N}\right) - T\left(\overline{L}^{\otimes n}\right) + T\left(\overline{L}|_T^{\otimes n}\right) &= A(n-N)^l\log(n-N) - An^l\log(n) + o(n^l) \\ &= An^l\log(1-N/n) + o(n^l) = o(n^l).\end{aligned}$$

Hence

$$\sigma(n) = -n^l \int_M \log|s|^2 \frac{c_1(\overline{L})^l}{l!} + o(n^l)$$

is obtained, so that the theorem follows. □

PROOF OF THEOREM 5.36. Let us start the proof of Theorem 5.36. As

$$\log \operatorname{vol}\{(x_1, \ldots, x_r) \in \mathbb{R}^r \mid x_1^2 + \cdots + x_r^2 \leq 1\} = O(r \log r),$$

by Proposition 3.12, we can see that

$$\widehat{\chi}\left(H^0(X, L^{\otimes n}), \|\cdot\|_{L^2}\right) = \widehat{\deg}\left(H^0(X, L^{\otimes n}), \langle , \rangle_n\right) + o(n^d).$$

Therefore, it is sufficient to show that

(5.19)

$$\widehat{\deg}\left(H^0(X, L^{\otimes n}), \langle , \rangle_n\right) + \sum_{i>0}(-1)^i \log \#\left(H^i(X, L^{\otimes n})\right) = \frac{\widehat{\deg}\left(\widehat{c}_1(\overline{L})^{\cdot d}\right)}{d!} n^d + o(n^d).$$

We prove it by induction on d. If $d = 1$, it follows from Theorem 3.14 and Lemma 3.18. We assume that $d > 1$. Since L is ample over $\mathbb{Q}$, for a sufficiently large integer N, there is a non-zero element of $s \in H^0(X, L^{\otimes N})$ such that $Y = \operatorname{div}(s)$ is smooth over $\mathbb{Q}$. If $d = 2$, then Y is not necessarily irreducible over $\mathbb{Q}$. However, if $d \geq 3$, then Y is irreducible over $\mathbb{Q}$. Moreover, there is n_0 such that, for all $n \geq n_0$ and $i > 0$, $H^i(X, L^{\otimes n})$ is a torsion module as a $\mathbb{Z}$-module. Here, considering an exact sequence

$$\begin{aligned} \overline{\mathbf{E}}_n : 0 \to (H^0(X(\mathbb{C}), L^{\otimes n-N}), \langle , \rangle_{n-N}) &\to (H^0(X(\mathbb{C}), L^{\otimes n}), \langle , \rangle_n) \\ &\to (H^0(Y(\mathbb{C}), L|_Y^{\otimes n}), \langle , \rangle_{n,Y}) \to 0 \end{aligned}$$

with metrics, we set $\gamma_n = \gamma(\overline{\mathbf{E}}_n)$. Then, using Theorem 5.37 in the case where $M = X(\mathbb{C})$, we have

$$\gamma_n = \frac{n^{d-1}}{(d-1)!} \int_{X(\mathbb{C})} \log |s| c_1(\overline{L})^{d-1} + o(n^{d-1}). \tag{5.20}$$

First we would like to show that if $n \geq n_0 + N$, then the following holds:

$$\begin{aligned} &\left(\widehat{\deg}\left(H^0(X, L^{\otimes n}), \langle , \rangle_n\right) + \sum_{i\geq 1}(-1)^i \log \#(H^i(X, L^{\otimes n}))\right) \\ &- \left(\widehat{\deg}\left(H^0(X, L^{\otimes n-N}), \langle , \rangle_{n-N}\right) + \sum_{i\geq 1}(-1)^i \log \#(H^i(X, L^{\otimes n-N}))\right) \\ &= -\gamma_n + \left(\widehat{\deg}\left(H^0(Y, L|_Y^{\otimes n}), \langle , \rangle_{n,Y}\right) + \sum_{i\geq 1}(-1)^i \log \#(H^i(Y, L|_Y^{\otimes n}))\right). \end{aligned} \tag{5.21}$$

We set $\Gamma_n = \operatorname{Image}\left(H^0(X, L^{\otimes n}) \to H^0(Y, L|_Y^{\otimes n})\right)$. Then since $H^0(Y, L|_Y^{\otimes n})/\Gamma_n$ is a torsion module, the inner product $\langle , \rangle_{n,Y}$ on $H^0(Y, L|_Y^{\otimes n})$ yields an inner product on Γ_n. By Proposition 3.21,

$$\widehat{\deg}\left(H^0(X, L^{\otimes n}), \langle , \rangle_n\right) - \widehat{\deg}\left(H^0(X, L^{\otimes n-N}), \langle , \rangle_{n-N}\right) = \widehat{\deg}(\Gamma_n, \langle , \rangle_{n,Y}) - \gamma_n$$

holds. Moreover, we have

$$\widehat{\deg}(\Gamma_n, \langle , \rangle_{n,Y}) = \widehat{\deg}\left(H^0(Y, L|_Y^{\otimes n}), \langle , \rangle_{n,Y}\right) - \log \#(H^0(Y, L|_Y^{\otimes n})/\Gamma_n).$$

On the other hand, by the following long exact sequence,

$$\begin{array}{lllllll} 0 \longrightarrow H^0(Y, L|_Y^{\otimes n})/\Gamma_n & & & & & & \\ \longrightarrow H^1(X, L^{\otimes n-N}) & \longrightarrow & H^1(X, L^{\otimes n}) & \longrightarrow & H^1(Y, L|_Y^{\otimes n}) & \longrightarrow & \\ \vdots & & \vdots & & \vdots & & \\ \longrightarrow H^i(X, L^{\otimes n-N}) & \longrightarrow & H^i(X, L^{\otimes n}) & \longrightarrow & H^i(Y, L|_Y^{\otimes n}) & \longrightarrow & \\ \vdots & & \vdots & & \vdots & & \end{array}$$

we obtain

$$\begin{aligned} \log \#(H^0(Y, L|_Y^{\otimes n})/\Gamma_n) = -\sum_{i\geq 1}(-1)^i \log \#(H^i(X, L^{\otimes n-N})) \\ + \sum_{i\geq 1}(-1)^i \log \#(H^i(X, L^{\otimes n})) \\ - \sum_{i\geq 1}(-1)^i \log \#(H^i(Y, L|_Y^{\otimes n})). \end{aligned}$$

Thus (5.21) follows.

Let us consider an embedding $X \hookrightarrow \mathbb{P}^r_{\mathbb{Z}}$. Let R_X be the homogeneous coordinate ring of X with respect to the above embedding. In the same way, let R_Y be the homogeneous coordinate ring of Y with respect to the embedding $Y \hookrightarrow X \hookrightarrow \mathbb{P}^r_{\mathbb{Z}}$. Let I be the kernel of the natural surjective ring homomorphism $R_X \to R_Y$. Note that I is a homogeneous ideal of R_X consisting of defining equations of Y in X. We set $S = \mathbb{Z} \setminus \{0\}$. We define J by $J = I_S \cap R_X$. Let Y' be the subscheme given by J, that is, Y' is the Zariski closure of $Y_{\mathbb{Q}}$ in X.

CLAIM 7. *There is a homogeneous ideal K of R_X such that $I = J \cap K$ and the subscheme Y'' defined by K is vertical.*

PROOF. By [**44**, (10.B) Proposition], there are homogeneous primary ideals $Q_1, \ldots, Q_l$ with $I = Q_1 \cap \cdots \cap Q_l$. Renumbering $Q_1, \ldots, Q_l$, we may assume that

$$\begin{cases} \sqrt{Q_i} \cap S = \emptyset & (i = 1, \ldots, m), \\ \sqrt{Q_i} \cap S \neq \emptyset & (i = m+1, \ldots, l). \end{cases}$$

Then, by using Atiyah-Macdonald's book [**3**, Corollary 3.4 and Proposition 4], we can see that

$$\begin{cases} (Q_1 \cap \cdots \cap Q_l)_S = (Q_1)_S \cap \cdots \cap (Q_l)_S, \\ (Q_i)_S = (R_X)_S \quad (i = m+1, \ldots, l), \\ (Q_i)_S \cap R_X = Q_i \quad (i = 1, \ldots, m). \end{cases}$$

Therefore, we obtain

$$\begin{aligned} J = I_S \cap R_X &= (Q_1)_S \cap \cdots \cap (Q_l)_S \cap R_X = (Q_1)_S \cap \cdots \cap (Q_m)_S \cap R_X \\ &= ((Q_1)_S \cap R_X) \cap \cdots \cap ((Q_m)_S \cap R_X) = Q_1 \cap \cdots \cap Q_m. \end{aligned}$$

Thus, if we set $K = Q_{m+1} \cap \cdots \cap Q_l$, then $I = J \cap K$ and the subscheme defined by K is vertical. ☐

Let $\mathcal{I}, \mathcal{J}, \mathcal{K}$ be the ideal sheaves of $\mathcal{O}_X$ given by I, J, K, respectively. Let us show the following:

$$
\begin{aligned}
(5.22)\quad \widehat{\deg}\left(H^0(Y, L|_Y^{\otimes n}), \langle\,,\rangle_{n,Y}\right) &+ \sum_{i\geq 1}(-1)^i \log \#(H^i(Y, L|_Y^{\otimes n})) \\
= \widehat{\deg}\left(H^0(Y', L|_{Y'}^{\otimes n}), \langle\,,\rangle_{n,Y'}\right) &+ \sum_{i\geq 1}(-1)^i \log \#(H^i(Y', L|_{Y'}^{\otimes n})) \\
&+ \sum_{i\geq 0}(-1)^i \log \#(H^i(Y'', L|_{Y''}^{\otimes n})) \\
&- \sum_{i\geq 0}(-1)^i \log \#(H^i(Y'\cap Y'', L|_{Y'\cap Y''}^{\otimes n})).
\end{aligned}
$$

First, considering a long exact sequence arising from the following exact sequence

$$
0 \to L^{\otimes n} \otimes \mathcal{J}/\mathcal{I} \to L^{\otimes n} \otimes \mathcal{O}_Y \to L^{\otimes n} \otimes \mathcal{O}_{Y'} \to 0,
$$

we have

$$
\begin{aligned}
\widehat{\deg}\left(H^0(Y, L|_Y^{\otimes n}), \langle\,,\rangle_{n,Y}\right) &+ \sum_{i\geq 1}(-1)^i \log \#(H^i(Y, L|_Y^{\otimes n})) \\
= \widehat{\deg}\left(H^0(Y', L|_{Y'}^{\otimes n}), \langle\,,\rangle_{n,Y'}\right) &+ \sum_{i\geq 1}(-1)^i \log \#(H^i(Y', L|_{Y'}^{\otimes n})) \\
&+ \sum_{i\geq 0}(-1)^i \log \#(H^i(L^{\otimes n} \otimes \mathcal{J}/\mathcal{I})).
\end{aligned}
$$

On the other hand, as $\mathcal{J}/\mathcal{I} = \mathcal{J}/\mathcal{J}\cap\mathcal{K} = (\mathcal{J}+\mathcal{K})/\mathcal{K}$, we obtain an exact sequence

$$
0 \to L^{\otimes n} \otimes \mathcal{J}/\mathcal{I} \to L^{\otimes n} \otimes \mathcal{O}_{Y''} \to L^{\otimes n} \otimes \mathcal{O}_{Y'\cap Y''} \to 0.
$$

Therefore, we get

$$
\begin{aligned}
\sum_{i\geq 0}(-1)^i \log \#(H^i(L^{\otimes n} \otimes \mathcal{J}/\mathcal{I})) &= \sum_{i\geq 0}(-1)^i \log \#(H^i(Y'', L|_{Y''}^{\otimes n})) \\
&\quad - \sum_{i\geq 0}(-1)^i \log \#(H^i(Y'\cap Y'', L|_{Y'\cap Y''}^{\otimes n})),
\end{aligned}
$$

so that (5.22) follows.

Here, applying (4) in Proposition 5.23 and Snapper's theorem [**35**, p. 295] of algebraic schemes,

$$
\begin{cases}
\displaystyle\sum_{i\geq 0}(-1)^i \log \#(H^i(Y'', L|_{Y''}^{\otimes n})) = \frac{\widehat{\deg}(\widehat{c}_1(\overline{L})^{d-1}\cdot(Y'',0))}{(d-1)!}n^{d-1} + o(n^{d-1}), \\
\displaystyle\sum_{i\geq 0}(-1)^i \log \#(H^i(Y'\cap Y'', L|_{Y'\cap Y''}^{\otimes n})) = o(n^{d-1})
\end{cases}
$$

follow. Moreover, by the hypothesis of induction,

$$\widehat{\deg}\left(H^0(Y', L|_{Y'}^{\otimes n}), \langle\,,\rangle_{n,Y'}\right) + \sum_{i\geq 1}(-1)^i \log \#(H^i(Y', L|_{Y'}^{\otimes n})) = \frac{\widehat{\deg}(\widehat{c}_1(\overline{L})^{d-1}\cdot(Y',0))}{(d-1)!}n^{d-1} + o(n^{d-1}).$$

Here we note the following: In the case where $d = 2$, Y' is not necessarily irreducible, so that we cannot use the hypothesis of induction. However, Theorem 3.14 and Lemma 3.18 hold on reduced arithmetic curves, so that we have no problem to proceed the induction.

Note that

$$\widehat{\deg}(\widehat{c}_1(\overline{L})^{d-1}\cdot(Y',0)) + \widehat{\deg}(\widehat{c}_1(\overline{L})^{d-1}\cdot(Y'',0)) - \int_{X(\mathbb{C})} \log|s| c_1(\overline{L})^{d-1} = \widehat{\deg}\left(\widehat{c}_1(\overline{L})^{d-1}\cdot(Y, -\log|s|^2)\right) = N\widehat{\deg}(\widehat{c}_1(\overline{L})^d)$$

and that

$$N\frac{n^{d-1}}{(d-1)!} = \frac{n^d}{d!} - \frac{(n-N)^d}{d!} + o(n^{d-1}).$$

Therefore, using (5.20), (5.21) and (5.22), for $n \geq n_0 + N$, we obtain

$$\begin{aligned}(5.23)\quad &\left(\widehat{\deg}\left(H^0(X, L^{\otimes n}), \langle\,,\rangle_n\right) + \sum_{i\geq 1}(-1)^i \log\#(H^i(X, L^{\otimes n}))\right) \\ &- \left(\widehat{\deg}\left(H^0(X, L^{\otimes n-N}), \langle\,,\rangle_{n-N}\right) + \sum_{i\geq 1}(-1)^i \log\#(H^i(X, L^{\otimes n-N}))\right) \\ &= \frac{\widehat{\deg}(\widehat{c}_1(\overline{L})^d)}{d!}(n^d - (n-N)^d) + o(n^{d-1}).\end{aligned}$$

Thus (5.19) follows from the difference equation (5.23). □

5.10. Several kinds of positivity of C^∞-hermitian invertible sheaves

In this section, we would like to introduce several kinds of positivity of continuous hermitian invertible sheaves. Moreover, throughout this section, the tensor product of invertible sheaves is denoted additively.

DEFINITION 5.38. Let X be a d-dimensional projective arithmetic variety and $\overline{L}$ a continuous hermitian invertible sheaf of real type on X, that is, the metric of $\overline{L}$ is continuous.

(1) Let s be a global section of L, that is, $s \in H^0(X, L)$. If $\|s\|_{\sup} \leq 1$, then s is called a *small section* of L. Moreover, if $\|s\|_{\sup} < 1$, then s is called a *strictly small section* of L.

(2) If $\overline{L}$ is C^∞, L is ample on X and $\overline{L}$ is positive on $X(\mathbb{C})$, then $\overline{L}$ is said to be *vertically ample*. Moreover, if $H^0(X, nL)$ is generated by strictly small sections of nL as a $\mathbb{Z}$-module for a sufficiently large n, then the vertical ample C^∞-hermitian invertible sheaf $\overline{L}$ is said to be *ample*.

(3) We assume that $\overline{L}$ is C^∞. We say that $\overline{L}$ is *vertically nef* if L is nef on all fibers of $X \to \operatorname{Spec}(\mathbb{Z})$ and $\overline{L}$ is semipositive on $X(\mathbb{C})$. Moreover, we say that $\overline{L}$ is *nef* if $\overline{L}$ is vertically nef and $\widehat{\deg}\left(\overline{L}|_C\right) = \widehat{\deg}\left(\widehat{c}_1(\overline{L}) \cdot (C, 0)\right) \geq 0$ for all 1-dimensional subschemes C of X.

(4) If there is a non-zero small section of L, then $\overline{L}$ is said to be *effective*. Further, $\overline{L}$ is said to be $\mathbb{Q}$-*effective* if $n\overline{L}$ is effective for some positive integer n. For two C^∞-hermitian invertible sheaves $\overline{L}_1$ and $\overline{L}_2$ of real type, if $\overline{L}_1 - \overline{L}_2$ is effective, then we denote it by $\overline{L}_2 \leq \overline{L}_1$ or $\overline{L}_1 \geq \overline{L}_2$. Similarly, if $\overline{L}_1 - \overline{L}_2$ is $\mathbb{Q}$-effective, then it is denoted by $\overline{L}_2 \leq_{\mathbb{Q}} \overline{L}_1$ or $\overline{L}_1 \geq_{\mathbb{Q}} \overline{L}_2$.

(5) We say that $\overline{L}$ is *big* if $L_{\mathbb{Q}}$ is big on $X_{\mathbb{Q}}$, that is,

$$\operatorname{vol}(L_{\mathbb{Q}}) = \lim_{m \to \infty} \frac{\operatorname{rk} H^0(X, mL)}{m^{d-1}} > 0,$$

and $n\overline{L}$ has a strictly small section for a sufficiently large n.

REMARK 5.2. A small section is an arithmetic analogue of a global section of an invertible sheaf on an algebraic variety. In order to see this, let us consider an algebraic variety over a finite field. Namely, let X be a normal projective algebraic variety over a finite field $\mathbb{F}_q$. Let L be an invertible sheaf on X. For a prime divisor Γ on X and a rational section s of L, we define $\|s\|_\Gamma$ as follows: Let ω_Γ be a local basis of L at Γ and $s = f_\Gamma \omega_\Gamma$ (f_Γ is a rational function on X). Then

$$\|s\|_\Gamma = \begin{cases} \exp(-\operatorname{ord}_\Gamma(f_\Gamma)) & \text{if } f_\Gamma \neq 0, \\ 0 & \text{if } f_\Gamma = 0. \end{cases}$$

$\|\cdot\|_\Gamma$ gives rise to a norm on L_η (η is the generic point of X), that is, the set of all rational sections of L. An important point of our consideration is the following formula:

$$H^0(X, L) = \{s \in L_\eta \mid \|s\|_\Gamma \leq 1 \quad (\forall \Gamma)\}.$$

In particular,

$$\dim_{\mathbb{F}_q} H^0(X, L) = \frac{\log\left(\#\{s \in L_\eta \mid \|s\|_\Gamma \leq 1 \quad (\forall \Gamma)\}\right)}{\log(q)},$$

which will give an important view point in the arithmetic setting.

Several kinds of positivity of C^∞-hermitian invertible sheaves have a lot of properties as in Remark 5.3. To treat them completely, we need to prepare many technical results, so in this section, we only consider Proposition 5.39, Proposition 5.41, Proposition 5.40 and Proposition 5.43.

PROPOSITION 5.39. *If $\overline{L}$ is an ample C^∞-hermitian invertible sheaf of real type, then* $\widehat{\deg}(\widehat{c}_1(\overline{L})^{\cdot d}) > 0$.

PROOF. We prove it by induction on d. If $d = 1$, the it follows from Corollary 3.23, so that we assume $d > 1$. Since $\overline{L}$ is ample, nL has a non-zero strictly small section s for some positive integer n. Let $\operatorname{div}(s) = a_1 Y_1 + \cdots + a_l Y_l$ be the decomposition as a cycle, where $a_i > 0$ for all $i = 1, \ldots, l$. Then

$$\begin{aligned} n\widehat{\deg}(\widehat{c}_1(\overline{L})^d) &= \widehat{\deg}(\widehat{c}_1(\overline{L})^{d-1} \cdot \widehat{c}_1(n\overline{L})) \\ &= \sum_i a_i \widehat{\deg}(\widehat{c}_1(\overline{L})^{d-1}(Y_i, 0)) - \int_{X(\mathbb{C})} \log|s| c_1(\overline{L})^{d-1}. \end{aligned}$$

Thus, by using the hypothesis of induction, we have $\widehat{\deg}(\widehat{c}_1(\overline{L})^d) > 0$. □

PROPOSITION 5.40. *Let $\overline{L}$, $\overline{L}'$, $\overline{L}''$, $\overline{M}$ and $\overline{M}'$ be continuous hermitian invertible sheaves of real type. Then we have the following.*

(1) *$\overline{L} \le \overline{M}$ if and only if there is a non-zero homomorphism $\phi : L \to M$ such that $|\phi(\cdot)|_M \le |\cdot|_L$ holds, where $|\cdot|_L$ and $|\cdot|_M$ are norms of $\overline{L}$ and $\overline{M}$.*
(2) *If $\overline{L} \le \overline{M}$, then $\hat{h}^0\left(H^0(X,L), \|\cdot\|_{\sup}^{\overline{L}}\right) \le \hat{h}^0\left(H^0(X,M), \|\cdot\|_{\sup}^{\overline{M}}\right)$ (see Section 2.5 for the definition of $\hat{h}^0$).*
(3) *$\overline{L} \le \overline{L}'$ and $\overline{L}' \le \overline{L}'' \implies \overline{L} \le \overline{L}''$.*
(4) *$\overline{L} \le \overline{M} \iff -\overline{M} \le -\overline{L}$.*
(5) *$\overline{L} \le \overline{M}$ and $\overline{L}' \le \overline{M}' \implies \overline{L} + \overline{L}' \le \overline{M} + \overline{M}'$.*
(6) *$\overline{L} \le_{\mathbb{Q}} \overline{M} \iff$ there is a positive integer n with $n\overline{L} \le n\overline{M}$.*
(7) *$\overline{L} \le_{\mathbb{Q}} \overline{L}'$ and $\overline{L}' \le_{\mathbb{Q}} \overline{L}'' \implies \overline{L} \le_{\mathbb{Q}} \overline{L}''$.*
(8) *$\overline{L} \le_{\mathbb{Q}} \overline{M} \iff -\overline{M} \le_{\mathbb{Q}} -\overline{L}$.*
(9) *$\overline{L} \le_{\mathbb{Q}} \overline{M}$ and $\overline{L}' \le_{\mathbb{Q}} \overline{M}' \implies \overline{L} + \overline{L}' \le_{\mathbb{Q}} \overline{M} + \overline{M}'$.*
(10) *$\overline{L} \le_{\mathbb{Q}} \overline{M} \iff m\overline{L} \le_{\mathbb{Q}} m\overline{M}$ for all positive integers m.*

PROOF. (1) ($\Longrightarrow$) : By assumption, $M - L$ has a non-zero small section s. Let $\phi : L \to M$ be a homomorphism given by $\phi(l) = s \otimes l$. Then

$$|\phi(l)|_M = |s \otimes l|_M = |s|_{M-L}|l|_L \le |l|_L.$$

($\Longleftarrow$) : The homomorphism $\phi : L \to M$ yields a homomorphism $\mathcal{O}_X \to M - L$, so that it gives rise to a non-zero global section s of $M - L$. Then $\phi(l) = s \otimes l$. Let ω be a local basis of L at $x \in X$. Then

$$|s|_{M-L}|\omega|_L = |s \otimes \omega|_M = |\phi(\omega)|_M \le |\omega|_L.$$

Thus $|s|_{M-L} \le 1$ around x:

(2) follows from (1).

(3) is obvious by definition.

(4) is obvious because $\overline{M} - \overline{L} = -\overline{L} - (-\overline{M})$.

(5) By assumption, $M - L$ and $M' - L'$ have non-zero small sections s and s', respectively. Then $s \otimes s' \in H^0(X, (M + M') - (L + L'))$ and

$$\|s \otimes s'\|_{\sup} \le \|s\|_{\sup}\|s'\|_{\sup} \le 1.$$

Thus (5) follows.

(6) is obvious by definition. (7), (8) and (9) follow from (3)+(6), (4)+(6) and (5)+(6), respectively. Finally (10) is obvious by (9) and (6). □

PROPOSITION 5.41. *Let X be a generically smooth projective arithmetic variety and let $\overline{L}$ be an ample C^∞-hermitian invertible sheaf of real type on X. Then we have the following:*

(1) *There is a positive integer m_1 such that, for all $m \ge m_1$, $H^0(X, mL)$ has a free basis consisting of strictly small sections.*
(2) $\hat{h}^0\left(H^0(X, mL), \|\cdot\|_{\sup}^{m\overline{L}}\right) = \dfrac{\widehat{\deg}(\widehat{c}_1(\overline{L})^{\cdot d})}{d!} m^d + o(m^d) \quad (m \gg 1).$

PROOF. (1) Since $\overline{L}$ is ample, we can find a positive integer m_0 such that $H^0(X, m_0L)$ is generated by strictly small sections $s_1, \ldots, s_l$ and

$$\operatorname{Sym}^m(H^0(X, m_0L)) \to H^0(X, mm_0L)$$

is surjective for every $m \geq 1$. We put $\upsilon = \min\{\|s_1\|_{\sup}, \ldots, \|s_l\|_{\sup}\} < 1$. Then, as $H^0(X, mm_0L)$ is generated by $\{s_1^{a_1} \cdots s_l^{a_l}\}_{\substack{a_1,\ldots,a_l \in \mathbb{Z}_{\geq 0} \\ a_1+\cdots+a_l=m}}$, we have

$$\lambda_{\mathbb{Q}}(H^0(X, mm_0L), \|\cdot\|_{\sup}) \leq \upsilon^m$$

for $m \geq 1$. Note that, by Corollary 1.25 and Lemma 5.44, if we set

$$R = \bigoplus_{m \geq 0} H^0(X, mL) \quad \text{and} \quad R^{(m_0)} = \bigoplus_{m \geq 0} H^0(X, mm_0L),$$

then $R^{(m_0)}$ is noetherian and R is a finitely generated m_0-graded $R^{(m_0)}$-module, and hence by Proposition 5.45, there is a positive constant A such that

$$\lambda_{\mathbb{Q}}(H^0(X, mL), \|\cdot\|_{\sup}) \leq A\upsilon^{m/m_0}$$

for $m \geq 0$. Therefore, by Lemma 2.18,

$$\lambda_{\mathbb{Z}}(H^0(X, mL), \|\cdot\|_{\sup}) \leq A \operatorname{rk}(H^0(X, mL))\upsilon^{m/m_0}$$

for $m \geq 0$, as required.

(2) Since L is ample, for a sufficiently large integer m, $H^i(X, mL) = \{0\}$ for all $i > 0$. Moreover,

$$\operatorname{rk} H^0(X, mL) = \frac{\deg(L_{\mathbb{Q}}^{d-1})}{(d-1)!} m^{d-1} + o(m^{d-1}) \quad (m \gg 1).$$

Thus, by Theorem 5.36,

$$\widehat{\chi}\left(H^0(X, mL), \|\cdot\|_{\sup}^{m\overline{L}}\right) = \frac{\widehat{\deg}(\widehat{c}_1(\overline{L})^{\cdot d})}{d!} m^d + o(m^d) \quad (m \gg 1).$$

Therefore, by (2) in Proposition 2.17,

$$\hat{h}^0\left(H^0(X, mL), \|\cdot\|_{\sup}^{m\overline{L}}\right) - \hat{h}^1\left(H^0(X, mL), \|\cdot\|_{\sup}^{m\overline{L}}\right) = \frac{\widehat{\deg}(\widehat{c}_1(\overline{L})^{\cdot d})}{d!} m^d + o(m^d).$$

On the other hand, by (1), $H^0(X, mL)$ has a free basis consisting of strictly small sections for $m \gg 1$. Therefore, by (5) in Proposition 2.17,

$$\hat{h}^1\left(H^0(X, mL), \|\cdot\|_{\sup}^{m\overline{L}}\right) = o(m^d).$$

Thus (2) follows. □

COROLLARY 5.42. *Let X be a d-dimensional projective arithmetic variety and let $\overline{L}$ be a continuous hermitian invertible sheaf of real type on X. Then we have the following.*

(1) *There is a constant C such that*

$$\hat{h}^0\left(H^0(X, mL), \|\cdot\|_{\sup}^{m\overline{L}}\right) \leq Cm^d \quad (m \geq 1).$$

(2) *We assume that $\overline{L}$ is C^∞. Let $\mu : Y \to X$ be a generic resolution of singularities of X. Let Φ be a C^∞-volume form of real type on $Y(\mathbb{C})$. Here we give an L^2-norm of $H^0(X, mL)$ as follows: For $t \in H^0(X, mL)$,*

$$\|t\|^{m\overline{L}}_{L^2,\Phi} := \left(\int_{Y(\mathbb{C})} \mu^*(|t|^2_{m\overline{L}})\Phi \right)^{1/2},$$

where $|\cdot|_{m\overline{L}}$ is the norm of $m\overline{L}$. Then

$$\hat{h}^0\left(H^0(X, mL), \|\cdot\|^{m\overline{L}}_{L^2,\Phi}\right) \le Cm^d \quad (m \ge 1).$$

PROOF. (1) Let $\mu : Y \to X$ be a generic resolution of singularities of X. Then

$$\hat{h}^0\left(H^0(X, mL), \|\cdot\|^{m\overline{L}}_{\sup}\right) \le \hat{h}^0\left(H^0(Y, m\mu^*(L)), \|\cdot\|^{m\mu^*(\overline{L})}_{\sup}\right).$$

Thus we may assume that X is generically smooth. Let $\overline{A}$ be an ample C^∞-hermitian invertible sheaf of real type on X. Then, for a sufficiently large n, $\overline{L} \le n\overline{A}$, so that $m\overline{L} \le mn\overline{A}$ $(m \ge 1)$. Thus (1) follows from Proposition 5.41.

(2) Using Gromov's inequality on $Y(\mathbb{C})$, there is a constant C_1 such that

$$\|\cdot\|^{m\overline{L}}_{\sup} \le C_1 m^{d-1}\|\cdot\|^{m\overline{L}}_{L^2,\Phi} \quad (m \gg 1).$$

Thus

$$\hat{h}^0\left(H^0(X, mL), C_1 m^{d-1}\|\cdot\|^{m\overline{L}}_{L^2,\Phi}\right) \le \hat{h}^0\left(H^0(X, mL), \|\cdot\|^{m\overline{L}}_{\sup}\right).$$

On the other hand, by (3) in Proposition 2.17,

$$\hat{h}^0\left(H^0(X, mL), C_1 m^{d-1}\|\cdot\|^{m\overline{L}}_{L^2,\Phi}\right) = \hat{h}^0\left(H^0(X, mL), \|\cdot\|^{m\overline{L}}_{L^2,\Phi}\right) + o(m^d).$$

Hence (2) follows. □

PROPOSITION 5.43. *Let $\overline{L}$ be a big continuous hermitian invertible sheaf of real type. Then, for any continuous hermitian invertible sheaf $\overline{M}$ of real type, there is a positive integer n such that $nL + M$ has a non-zero strictly small section.*

PROOF. By our assumption, there is a positive integer n_1 such that $n_1 L$ has a non-zero strictly small section s_1. Since $L_{\mathbb{Q}}$ is big on $X_{\mathbb{Q}}$, there are a sufficiently large n_2 and a non-zero section $s_2 \in H^0(X, L^{\otimes n_2} \otimes M)$. Let n_3 be a positive integer with $(\|s_1\|_{\sup})^{n_3}\|s_2\|_{\sup} < 1$. If we set $s = s_1^{\otimes n_3} \otimes s_2$, then $s \in H^0(X, L^{\otimes n_3 n_1 + n_2} \otimes M)$ and

$$\|s\|_{\sup} \le (\|s_1\|_{\sup})^{n_3}\|s_2\|_{\sup} < 1,$$

as required. □

REMARK 5.3. In later sections, we will show the following properties.
(1) (Corollary 6.14) If $\overline{L}$ is vertically nef and $\widehat{\deg}(\widehat{c}_1(\overline{L})^d)) > 0$, then $\overline{L}$ is big.

(2) (Theorem 6.15) If $\overline{L}_1, \ldots, \overline{L}_d$ are nef C^∞-hermitian invertible sheaves of real type, then

$$\widehat{\deg}\left(\widehat{c}_1(\overline{L}_1)\cdots\widehat{c}_1(\overline{L}_d)\right) \ge 0.$$

(3) (Theorem 7.1, Nakai-Moishezon's criterion on arithmetic varieties) We assume that X is generically smooth and $\overline{L}$ is vertically ample. If

$$\widehat{\deg}\left(\widehat{c}_1(\overline{L})^{\cdot \dim Y} \cdot (Y, 0)\right) > 0$$

for any horizontal integral subscheme Y of X (i.e. Y is flat over $\mathbb{Z}$), then $\overline{L}$ is ample as a C^∞-hermitian invertible sheaf.

(4) (Proposition 7.9) We assume that X is generically smooth, $\overline{L}$ is nef and $\overline{A}$ is ample, then $\overline{A} + \overline{L}$ is ample, that is, a nef C^∞-hermitian invertible sheaf is the limit of ample C^∞-hermitian $\mathbb{Q}$-invertible sheaves.

5.11. Estimation of $\lambda_{\mathbb{Q}}$ for a normed graded ring

In this section, we consider technical facts treated in [**54**] and [**76**]. These results are used in other sections.

Let $R = \bigoplus_{n=0}^{\infty} R_n$ be a graded ring over a commutative ring k with unity. Let h be a positive integer. We define $R^{(h)}$ to be $R^{(h)} = \bigoplus_{n=0}^{\infty} R_{nh}$. Let us begin with the following lemma.

LEMMA 5.44. *If R is noetherian, then R is a finitely generated $R^{(h)}$-module. In particular, $R^{(h)}$ is also noetherian.*

PROOF. Since R is noetherian, by [**3**, Proposition 10.7], R_0 is a noetherian and there are homogeneous elements $x_1, \ldots, x_l$ such that $R = R_0[x_1, \ldots, x_l]$. Let a_i be the degree of x_i. For $(t_1, \ldots, t_l) \in \mathbb{Z}^l_{\geq 0}$, we set $t_i = hq_i + r_i$ $(0 \leq r_i < h)$ for each i. Then $x_1^{t_1} \cdots x_l^{t_l} = (x_1^h)^{q_1} \cdots (x_l^h)^{q_l} x_1^{r_1} \cdots x_l^{r_l}$. Notice that $x_1^h, \ldots, x_l^h \in R^{(h)}$. Thus R is generated by a finite set $\{x_1^{r_1} \cdots x_l^{r_l}\}_{(r_1,\ldots,r_l) \in \{0,1,\ldots,h-1\}^l}$ as a $R^{(h)}$-module. We put

$$\Delta_h = \left\{(r_1, \ldots, r_l) \in \mathbb{Z}^l_{\geq 0} \mid 0 \leq r_i < h \ (\forall i) \text{ and } h \mid r_1 a_1 + \cdots + r_l a_l\right\},$$

which is a finite set. Note that, if $h \mid t_1 a_1 + \cdots + t_l a_l$, then $h \mid r_1 a_1 + \cdots + r_l a_l$. This means that $R^{(h)}$ is generated by $\{x_1^{r_1} \cdots x_l^{r_l}\}_{(r_1,\ldots,r_l) \in \Delta_h}$ as a $R_0[x_1^h, \ldots, x_l^h]$-module. Thus $R^{(h)}$ is noetherian. $\square$

Let M be a R-module. We say that M is an *h-graded R-module* if M has the following properties:

(i) $M = \bigoplus_{n=-\infty}^{\infty} M_n$ as k-modules.
(ii) For all $n \in \mathbb{Z}_{\geq 0}$ and $n' \in \mathbb{Z}$, $x \cdot m \in M_{hn+n'}$ whenever $x \in R_n$ and $m \in M_{n'}$.

For example, R is an h-graded $R^{(h)}$-module.

We assume the following:

(a) $k = \mathbb{Z}$.
(b) R_n is a finitely generated $\mathbb{Z}$-module for every $n \in \mathbb{Z}_{\geq 0}$.
(c) M is an h-graded R-module.
(d) M_n is a finitely generated $\mathbb{Z}$-module for every $n \in \mathbb{Z}$.

Let $\mathbb{K}$ be either $\mathbb{Q}$ or $\mathbb{R}$. If we set $R_{\mathbb{K}} = R \otimes_{\mathbb{Z}} \mathbb{K}$ and $M_{\mathbb{K}} = M \otimes_{\mathbb{Z}} \mathbb{K}$, then

$$R_{\mathbb{K}} = \bigoplus_{n=0}^{\infty} (R_n)_{\mathbb{K}} \quad \text{and} \quad M_{\mathbb{K}} = \bigoplus_{n=-\infty}^{\infty} (M_n)_{\mathbb{K}},$$

where $(R_n)_{\mathbb{K}} = R_n \otimes_{\mathbb{Z}} \mathbb{K}$ and $(M_n)_{\mathbb{K}} = M_n \otimes_{\mathbb{Z}} \mathbb{K}$. Note that $R_{\mathbb{K}}$ is a graded ring over $\mathbb{K}$ and $M_{\mathbb{K}}$ is an h-graded $R_{\mathbb{K}}$-module. Let $\|\cdot\|_n$ be a norm of $(R_n)_{\mathbb{R}}$ for each $n \in \mathbb{Z}_{\geq 0}$. We say that

$$(R, \|\cdot\|) = \bigoplus_{n=0}^{\infty} (R_n, \|\cdot\|_n)$$

is a *normed graded ring over* $\mathbb{Z}$ if $\|x \cdot x'\|_{n+n'} \le \|x\|_n \|x'\|_{n'}$ for all $x \in (R_n)_{\mathbb{R}}$ and $x' \in (R_{n'})_{\mathbb{R}}$. Moreover, let $\|\cdot\|_{M_n}$ be a norm of $(M_n)_{\mathbb{R}}$ for each $n \in \mathbb{Z}$. Then,

$$(M, \|\cdot\|_M) = \bigoplus_{n=-\infty}^{\infty} (M_n, \|\cdot\|_{M_n})$$

is called a *normed h-graded $(R, \|\cdot\|)$-module* if $\|x \cdot m\|_{M_{hn+n'}} \le \|s\|_n \|m\|_{M_{n'}}$ for all $x \in (R_n)_{\mathbb{R}}$ and $m \in (M_{n'})_{\mathbb{R}}$.

PROPOSITION 5.45. *Let $(R, \|\cdot\|)$ be a normed graded ring over $\mathbb{Z}$ and $(M, \|\cdot\|_M)$ a normed h-graded $(R, \|\cdot\|)$-module. We assume that $M_{\mathbb{Q}}$ is a finitely generated $R_{\mathbb{Q}}$-module. If there are $A, e, v \in \mathbb{R}_{\ge 0}$ such that $\lambda_{\mathbb{Q}}(R_n, \|\cdot\|_n) \le A(n+1)^e v^n$ for all $n \ge 0$, then there is $B \in \mathbb{R}_{\ge 0}$ such that $\lambda_{\mathbb{Q}}(M_n, \|\cdot\|_{M_n}) \le B(n+1)^e v^{n/h}$ for all $n \ge 0$.*

PROOF. By our assumption, for each $n \ge 0$, we can choose a basis $x_{n,1}, \dots, x_{n,t_n}$ of $(R_n)_{\mathbb{Q}}$ over $\mathbb{Q}$ such that $x_{n,1}, \dots, x_{n,t_n} \in R_n$ and $\|x_{n,j}\|_n \le A(n+1)^e v^n$ for all $j = 1, \dots, t_n$. Moreover, let $m_1, \dots, m_l$ be homogeneous generators of $M_{\mathbb{Q}}$ over $R_{\mathbb{Q}}$. Let d_i denote the degree of m_i. By replacing m_i with am_i for some $a \in \mathbb{Z}_{>0}$, we may assume that $m_i \in M_{d_i}$. Then $(M_n)_{\mathbb{Q}}$ is generated by elements of the form $x_{i,j} m_k$ with $ih + d_k = n$. We write

$$C = \max_{k=1,\dots,l} \{\|m_k\|_{M_{d_k}} v^{-d_k/h}\}.$$

In addition, we choose a positive number D such that

$$\frac{n - d_k}{h} + 1 \le D(n+1)$$

for all $n \ge 0$ and $k = 1, \dots, l$. Since $x_{i,j} m_k \in M_n$ and

$$\begin{aligned} \|x_{i,j} m_k\|_{M_n} &\le \|x_{i,j}\|_i \|m_k\|_{M_{d_k}} \le A(i+1)^e v^i \|m_k\|_{M_{d_k}} \\ &= A\left(\frac{n-d_k}{h} + 1\right)^e v^{(n-d_k)/h} \|m_k\|_{M_{d_k}} \le ACD^e (n+1)^e v^{n/h}, \end{aligned}$$

we have $\lambda_{\mathbb{Q}}(M_n, \|\cdot\|_{M_n}) \le ACD^e (n+1)^e v^{n/h}$ for all $n \ge 0$. □

Let I be a homogeneous ideal of R and let $R' = R/I$. Let $Q = \bigoplus_{n \in \mathbb{Z}} Q_n$ be an h-graded R-module with $I \cdot Q = 0$, and let $f : M \to Q$ be a surjective homomorphism of h-graded R-modules of degree 0, that is, $f(M_n) = Q_n$ for all $n \in \mathbb{Z}$. Note that Q is an h-graded R'-module. Let $\|\cdot\|'_n = \|\cdot\|_{n, R_n \twoheadrightarrow R'_n}$ and $\|\cdot\|_{Q_n} = \|\cdot\|_{M_n, M_n \twoheadrightarrow Q_n}$ (cf. Section 1.2). Then

$$\|x' \cdot q\|_{Q_{hn+n'}} \le \|x'\|'_n \|q\|_{Q_{n'}}$$

holds for all $x' \in (R'_n)_{\mathbb{R}}$ and $q \in (Q_{n'})_{\mathbb{R}}$. Indeed, choose $x \in (R_n)_{\mathbb{R}}$ and $m \in (M_{n'})_{\mathbb{R}}$ such that the classes of x in $R'_{\mathbb{R}}$ is x' and $f(m) = q$, and that $\|x\|_n = \|x'\|'_n$ and $\|m\|_{M_{n'}} = \|q\|_{Q_{n'}}$. As $f(x \cdot m) = x' \cdot q$,

$$\|x' \cdot q\|_{Q_{hn+n'}} \le \|x \cdot m\|_{M_{hn+n'}} \le \|x\|_n \|m\|_{M_{n'}} = \|x'\|'_n \|q\|_{Q_{n'}}.$$

Therefore, $(R', \|\cdot\|') = \bigoplus_{n=0}^{\infty} (R'_n, \|\cdot\|'_n)$ is a normed graded ring and $(Q, \|\cdot\|_Q) = \bigoplus_{n=-\infty}^{\infty} (Q_n, \|\cdot\|_{Q_n})$ is a normed h-graded $(R', \|\cdot\|')$-module.

COROLLARY 5.46. *Let I, J and K be homogeneous ideals of R such that $J \subseteq K$ and $I \cdot K \subseteq J$. Let $R' = R/I$ and $Q = K/J$, and let $\|\cdot\|_{K_n} = \|\cdot\|_{n, K_n \hookrightarrow R_n}$ and $\|\cdot\|_{Q_n} = \|\cdot\|_{K_n, K_n \twoheadrightarrow Q_n}$. We assume the following:*

(1) *$R_{\mathbb{Q}}$ is noetherian.*
(2) *There are $A, e, v \in \mathbb{R}_{\geq 0}$ such that $\lambda_{\mathbb{Q}}(R'_n, \|\cdot\|'_n) \leq A(n+1)^e v^n$ for all $n \geq 0$.*

Then there is $B \in \mathbb{R}_{\geq 0}$ such that $\lambda_{\mathbb{Q}}(Q_n, \|\cdot\|_{Q_n}) \leq B(n+1)^e v^n$ for all $n \geq 0$.

PROOF. Clearly, $(K, \|\cdot\|_K) = \bigoplus_{n=0}^{\infty}(K_n, \|\cdot\|_{K_n})$ is a normed 1-graded $(R, \|\cdot\|)$-module. Note that $(Q, \|\cdot\|_Q)$ is a normed 1-graded $(R', \|\cdot\|')$-module because $I \cdot Q = 0$. Further, $K_{\mathbb{Q}}$ is finitely generated as a $R_{\mathbb{Q}}$-module because $K_{\mathbb{Q}}$ is an ideal of a noetherian ring $R_{\mathbb{Q}}$. Thus $Q_{\mathbb{Q}}$ is also finitely generated as a $R'_{\mathbb{Q}}$-module, and hence the corollary follows from Proposition 5.45. □

Let X be a d-dimensional projective arithmetic variety. Let

$$(R, \|\cdot\|) = \bigoplus_{n=0}^{\infty}(R_n, \|\cdot\|_n)$$

is a normed graded ring over $\mathbb{Z}$ such that $R = \bigoplus_{n=0}^{\infty} R_n$ is a graded subring of $\bigoplus_{n=0}^{\infty} H^0(X, nL)$ for some invertible sheaf L on X. For an ideal sheaf $\mathcal{I}$ of X, we define $I_n(R; \mathcal{I})$ and $I(R; \mathcal{I})$ to be

$$I_n(R; \mathcal{I}) = H^0(X, nL \otimes \mathcal{I}) \cap R_n, \quad \text{and} \quad I(R; \mathcal{I}) = \bigoplus_{n=0}^{\infty} I_n(R; \mathcal{I}).$$

Note that $I(R; \mathcal{I})$ is a homogeneous ideal of R, so that we have a graded ring $R_{\mathcal{I}}$ given by $R_{\mathcal{I}} := R/I(R; \mathcal{I})$. Let $\|\cdot\|_{(R_{\mathcal{I}})_n}$ be the quotient norm of $(R_{\mathcal{I}})_n$ induced by $R_n \twoheadrightarrow (R_{\mathcal{I}})_n$ and the norm $\|\cdot\|_n$ of R_n. Let Y be an arithmetic subvariety of X, that is, Y is an integral subscheme flat over $\mathbb{Z}$, and let $\mathcal{I}_Y$ be the defining ideal sheaf of Y. For simplicity, $R_{\mathcal{I}_Y}$, $\|\cdot\|_{R_{\mathcal{I}_Y}}$, $(R_Y)_n$, and $\|\cdot\|_{(R_Y)_n}$ are denoted by R_Y, $\|\cdot\|_Y$, $R_{Y,n}$ and $\|\cdot\|_{Y,n}$, respectively. Notice that

$$R_{Y,n} \subseteq H^0(X, nL)/H^0(X, nL \otimes \mathcal{I}_Y) \hookrightarrow H^0(Y, nL|_Y),$$

that is, $\bigoplus_{n=0}^{\infty} R_{Y,n}$ is isomorphic to a graded subring of $\bigoplus_{n=0}^{\infty} H^0(Y, nL|_Y)$. In particular, R_Y is an integral domain. Let $\mathfrak{S}_X$ denote the set of all arithmetic subvarieties of X.

THEOREM 5.47 (cf. [**54**, Theorem 3.1]). *Let $v : \mathfrak{S}_X \to \mathbb{R}_{>0}$ be a map such that $(R, \|\cdot\|)$ and v satisfy the following conditions:*

(1) *$R_{\mathbb{Q}}$ is noetherian.*
(2) *For each $Y \in \mathfrak{S}_X$, there is a positive integer n_0 such that, for all $n \geq n_0$, $(R_{Y,n})_{\mathbb{Q}} = H^0\left(Y_{\mathbb{Q}}, nL_{\mathbb{Q}}|_{Y_{\mathbb{Q}}}\right)$.*
(3) *For each $Y \in \mathfrak{S}_X$, there are a positive integer n_1 and a non-zero $s \in R_{Y,n_1}$ with $\|s\|_{Y,n_1} \leq v(Y)^{n_1}$.*

Then there are a positive constant B and a finite subset S of $\mathfrak{S}_X$ such that

$$\lambda_{\mathbb{Q}}(R_n, \|\cdot\|_n) \leq B(n+1)^{d(d-1)/2} \left(\max\{v(Y) \mid Y \in S\}\right)^n$$

for all $n \geq 0$.

PROOF. For a positive integer h, we set $\|\cdot\|_n^{(h)} = \|\cdot\|_{hn}$ for $n \geq 0$. Let us begin with the following claim:

CLAIM 8. (i) $(R^{(h)}, \|\cdot\|^{(h)})$ *and* v^h *satisfy the above conditions* (1), (2) *and* (3).

(ii) *If the theorem holds for* $(R^{(h)}, \|\cdot\|^{(h)})$ *and* v^h *for a positive integer* h, *then it also holds for* $(R, \|\cdot\|)$ *and* v.

PROOF. (i) $R_{\mathbb{Q}}^{(h)}$ is noetherian by Lemma 5.44, so that (1) holds. Obviously, (2) is satisfied. Let us consider (3). For $Y \in \mathfrak{S}_X$, we choose a positive integer n_1 and a non-zero $t \in R_{Y,n_1}$ with $\|t\|_{Y,n_1} \leq v(Y)^{n_1}$, so that $t^h \in R_{Y,hn_1} \setminus \{0\}$ and $\|t^h\|_{Y,hn_1} \leq (\|t\|_{Y,n_1})^h \leq (v(Y)^h)^{n_1}$, as required.

(ii) By our assumption, there are $B \in \mathbb{R}_{>0}$ and a finite subset S of $\mathfrak{S}_X$ such that

$$\lambda_{\mathbb{Q}}(R_{nh}, \|\cdot\|_{nh}) \leq B(n+1)^{d(d-1)/2} \left(\max\{v(Y)^h \mid Y \in S\}\right)^n$$

for all $n \geq 0$. On the other hand, by Lemma 5.44, $R_{\mathbb{Q}}$ is a finitely generated $R_{\mathbb{Q}}^{(h)}$-module. Thus, by Proposition 5.45, there is $B' \in \mathbb{R}_{>0}$ such that

$$\lambda_{\mathbb{Q}}(R_n, \|\cdot\|_n) \leq B'(n+1)^{d(d-1)/2} \left(\max\{v(Y)^h \mid Y \in S\}\right)^{n/h}$$

for all $n \geq 0$, so that the claim follows. □

We prove this theorem by induction on d. By Claim 8, we may assume that there is a non-zero $s \in R_1$ with $\|s\|_1 \leq v(X)$. Let us consider the case where $d = 1$. As $R_n \xrightarrow{\cdot s} R_{n+1}$ is injective, we have

$$\mathrm{rk}_{\mathbb{Z}} R_1 \leq \cdots \leq \mathrm{rk}_{\mathbb{Z}} R_n \leq \mathrm{rk}_{\mathbb{Z}} R_{n+1} \leq \cdots \leq \mathrm{rk}_{\mathbb{Z}} L_{\mathbb{Z}}.$$

Thus there is a positive integer n_0 such that $R_{n_0} \xrightarrow{\cdot s^n} R_{n_0+n}$ yields an isomorphism over $\mathbb{Q}$. Hence, by (1) in Lemma 2.19,

$$\lambda_{\mathbb{Q}}(R_{n+n_0}, \|\cdot\|_{n+n_0}) \leq \|s\|_1^n \lambda_{\mathbb{Q}}(R_{n_0}, \|\cdot\|_{n_0}) \leq v(X)^n \lambda_{\mathbb{Q}}(R_{n_0}, \|\cdot\|_{n_0}),$$

as required.

We assume that $d > 1$. Let $\mathcal{I}$ be the ideal sheaf of $\mathcal{O}_X$ given by

$$\mathcal{I} = \mathrm{Image}\left(L^{-1} \xrightarrow{\otimes s} \mathcal{O}_X\right).$$

CLAIM 9. *We can find a sequence* $\mathcal{I}_0 = \mathcal{I} \subsetneq \mathcal{I}_1 \subsetneq \cdots \subsetneq \mathcal{I}_m = \mathcal{O}_X$ *of ideal sheaves and proper integral subschemes* $D_1, \ldots, D_m$ *of* X *given by the ideal sheaves* $\mathcal{I}_{D_1}, \ldots, \mathcal{I}_{D_m}$, *respectively, such that* $\mathcal{I}_{D_r} \cdot \mathcal{I}_r \subseteq \mathcal{I}_{r-1}$ *for all* $r = 1, \ldots, m$.

PROOF. We would like to build $\mathcal{I}_0, \mathcal{I}_1, \ldots$ and $D_1, \ldots$ inductively, so that we assume that $\mathcal{I}_0, \ldots, \mathcal{I}_{i-1}$ and $D_1, \ldots, D_{i-1}$ have been constructed. We choose an $\mathcal{O}_X$-submodule $\mathcal{F}_i$ of $\mathcal{O}_X/\mathcal{I}_{i-1}$ such that the support of any non-zero $\mathcal{O}_X$-submodule of $\mathcal{F}_i$ is $\mathrm{Supp}(\mathcal{F}_i)$. Let D_i be the support of $\mathcal{F}_i$. Then D_i is irreducible and $D_i \subsetneq X$. Let n_i be the maximal non-negative integer with $\mathcal{G}_i := \mathcal{I}_{D_i}^{n_i} \cdot \mathcal{F}_i \neq 0$, and let $\mathcal{I}_i$ be the ideal sheaf with $\mathcal{I}_i/\mathcal{I}_{i-1} = \mathcal{G}_i$. Then, as $\mathcal{I}_{D_i} \cdot \mathcal{G}_i = 0$, we have $\mathcal{I}_{D_i} \cdot \mathcal{I}_i \subseteq \mathcal{I}_{i-1}$. Since $\mathcal{O}_X$ is noetherian, this precess must be stopped after finitely many repetitions. □

Fix a positive integer n_0 such that $(R_n)_{\mathbb{Q}} = H^0(X_{\mathbb{Q}}, nL_{\mathbb{Q}})$ for all $n \geq n_0$. Let $\|\cdot\|_{n,r}'$ be the quotient norm of $I_n(R;\mathcal{I}_r)/I_n(R;\mathcal{I}_{r-1})$ induced by $I_n(R;\mathcal{I}_r) \twoheadrightarrow I_n(R;\mathcal{I}_r)/I_n(R;\mathcal{I}_{r-1})$ and $\|\cdot\|_{n,r}$ of $I_n(R;\mathcal{I}_r)$. Let us consider the following claim.

CLAIM 10. (a) *For each* $1 \leq r \leq m$, *there are* $B_r \in \mathbb{R}_{>0}$ *and a finite subset* S_r *of* $\mathfrak{S}_X$ *such that*

$$\lambda_{\mathbb{Q}}\left(I_n(R;\mathcal{I}_r)/I_n(R;\mathcal{I}_{r-1}),\ \|\cdot\|'_{n,r}\right) \leq B_r(n+1)^{(d-1)(d-2)/2}\left(\max\{v(Y) \mid Y \in S_r\}\right)^n.$$

for all $n \geq 0$.

(b) *There is* $C_1 \in \mathbb{R}_{>0}$ *such that*

$$\operatorname{rk}(I_n(R;\mathcal{I}_r)/I_n(R;\mathcal{I}_{r-1})) \leq C_1(n+1)^{d-2}$$

for all $n \geq 0$ *and* $r = 1, \ldots, m$.

(c) $\operatorname{rk}(I_n(R;\mathcal{I}_0)/R_{n-1}s) = 0$ *for all* $n \geq n_0 + 1$.

PROOF. (a) In the case where D_r is vertical, $I_n(R;\mathcal{I}_r)/I_n(R;\mathcal{I}_{r-1})$ is a torsion module for all $n \geq 0$, so that, if we set $S_r = \{X\}$ and $B_r = 1$, then the assertion is obvious. Otherwise, the assertion follows from Corollary 5.46 and the hypothesis of induction because $I(R;\mathcal{I}_{D_r}) \cdot I(R;\mathcal{I}_r) \subseteq I(R;\mathcal{I}_{r-1})$.

(b) is obvious because $I_n(R;\mathcal{I}_r)/I_n(R;\mathcal{I}_{r-1}) \hookrightarrow H^0(D_r, nL \otimes \mathcal{I}_r/\mathcal{I}_{r-1})$.

(c) follows from

$$(R_{n-1})_{\mathbb{Q}}s = H^0(X_{\mathbb{Q}}, (n-1)L_{\mathbb{Q}})s = H^0(X_{\mathbb{Q}}, (nL \otimes \mathcal{I})_{\mathbb{Q}}) = I_n(R;\mathcal{I})_{\mathbb{Q}}. \qquad \square$$

Let $\overline{R}_n$ and $\overline{I}_n(R;\mathcal{I}_r)$ denote $(R_n, \|\cdot\|_n)$ and $(I_n(R;\mathcal{I}_r), \|\cdot\|_{n,r})$, respectively, where $\|\cdot\|_{n,r} = \|\cdot\|_{n, I_n(R;\mathcal{I}_r)\hookrightarrow R_n}$. By the definition, $\overline{R}_n = \overline{I}_n(R;\mathcal{I}_m)$.

Using (c) in Claim 10 and applying Lemma 2.20 to the sequence

$$\begin{array}{cccccccc}
\overline{R}_{n_0} & \xrightarrow{\cdot s} & \overline{I}_{n_0+1}(R;\mathcal{I}_0) & \hookrightarrow \cdots \hookrightarrow & \overline{I}_{n_0+1}(R;\mathcal{I}_r) & \hookrightarrow \cdots \hookrightarrow & \overline{I}_{n_0+1}(R;\mathcal{I}_m) \\
& \xrightarrow{\cdot s} & \overline{I}_{n_0+2}(R;\mathcal{I}_0) & \hookrightarrow \cdots \hookrightarrow & \overline{I}_{n_0+2}(R;\mathcal{I}_r) & \hookrightarrow \cdots \hookrightarrow & \overline{I}_{n_0+2}(R;\mathcal{I}_m) \\
& \vdots & \vdots & \vdots & \vdots & \vdots & \vdots \\
& \xrightarrow{\cdot s} & \overline{I}_{n}(R;\mathcal{I}_0) & \hookrightarrow \cdots \hookrightarrow & \overline{I}_{n}(R;\mathcal{I}_r) & \hookrightarrow \cdots \hookrightarrow & \overline{I}_{n}(R;\mathcal{I}_m),
\end{array}$$

we obtain

$$\begin{aligned}
&\lambda_{\mathbb{Q}}(R_n, \|\cdot\|_n) \\
&\leq \sum_{i=n_0+1}^{n}\left(\sum_{r=1}^{m}\|s\|_1^{n-i}\lambda_{\mathbb{Q}}\left(I_i(R;\mathcal{I}_r)/I_i(R;\mathcal{I}_{r-1}),\ \|\cdot\|'_{i,r}\right)\operatorname{rk}(I_i(R;\mathcal{I}_r)/I_i(R;\mathcal{I}_{r-1}))\right) \\
&\qquad\qquad + \|s\|_1^{n-n_0}\lambda_{\mathbb{Q}}(R_{n_0}, \|\cdot\|_{n_0})\operatorname{rk}(R_{n_0})
\end{aligned}$$

for $n \geq n_0 + 1$, so that, if we set $S = S_1 \cup \cdots \cup S_r \cup \{X\}$, $v = \max\{v(Y) \mid Y \in S\}$, $C_2 = \max\{B_1, \ldots, B_r\}$ and $C_3 = \|s\|_1^{-n_0}\lambda_{\mathbb{Q}}(R_{n_0}, \|\cdot\|_{n_0})\operatorname{rk}(R_{n_0})$, then, using (a) and (b) in Claim 10, we have

$$\begin{aligned}
\lambda_{\mathbb{Q}}(R_n, \|\cdot\|_n) &\leq \sum_{i=n_0+1}^{n} \left(\sum_{r=1}^{m} v^{n-i} \cdot (C_2(i+1)^{(d-1)(d-2)/2} v^i) \cdot (C_1(i+1)^{d-2}) \right) + C_3 v^n \\
&\leq \sum_{i=n_0+1}^{n} \left(mC_1C_2 v^n (i+1)^{(d-1)(d-2)/2+d-2} \right) + C_3 v^n \\
&\leq \sum_{i=n_0+1}^{n} \left(mC_1C_2 v^n (n+1)^{(d-1)(d-2)/2+d-2} \right) + C_3 v^n \\
&\leq mC_1C_2 v^n (n+1)^{(d-1)(d-2)/2+d-2+1} + C_3 v^n \\
&= \left(mC_1C_2 (n+1)^{d(d-1)/2} + C_3 \right) v^n,
\end{aligned}$$

as required. □

COROLLARY 5.48. *Let X be a d-dimensional projective arithmetic variety and let $\overline{L}$ be a continuous hermitian invertible sheaf on X. For $Y \in \mathfrak{S}_X$, let $H^0(X|Y, nL)$ denote*

$$\operatorname{Image}\left(H^0(X, nL) \to H^0(Y, nL|_Y) \right).$$

Let $\|\cdot\|_{n,Y}$ be the quotient norm of $H^0(X|Y, nL)$ induced by the supremum norm $\|\cdot\|_n$ of $H^0(X, nL)_{\mathbb{C}}$ and the surjective homomorphism $H^0(X, nL)_{\mathbb{C}} \to H^0(X|Y, nL)_{\mathbb{C}}$. Here we assume the following:

1. *$\bigoplus_{n=0}^{\infty} H^0(X, nL)_{\mathbb{Q}}$ is noetherian.*
2. *For each $Y \in \mathfrak{S}_X$, there is a positive integer n_0 such that, for all $n \geq n_0$, $H^0(X|Y, nL)_{\mathbb{Q}} = H^0(Y, nL|_Y)_{\mathbb{Q}}$.*
3. *For each $Y \in \mathfrak{S}_X$, there are a positive integer n_1 and a non-zero $s \in H^0(X|Y, n_1 L)$ such that $\|s\|_{n_1,Y} < 1$.*

Then $H^0(X, nL)$ has a free basis consisting of strictly small sections for a sufficiently large n.

PROOF. By using the above (3), for each arithmetic subvariety Y of X, we set $v(Y) = \|s\|_{n_1,Y}^{1/n_1} < 1$. Then, by Theorem 5.47 and Lemma 2.18, there are a positive constant B and a finite subset S of $\mathfrak{S}_X$ such that

$$\lambda_{\mathbb{Z}}(H^0(X, nL), \|\cdot\|_{\sup}) \leq B \operatorname{rk}(H^0(X, nL))(n+1)^{d(d-1)/2} \left(\max\{v(Y) \mid Y \in S\}\right)^n$$

for all $n \geq 0$. Note that $\max\{v(Y) \mid Y \in S\} < 1$, so that the corollary follows. □

CHAPTER 6

Arithmetic Volume Function and Its Continuity

In this section, we introduce the volume of a continuous hermitian invertible sheaf on a projective arithmetic variety and investigate its several properties. The contexts of this section consist largely of Moriwaki's paper [**51**]. The most important property is the continuity of volumes due to Moriwaki [**51**]. The fundamental estimate for it is Theorem 6.17 whose proof is very technical, so it will be proved in the last part of this section. As an application of the continuity of volumes, we will show the generalized Hodge index theorem. This is one of the fundamental theorems in birational Arakelov geometry. This is a generalization of several results due to Zhang [**76**]. For example, we can see that a nef C^∞-hermitian invertible sheaf on an projective arithmetic variety is big if and only if its arithmetic self-intersection number is positive, which is a well-known fact in the geometric case. Moreover, the arithmetic Bogomolov inequality in Section 8 is also a consequence of the generalized Hodge index theorem.

6.1. Arithmetic volume function

Let X be a d-dimensional projective arithmetic variety. Let $\overline{L}$ be a continuous hermitian invertible sheaf of real type on X. We define $\widehat{\mathrm{vol}}(\overline{L})$ to be

$$\widehat{\mathrm{vol}}(\overline{L}) := \limsup_{m\to\infty} \frac{\hat{h}^0(H^0(X, mL), \|\cdot\|_{\sup}^{m\overline{L}})}{m^d/d!},$$

called the *arithmetic volume* of $\overline{L}$. In addition, $\widehat{\mathrm{vol}}_\flat(\overline{L})$ is defined to be

$$\widehat{\mathrm{vol}}_\flat(\overline{L}) := \liminf_{m\to\infty} \frac{\hat{h}^0(H^0(X, mL), \|\cdot\|_{\sup}^{m\overline{L}})}{m^d/d!}.$$

For example, if X is generically smooth and $\overline{L}$ is ample, then

$$\widehat{\mathrm{vol}}(\overline{L}) = \widehat{\mathrm{vol}}_\flat(\overline{L}) = \widehat{\deg}(\widehat{c}_1(\overline{L})^{\cdot d})$$

by Proposition 5.39 and Proposition 5.41.

PROPOSITION 6.1. *$0 \leq \widehat{\mathrm{vol}}_\flat(\overline{L}) \leq \widehat{\mathrm{vol}}(\overline{L}) < \infty$ holds.*

PROOF. It follows from Corollary 5.42. □

REMARK 6.1. We can expect that $\widehat{\mathrm{vol}}(\overline{L}) = \widehat{\mathrm{vol}}_\flat(\overline{L})$, that is, $\widehat{\mathrm{vol}}(\overline{L})$ is given by lim instead of lim sup. This was actually proved by H. Chen [**12**]. In this book, we will not use this fact.

In order to discuss $\widehat{\mathrm{vol}}(\overline{L})$ and $\widehat{\mathrm{vol}}_\flat(\overline{L})$ simultaneously, $\widehat{\mathrm{vol}}(\overline{L})$ is sometimes denoted by $\widehat{\mathrm{vol}}_\sharp(\overline{L})$. Let ϕ be an F_∞-invariant continuous function on $X(\mathbb{C})$. We define $\overline{\mathscr{O}}(\phi)$ to be $\overline{\mathscr{O}}(\phi) := (\mathscr{O}_X, \exp(-\phi)|\cdot|_{can})$.

PROPOSITION 6.2. *Let $\overline{L}$ and $\overline{M}$ be continuous hermitian invertible sheaves of real type on X. If $\natural$ is either $\sharp$ or $\flat$, then we have the following.*

(1) *If $\overline{L} \leq \overline{M}$, then $\widehat{\mathrm{vol}}_{\natural}(\overline{L}) \leq \widehat{\mathrm{vol}}_{\natural}(\overline{M})$.*
(2) $|\widehat{\mathrm{vol}}_{\natural}(\overline{L} + \overline{\mathscr{O}}(\phi)) - \widehat{\mathrm{vol}}_{\natural}(\overline{L})| \leq d\|\phi\|_{\sup} \mathrm{vol}(L_{\mathbb{Q}})$, *where*
$$\mathrm{vol}(L_{\mathbb{Q}}) = \lim_{m\to\infty} \frac{\dim_{\mathbb{Q}} H^0(X_{\mathbb{Q}}, mL_{\mathbb{Q}})}{m^{d-1}/(d-1)!}$$
(for details, see [**40**, Subsection 2.2.C and Subsection 11.4.A]).

PROOF. (1) Since $\overline{L} \leq \overline{M}$, by (5) of Proposition 5.40, $m\overline{L} \leq m\overline{M}$ $(m \geq 1)$. Thus, by (2) in Proposition 5.40,
$$\hat{h}^0\left(H^0(X, mL), \|\cdot\|_{\sup}^{m\overline{L}}\right) \leq \hat{h}^0\left(H^0(X, mM), \|\cdot\|_{\sup}^{m\overline{M}}\right)$$
for $m \geq 1$. Thus (1) follows.

(2) First we consider the case where ϕ is a constant λ. Note that
$$\|\cdot\|_{\sup}^{m(\overline{L}+\overline{\mathscr{O}}(\lambda))} = \exp(-m\lambda)\|\cdot\|_{\sup}^{m\overline{L}}.$$
Thus, if $\lambda \geq 0$, then, by (3) in Proposition 2.17, there is a constant C' such that
$$\begin{aligned} 0 \leq \hat{h}^0(H^0(X, mL), \|\cdot\|_{\sup}^{m(\overline{L}+\overline{\mathscr{O}}(\lambda))}) - \hat{h}^0(H^0(X, mL), \|\cdot\|_{\sup}^{m\overline{L}}) \\ \leq \lambda m \dim H^0(X_{\mathbb{Q}}, mL_{\mathbb{Q}}) + C' m^{d-1}\log(m) \end{aligned}$$
for $m \gg 1$, and hence
$$\widehat{\mathrm{vol}}_{\natural}(\overline{L}) \leq \widehat{\mathrm{vol}}_{\natural}(\overline{L} + \overline{\mathscr{O}}(\lambda)) \leq \widehat{\mathrm{vol}}_{\natural}(\overline{L}) + d\lambda \,\mathrm{vol}(L_{\mathbb{Q}}).$$
If $\lambda < 0$, then, applying the above inequalities to the case where $\overline{L}$ is given as $\overline{L} + \overline{\mathscr{O}}(\lambda)$, we obtain
$$\begin{aligned} \widehat{\mathrm{vol}}_{\natural}(\overline{L} + \overline{\mathscr{O}}(\lambda)) \leq \widehat{\mathrm{vol}}_{\natural}(\overline{L}) = \widehat{\mathrm{vol}}_{\natural}((\overline{L} + \overline{\mathscr{O}}(\lambda)) + \overline{\mathscr{O}}(-\lambda)) \\ \leq \widehat{\mathrm{vol}}_{\natural}(\overline{L} + \overline{\mathscr{O}}(\lambda)) + d(-\lambda)\,\mathrm{vol}(L_{\mathbb{Q}}). \end{aligned}$$
Therefore, we have the assertion in the case where ϕ is a constant λ.

In general, as $-\|\phi\|_{\sup} \leq \phi \leq \|\phi\|_{\sup}$, we have
$$\overline{\mathscr{O}}(-\|\phi\|_{\sup}) \leq \overline{\mathscr{O}}(\phi) \leq \overline{\mathscr{O}}(\|\phi\|_{\sup}),$$
and hence
$$\widehat{\mathrm{vol}}_{\natural}(\overline{L} + \overline{\mathscr{O}}(-\|\phi\|_{\sup})) \leq \widehat{\mathrm{vol}}_{\natural}(\overline{L} + \overline{\mathscr{O}}(\phi)) \leq \widehat{\mathrm{vol}}_{\natural}(\overline{L} + \overline{\mathscr{O}}(\|\phi\|_{\sup})).$$
Therefore, the assertion follows from the case where ϕ is a constant. $\square$

COROLLARY 6.3. *Let $\overline{L}$ be a continuous hermitian invertible sheaf of real type on X. Then*
$$\widehat{\mathrm{vol}}(\overline{L}) = \limsup_{m\to\infty} \frac{\log \#\{s \in H^0(X, mL) \mid \|s\|_{\sup}^{m\overline{L}} < 1\}}{m^d/d!}$$
and
$$\widehat{\mathrm{vol}}_{\flat}(\overline{L}) = \liminf_{m\to\infty} \frac{\log \#\{s \in H^0(X, mL) \mid \|s\|_{\sup}^{m\overline{L}} < 1\}}{m^d/d!}.$$

PROOF. Since $\|\cdot\|_{\sup}^{m(\overline{L}+\overline{\mathcal{O}}(-\lambda))} = \exp(m\lambda)\|\cdot\|_{\sup}^{m\overline{L}}$ for a positive real number λ, we have

$$\hat{H}^0\left(H^0(X, mL), \|\cdot\|_{\sup}^{m(\overline{L}+\overline{\mathcal{O}}(-\lambda))}\right) \subseteq \left\{s \in H^0(X, mL) \mid \|s\|_{\sup}^{m\overline{L}} < 1\right\} \subseteq \hat{H}^0\left(H^0(X, mL), \|\cdot\|_{\sup}^{m\overline{L}}\right).$$

Thus, by (2) in Proposition 6.2,

$$\widehat{\mathrm{vol}}(\overline{L}) - d\lambda\,\mathrm{vol}(L_{\mathbb{Q}}) \le \limsup_{m\to\infty} \frac{\log \#\{s \in H^0(X, mL) \mid \|s\|_{\sup}^{m\overline{L}} < 1\}}{m^d/d!} \le \widehat{\mathrm{vol}}(\overline{L})$$

and

$$\widehat{\mathrm{vol}}_{\flat}(\overline{L}) - d\lambda\,\mathrm{vol}(L_{\mathbb{Q}}) \le \liminf_{m\to\infty} \frac{\log \#\{s \in H^0(X, mL) \mid \|s\|_{\sup}^{m\overline{L}} < 1\}}{m^d/d!} \le \widehat{\mathrm{vol}}_{\flat}(\overline{L}).$$

Thus the corollary follows. □

Here let us consider the following theorem.

THEOREM 6.4. *Let $\overline{L}$ and $\overline{M}$ be C^∞-hermitian invertible sheaves of real type on X.*

(1) *The following estimation holds:*
$$\left|\hat{h}^0\left(H^0(X, mL + M), \|\cdot\|_{\sup}^{m\overline{L}+\overline{M}}\right) - \hat{h}^0\left(H^0(X, mL), \|\cdot\|_{\sup}^{m\overline{L}}\right)\right| \le o(m^d).$$

(2) *If $\pi : X' \to X$ is a birational morphism of projective arithmetic varieties, then*
$$\left|\hat{h}^0\left(H^0(X', \pi^*(mL + M)), \|\cdot\|_{\sup}^{\pi^*(m\overline{L}+\overline{M})}\right) - \hat{h}^0\left(H^0(X, mL + M), \|\cdot\|_{\sup}^{m\overline{L}+\overline{M}}\right)\right| \le o(m^d).$$

PROOF. Let us begin with the following claim:

CLAIM 1. *If X is normal, then*
$$\left|\hat{h}^0\left(H^0(X, mL + M), \|\cdot\|_{\sup}^{m\overline{L}+\overline{M}}\right) - \hat{h}^0\left(H^0(X, mL), \|\cdot\|_{\sup}^{m\overline{L}}\right)\right| \le o(m^d).$$

PROOF. We assume the technical result Theorem 6.17, which will be proved in Section 6.5. First we consider the case where X is generically smooth. Applying Theorem 6.17 to the case where $\overline{A} = \overline{M}$, $b = 1$ and $c = 0$, if $m \gg 1$, then
$$\hat{h}^0\left(H^0(X, mL + M), \|\cdot\|_{\sup}^{m\overline{L}+\overline{M}}\right) \le \hat{h}^0\left(H^0(X, mL), \|\cdot\|_{\sup}^{m\overline{L}}\right) + o(m^d).$$
Moreover, applying Theorem 6.17 to the case where $\overline{A} = -\overline{M}$, and $b = c = 1$, if $m \gg 1$, then
$$\hat{h}^0\left(H^0(X, mL), \|\cdot\|_{\sup}^{m\overline{L}}\right) \le \hat{h}^0\left(H^0(X, mL + M), \|\cdot\|_{\sup}^{m\overline{L}+\overline{M}}\right) + o(m^d).$$
Therefore, the claim follows in this case.

Next we consider the case where X is normal. Let $\pi : X' \to X$ be a generic resolution of singularities of X (cf. Theorem 5.1). We may assume that X' is normal. Then, as $\pi_*(\mathscr{O}_{X'}) = \mathscr{O}_X$, we have $H^0(X, mL) = H^0(X', \pi^*(mL))$ and $H^0(X, mL+M) = H^0(X', \pi^*(mL+M))$. Thus the claim follows. $\square$

CLAIM 2. *If $\pi_*(\mathscr{O}_{X'}) = \mathscr{O}_X$, then (2) holds.*

PROOF. In this case, we have $H^0(X, mL+M) = H^0(X', \pi^*(mL+M))$, so that the claim is obvious. $\square$

CLAIM 3. *If $\pi : X' \to X$ is the normalization of X, then (2) holds.*

PROOF. Clearly,

$$\hat{h}^0\left(H^0(X, mL+M), \|\cdot\|_{\sup}^{m\overline{L}+\overline{M}}\right) \leq \hat{h}^0\left(H^0(X', \pi^*(mL+M)), \|\cdot\|_{\sup}^{\pi^*(m\overline{L}+\overline{M})}\right).$$

Let $\mathcal{I}_{X'/X}$ be the conductor ideal of $\pi : X' \to X$, that is, $\mathcal{I}_{X'/X}$ is an ideal sheaf of $\mathscr{O}_{X'}$ with $\pi_*(\mathcal{I}_{X'/X}) = \mathscr{H}\!om_{\mathscr{O}_X}(\pi_*(\mathscr{O}_{X'}), \mathscr{O}_X)$. Let H be an ample invertible sheaf on X' such that $H^0(X', H \otimes \mathcal{I}_{X'/X}) \neq 0$. Let s be a non-zero element of $H^0(X', H \otimes \mathcal{I}_{X'/X})$. Let us take a C^∞-hermitian norm $|\cdot|$ of H such that $|\cdot|$ is of real type and $\|s\|_{\sup} \leq 1$. We set $\overline{H} = (H, |\cdot|)$. Here let us see the following.

$$(6.1)\quad \operatorname{Image}\left(H^0(X', \pi^*(mL+M) - H) \xrightarrow{s} H^0(X', \pi^*(mL+M))\right) \subseteq H^0(X, mL+M).$$

Indeed, as $s \in H^0(X', H \otimes \mathcal{I}_{X'/X})$, if $t \in H^0(X', \pi^*(mL+M) - H)$, then

$$ts \in H^0(X', \pi^*(mL+M) \otimes \mathcal{I}_{X'/X}).$$

Since $\pi_*(\mathcal{I}_{X'/X}) \subseteq \mathscr{O}_X$, we have

$$\begin{aligned} H^0(X', \pi^*(mL+M) \otimes \mathcal{I}_{X'/X}) &= H^0(X, (mL+M) \otimes \pi_*(\mathcal{I}_{X'/X})) \\ &\subseteq H^0(X, mL+M). \end{aligned}$$

Thus (6.1) follows. On the other hand, as $\|s\|_{\sup} \leq 1$, we obtain

$$\begin{aligned} &\hat{h}^0\left(H^0(X', \pi^*(mL+M) - H), \|\cdot\|_{\sup}^{\pi^*(m\overline{L}+\overline{M})-\overline{H}}\right) \\ &\qquad \leq \hat{h}^0\left(H^0(X, mL+M), \|\cdot\|_{\sup}^{m\overline{L}+\overline{M}}\right). \end{aligned}$$

Moreover, by Claim 1,

$$\begin{aligned} &\hat{h}^0\left(H^0(X', \pi^*(mL+M)), \|\cdot\|_{\sup}^{\pi^*(m\overline{L}+\overline{M})}\right) \\ &\qquad \leq \hat{h}^0\left(H^0(X', \pi^*(mL)), \|\cdot\|_{\sup}^{\pi^*(m\overline{L})}\right) + o(m^d) \\ &\qquad \leq \hat{h}^0\left(H^0(X', \pi^*(mL+M) - H), \|\cdot\|_{\sup}^{\pi^*(m\overline{L}+\overline{M})-\overline{H}}\right) + o(m^d). \end{aligned}$$

Therefore, the claim follows. $\square$

The proof of (2): Let $\mu : Y \to X$ and $\nu : Y' \to X'$ be the normalization of X and X', respectively. Then we have the following commutative diagram

$$\begin{array}{ccc} Y & \xleftarrow{\pi'} & Y' \\ {\scriptstyle\mu}\downarrow & & \downarrow{\scriptstyle\nu} \\ X & \xleftarrow{\pi} & X', \end{array}$$

where $\pi' : Y' \to Y$ is the induced birational morphism. Thus, by Claim 2 and Claim 3, (2) follows.

The proof of (1): (1) follows from (2) and Claim 1. □

COROLLARY 6.5. *If $\overline{L}$ and $\overline{M}$ are C^∞-hermitian invertible sheaves on X, then*

$$\limsup_{m\to\infty} \frac{\hat{h}^0(H^0(X, mL+M), \|\cdot\|_{\sup}^{m\overline{L}+\overline{M}})}{m^d/d!} = \widehat{\mathrm{vol}}(\overline{L})$$

and

$$\liminf_{m\to\infty} \frac{\hat{h}^0(H^0(X, mL+M), \|\cdot\|_{\sup}^{m\overline{L}+\overline{M}})}{m^d/d!} = \widehat{\mathrm{vol}}_\flat(\overline{L}).$$

PROOF. It follows from Theorem 6.4. □

As a consequence, we have the following theorem:

THEOREM 6.6. *Let X be a d-dimensional projective arithmetic variety and let $\overline{L}$ be a continuous hermitian invertible sheaf $\overline{L}$ of real type on X. If $\natural$ is either $\sharp$ or $\flat$, then we have the following:*

1. *$\widehat{\mathrm{vol}}_\natural(p\overline{L}) = p^d\,\widehat{\mathrm{vol}}_\natural(\overline{L})$ for a non-negative integer p.*
2. *If $\pi : X' \to X$ is a birational morphism of projective arithmetic varieties, then $\widehat{\mathrm{vol}}_\natural(\pi^*(\overline{L})) = \widehat{\mathrm{vol}}_\natural(\overline{L})$.*
3. *The following are equivalent:*
   - (3.1) *$\overline{L}$ is big.*
   - (3.2) *$\widehat{\mathrm{vol}}_\flat(\overline{L}) > 0$.*
   - (3.3) *$\widehat{\mathrm{vol}}(\overline{L}) > 0$.*

PROOF. (1) and (2): First we discuss (1) and (2) under the assumption that $\overline{L}$ is C^∞. In this case, (2) follows from (2) in Theorem 6.4, so let us consider (1). For an integer i with $0 \le i < p$, by (1) in Theorem 6.4,

$$\left|\hat{h}^0\left(H^0(X, pnL), \|\cdot\|_{\sup}^{pn\overline{L}}\right) - \hat{h}^0\left(H^0(X, (pn+i)L), \|\cdot\|_{\sup}^{(pn+i)\overline{L}}\right)\right| \le o(n^d).$$

Thus, since $\lim_{n\to\infty}(pn+i)^d/(pn)^d = 1$, we have

$$\limsup_{n\to\infty} \frac{\hat{h}^0\left(H^0(X, pnL), \|\cdot\|_{\sup}^{pn\overline{L}}\right)}{(pn)^d} = \limsup_{n\to\infty} \frac{\hat{h}^0\left(H^0(X, (pn+i)L), \|\cdot\|_{\sup}^{(pn+i)\overline{L}}\right)}{(pn+i)^d}$$

and

$$\liminf_{n\to\infty} \frac{\hat{h}^0\left(H^0(X, pnL), \|\cdot\|_{\sup}^{pn\overline{L}}\right)}{(pn)^d} = \liminf_{n\to\infty} \frac{\hat{h}^0\left(H^0(X, (pn+i)L), \|\cdot\|_{\sup}^{(pn+i)\overline{L}}\right)}{(pn+i)^d},$$

so that

$$\limsup_{n\to\infty} \frac{\hat{h}^0\left(H^0(X, pnL), \|\cdot\|_{\sup}^{pn\overline{L}}\right)}{(pn)^d} = \limsup_{m\to\infty} \frac{\hat{h}^0\left(H^0(X, mL), \|\cdot\|_{\sup}^{(pn+i)\overline{L}}\right)}{m^d}$$

and

$$\liminf_{n\to\infty} \frac{\hat{h}^0\left(H^0(X, pnL), \|\cdot\|_{\sup}^{pn\overline{L}}\right)}{(pn)^d} = \liminf_{m\to\infty} \frac{\hat{h}^0\left(H^0(X, mL), \|\cdot\|_{\sup}^{(pn+i)\overline{L}}\right)}{m^d},$$

which proves (1) in the case where $\overline{L}$ is C^∞.

In general, for a given positive number ϵ, by the Stone-Weierstrass theorem, there is an F_∞-invariant continuous function ϕ such that $\|\phi\|_{\sup} \leq \epsilon$ and that $\overline{L} + \overline{\mathscr{O}}(\phi)$ is C^∞. Then, by the previous consideration,

$$\widehat{\mathrm{vol}}_\natural(p(\overline{L} + \overline{\mathscr{O}}(\phi))) = p^d \,\widehat{\mathrm{vol}}_\natural(\overline{L} + \overline{\mathscr{O}}(\phi)) \text{ and } \widehat{\mathrm{vol}}_\natural(\pi^*(\overline{L} + \overline{\mathscr{O}}(\phi))) = \widehat{\mathrm{vol}}_\natural(\overline{L} + \overline{\mathscr{O}}(\phi)).$$

On the other hand, by (2) in Proposition 6.2, we have

$$\begin{cases} \left|\widehat{\mathrm{vol}}_\natural(\overline{L} + \overline{\mathscr{O}}(\phi)) - \widehat{\mathrm{vol}}_\natural(\overline{L})\right| \leq d\|\phi\|_{\sup} \mathrm{vol}(L_{\mathbb{Q}}) \leq (d \,\mathrm{vol}(L_{\mathbb{Q}}))\epsilon, \\ \\ \left|\widehat{\mathrm{vol}}_\natural(p(\overline{L} + \overline{\mathscr{O}}(\phi))) - \widehat{\mathrm{vol}}_\natural(p\overline{L})\right| \leq d\|p\phi\|_{\sup} \mathrm{vol}(pL_{\mathbb{Q}}) \leq (p^d d \,\mathrm{vol}(L_{\mathbb{Q}}))\epsilon, \\ \\ \left|\widehat{\mathrm{vol}}_\natural(\pi^*(\overline{L} + \overline{\mathscr{O}}(\phi))) - \widehat{\mathrm{vol}}_\natural(\pi^*(\overline{L}))\right| \leq d\|\pi^*(\phi)\|_{\sup} \mathrm{vol}(\pi^*(L_{\mathbb{Q}})) \leq (d \,\mathrm{vol}(L_{\mathbb{Q}}))\epsilon, \end{cases}$$

and hence,

$$\left|\widehat{\mathrm{vol}}_\natural(p\overline{L}) - p^d \,\widehat{\mathrm{vol}}_\natural(\overline{L})\right| \leq (2p^d d \,\mathrm{vol}(L_{\mathbb{Q}}))\epsilon$$

and

$$\left|\widehat{\mathrm{vol}}_\natural(\pi^*(\overline{L})) - \widehat{\mathrm{vol}}_\natural(\overline{L})\right| \leq (2d \,\mathrm{vol}(L_{\mathbb{Q}}))\epsilon.$$

Thus (1) and (2) follow.

(3) Let $\overline{A}$ be an ample C^∞-hermitian invertible sheaf of real type on X. Now we see that (3.1) implies (3.2). Since $\overline{L}$ is big, there is a positive integer p with $\overline{A} \leq p\overline{L}$ by Proposition 5.43. Note that $\widehat{\mathrm{vol}}_\flat(\overline{A}) > 0$ by Proposition 5.39 and Proposition 5.41. Therefore, by (1) in Proposition 6.2,

$$0 < \widehat{\mathrm{vol}}_\flat(\overline{A}) \leq \widehat{\mathrm{vol}}_\flat(p\overline{L}) = p^d \,\widehat{\mathrm{vol}}_\flat(\overline{L}),$$

as required.

Clearly, (3.2) implies (3.3). Finally, we assume (3.3), that is, $\widehat{\mathrm{vol}}(\overline{L}) > 0$. By Corollary 6.3, there is a positive integer m such that mL has a non-zero strictly small section. Therefore, it is sufficient to show that $L_{\mathbb{Q}}$ is big on $X_{\mathbb{Q}}$. For this purpose, by using the Stone-Weierstrass theorem together with (2) in Proposition 6.2, we may assume that $\overline{L}$ is C^∞. By (1) in Theorem 6.4,

$$\limsup_{m\to\infty} \frac{\hat{h}^0\left(H^0(X, mL - A), \|\cdot\|_{\sup}^{m\overline{L}-\overline{A}}\right)}{m^d} = \frac{\widehat{\mathrm{vol}}(\overline{L})}{d!} > 0,$$

so that there is a positive integer n with $\hat{h}^0\left(H^0(X, nL - A), \|\cdot\|_{\sup}^{n\overline{L}-\overline{A}}\right) \neq 0$. Therefore, $n\overline{L} \geq \overline{A}$. In particular, $L_{\mathbb{Q}}$ is big on $X_{\mathbb{Q}}$. □

6.2. Extension of volume function over $\mathbb{Q}$

Let X be a d-dimensional projective arithmetic variety. We denote the group of isomorphism classes of C^∞-hermitian (resp. continuous hermitian) invertible sheaves of real type on X by $\widehat{\mathrm{Pic}}(X)$ (resp. $\widehat{\mathrm{Pic}}_{C^0}(X)$). An element of $\widehat{\mathrm{Pic}}(X) \otimes \mathbb{Q}$ (resp. $\widehat{\mathrm{Pic}}_{C^0}(X) \otimes \mathbb{Q}$) is called a *$C^\infty$-hermitian $\mathbb{Q}$-invertible sheaf of real type* (resp. *continuous hermitian $\mathbb{Q}$-invertible sheaf of real type*) on X. For $\overline{M} \in \widehat{\mathrm{Pic}}_{C^0}(X)$, an element $\overline{M} \otimes 1$ of $\widehat{\mathrm{Pic}}_{C^0}(X) \otimes \mathbb{Q}$ is denoted by $[\overline{M}]$. Note that $[\overline{M}] = [(\mathscr{O}_X, |\cdot|_{can})]$ if and only if $\overline{M}$ is a torsion element of $\widehat{\mathrm{Pic}}_{C^0}(X)$, that is, there is a positive integer n such that $n\overline{M} = (\mathscr{O}_X, |\cdot|_{can})$. We say that a continuous hermitian $\mathbb{Q}$-invertible sheaf $\overline{L}$ of real type is represented by $\overline{M} \in \widehat{\mathrm{Pic}}_{C^0}(X)$ if $[\overline{M}] = \overline{L}$. Note that if $\overline{M}$ is C^∞, then $\overline{L} \in \widehat{\mathrm{Pic}}(X) \otimes \mathbb{Q}$. Moreover, a continuous hermitian $\mathbb{Q}$-invertible sheaf of real type on X is said to be ample if there is a positive integer n such that $n\overline{L}$ is represented by an ample C^∞-hermitian invertible sheaf of real type on X. Similarly, $\overline{L}$ is said to be nef (resp. big) if $n\overline{L}$ is represented by a nef C^∞-hermitian (resp. big continuous hermitian) invertible sheaf of real type on X.

Let $\natural$ be either $\sharp$ or $\flat$. By (1) in Theorem 6.6 and the following lemma,

$$\widehat{\mathrm{vol}}_\natural : \widehat{\mathrm{Pic}}_{C^0}(X) \to \mathbb{R}$$

extends to a positively homogeneous function

$$\widehat{\mathrm{vol}}_\natural : \widehat{\mathrm{Pic}}_{C^0}(X) \otimes \mathbb{Q} \to \mathbb{R}$$

of degree d, that is,

$$\widehat{\mathrm{vol}}_\natural([\overline{M}]) = \widehat{\mathrm{vol}}_\natural(\overline{M}) \quad \text{and} \quad \widehat{\mathrm{vol}}_\natural(a\overline{L}) = a^d \widehat{\mathrm{vol}}_\natural(\overline{L})$$

for $\overline{M} \in \widehat{\mathrm{Pic}}_{C^0}(X)$, $\overline{L} \in \widehat{\mathrm{Pic}}_{C^0}(X) \otimes \mathbb{Q}$ and $a \in \mathbb{Q}_{\geq 0}$.

LEMMA 6.7. *Let $f : M \to \mathbb{R}$ be a function on a $\mathbb{Z}$-module M such that there is a positive number d with $f(px) = p^d f(x)$ for all $x \in M$ and $p \in \mathbb{Z}_{\geq 0}$. Then there is a unique $\tilde{f} : M \otimes \mathbb{Q} \to \mathbb{R}$ such that $\tilde{f}(x \otimes 1) = f(x)$ for all $x \in M$ and that $\tilde{f}(ay) = a^d \tilde{f}(y)$ for all $y \in M \otimes \mathbb{Q}$ and $a \in \mathbb{Q}_{\geq 0}$.*

PROOF. First we consider the uniqueness of $\tilde{f}$. As any element of $M \otimes \mathbb{Q}$ can be written by a form $x \otimes a$ ($x \in M$, $a \in \mathbb{Q}_{\geq 0}$), $\tilde{f}(x \otimes a) = a^d \tilde{f}(x \otimes 1) = a^d f(x)$, which shows the uniqueness.

In order to prove the existence, let us see that if $x_1 \otimes a_1 = x_2 \otimes a_2$ in $M \otimes \mathbb{Q}$ for $x_1, x_2 \in M$ and $a_1, a_2 \in \mathbb{Q}_{\geq 0}$, then $a_1^d f(x_1) = a_2^d f(x_2)$. Indeed, let m be a positive integer with $ma_1, ma_2 \in \mathbb{Z}$. Then

$$(ma_1x_1) \otimes 1 = m(x_1 \otimes a_1) = m(x_2 \otimes a_2) = (ma_2x_2) \otimes 1,$$

so that there is a positive integer n with $nma_1x_1 = nma_2x_2$. Thus $(nma_1)^d f(x_1) = (nma_2)^d f(x_2)$ follows, as required.

By the above discussion, we define $\tilde{f}$ to be $\tilde{f}(x \otimes a) := a^d f(x)$ for $x \otimes a$ ($x \in M$, $a \in \mathbb{Q}_{\geq 0}$). It is easy to see that $\tilde{f}$ satisfies all condition of the lemma. □

The order $\leq_{\mathbb{Q}}$ on $\widehat{\mathrm{Pic}}_{C^0}(X)$ defined in (4) of Definition 5.38 can be extended to an order on $\widehat{\mathrm{Pic}}_{C^0}(X) \otimes \mathbb{Q}$ as follows. For $\overline{L}, \overline{M} \in \widehat{\mathrm{Pic}}_{C^0}(X) \otimes \mathbb{Q}$, $\overline{L} \leq_{\mathbb{Q}} \overline{M}$ means that there is a positive integer n such that $n\overline{L}$ and $n\overline{M}$ are represented by continuous hermitian invertible sheaves $\overline{L}'$ and $\overline{M}'$ of real type, respectively, with $\overline{L}' \leq_{\mathbb{Q}} \overline{M}'$. Then, by using Proposition 5.40, we can see the following:

(1) If $\overline{L} \leq_{\mathbb{Q}} \overline{L}'$ and $\overline{L}' \leq_{\mathbb{Q}} \overline{L}''$, then $\overline{L} \leq_{\mathbb{Q}} \overline{L}''$.
(2) $\overline{L} \leq_{\mathbb{Q}} \overline{M}$ if and only if $-\overline{M} \leq_{\mathbb{Q}} -\overline{L}$.
(3) If $\overline{L} \leq_{\mathbb{Q}} \overline{M}$ and $\overline{L}' \leq_{\mathbb{Q}} \overline{M}'$, then $\overline{L} + \overline{L}' \leq_{\mathbb{Q}} \overline{M} + \overline{M}'$.
(4) For a positive rational number a, $\overline{L} \leq_{\mathbb{Q}} \overline{M}$ if and only if $a\overline{L} \leq_{\mathbb{Q}} a\overline{M}$.

Let $\pi : X' \to X$ be a birational morphism of projective arithmetic varieties. We can easily see that the homomorphism $\pi^* : \widehat{\mathrm{Pic}}_{C^0}(X) \to \widehat{\mathrm{Pic}}_{C^0}(X')$ extends to the homomorphism $\widehat{\mathrm{Pic}}_{C^0}(X) \otimes \mathbb{Q} \to \widehat{\mathrm{Pic}}_{C^0}(X') \otimes \mathbb{Q}$. By abuse of notation, we also denote it by π^*. Then we have the following proposition.

PROPOSITION 6.8. *For $\overline{L}, \overline{M} \in \widehat{\mathrm{Pic}}_{C^0}(X) \otimes \mathbb{Q}$, the following assertions hold.*

(1) *If $\overline{L} \leq_{\mathbb{Q}} \overline{M}$, then $\widehat{\mathrm{vol}}_{\natural}(\overline{L}) \leq \widehat{\mathrm{vol}}_{\natural}(\overline{M})$ holds.*
(2) $\left|\widehat{\mathrm{vol}}_{\natural}(\overline{L} + \overline{\mathscr{O}}(\phi)) - \widehat{\mathrm{vol}}_{\natural}(\overline{L})\right| \leq d\,\mathrm{vol}(L_{\mathbb{Q}})\|\phi\|_{\sup}$ *for $\phi \in C^0(X)$.*
(3) $\widehat{\mathrm{vol}}_{\natural}(\pi^*(\overline{L})) = \widehat{\mathrm{vol}}_{\natural}(\overline{L})$.
(4) *$\overline{L}$ is big if and only if $\widehat{\mathrm{vol}}_{\natural}(\overline{L}) > 0$.*

PROOF. (1) We choose a positive integer n such that $n\overline{L}$ and $n\overline{M}$ are represented by continuous hermitian invertible sheaves $\overline{L}'$ and $\overline{M}'$ of real type, respectively, with $\overline{L}' \leq_{\mathbb{Q}} \overline{M}'$. By Proposition 5.40, there is a positive integer m with $m\overline{L}' \leq m\overline{M}'$. Thus, by (1) in Proposition 6.2, $\widehat{\mathrm{vol}}_{\natural}(m\overline{L}') \leq \widehat{\mathrm{vol}}_{\natural}(m\overline{M}')$ holds. Note that $[m\overline{L}'] = nm\overline{L}$ and $[m\overline{M}'] = nm\overline{M}$. Thus (1) follows from the homogeneity of $\widehat{\mathrm{vol}}_{\natural}$.

(2) We choose a positive integer n such that $n\overline{L}$ is represented by a continuous hermitian invertible sheaf $\overline{L}'$. Then, by (2) in Proposition 6.2,

$$\left|\widehat{\mathrm{vol}}_{\natural}(\overline{L}' + \overline{\mathscr{O}}(n\phi)) - \widehat{\mathrm{vol}}_{\natural}(\overline{L}')\right| \leq d\,\mathrm{vol}(L'_{\mathbb{Q}})\|n\phi\|_{\sup}.$$

Thus, by the homogeneity of $\widehat{\mathrm{vol}}_{\natural}$, we have (2).

(3) and (4) are consequences of (2) and (3) in Theorem 6.6, respectively. □

6.3. Continuity of volume function

Throughout this section, let X be a d-dimensional projective arithmetic variety and let $\natural$ be either $\sharp$ or $\flat$. The following theorem due to Moriwaki [**51**] is the main theorem of this section.

THEOREM 6.9 (Continuity of arithmetic volume function). *Let $\overline{L}, \overline{A}_1, \ldots, \overline{A}_n$ be C^∞-hermitian $\mathbb{Q}$-invertible sheaves of real type on X. There is a positive constant C depending only on X and $\overline{L}, \overline{A}_1, \ldots, \overline{A}_n$ such that, for any rational numbers $\epsilon_1, \ldots, \epsilon_n$ with $|\epsilon_1| + \cdots + |\epsilon_n| \leq 1$,*

$$\left|\widehat{\mathrm{vol}}_{\natural}(\overline{L} + \epsilon_1\overline{A}_1 + \cdots + \epsilon_n\overline{A}_n) - \widehat{\mathrm{vol}}_{\natural}(\overline{L})\right| \leq C\left(|\epsilon_1| + \cdots + |\epsilon_n|\right).$$

PROOF. In this proof, we use Theorem 6.17 which will be proved Section 6.5.

First we consider the case where $n = 1$. We set $\overline{A} = \overline{A}_1$ and $\epsilon = \epsilon_1$. By virtue of the homogeneity of $\mathrm{vol}_{\natural}$,

$$\widehat{\mathrm{vol}}_{\natural}(a\overline{L} + \epsilon a\overline{A}) - \widehat{\mathrm{vol}}_{\natural}(a\overline{L}) = a^d\left(\widehat{\mathrm{vol}}_{\natural}(\overline{L} + \epsilon\overline{A}) - \widehat{\mathrm{vol}}_{\natural}(\overline{L})\right),$$

for $a \in \mathbb{Q}_{>0}$, so that we may assume that $\overline{L}$ and $\overline{A}$ are C^∞-hermitian invertible sheaves of real type on X. Let $\mu : X' \to X$ be a generic resolution of singularities of X. Then, by (3) in Proposition 6.8, we have $\widehat{\mathrm{vol}}_{\natural}(\overline{L}+\epsilon\overline{A}) = \widehat{\mathrm{vol}}_{\natural}(\mu^*(\overline{L})+\epsilon\mu^*(\overline{A}))$ and $\widehat{\mathrm{vol}}_{\natural}(\overline{L}) = \widehat{\mathrm{vol}}_{\natural}(\mu^*(\overline{L}))$. Therefore, we may further assume that X is generically smooth.

By Theorem 6.17, there are constants a_0', C' and D' depending only on X, $\overline{L}$ and $\overline{A}$ such that

$$\hat{h}^0\left(H^0(aL+(b-c)A), \|\cdot\|_{\sup}^{a\overline{L}+(b-c)\overline{A}}\right) \leq \hat{h}^0\left(H^0(aL-cA), \|\cdot\|_{\sup}^{a\overline{L}-c\overline{A}}\right) + C'ba^{d-1} + D'a^{d-1}\log(a)$$

holds for all integers a, b, c with $a \geq b \geq c \geq 0$ and $a \geq a_0'$. We assume $0 \leq \epsilon \leq 1$. Let m_0 be a positive integer with $m_0\epsilon \in \mathbb{Z}$. Applying the above inequality to the case where $a = mm_0$, $b = mm_0\epsilon$ and $c = 0$, we obtain

$$\begin{aligned}\hat{h}^0&\left(H^0(mm_0L+mm_0\epsilon A), \|\cdot\|_{\sup}^{mm_0\overline{L}+mm_0\epsilon\overline{A}}\right)\\ &\leq \hat{h}^0\left(H^0(mm_0L), \|\cdot\|_{\sup}^{mm_0\overline{L}}\right)\\ &\qquad + C'\epsilon(mm_0)^d + D'(mm_0)^{d-1}\log(mm_0)\end{aligned}$$

for $m \gg 1$, which implies

$$\widehat{\mathrm{vol}}_\natural(m_0\overline{L}+m_0\epsilon\overline{A}) \leq \widehat{\mathrm{vol}}_\natural(m_0\overline{L}) + d!\epsilon C'm_0^d.$$

Therefore, using the homogeneity of $\widehat{\mathrm{vol}}_\natural$, we have

$$\widehat{\mathrm{vol}}_\natural(\overline{L}+\epsilon\overline{A}) \leq \widehat{\mathrm{vol}}_\natural(\overline{L}) + d!C'\epsilon.$$

Next, applying the above inequality to the case where $a = mm_0$ and $b = c = mm_0\epsilon$,

$$\begin{aligned}\hat{h}^0&\left(H^0(mm_0\overline{L}), \|\cdot\|_{\sup}^{mm_0\overline{L}}\right)\\ &\leq \hat{h}^0\left(H^0(mm_0L-mm_0\epsilon\overline{A}), \|\cdot\|_{\sup}^{mm_0\overline{L}-mm_0\epsilon\overline{A}}\right)\\ &\qquad + C'\epsilon(mm_0)^d + D'(mm_0)^{d-1}\log(mm_0)\end{aligned}$$

holds for $m \gg 1$. Thus, in the same way as before, we have

$$d!C'(-\epsilon) + \widehat{\mathrm{vol}}_\natural(\overline{L}) \leq \widehat{\mathrm{vol}}_\natural(\overline{L}-\epsilon\overline{A}).$$

Therefore, there is a positive constant C' depending only on X, $\overline{L}$ and $\overline{A}$ such that

$$\begin{cases}\widehat{\mathrm{vol}}_\natural(\overline{L}+\epsilon\overline{A}) \leq \widehat{\mathrm{vol}}_\natural(\overline{L}) + d!C'\epsilon,\\ d!C'(-\epsilon) + \widehat{\mathrm{vol}}_\natural(\overline{L}) \leq \widehat{\mathrm{vol}}_\natural(\overline{L}-\epsilon\overline{A}).\end{cases}$$

We can do the same argument if we replace $\overline{A}$ by $-\overline{A}$, so that there is a positive constant C'' depending only on X, $\overline{L}$ and $\overline{A}$ such that

$$\begin{cases}\widehat{\mathrm{vol}}_\natural(\overline{L}-\epsilon\overline{A}) \leq \widehat{\mathrm{vol}}_\natural(\overline{L}) + d!C''\epsilon,\\ d!C''(-\epsilon) + \widehat{\mathrm{vol}}(\overline{L}) \leq \widehat{\mathrm{vol}}_\natural(\overline{L}+\epsilon\overline{A}).\end{cases}$$

Thus, if we set $C = d!\max\{C', C''\}$, then we obtain the case where $n = 1$.

Next we consider a general case. Let us choose an ample C^∞-hermitian $\mathbb{Q}$-invertible sheaf $\overline{B}$ of real type such that $-\overline{B} \leq_{\mathbb{Q}} \overline{A}_i \leq_{\mathbb{Q}} \overline{B}$ $(i = 1, \ldots, n)$. Then, for each $i = 1, \ldots, n$,

$$-|\epsilon_i|\overline{B} \leq_{\mathbb{Q}} \epsilon_i\overline{A}_i \leq_{\mathbb{Q}} \leq_{\mathbb{Q}} |\epsilon_i|\overline{B},$$

which implies

$$\overline{L} - (|\epsilon_1| + \cdots + |\epsilon_n|)\overline{B} \leq_{\mathbb{Q}} \overline{L} + \epsilon_1\overline{A}_1 + \cdots + \epsilon_n\overline{A}_n \leq_{\mathbb{Q}} \overline{L} + (|\epsilon_1| + \cdots + |\epsilon_n|)\overline{B},$$

and hence,

$$\widehat{\mathrm{vol}}_\natural(\overline{L} - (|\epsilon_1| + \cdots + |\epsilon_n|)\overline{B}) \\ \leq \widehat{\mathrm{vol}}_\natural(\overline{L} + \epsilon_1\overline{A}_1 + \cdots + \epsilon_n\overline{A}_n) \\ \leq \widehat{\mathrm{vol}}_\natural(\overline{L} + (|\epsilon_1| + \cdots + |\epsilon_n|)\overline{B}).$$

Thus the general case follows from the case where $n = 1$. □

The arithmetic volume function $\widehat{\mathrm{vol}}_\natural : \widehat{\mathrm{Pic}}_{C^0}(X) \otimes \mathbb{Q} \to \mathbb{R}$ is continuous in the following sense.

COROLLARY 6.10. *Let $\overline{L}, \overline{A}_1, \ldots, \overline{A}_n$ be continuous hermitian $\mathbb{Q}$-invertible sheaves of real type on X. For a positive number ϵ, there is a positive number δ such that*

$$\left| \widehat{\mathrm{vol}}_\natural\left(\overline{L} + \sum_{j=1}^n \delta_j \overline{A}_j\right) - \widehat{\mathrm{vol}}_\natural\left(\overline{L}\right) \right| \leq \epsilon$$

for all $\delta_1, \ldots, \delta_n \in \mathbb{Q}$ with $|\delta_1| + \cdots + |\delta_n| \leq \delta$.

PROOF. It is easy to find a positive constant C_1 such that

$$\mathrm{vol}\left(L_{\mathbb{Q}} + \sum_{j=1}^n \delta_j (A_j)_{\mathbb{Q}}\right) \leq C_1$$

for all $\delta_1, \ldots, \delta_n \in \mathbb{Q}$ with $|\delta_1| + \cdots + |\delta_n| \leq 1$. By using the Stone-Weierstrass theorem, we can find F_∞-invariant continuous functions $f, g_1, \ldots, g_n$ on $X(\mathbb{C})$ such that

$$\overline{L}' := \overline{L} + \overline{\mathscr{O}}(f) \quad \text{and} \quad \overline{A}_j' := \overline{A}_j + \overline{\mathscr{O}}(g_j) \quad (j = 1, \ldots, n)$$

are C^∞ and

$$\|f\| \leq \frac{\epsilon}{8dC_1} \quad \text{and} \quad \|g_j\| \leq \frac{\epsilon}{8dC_1} \quad (j = 1, \ldots, n).$$

Here we consider the following two claims:

CLAIM 4. *If $\delta_1, \ldots, \delta_n \in \mathbb{Q}$ and $|\delta_1| + \cdots + |\delta_n| \leq 1$, then*

$$\left| \widehat{\mathrm{vol}}_\natural\left(\overline{L}' + \sum_{j=1}^n \delta_j \overline{A}_j'\right) - \widehat{\mathrm{vol}}_\natural\left(\overline{L} + \sum_{j=1}^n \delta_j \overline{A}_j\right) \right| \leq \frac{\epsilon}{4}.$$

PROOF. As

$$\left(\overline{L}' + \sum_{j=1}^n \delta_j \overline{A}_j'\right) - \left(\overline{L} + \sum_{j=1}^n \delta_j \overline{A}_j\right) = \overline{\mathscr{O}}\left(f + \sum_{j=1}^n \delta_j g_j\right),$$

by (2) in Proposition 6.8,

$$\begin{aligned}\left|\widehat{\mathrm{vol}}_{\natural}\left(\overline{L}' + \sum_{j=1}^{n}\delta_j \overline{A}'_j\right) - \widehat{\mathrm{vol}}_{\natural}\left(\overline{L} + \sum_{j=1}^{n}\delta_j \overline{A}_j\right)\right| \\ \leq d\,\mathrm{vol}\left(L_{\mathbb{Q}} + \sum_{j=1}^{n}\delta_j (A_j)_{\mathbb{Q}}\right)\left\| f + \sum_{j=1}^{n}\delta_j g_j\right\|_{\sup} \\ \leq dC_1\left(\|f\|_{\sup} + \sum_{j=1}^{n}|\delta_j|\|g_j\|_{\sup}\right) \leq dC_1\frac{\epsilon}{8dC_1}\left(1 + \sum_{j=1}^{n}|\delta_j|\right) \leq \frac{\epsilon}{4},\end{aligned}$$

as required. □

CLAIM 5. *There is a positive number δ such that*

$$\left|\widehat{\mathrm{vol}}_{\natural}\left(\overline{L}' + \sum_{j=1}^{n}\delta_j \overline{A}'_j\right) - \widehat{\mathrm{vol}}_{\natural}\left(\overline{L}'\right)\right| \leq \epsilon/2$$

for all $\delta_1, \ldots, \delta_n \in \mathbb{Q}$ with $|\delta_1| + \cdots + |\delta_n| \leq \delta$.

PROOF. By Theorem 6.9, there is a constant C depending only on X and $\overline{L}', \overline{A}'_1, \ldots, \overline{A}'_n$ such that

$$\left|\widehat{\mathrm{vol}}_{\natural}\left(\overline{L}' + \sum_{j=1}^{n}\delta_j \overline{A}'_j\right) - \widehat{\mathrm{vol}}_{\natural}\left(\overline{L}'\right)\right| \leq C\left(\sum_{j=1}^{n}|\delta_j|\right)$$

for all $\delta_1, \ldots, \delta_n \in \mathbb{Q}$ with $|\delta_1| + \cdots + |\delta_n| \leq 1$, and hence the claim follows. □

Let us go back to the proof of Corollary 6.10. Note that

$$\begin{aligned}\left|\widehat{\mathrm{vol}}_{\natural}\left(\overline{L} + \sum_{j=1}^{n}\delta_j \overline{A}_j\right) - \widehat{\mathrm{vol}}_{\natural}\left(\overline{L}\right)\right| \leq \left|\widehat{\mathrm{vol}}_{\natural}\left(\overline{L}' + \sum_{j=1}^{n}\delta_j \overline{A}'_j\right) - \widehat{\mathrm{vol}}_{\natural}\left(\overline{L}'\right)\right| \\ + \left|\widehat{\mathrm{vol}}_{\natural}\left(\overline{L}' + \sum_{j=1}^{n}\delta_j \overline{A}'_j\right) - \widehat{\mathrm{vol}}_{\natural}\left(\overline{L} + \sum_{j=1}^{n}\delta_j \overline{A}_j\right)\right| + \left|\widehat{\mathrm{vol}}_{\natural}\left(\overline{L}'\right) - \widehat{\mathrm{vol}}_{\natural}\left(\overline{L}\right)\right|.\end{aligned}$$

Thus the theorem follows from Claim 4 and Claim 5. □

REMARK 6.2. Let $\overline{L}_1, \ldots, \overline{L}_r, \overline{A}_1, \ldots, \overline{A}_n$ be continuous hermitian $\mathbb{Q}$-invertible sheaves of real type on X. In [**52**], the following continuity is proved: for a compact subset B in $\mathbb{R}^r$ and a positive number ϵ, there are positive numbers δ and δ' such that

$$\left|\widehat{\mathrm{vol}}_{\natural}\left(\sum_{i=1}^{r}a_i \overline{L}_i + \sum_{j=1}^{n}\delta_j \overline{A}_j + \overline{\mathscr{O}}(\phi)\right) - \widehat{\mathrm{vol}}_{\natural}\left(\sum_{i=1}^{r}a_i \overline{L}_i\right)\right| \leq \epsilon$$

for all $a_1, \ldots, a_r, \delta_1, \ldots, \delta_n \in \mathbb{Q}$ and $\phi \in C^0(X)$ with

$$(a_1, \ldots, a_r) \in B, \quad |\delta_1| + \cdots + |\delta_n| \leq \delta \quad \text{and} \quad \|\phi\|_{\sup} \leq \delta'.$$

6.4. Generalized Hodge index theorem

In this section, let us consider the generalized Hodge index theorem as an application of the continuity of arithmetic volume functions. Let us begin with the following technical definition.

Let X be a d-dimensional projective arithmetic variety. Let L be an invertible sheaf on X such that L is nef on the generic fiber $X_{\mathbb{Q}}$ of $X \to \operatorname{Spec}(\mathbb{Z})$. We say that L has *moderate growth of positive even cohomologies* if there are a genetic resolution $\mu : Y \to X$ of singularities of X and an ample invertible sheaf A on Y such that, for any positive integer n, we can find a positive integer m_0 with

$$\log \#(H^{2i}(Y, m(n\mu^*(L) + A))) = o(m^d)$$

for all $m \geq m_0$ and all $i > 0$. Let us consider several examples having moderate growth of positive even cohomologies.

EXAMPLE 6.11. (1) We assume $d = 2$. Then L has moderate growth of positive even cohomologies. In fact, $H^i(X, F) = 0$ $(i \geq 2)$ for any coherent sheaf F on X.

(2) We assume that L is nef on every fiber of $X \to \operatorname{Spec}(\mathbb{Z})$. Then L has moderate growth of positive even cohomologies. Indeed, let $\mu : Y \to X$ be a generic resolution of singularities of X and A an ample invertible sheaf on Y. Then, for $n \geq 1$, $n\mu^*(L) + A$ is ample. Thus $H^i(Y, m(n\mu^*(L) + A)) = 0$ for $m \gg 1$ and $i > 0$.

The following theorem is the generalized Hodge index theorem due to Moriwaki [**51**].

THEOREM 6.12 (Generalized Hodge index theorem). *Let X be a d-dimensional projective arithmetic variety and let $\overline{L}$ be a C^∞-hermitian invertible sheaf of real type on X with the following properties.*

(1) *$L_{\mathbb{Q}}$ is nef on $X_{\mathbb{Q}}$.*
(2) *$\overline{L}$ is semipositive on $X(\mathbb{C})$.*
(3) *L has moderate growth of positive even cohomologies.*

Then an inequality $\widehat{\operatorname{vol}}(\overline{L}) \geq \widehat{\deg}(\widehat{c}_1(\overline{L})^{\cdot d})$ holds.

PROOF. First we assume that X is generically smooth and the following conditions (a), (b), (c) instead of (1), (2), (3) as above.

(a) $L_{\mathbb{Q}}$ is ample $X_{\mathbb{Q}}$.
(b) $\overline{L}$ is positive on $X(\mathbb{C})$.
(c) There is a positive integer m_0 such that $\log \#(H^{2i}(X, mL)) = o(m^d)$ holds for all $m \geq m_0$ and all $i > 0$.

Then we have the following.

CLAIM 6. $\widehat{\operatorname{vol}}(\overline{L}) \geq \widehat{\deg}(\widehat{c}_1(\overline{L})^{\cdot d})$.

PROOF. By the assumptions (a) and (b), using Theorem 5.36,

$$\begin{aligned}\widehat{\chi}(H^0(X, mL), \|\cdot\|_{\sup}^{m\overline{L}}) &\\ + \sum_{i \geq 1} \log \#(H^{2i}(X, mL)) - \sum_{i \geq 1} \log \#(H^{2i-1}(X, mL)) &\\ &= \frac{\widehat{\deg}(\widehat{c}_1(\overline{L})^{\cdot d})}{d!} m^d + o(m^d)\end{aligned}$$

for $m \gg 1$. Thus, by the assumption (c) and (1) in Proposition 2.17, we obtain

$$\hat{h}^0(H^0(X, mL), \|\cdot\|_{\sup}^{m\overline{L}}) \geq \frac{\widehat{\deg}(\widehat{c}_1(\overline{L})^{\cdot d})}{d!} m^d + o(m^d)$$

for $m \gg 1$. Thus the claim follows. □

Let us consider a general case. Since L has moderate growth of positive even cohomologies, there are a generic resolution $\mu : Y \to X$ of singularities and an ample invertible sheaf A on Y such that, for all n, we can find a positive integer m_0 with

$$\log \#(H^{2i}(Y, m(n\mu^*(L) + A))) = o(m^d)$$

for all $m \geq m_0$ and all $i > 0$. We give a C^∞-hermitian norm $|\cdot|_A$ to A such that $\overline{A} = (A, |\cdot|_A)$ is an ample C^∞-hermitian invertible sheaf of real type on Y. Then, by Claim 6,

$$\widehat{\mathrm{vol}}(n\mu^*(\overline{L}) + \overline{A}) \geq \widehat{\deg}(\widehat{c}_1(n\mu^*(\overline{L}) + \overline{A})^{\cdot d}),$$

which means

$$\widehat{\mathrm{vol}}(\mu^*(\overline{L}) + (1/n)\overline{A}) \geq \widehat{\deg}((\widehat{c}_1(\mu^*(\overline{L})) + (1/n)\widehat{c}_1(\overline{A}))^{\cdot d})$$

by (1) in Theorem 6.6. Therefore, by the continuity of the arithmetic volume function, we obtain

$$\widehat{\mathrm{vol}}(\mu^*(\overline{L})) \geq \widehat{\deg}(\widehat{c}_1(\mu^*(\overline{L}))^{\cdot d}).$$

Thus the theorem follows from (2) in Theorem 6.6 and the projection formula. □

By (1) and (2) in Example 6.11, we have the following.

COROLLARY 6.13. *Let X be a projective arithmetic surface and let $\overline{L}$ be a C^∞-hermitian invertible sheaf of real type on X such that L is nef on the generic fiber $X \to \mathrm{Spec}(\mathbb{Z})$ and $\overline{L}$ is semipositive on $X(\mathbb{C})$. Then*

$$\widehat{\mathrm{vol}}(\overline{L}) \geq \widehat{\deg}(\widehat{c}_1(\overline{L})^{\cdot 2}).$$

COROLLARY 6.14. *Let X be a d-dimensional projective arithmetic variety and let $\overline{L}$ be a C^∞-hermitian invertible sheaf of real type on X such that $\overline{L}$ is vertically nef, that is, L is nef on every fiber of $X \to \mathrm{Spec}(\mathbb{Z})$ and $\overline{L}$ is semipositive on $X(\mathbb{C})$. Then*

$$\widehat{\mathrm{vol}}(\overline{L}) \geq \widehat{\deg}(\widehat{c}_1(\overline{L})^{\cdot d})$$

holds. In particular, if $\widehat{\deg}(\widehat{c}_1(\overline{L})^{\cdot d}) > 0$, then $\overline{L}$ is big.

REMARK 6.3. Here we would like to give several remarks on the existence theorem of small sections. Let X be a projective arithmetic variety and let $\overline{L}$ be a C^∞-hermitian invertible sheaf of real type on X with $\widehat{\deg}(\widehat{c}_1(\overline{L})^{\cdot \dim X}) > 0$. If $\overline{L}$ is vertically nef (in the case where X is an arithmetic surface, the assumption that $L_{\mathbb{Q}}$ is nef and $\overline{L}$ is semipositive on $X(\mathbb{C})$ is sufficient), then, by Corollary 6.13 or Corollary 6.14, nL has a non-zero strictly small section for a sufficiently large n. As we point out in the introduction, the existence theorem of small sections is one of the main themes of Arakelov geometry; here we give its brief history. As an application of the arithmetic Riemann-Roch theorem on an arithmetic surface, Faltings **[18]** proved the existence theorem of small sections in the case where X is a regular arithmetic surface, $L_{\mathbb{Q}}$ is ample and $\overline{L}$ is admissible on $X(\mathbb{C})$. Gillet-Soulé **[26]** established the existence theorem of small sections as an application of their arithmetic Riemann-Roch theorem under the assumption that X is regular and $\overline{L}$ is vertically ample. Abbes-Bouche [**1**] gave the same result without using the arithmetic Riemann-Roch theorem. Moreover, Zhang [**76**] showed that the assumption that $L_{\mathbb{Q}}$ is ample and $\overline{L}$ is vertically nef is sufficient for the existence of small sections. The assumption that $L_{\mathbb{Q}}$ is ample seems to be geometrically very strong. However, Moriwaki [**51**] showed the continuity of the arithmetic volume function and proved the above result as its application.

REMARK 6.4. In Corollary 6.13, if $\deg(L_{\mathbb{Q}}) = 0$, then $\widehat{\deg}(\widehat{c}_1(\overline{L})^{\cdot 2}) \leq 0$. This is nothing more than the Hodge index theorem due to Faltings-Hriljac (see Remark 9.3). In this sense, Theorem 6.12 is called the generalized Hodge index theorem.

REMARK 6.5. As a similar inequality, Yuan [**73**] proved the arithmetic Siu's inequality as follows. Let $\overline{L}$ and $\overline{M}$ be nef C^{∞}-hermitian $\mathbb{Q}$-invertible sheaves of real type on X on a d-dimensional projective arithmetic variety. Then

$$\widehat{\mathrm{vol}}(\overline{L} - \overline{M}) \geq \widehat{\deg}(\widehat{c}_1(\overline{L})^d) - d \cdot \widehat{\deg}(\widehat{c}_1(\overline{L})^{(d-1)} \cdot \widehat{c}_1(\overline{M})).$$

Finally, let us consider the following as an application of the generalized Hodge index theorem.

THEOREM 6.15. *Let $\overline{L}_1, \ldots, \overline{L}_d$ be nef C^{∞}-hermitian invertible sheaves of real type on X. Then*

$$\widehat{\deg}\left(\widehat{c}_1(\overline{L}_1) \cdots \widehat{c}_1(\overline{L}_d)\right) \geq 0.$$

Let us begin with the following lemma.

LEMMA 6.16. *Let $\overline{L}$ be a nef C^{∞}-hermitian invertible sheaf of real type on X and let $\overline{M}$ be a vertically nef C^{∞}-hermitian invertible sheaf of real type on X such that*

$$\widehat{\deg}\left(\widehat{c}_1(\overline{M})^{\dim(\Gamma)} \cdot (\Gamma, 0)\right) > 0$$

for any horizontal integral subscheme Γ of X (i.e. Γ is flat over $\mathbb{Z}$). Then

$$\widehat{\deg}\left(\widehat{c}_1(\overline{L})^i \cdot \widehat{c}_1(\overline{M})^{d-i}\right) \geq 0$$

for all $0 \leq i \leq d$.

PROOF. We prove it by induction on d. In the case where $d = 1$, the assertion is obvious, so that we assume $d > 1$. First we consider the case where $0 \leq i < d$. Since $\widehat{\deg}\left(\widehat{c}_1(\overline{M})^d\right) > 0$, replacing $\overline{M}$ by $\overline{M}^{\otimes n}$ $(n > 0)$ if necessarily, by Corollary 6.14, we may assume that M has a non-zero strictly small section s. Let $\mathrm{div}(s) = a_1\Gamma_1 + \cdots + a_e\Gamma_e$ be the decomposition as cycles, where $a_j > 0$ for all j. Then

$$\widehat{\deg}\left(\widehat{c}_1(\overline{L})^i \cdot \widehat{c}_1(\overline{M})^{d-i}\right) = \sum_j a_j \widehat{\deg}\left(\widehat{c}_1(\overline{L})^i \cdot \widehat{c}_1(\overline{M})^{d-1-i} \cdot (\Gamma_j, 0)\right)$$
$$- \int_{X(\mathbb{C})} \log(|s|) c_1(\overline{L})^i \cdot \widehat{c}_1(\overline{M})^{d-1-i},$$

which is non-negative by the hypothesis of induction.

Next we consider the case where $i = d$. We set $\overline{L}_t = \overline{L} + t\overline{M}$ for $t \in \mathbb{Q}$. Let us consider the polynomial $P(t)$ given by $P(t) = \widehat{\deg}\left(\widehat{c}_1(\overline{L}_t)^d\right)$.

CLAIM 7. *Let t be a positive rational number with $P(t) > 0$. Then*

$$P(t) \geq t^d \widehat{\deg}\left(\widehat{c}_1(\overline{M})^d\right).$$

PROOF. Let Γ be a subscheme of X such that Γ is flat over $\mathbb{Z}$. If $\Gamma \neq X$, then

$$\widehat{\deg}\left(\widehat{c}_1(\overline{L}_t)^{\dim(\Gamma)} \cdot (\Gamma, 0)\right) = t^{\dim \Gamma}\widehat{\deg}\left(\widehat{c}_1(\overline{M})^{\dim(\Gamma)} \cdot (\Gamma, 0)\right) + \sum_{j=1}^{\dim \Gamma} \binom{\dim \Gamma}{j} t^{\dim \Gamma - j}\widehat{\deg}\left(\widehat{c}_1(\overline{L}\big|_{\Gamma})^j \cdot \widehat{c}_1(\overline{M}\big|_{\Gamma})^{\dim \Gamma - j}\right).$$

Thus, by hypothesis of induction, we have

$$\widehat{\deg}\left(\widehat{c}_1(\overline{L}_t)^{\dim(\Gamma)} \cdot (\Gamma, 0)\right) > 0.$$

If $\Gamma = X$, then

$$\widehat{\deg}\left(\widehat{c}_1(\overline{L}_t)^{\dim(\Gamma)} \cdot (\Gamma, 0)\right) = P(t) > 0.$$

Therefore, in the same argument as above, we obtain

$$\widehat{\deg}\left(\widehat{c}_1(\overline{L}) \cdot \widehat{c}_1(\overline{L}_t)^{d-1}\right) \geq 0,$$

and hence,

$$\begin{aligned} P(t) &= \widehat{\deg}\left((\widehat{c}_1(\overline{L}) + t\widehat{c}_1(\overline{M})) \cdot \widehat{c}_1(\overline{L}_t)^{d-1}\right) \\ &\geq t\widehat{\deg}\left(\widehat{c}_1(\overline{M}) \cdot \widehat{c}_1(\overline{L}_t)^{d-1}\right) \\ &\geq t^d\widehat{\deg}\left(\widehat{c}_1(\overline{M})^d\right), \end{aligned}$$

as required. □

Let us go back to the proof of the lemma. We set $t_0 = \max\{t \in \mathbb{R} \mid P(t) = 0\}$. We assume $t_0 > 0$. Then, by the previous claim, for all $t > t_0$,

$$P(t) \geq t^d\widehat{\deg}\left(\widehat{c}_1(\overline{M})^d\right).$$

As t goes to t_0, we have

$$0 = P(t_0) \geq t_0^d\widehat{\deg}\left(\widehat{c}_1(\overline{M})^d\right) > 0.$$

This is a contradiction, that is, $t_0 \leq 0$. In particular, $P(0) \geq 0$. Thus the lemma follows. □

PROOF OF THEOREM 6.15. Let us start the proof of the theorem. We also prove it by induction on d. If $d = 1$, then the assertion is obvious, so that we assume $d > 1$. Let $\overline{M}$ be an ample C^∞-hermitian invertible sheaf of real type on X. By Proposition 5.39,

$$\widehat{\deg}\left(\widehat{c}_1(\overline{M})^{\dim(\Gamma)} \cdot (\Gamma, 0)\right) > 0$$

holds for all horizontal integral subschemes Γ. Let t be a positive rational number. We consider $\overline{L}_d + t\overline{M}$. Then, by Lemma 6.16,

$$\widehat{\deg}\left((\widehat{c}_1(\overline{L}_d) + t\widehat{c}_1(\overline{M}))^d\right) > 0.$$

Thus, by Corollary 6.14, for a sufficiently large integer n, $n(L_d+tM)$ has a non-zero strictly small section s. Let $\operatorname{div}(s) = a_1\Gamma_1 + \cdots + a_e\Gamma_e$ be the decomposition as cycles as in the previous lemma. Then

$$n\widehat{\deg}\left(\widehat{c}_1(\overline{L}_1)\cdots\widehat{c}_1(\overline{L}_{d-1})\cdot\widehat{c}_1(\overline{L}_d+t\widehat{c}_1(\overline{M}))\right) = \sum_j a_j\widehat{\deg}\left(\widehat{c}_1(\overline{L}_1)\cdots\widehat{c}_1(\overline{L}_{d-1})\cdot(\Gamma_j,0)\right) - \int_{X(\mathbb{C})}\log(|s|)c_1(\overline{L}_1)\cdots c_1(\overline{L}_{d-1}).$$

Thus, by the hypothesis of induction, we have

$$\widehat{\deg}\left(\widehat{c}_1(\overline{L}_1)\cdots\widehat{c}_1(\overline{L}_{d-1})\cdot\widehat{c}_1(\overline{L}_d+t\widehat{c}_1(\overline{M}))\right)\geq 0.$$

Therefore, the theorem follows because t is an arbitrary positive rational number. □

6.5. Estimate of the number of small sections

The purpose of this section is to prove the following fundamental estimate due to Moriwaki [**51**, Theorem 3.1] and [**52**, Theorem 1.3.1 and Theorem 1.3.2]. This theorem was used in the previous sections to establish the continuity of the volume function.

THEOREM 6.17. *Let X be a d-dimensional generically smooth projective arithmetic variety. We fix a volume form Φ of real type on $X(\mathbb{C})$. Let $\overline{L}$ and $\overline{A}$ be C^∞-hermitian invertible sheaves of real type on X. Then there are constants a_0, C, D depending only on X, Φ, $\overline{L}$ and $\overline{A}$ such that*

$$\hat{h}^0\left(H^0(aL+(b-c)A),\|\cdot\|_{L^2}^{a\overline{L}+(b-c)\overline{A}}\right)\leq\hat{h}^0\left(H^0(aL-cA),\|\cdot\|_{L^2}^{a\overline{L}-c\overline{A}}\right) + Cba^{d-1}+Da^{d-1}\log(a)$$

holds for all integers a,b,c with $a\geq b\geq c\geq 0$ and $a\geq a_0$, where the L^2-norms $\|\cdot\|_{L^2}^{a\overline{L}+(b-c)\overline{A}}$ and $\|\cdot\|_{L^2}^{a\overline{L}-c\overline{A}}$ are given by Φ. The above inequality is called the L^2-version. In addition, the sup*-version also holds, that is, there are constants a_0', C' and D' depending only on X, $\overline{L}$ and $\overline{A}$ such that*

$$\hat{h}^0\left(H^0(aL+(b-c)A),\|\cdot\|_{\sup}^{a\overline{L}+(b-c)\overline{A}}\right)\leq\hat{h}^0\left(H^0(aL-cA),\|\cdot\|_{\sup}^{a\overline{L}-c\overline{A}}\right) + C'ba^{d-1}+D'a^{d-1}\log(a)$$

holds for all integers a,b,c with $a\geq b\geq c\geq 0$ and $a\geq a_0'$.

6.5.1. Distorsion function and its estimate. Let X be a non-singular complex projective variety such that every connected component of X has the fixed dimension n. Let Φ be a volume form of X. Let $\overline{H}=(H,h)$ be a C^∞-hermitian invertible sheaf on X. For $s,s'\in H^0(X,H)$, we define $\langle s,s'\rangle_{\overline{H},\Phi}$ to be

$$\langle s,s'\rangle_{\overline{H},\Phi} := \int_X h(s,s')\Phi.$$

Let $s_1,\ldots,s_N$ be an orthonormal basis of $H^0(X,H)$ with respect to $\langle\ ,\ \rangle_{\overline{H},\Phi}$. We define the *distorsion function* $\operatorname{dist}(\overline{H},\Phi)$ of $\overline{H}$ with respect to Φ to be

$$\operatorname{dist}(\overline{H},\Phi)(x) := \sum_{i=1}^N h(s_i,s_i)(x).$$

The above function does not depend on the choice of the orthonormal basis of $H^0(X, H)$. In the case where $H^0(X, H) = \{0\}$, $\operatorname{dist}(\overline{H}, \Phi)$ is defined to be the constant function 0. Let λ be a positive number. We can easily see that $\operatorname{dist}(\overline{H}, \lambda\Phi) = \lambda^{-1} \operatorname{dist}(\overline{H}, \Phi)$. In addition, if $X_1, \dots, X_l$ are connected components of X, then

$$\operatorname{dist}(\overline{H}, \Phi) = \operatorname{dist}\left(\overline{H}\big|_{X_1}, \Phi|_{X_1}\right) + \cdots + \operatorname{dist}\left(\overline{H}\big|_{X_l}, \Phi|_{X_l}\right).$$

If X is connected and $\overline{A}$ is a positive C^∞-hermitian invertible sheaf on X, then, by Bouche [**10**] and Tian [**67**], it is known that

$$\lim_{a\to\infty} \sup_{x\in X} \left| \frac{\operatorname{dist}(a\overline{A}, \Phi(\overline{A}))(x)}{\dim H^0(aA)} - 1 \right| = 0,$$

where

$$\Phi(\overline{A}) = \frac{c_1(\overline{A})^n}{\displaystyle\int_X c_1(\overline{A})^n}.$$

We have the following theorem due to Yuan [**73**].

THEOREM 6.18. *Let $\overline{A} = (A, h_A)$ and $\overline{B} = (B, h_B)$ be positive C^∞-hermitian invertible sheaves on X. Then, for any real number ϵ with $0 < \epsilon < 1$, there are positive numbers a_0 and b_0 such that*

$$\operatorname{dist}(a\overline{A} - b\overline{B}, c_1(\overline{A})^n)(x) \le \frac{(1+\epsilon)^2}{1-\epsilon} \dim H^0(aA)$$

for all $x \in X$ and $a \ge a_0$, $b \ge b_0$.

PROOF. Clearly, it is sufficient to show the theorem in the case where X is connected. The above theorem due to Bouche and Tian gives rise to positive constants a_0 and b_0 such that

$$\dim H^0(aA)\,(1-\epsilon) \le \operatorname{dist}(a\overline{A}, \Phi(A))(z) \le \dim H^0(aA)\,(1+\epsilon)$$

and

$$\dim H^0(bB)\,(1-\epsilon) \le \operatorname{dist}(b\overline{B}, \Phi(B))(z) \le \dim H^0(bB)\,(1+\epsilon)$$

for all $z \in X$, $a \ge a_0$ and $b \ge b_0$.

Let x be an arbitrary fixed point of X. For $b \ge b_0$, let $s_1, \dots, s_N$ be an orthonormal basis of $H^0(bB)$ with respect to $\langle\ ,\ \rangle_{b\overline{B},\Phi(\overline{B})}$ such that $s_1(x) \ne 0$ and $s_2(x) = \cdots = s_N(x) = 0$. Then

$$h_{b\overline{B}}(s_1, s_1)(x) = \operatorname{dist}(b\overline{B}, \Phi(\overline{B}))(x) \ge (1-\epsilon)\dim H^0(bB).$$

In addition, note that

$$\|s_1\|_{\sup}^2 \le \sup_{z\in X} \operatorname{dist}(b\overline{B}, \Phi(B))(z) \le (1+\epsilon)\dim H^0(bB),$$

so that

$$\frac{h_{b\overline{B}}(s_1, s_1)(x)}{\|s_1\|_{\sup}^2} \ge \frac{1-\epsilon}{1+\epsilon},$$

for $b \ge b_0$.

For $a \ge a_0$ and $b \ge b_0$, let us take an orthonormal basis $t_1, \dots, t_r$ of $H^0(aA - bB)$ with respect to $\langle\ ,\ \rangle_{a\overline{A}-b\overline{B},\Phi(\overline{A})}$ such that

$$s_1t_1, \dots, s_1t_r \in H^0(aA)$$

are orthogonal with respect to $\langle\ ,\ \rangle_{a\overline{A},\Phi(\overline{A})}$. The existence of such a basis $t_1,\ldots,t_r$ follows from the fact that any hermitian matrix is diagonalizable by a unitary matrix. Then, as

$$\left\{s_1t_i/\|s_1t_i\|_{a\overline{A},\Phi(\overline{A})}\right\}_{i=1,\ldots,r}$$

forms a part of an orthonormal basis of $H^0(aA)$, we have

$$\sum_{i=1}^{r}\frac{h_{a\overline{A}}(s_1t_i,s_1t_i)(x)}{\|s_1t_i\|^2_{a\overline{A},\Phi(\overline{A})}}\leq \operatorname{dist}(aA,\Phi(\overline{A}))(x)\leq (1+\epsilon)\dim H^0(aA).$$

Moreover,

$$\|s_1t_i\|^2_{a\overline{A},\Phi(\overline{A})}\leq \|s_1\|^2_{\sup}\qquad (i=1,\ldots,r),$$

so that, if we set $\lambda=\int_X c_1(\overline{A})^n\geq 1$, then

$$\operatorname{dist}(a\overline{A}-b\overline{B},c_1(\overline{A})^n)=\lambda^{-1}\operatorname{dist}(a\overline{A}-b\overline{B},\Phi(\overline{A}))\leq \operatorname{dist}(a\overline{A}-b\overline{B},\Phi(\overline{A})),$$

and hence

$$\begin{aligned}\frac{1-\epsilon}{1+\epsilon}\operatorname{dist}(a\overline{A}-b\overline{B},c_1(\overline{A})^n)(x) &\leq \frac{h_{b\overline{B}}(s_1,s_1)(x)}{\|s_1\|^2_{\sup}}\sum_{i=1}^{r}h_{a\overline{A}-b\overline{B}}(t_i,t_i)(x)\\ &\leq \sum_{i=1}^{r}\frac{h_{b\overline{B}}(s_1,s_1)(x)}{\|s_1t_i\|^2_{a\overline{A},\Phi(\overline{A})}}h_{a\overline{A}-b\overline{B}}(t_i,t_i)(x)\\ &=\sum_{i=1}^{r}\frac{h_{a\overline{A}}(s_1t_i,s_1t_i)(x)}{\|s_1t_i\|^2_{a\overline{A},\Phi(\overline{A})}}\leq (1+\epsilon)\dim H^0(aA),\end{aligned}$$

as required. □

Let $\overline{L}$ and $\overline{A}$ be C^∞-hermitian invertible sheaves on X such that $\overline{A}$ and $\overline{L}+\overline{A}$ are positive. We set $\Phi=c_1(\overline{L}+\overline{A})^n$. Let a,b,c be non-negative integers. Let s be an element of $H^0(bA)$ such that s is not zero on each connected component and $\|s\|_{\sup}\leq 1$. Let $\langle\ ,\ \rangle_{a\overline{L}-c\overline{A}}$ and $\langle\ ,\ \rangle_{a\overline{L}+(b-c)\overline{A}}$ be the natural inner products of $H^0(aL-cA)$ and $H^0(aL+(b-c)A)$ with respect to Φ. Let $\langle\ ,\ \rangle_{a\overline{L}+(b-c)\overline{A},s,\mathrm{sub}}$ be an inner product of $H^0(aL-cA)$ given by

$$\langle t,t'\rangle_{a\overline{L}+(b-c)\overline{A},s,\mathrm{sub}}=\langle st,st'\rangle_{a\overline{L}+(b-c)\overline{A}}\qquad (t,t'\in H^0(aL-cA)).$$

Then we have the following.

COROLLARY 6.19. *Let $\gamma(a,c,s)$ be the volume difference (see Section 3.7) of*

$$0\to\left(H^0(aL-cA),\langle,\rangle_{a\overline{L}-c\overline{A}}\right)\to\left(H^0(aL-cA),\langle,\rangle_{a\overline{L}+(b-c)\overline{A},s,\mathrm{sub}}\right)\to(0,0)\to 0.$$

Then, for a real number ϵ with $0<\epsilon<1$, there is a positive number a_0 such that

$$\gamma(a,c,s)\geq\frac{(1+\epsilon)^2}{1-\epsilon}\dim H^0(a(L+A))\left(\int_X\log(|s|)\Phi\right)$$

holds for all $a\geq a_0$ and $c\geq 0$.

PROOF. Note that $a\overline{L}-c\overline{A}=a(\overline{L}+\overline{A})-(a+c)\overline{A}$, so that Theorem 6.18 ensures a positive constant a_0 such that

$$\operatorname{dist}(a\overline{L}-c\overline{A},\Phi)(x)\leq\frac{(1+\epsilon)^2}{1-\epsilon}\dim H^0(a(L+A))$$

for all $x\in X$ and all $a\geq a_0$ and $c\geq 0$.

We choose an orthonormal basis $t_1, \ldots, t_r$ of $H^0(aL - cA)$ with respect to $\langle\ ,\ \rangle_{a\overline{L}-c\overline{A}}$ such that $st_1, \ldots, st_r$ are orthogonal with respect to $\langle\ ,\ \rangle_{a\overline{L}+(b-c)\overline{A}}$. Then, as

$$\gamma(a,c,s) = -\frac{1}{2}\log\left(\frac{\det(\langle t_i, t_j\rangle_{a\overline{L}-c\overline{A}})}{\det(\langle st_i, st_j\rangle_{a\overline{L}-c\overline{A}})}\right) = \frac{1}{2}\sum_{i=1}^{r}\log\int_X |s|^2|t_i|^2\Phi,$$

by virtue of Jensen's inequality,

$$\begin{aligned}\gamma(a,c,s) &\geq \frac{1}{2}\sum_{i=1}^{r}\int_X \log(|s|^2)|t_i|^2\Phi = \frac{1}{2}\int_X \log(|s|^2)\operatorname{dist}(a\overline{L}-c\overline{A},\Phi)\Phi \\ &\geq \frac{(1+\epsilon)^2}{1-\epsilon}\dim H^0(a(L+A))\left(\int_X \log(|s|)\Phi\right)\end{aligned}$$

for all $a \geq a_0$ and $c \geq 0$, as required. □

6.5.2. Proof of Theorem 6.17. Let us start the proof of Theorem 6.17. First we have the following claim.

CLAIM 8. *Let $\overline{A}'$ be another C^∞-hermitian invertible sheaf of real type on X with $\overline{A} \leq \overline{A}'$. If the L^2-version and the* sup*-version hold for $\overline{L}$ and $\overline{A}'$, then the L^2-version and the* sup*-version also hold for $\overline{L}$ and $\overline{A}$.*

PROOF. Since $a\overline{L} + (b-c)\overline{A} \leq a\overline{L} + (b-c)\overline{A}'$ and $a\overline{L} - c\overline{A}' \leq a\overline{L} - c\overline{A}$, the assertion is obvious. □

By this claim, we may replace $\overline{A}$ by a sufficiently large C^∞-hermitian invertible sheaf of real type. Thus we may assume the following properties for $\overline{A}$.

(1) A is very ample on X.
(2) $\overline{A}$ and $\overline{L} + \overline{A}$ is positive on $X(\mathbb{C})$.

Under the above assumption, we fix constants C_1 and C_2 such that

$$\begin{cases}\operatorname{rk} H^0(a(L+A)) \leq C_1 a^{d-1} \qquad (a \geq 1), \\ \log(18)\operatorname{rk} H^0(a(L+A)) + 2\log\left((\operatorname{rk} H^0(a(L+A))!\right) \\ \qquad\qquad\qquad\qquad \leq C_2 a^{d-1}\log(a) \qquad (a \geq 2).\end{cases}$$

CLAIM 9. *If the L^2-version holds for a volume form of real type, then the L^2-version holds for any volume form of real type.*

PROOF. We assume that the L^2-version holds for a volume form Φ of real type. Let Φ' be another volume form of real type. Then there are constants $0 < \sigma_0 \leq 1$ and $\sigma_1 \geq 1$ with $\sigma_0^2\Phi' \leq \Phi \leq \sigma_1^2\Phi'$. Thus we have

$$\|\cdot\|_{L^2}^{a\overline{L}+(b-c)\overline{A},\Phi} \leq \sigma_1\|\cdot\|_{L^2}^{a\overline{L}+(b-c)\overline{A},\Phi'} \quad\text{and}\quad \sigma_0\|\cdot\|_{L^2}^{a\overline{L}-c\overline{A},\Phi'} \leq \|\cdot\|_{L^2}^{a\overline{L}-c\overline{A},\Phi},$$

so that

$$\begin{aligned}&\hat{h}^0\left(H^0(aL+(b-c)A), \sigma_1\|\cdot\|_{L^2}^{a\overline{L}+(b-c)\overline{A},\Phi'}\right) \\ &\qquad \leq \hat{h}^0\left(H^0(aL+(b-c)A), \|\cdot\|_{L^2}^{a\overline{L}+(b-c)\overline{A},\Phi}\right)\end{aligned}$$

and

$$\hat{h}^0\left(H^0(aL-cA), \|\cdot\|_{L^2}^{a\overline{L}-c\overline{A},\Phi}\right) \leq \hat{h}^0\left(H^0(aL-cA), \sigma_0\|\cdot\|_{L^2}^{a\overline{L}-c\overline{A},\Phi'}\right).$$

On the other hand, note that

$$\operatorname{rk} H^0(aL + (b-c)A) \leq \operatorname{rk} H^0(a(L+A))$$

and

$$\operatorname{rk} H^0(aL - cA) \leq \operatorname{rk} H^0(a(L+A)).$$

Then, by (3) in Proposition 2.17,

$$\begin{aligned}\hat{h}^0 &\left(H^0(aL + (b-c)A), \|\cdot\|_{L^2}^{a\overline{L}+(b-c)\overline{A},\Phi'}\right) \\ &\leq \hat{h}^0\left(H^0(aL + (b-c)A), \sigma_1\|\cdot\|_{L^2}^{a\overline{L}+(b-c)\overline{A},\Phi'}\right) \\ &\quad + \log(\sigma_1)C_1 a^{d-1} + C_2\log(a)a^{d-1},\end{aligned}$$

and

$$\begin{aligned}\hat{h}^0\left(H^0(aL - cA), \sigma_0\|\cdot\|_{L^2}^{a\overline{L}-c\overline{A},\Phi'}\right) &\leq \hat{h}^0\left(H^0(aL - cA), \|\cdot\|_{L^2}^{a\overline{L}-c\overline{A},\Phi'}\right) \\ &\quad + \log(\sigma_0^{-1})C_1 a^{d-1} + C_2\log(a)a^{d-1}. \quad \square\end{aligned}$$

Thus the claim follows.

Therefore, we may assume that the volume form Φ to define the L^2-norms is given by

$$\Phi(\overline{L} + \overline{A}) =: \frac{c_1(\overline{L}+\overline{A})^{d-1}}{\int_{X(\mathbb{C})} c_1(\overline{L}+\overline{A})^{d-1}}.$$

CLAIM 10. *The* sup*-version follows from the* L^2*-version.*

PROOF. First, as $\|\cdot\|_{\sup}^{a\overline{L}+(b-c)\overline{A}} \geq \|\cdot\|_{L^2}^{a\overline{L}+(b-c)\overline{A}}$, we obtain

$$\hat{h}^0\left(H^0(aL + (b-c)A), \|\cdot\|_{\sup}^{a\overline{L}+(b-c)\overline{A}}\right) \leq \hat{h}^0\left(H^0(aL + (b-c)A), \|\cdot\|_{L^2}^{a\overline{L}+(b-c)\overline{A}}\right).$$

In addition, by Gromov's inequality (Proposition 5.32), there is a constant $D \geq 1$ such that

$$\|\cdot\|_{L^2}^{a\overline{L}-c\overline{A}} \geq D^{-1}(a+c+1)^{-(d-1)}\|\cdot\|_{\sup}^{a\overline{L}-c\overline{A}}$$

for all $a, c \geq 0$, so that

$$\begin{aligned}\hat{h}^0 &\left(H^0(aL - cA), \|\cdot\|_{L^2}^{a\overline{L}-c\overline{A}}\right) \\ &\leq \hat{h}^0\left(H^0(aL - cA), D^{-1}(a+c+1)^{-(d-1)}\|\cdot\|_{\sup}^{a\overline{L}-c\overline{A}}\right) \\ &\leq \hat{h}^0\left(H^0(aL - cA), \|\cdot\|_{\sup}^{a\overline{L}-c\overline{A}}\right) + \log(D(2a+1)^{d-1})C_1 a^{d-1} + C_2 a^{d-1}\log(a)\end{aligned}$$

for all $a \geq c \geq 0$. Thus the assertion of the claim follows. $\square$

Gathering all of the above arguments, it is sufficient to see the L^2-version under the assumptions that $\overline{L}$ and $\overline{A}$ satisfy conditions (1) and (2) as above and the volume form Φ to define the L^2-norms is given by $\Phi(\overline{L}+\overline{A})$.

Here we fix several notations. Let V be a compact complex manifold and Φ a volume form. Let $\overline{M} = (M, |\cdot|_M)$ and $\overline{N} = (N, |\cdot|_N)$ be C^∞-hermitian invertible sheaves on V. Let $t \in H^0(V, N)$ be a section such that t is non-zero on each connected component of V. The subnorm of $H^0(V, M-N)$ induced by the natural injective homomorphism

$$H^0(V, M-N) \xrightarrow{\otimes t} H^0(V, M),$$

$|\cdot|_M$ and Φ is denoted by $\|\cdot\|^{\overline{M}, M-N}_{L^2,t,\mathrm{sub}}$, that is,

$$\|s\|^{\overline{M}, M-N}_{L^2,t,\mathrm{sub}} := \left(\int_V |s\otimes t|^2_M \Phi\right)^{1/2} \qquad (s \in H^0(V, M-N)).$$

The norm $\|\cdot\|^{\overline{M}, M-N}_{L^2,t,\mathrm{sub}}$ is often denoted by $\|\cdot\|^{\overline{M}}_{L^2,t,\mathrm{sub}}$ for short. Let $|\cdot|_A$ be the L^2-norm of $\overline{A}$. Then, for a real number λ, we set

$$\overline{A}^{\lambda} = (A, \exp(-\lambda)|\cdot|_A).$$

CLAIM 11. *We may assume the following.*

(a) *There is a non-zero small section s of A such that* $\mathrm{div}(s)$ *is smooth over* $\mathbb{Q}$ *(if $d=1$, then* $\mathrm{div}(s)$ *is an empty set on $X_{\mathbb{Q}}$).*

(b) *There are a positive integer n and a non-zero section t of $nA-L$ such that t is not zero on each irreducible component of* $\mathrm{div}(s)$.

PROOF. Since A is very ample, there is a non-zero section $s \in H^0(X, A)$ such that $\mathrm{div}(s)$ is smooth over $\mathbb{Q}$. Moreover, by Proposition 1.22, there are a positive number n and a non-zero section $t \in H^0(nA-L)$ such that t is non-zero on each irreducible component of $\mathrm{div}(s)$. Let λ be a non-negative real number with $\exp(-\lambda)\|s\|_{\mathrm{sup}} \leq 1$ and $\exp(-n\lambda)\|t\|_{\mathrm{sup}} \leq 1$. Then s and t are small sections of $\overline{A}^{\lambda}$ and $n\overline{A}^{\lambda} - \overline{L}$, respectively. On the other hand, $\overline{A}^{\lambda}$ satisfies conditions (1) and (2), $\overline{A} \leq \overline{A}^{\lambda}$ and $\Phi(\overline{L}+\overline{A}) = \Phi(\overline{L}+\overline{A}^{\lambda})$. Thus, the claim follows from Claim 8. □

Let $\mathcal{F}$ be a coherent sheaf on X and let Z be a subscheme of X. We denote the image of the natural homomorphism $H^i(X, \mathcal{F}) \to H^i(Z, \mathcal{F}|_Z)$ by $I^i(Z, \mathcal{F}|_Z)$ or $I^i(\mathcal{F}|_Z)$ for short.

From here, we start the proof of Theorem 6.17. If $b=0$, then $c=0$, so that the theorem is obvious in this case. We assume $b \geq 1$. Let s be the section of A as in Claim 11, that is, $s \in H^0(X, A)$ is a non-zero small section and $Y := \mathrm{div}(s)$ is smooth over $\mathbb{Q}$.

In addition to C_1 and C_2, we fix constants C_3 and C_4 such that

$$\begin{cases} \mathrm{rk}\, H^0(Y, a(L+A)|_Y) \leq C_3 a^{d-2} & (a \geq 1), \\ \log(18)\,\mathrm{rk}\, H^0(Y, a(L+A)|_Y) + 2\log\left((\mathrm{rk}\, H^0(Y, a(L+A)|_Y))!\right) & \\ \qquad\qquad \leq C_4 a^{d-2}\log(a) & (a \geq 2). \end{cases}$$

Note that if $d=1$, then $\mathrm{rk}\, H^0(Y, a(L+A)|_Y) = 0$.

Here we consider a surjective homomorphism

$$H^0(aL-(b-c)A) \to I^0(aL+(b-c)A|_{bY})$$

and the L^2-norm $\|\cdot\|^{a\overline{L}+(b-c)\overline{A}}_{L^2}$ on $H^0(aL+(b-c)A)$. Let $\|\cdot\|^{a\overline{L}+(b-c)\overline{A}}_{L^2,\mathrm{quot}}$ be the quotient norm of $I^0(aL+(b-c)A|_{bY})$ induced by the norm $\|\cdot\|^{a\overline{L}+(b-c)\overline{A}}_{L^2}$. Then an exact sequence

$$0 \to H^0(aL - cA) \xrightarrow{s^b} H^0(aL + (b-c)A) \to I^0(aL + (b-c)A|_{bY}) \to 0$$

yields an exact sequence

$$\begin{aligned} 0 \to \left(H^0(aL - cA), \|\cdot\|_{L^2, s^b, \mathrm{sub}}^{a\overline{L}+(b-c)\overline{A}}\right) &\to \left(H^0(aL + (b-c)A), \|\cdot\|_{L^2}^{a\overline{L}+(b-c)\overline{A}}\right) \\ &\to \left(I^0(aL + (b-c)A|_{bY}), \|\cdot\|_{L^2, \mathrm{quot}}^{a\overline{L}+(b-c)\overline{A}}\right) \to 0 \end{aligned}$$

as normed $\mathbb{Z}$-modules. Therefore, since $\operatorname{rk} H^0(aL - cA) \leq \operatorname{rk} H^0(a(L+A))$, by using the above exact sequence and (4) in Proposition 2.17, we obtain

$$\begin{aligned} (6.2)\quad \hat{h}^0\left(H^0(aL + (b-c)A), \|\cdot\|_{L^2}^{a\overline{L}+(b-c)\overline{A}}\right) &\leq \hat{h}^0\left(H^0(aL - cA), \|\cdot\|_{L^2, s^b, \mathrm{sub}}^{a\overline{L}+(b-c)\overline{A}}\right) \\ &\quad + \hat{h}^0\left(I^0(aL + (b-c)A|_{bY}), \|\cdot\|_{L^2, \mathrm{quot}}^{a\overline{L}+(b-c)\overline{A}}\right) + C_2 a^{d-1}\log(a) \end{aligned}$$

for all integers a, b, c with $a \geq b \geq c \geq 0$, $a \geq 2$.

Here we have the following lemmas.

LEMMA 6.20. *We can find constants $a_0 = a_0(\overline{L}, \overline{A})$ and $C_5 = C_5(\overline{L}, \overline{A})$ satisfying the following inequality*

$$\begin{aligned} \hat{h}^0\left(H^0(aL - cA), \|\cdot\|_{L^2, s^b, \mathrm{sub}}^{a\overline{L}+(b-c)\overline{A}}\right) &\leq \hat{h}^0\left(H^0(aL - cA), \|\cdot\|_{L^2}^{a\overline{L}-c\overline{A}}\right) \\ &\quad + C_5 b a^{d-1} + C_2 a^{d-1}\log(a) \end{aligned}$$

for all integers a, b, c with $a \geq b \geq c \geq 0$, $a \geq a_0$.

LEMMA 6.21. *We can find constants $C_6 = C_6(\overline{L}, \overline{A})$ and $C_7 = C_7(\overline{L}, \overline{A})$ satisfying the following inequality*

$$\hat{h}^0\left(I^0((aL + (b-c)A)|_{bY}), \|\cdot\|_{L^2, \mathrm{quot}}^{a\overline{L}+(b-c)\overline{A}}\right) \leq C_6 b a^{d-1} + (C_4 + C_7) a^{d-1}\log(a)$$

for all integers a, b, c with $a \geq b \geq c \geq 0$, $a \geq 2$.

We will prove these lemmas later. We proceed with the proof of the theorem. By Lemma 6.20 and Lemma 6.21 together with (6.2), if we set $C = C_5 + C_6$ and $D = 2C_2 + C_4 + C_7$, then

$$\begin{aligned} \hat{h}^0\left(H^0(aL + (b-c)A), \|\cdot\|_{L^2}^{a\overline{L}+(b-c)\overline{A}}\right) \\ \leq \hat{h}^0\left(H^0(aL - cA), \|\cdot\|_{L^2}^{a\overline{L}-c\overline{A}}\right) + C b a^{d-1} + D a^{d-1}\log(a) \end{aligned}$$

holds for all integers a, b, c with $a \geq b \geq c \geq 0$, $a \geq a_0$. Thus the assertion of Theorem 6.17 follows.

Let us consider the proofs of Lemma 6.20 and Lemma 6.21. We use the same notation as before.

PROOF OF LEMMA 6.20. Since $\|\cdot\|^{a\overline{L}+(b-c)\overline{A}}_{L^2,s^b,\mathrm{sub}} \leq \|\cdot\|^{a\overline{L}-c\overline{A}}_{L^2}$, by (2) in Proposition 2.17, we obtain

$$\begin{aligned}\hat{h}^0\left(H^0(aL-cA),\|\cdot\|^{a\overline{L}-c\overline{A}}_{L^2}\right)-\hat{h}^0\left(H^0(aL-cA),\|\cdot\|^{a\overline{L}+(b-c)\overline{A}}_{L^2,s^b,\mathrm{sub}}\right)+C_2a^{d-1}\log(a)\\ \geq \widehat{\chi}\left(H^0(aL-cA),\|\cdot\|^{a\overline{L}-c\overline{A}}_{L^2}\right)-\widehat{\chi}\left(H^0(aL-cA),\|\cdot\|^{a\overline{L}+(b-c)\overline{A}}_{L^2,s^b,\mathrm{sub}}\right).\end{aligned}$$

As

$$\begin{aligned}&\widehat{\chi}\left(H^0(aL-cA),\|\cdot\|^{a\overline{L}-c\overline{A}}_{L^2}\right)-\widehat{\chi}\left(H^0(aL-cA),\|\cdot\|^{a\overline{L}+(b-c)\overline{A}}_{L^2,s^b,\mathrm{sub}}\right)\\ &\quad=\widehat{\deg}\left(H^0(aL-cA),\langle\,,\rangle^{a\overline{L}-c\overline{A}}_{L^2}\right)-\widehat{\deg}\left(H^0(aL-cA),\langle\,,\rangle^{a\overline{L}+(b-c)\overline{A}}_{L^2,s^b,\mathrm{sub}}\right)\end{aligned}$$

by Proposition 3.12 and

$$\int_{X(\mathbb{C})}\log|s^b|\Phi=b\int_{X(\mathbb{C})}\log|s|\Phi,$$

Lemma 6.20 follows from Corollary 6.19 and Proposition 3.21. □

PROOF OF LEMMA 6.21. Let k be an integer with $0 \leq k < b$. Let us begin with the following claim:

CLAIM 12. *There is an injective homomorphism*

$$\alpha_k : aL+(b-c-k)A|_Y \to aL+(b-c)A|_{(k+1)Y}$$

together with a commutative diagram

$$\begin{array}{ccccccccc}0 & \longrightarrow & aL+(b-c-k)A & \xrightarrow{s^k} & aL+(b-c)A & \longrightarrow & aL+(b-c)A|_{kY} & \longrightarrow & 0\\ & & \downarrow & & \downarrow & & \| & & \\ 0 & \longrightarrow & aL+(b-c-k)A|_Y & \xrightarrow{\alpha_k} & aL+(b-c)A|_{(k+1)Y} & \longrightarrow & aL+(b-c)A|_{kY} & \longrightarrow & 0\end{array}$$

such that two horizontal sequences are exact. In particular, we obtain an exact sequence

$$\begin{aligned}0\to I^0((aL+(b-c-k)A)|_Y)\to I^0((aL+(b-c)A)|_{(k+1)Y})\\ \to I^0((aL+(b-c)A)|_{kY})\to 0.\end{aligned}$$

PROOF. First we consider a commutative diagram

$$\begin{array}{ccccccccc}0 & \longrightarrow & -(k+1)A & \xrightarrow{s^{k+1}} & \mathscr{O}_X & \longrightarrow & \mathscr{O}_{(k+1)Y} & \longrightarrow & 0\\ & & \downarrow s & & \| & & \downarrow & & \\ 0 & \longrightarrow & -kA & \xrightarrow{s^k} & \mathscr{O}_X & \longrightarrow & \mathscr{O}_{kY} & \longrightarrow & 0\\ & & \downarrow & & & & & & \\ & & -kA|_Y & & & & & & \end{array}$$

which gives rise to an injective homomorphism $\alpha'_k : -kA|_Y \to \mathscr{O}_{(k+1)Y}$ and a commutative diagram

$$\begin{array}{ccccccccc} 0 & \longrightarrow & -kA & \xrightarrow{s^k} & \mathscr{O}_X & \longrightarrow & \mathscr{O}_{kY} & \longrightarrow & 0 \\ & & \downarrow & & \downarrow & & \| & & \\ 0 & \longrightarrow & -kA|_Y & \xrightarrow{\alpha'_k} & \mathscr{O}_{(k+1)Y} & \longrightarrow & \mathscr{O}_{kY} & \longrightarrow & 0. \end{array}$$

Note that two horizontal sequences are exact. Thus, by applying the tensor product with $aL + (b-c)A$, the assertion of the claim follows. $\square$

The surjective homomorphism

$$H^0(aL + (b-c-k)A) \to I^0(Y, aL + (b-c-k)A|_Y)$$

and the norm $\|\cdot\|_{L^2,s^k,\mathrm{sub}}^{a\overline{L}+(b-c)\overline{A}}$ on $H^0(aL + (b-c-k)A)$ induce the quotient norm

$$\|\cdot\|_{L^2,s^k,\mathrm{sub},\mathrm{quot}}^{a\overline{L}+(b-c)\overline{A}}$$

on $I^0(Y, aL + (b-c-k)A|_Y)$. In addition, for each $1 \le i \le b$, let $\|\cdot\|_{L^2,\mathrm{quot},i}^{a\overline{L}+(b-c)\overline{A}}$ be the quotient norm of $I^0((aL+(b-c)A)|_{iY})$ induced by the surjective homomorphism

$$H^0(aL + (b-c)A) \to I^0((aL+(b-c)A)|_{iY})$$

and the L^2-norm $\|\cdot\|_{L^2}^{a\overline{L}+(b-c)\overline{A}}$ of $H^0(aL+(b-c)A)$. We consider the following commutative diagram:

$$\begin{array}{ccc} H^0(aL + (b-c-k)A) & \xrightarrow{s^k} & H^0(aL+(b-c)A) \\ \downarrow & & \downarrow \\ I^0((aL+(b-c-k)A)|_Y) & \xrightarrow{\alpha_k} & I^0((aL+(b-c)A)|_{(k+1)Y}), \end{array}$$

Note that the two vertical homomorphisms have the same kernel. Thus, a sequence

$$\begin{aligned} 0 \to \left(I^0((aL+(b-c-k)A)|_Y), \|\cdot\|_{L^2,s^k,\mathrm{sub},\mathrm{quot}}^{a\overline{L}+(b-c)\overline{A}}\right) \\ \to \left(I^0((aL+(b-c)A)|_{(k+1)Y}), \|\cdot\|_{L^2,\mathrm{quot},k+1}^{a\overline{L}+(b-c)\overline{A}}\right) \\ \to \left(I^0((aL+(b-c)A)|_{kY}), \|\cdot\|_{L^2,\mathrm{quot},k}^{a\overline{L}+(b-c)\overline{A}}\right) \to 0 \end{aligned}$$

is an exact sequence as normed $\mathbb{Z}$-modules by virtue of Lemma 1.2. Therefore, since

$$\operatorname{rk} H^0(aL + (b-c-k)A|_Y) \le \operatorname{rk} H^0(a(L+A)|_Y),$$

by (4) in Proposition 2.17, we obtain

$$\begin{aligned} \hat{h}^0 &\left(I^0((aL+(b-c)A)|_{(k+1)Y}), \|\cdot\|_{L^2,\mathrm{quot},k+1}^{a\overline{L}+(b-c)\overline{A}}\right) \\ &\qquad - \hat{h}^0\left(I^0((aL+(b-c)A)|_{kY}), \|\cdot\|_{L^2,\mathrm{quot},k}^{a\overline{L}+(b-c)\overline{A}}\right) \\ &\le \hat{h}^0\left(I^0((aL+(b-c-k)A)|_Y), \|\cdot\|_{L^2,s^k,\mathrm{sub},\mathrm{quot}}^{a\overline{L}+(b-c)\overline{A}}\right) + C_4 a^{d-2}\log(a), \end{aligned}$$

which implies

$$
\begin{aligned}
&\hat{h}^0\left(I^0((aL+(b-c)A)|_{bY}), \|\cdot\|^{a\overline{L}+(b-c)\overline{A}}_{L^2,\mathrm{quot}}\right)\\
&\qquad\qquad - \hat{h}^0\left(I^0((aL+(b-c)A)|_{Y}), \|\cdot\|^{a\overline{L}+(b-c)\overline{A}}_{L^2,\mathrm{quot}}\right)\\
&=\sum_{k=1}^{b-1}\Big[\hat{h}^0\left(I^0((aL+(b-c)A)|_{(k+1)Y}), \|\cdot\|^{a\overline{L}+(b-c)\overline{A}}_{L^2,\mathrm{quot},k+1}\right)\\
&\qquad\qquad -\hat{h}^0\left(I^0((aL+(b-c)A)|_{kY}), \|\cdot\|^{a\overline{L}+(b-c)\overline{A}}_{L^2,\mathrm{quot},k}\right)\Big]\\
&\leq \sum_{k=1}^{b-1}\hat{h}^0\left(I^0((aL+(b-c-k)A)|_{Y}), \|\cdot\|^{a\overline{L}+(b-c)\overline{A}}_{L^2,s^k,\mathrm{sub},\mathrm{quot}}\right) + (b-1)C_4 a^{d-2}\log(a).
\end{aligned}
$$

Thus Lemma 6.21 follows from Sublemma 6.22 below. □

SUBLEMMA 6.22. *There are positive constants C_6 and C_7 depending only on $\overline{L}$ and $\overline{A}$ such that*

$$
\hat{h}^0\left(I^0(Y, aL+(b-c-k)A|_{Y}), \|\cdot\|^{a\overline{L}+(b-c)\overline{A}}_{L^2,s^k,\mathrm{sub},\mathrm{quot}}\right) \leq C_6 a^{d-1} + C_7 a^{d-2}\log(a)
$$

for all integers a, b, c, k with $a \geq b \geq c \geq 0$, $a \geq 2$ and $0 \leq k < b$.

PROOF. Let Y' be the horizontal component of Y, that is, Y' is the Zariski closure of $Y \cap X_{\mathbb{Q}}$ in X. Note that if $d = 1$, then $Y' = \emptyset$. Let I and I' be the defining ideal sheaves of Y and Y', respectively. As A is ample, there is a positive integer m_0 such that

$$
H^j(X, mA \otimes I'/I) = 0
$$

for all $m \geq m_0$ and $j > 0$. Let $\{x_1, \ldots, x_r\}$ denote $\mathrm{Ass}_{\mathcal{O}_X}(I'/I)$. By Proposition 1.22, there is a positive integer n_0 with the following properties:

(i) We can find a non-zero section l of $H^0(X, n_0 A - L)$ such that $l(x_i) \neq 0$ for all i.

(ii) $2(n_0 - 1) \geq m_0$.

By the above (i), we have an injective homomorphism

$$
(aL + (b-c-k)A) \otimes I'/I \to (n_0 a + b - c - k)A \otimes I'/I.
$$

Since

$$
n_0 a + b - c - k \geq (n_0 - 1)a + a + b - c - k \geq 2(n_0 - 1) \geq m_0,
$$

we have

$$
\begin{aligned}
\log \#H^0(X, &(aL + (b-c-k)A) \otimes I'/I)\\
&\leq \log \#H^0(X, (n_0 a + b - c - k)A \otimes I'/I)\\
&\qquad = \sum_{j \geq 0} (-1)^j \log \#H^j(X, (n_0 a + b - c - k)A \otimes I'/I).
\end{aligned}
$$

Note that $\mathrm{Supp}(I'/I)$ is contained in the fiber of $X \to \mathrm{Spec}(\mathbb{Z})$, so that, by Snapper's theorem (see [**35**, Theorem in Section 1]), there is a polynomial $P(T)$ of degree at most $d-1$ over $\mathbb{R}$ with

$$
\sum_{j \geq 0} (-1)^j \log \#H^j(X, (n_0 a + b - c - k)A \otimes I'/I) = P(n_0 a + b - c - k).
$$

Thus there is a constant C''_6 such that

$$\log \#H^0(X, (aL + (b-c-k)A) \otimes I'/I) \le C''_6 a^{d-1}$$

holds for all integers a, b, c, k with $a \ge b \ge c \ge 0$, $a \ge 2$ and $0 \le k < b$. Since the kernel of the natural surjective homomorphism

$$I^0(Y, aL + (b-c-k)A|_Y) \to I^0(Y', aL + (b-c-k)A|_{Y'})$$

is contained in $H^0(X, (aL + (b-c-k)A) \otimes I'/I)$, then we can see that $I^0(Y', aL + (b-c-k)A|_{Y'})$ has the same norm $\|\cdot\|^{a\overline{L}+(b-c)\overline{A}}_{L^2,s^k,\mathrm{sub},\mathrm{quot}}$ and

$$\begin{aligned}\hat{h}^0\left(I^0(Y, aL + (b-c-k)A|_Y), \|\cdot\|^{a\overline{L}+(b-c)\overline{A}}_{L^2,s^k,\mathrm{sub},\mathrm{quot}}\right)\\ \le \hat{h}^0\left(I^0(Y', aL + (b-c-k)A|_{Y'}), \|\cdot\|^{a\overline{L}+(b-c)\overline{A}}_{L^2,s^k,\mathrm{sub},\mathrm{quot}}\right) + C''_6 a^{d-1}.\end{aligned}$$

Therefore, it is sufficient to find constants C'_6 and C'_7 such that
(6.3)

$$\hat{h}^0\left(I^0(Y', aL + (b-c-k)A|_{Y'}), \|\cdot\|^{a\overline{L}+(b-c)\overline{A}}_{L^2,s^k,\mathrm{sub},\mathrm{quot}}\right) \le C'_6 a^{d-1} + C'_7 a^{d-2}\log(a)$$

for all integers a, b, c, k with $a \ge b \ge c \ge 0$, $a \ge 2$ and $0 \le k < b$.

If $d = 1$, then $I^0(Y', aL + (b-c-k)A|_{Y'}) = 0$, and hence (6.3) is obvious, so we assume that $d \ge 2$. Let U be an open set of $X(\mathbb{C})$ such that the closure of U does not intersect with $Y'(\mathbb{C})$ and U is non-empty on each connected component of $X(\mathbb{C})$. We apply Lemma 5.35 to the cases where $L_{\mathbb{C}}, A_{\mathbb{C}}$ and $L_{\mathbb{C}}, -A_{\mathbb{C}}$, so that we can find constants $D_1 \ge 1$ and $D'_1 \ge 1$ such that

$$D'_1 D_1^{l+|m|} \int_U |u|^2\Phi \ge \int_{X(\mathbb{C})} |u|^2\Phi$$

for all integers l, m with $l \ge 0$ and $u \in H^0(X(\mathbb{C}), lL + mA)$. As

$$(\text{the closure of } U) \cap Y'(\mathbb{C}) = \emptyset,$$

we can give D_2 by

$$D_2 := \frac{1}{\inf_{x\in U}\{|s|(x)\}}.$$

Note that $D_2 \ge 1$ because $\inf_{x\in U}\{|s|(x)\} \le 1$. Therefore, if we put $D_3 = \max\{D_2, D_1\}$, then

$$\begin{aligned}\int_{X(\mathbb{C})} |s^k \otimes u|^2\Phi \ge \int_U |s^k \otimes u|^2\Phi \ge \frac{1}{D_2^{2k}}\int_U |u|^2\Phi \ge \frac{1}{D_2^{2k} D'_1 D_1^{a+|b-c-k|}}\int_{X(\mathbb{C})} |u|^2\Phi\\ \ge \frac{1}{D'_1 D_3^{4a}}\int_{X(\mathbb{C})} |u|^2\Phi\end{aligned}$$

for $u \in H^0(X, aL + (b-c-k)A)$, and hence

$$\|\cdot\|^{a\overline{L}+(b-c)\overline{A}}_{L^2,s^k,\mathrm{sub}} \ge \frac{1}{\sqrt{D'_1 D_3^{2a}}}\|\cdot\|^{a\overline{L}+(b-c-k)\overline{A}}_{L^2}.$$

Let us consider a surjective homomorphism

$$H^0(X, aL + (b-c-k)A) \to I^0(Y', aL + (b-c-k)A|_{Y'}).$$

Let $\|\cdot\|_{L^2,\mathrm{quot}}^{a\overline{L}+(b-c-k)\overline{A}}$ be the quotient norm of $I^0(Y', aL+(b-c-k)A|_{Y'})$ induced by the above homomorphism. Then the above inequality of the norms implies

$$\|\cdot\|_{L^2,s^k,\mathrm{sub},\mathrm{quot}}^{a\overline{L}+(b-c)\overline{A}} \geq \frac{1}{\sqrt{D_1' D_3^{2a}}}\|\cdot\|_{L^2,\mathrm{quot}}^{a\overline{L}+(b-c-k)\overline{A}}.$$

Let us fix a volume form of Y' given by the C^∞-hermitian invertible sheaf $\overline{L}+\overline{A}|_{Y'}$. We apply Corollary 5.34 to the cases where $L_{\mathbb{C}}, A_{\mathbb{C}}$ and $L_{\mathbb{C}}, -A_{\mathbb{C}}$, so that there are constants $D_4, D_4' \geq 1$ such that

$$\begin{aligned}\|\cdot\|_{L^2,\mathrm{quot}}^{a\overline{L}+(b-c-k)\overline{A}} &\geq \frac{1}{\sqrt{D_4' D_4^{a+|b-c-k|}}}\|\cdot\|_{L^2}^{a\overline{L}+(b-c-k)\overline{A}|_{Y'}} \\ &\geq \frac{1}{\sqrt{D_4' D_4^{a}}}\|\cdot\|_{L^2}^{a\overline{L}+(b-c-k)\overline{A}|_{Y'}}\end{aligned}$$

on $I^0(Y', aL+(b-c-k)A|_{Y'})$. Here we use an inequality $e^x \geq x+1$ $(x \geq 0)$. Let $D_5 = \max\{D_3, D_4\}$ and $D_5' = \max\{D_1', D_4'\}$. Then we have

$$\|\cdot\|_{L^2,s^k,\mathrm{sub},\mathrm{quot}}^{a\overline{L}+(b-c)\overline{A}} \geq \frac{1}{D_5' D_5^{3a}}\|\cdot\|_{L^2}^{a\overline{L}+(b-c-k)\overline{A}|_{Y'}}$$

on $I^0(Y', aL+(b-c-k)A|_{Y'})$. Since

$$\operatorname{rk} H^0(aL+(b-c-k)A|_{Y'}) \leq \operatorname{rk} H^0(a(L+A)|_{Y'}),$$

by (3) in Proposition 2.17, we obtain

$$\begin{aligned}&\hat{h}^0\left(I^0(Y', aL+(b-c-k)A_{Y'}), \|\cdot\|_{L^2,s^k,\mathrm{sub},\mathrm{quot}}^{a\overline{L}+(b-c)\overline{A}}\right) \\ &\quad\leq \hat{h}^0\left(I^0(Y', aL+(b-c-k)A|_{Y'}), \|\cdot\|_{L^2}^{a\overline{L}+(b-c-k)\overline{A}|_{Y'}}\right) \\ &\qquad\qquad + \log(D_5' D_5^{3a})C_3 a^{d-2} + C_4 a^{d-2}\log(a) \\ &\quad\leq \hat{h}^0\left(H^0(Y', aL-(b-c-k)A|_{Y'}), \|\cdot\|_{L^2}^{a\overline{L}+(b-c-k)\overline{A}|_{Y'}}\right) \\ &\qquad\qquad + \log(D_5' D_5^{3a})C_3 a^{d-2} + C_4 a^{d-2}\log(a).\end{aligned}$$

Let $\widetilde{Y'}$ be the normalization of Y' and t the section as in Claim 11. Note that t yields an inequality $\overline{L}|_{\widetilde{Y'}} \leq n\overline{A}|_{\widetilde{Y'}}$, and hence, as

$$a\overline{L}+(b-c-k)\overline{A}|_{\widetilde{Y'}} \leq (an+b-c-k)\overline{A}|_{\widetilde{Y'}},$$

we have

$$\begin{aligned}&\hat{h}^0\left(H^0(Y', aL+(b-c-k)A|_{Y'}), \|\cdot\|_{L^2}^{a\overline{L}+(b-c-k)\overline{A}|_{Y'}}\right) \\ &\quad\leq \hat{h}^0\left(H^0(\widetilde{Y'}, aL+(b-c-k)A|_{\widetilde{Y'}}), \|\cdot\|_{L^2}^{a\overline{L}+(b-c-k)\overline{A}|_{\widetilde{Y'}}}\right) \\ &\qquad\leq \hat{h}^0\left(H^0(\widetilde{Y'}, (an+b-c-k)A|_{\widetilde{Y'}}), \|\cdot\|_{L^2}^{(an+b-c-k)\overline{A}|_{\widetilde{Y'}}}\right).\end{aligned}$$

Moreover, by Corollary 5.42, there is a constant D_6 such that

$$\hat{h}^0\left(H^0(\widetilde{Y'}, mA|_{\widetilde{Y'}}), \|\cdot\|_{L^2}^{m\overline{A}|_{\widetilde{Y'}}}\right) \leq D_6 m^{d-1}$$

for all $m \geq 1$. Thus (6.3) follows. □

CHAPTER 7

Nakai-Moishezon Criterion on an Arithmetic Variety

An invertible sheaf L on a projective scheme V over a field is ample if and only if $(L^{\dim W} \cdot W) > 0$ holds for any reduced and irreducible subscheme W of V with $\dim W > 0$. This is what is called Nakai-Moishezon's criterion. The purpose of this section is to establish the following analogue on arithmetic variety.

THEOREM 7.1. *Let X be a generically smooth projective arithmetic variety and let $\overline{L}$ be a C^∞-hermitian invertible sheaf of real type on X. If $c_1(\overline{L})$ is semipositive and $\widehat{\deg}\left(\widehat{c}_1(\overline{L})^{\cdot \dim Y} \cdot (Y, 0)\right) > 0$ for any integral subscheme Y of X, then L is ample and $H^0(X, nL)$ has a free basis consisting of strictly small sections for a sufficiently large n.*

For the proof of the above theorem, we use many methods due to Zhang [**76**]. Combining this theorem with the generalized Hodge index theorem in the last chapter, we can obtain the Hilbert-Samuel formula (Theorem 7.10) for a nef C^∞-hermitian invertible sheaf. Let us begin with several preparations

7.1. Endmorphism $\mathcal{N}$ and its basic properties

Let X be an n-dimensional complex manifold and $ds^2 = \sum g_{\alpha\beta} dz^\alpha \otimes d\bar{z}^\beta$ an hermitian metric of X. Moreover, let (E, h) be a C^∞-hermitian locally free coherent sheaf on X. We denote $\det \Omega_X^1$ by ω_X. Since $A^{n,q}(E \otimes \omega_X^{-1})$ is naturally identified with $A^{0,q}((E \otimes \omega_X^{-1}) \otimes \omega_X)$ and $\omega_X^{-1} \otimes \omega_X \simeq \mathcal{O}_X$, there is a natural isomorphism

$$\iota : A^{0,q}(E) \to A^{n,q}(E \otimes \omega_X^{-1}).$$

Concretely it can be described as follows: if an element η of $A^{0,q}(E)$ is locally written by

$$\eta = \sum_{i_1 < \ldots < i_q} \eta_{i_1, \ldots, i_q} d\bar{z}_{i_1} \wedge \cdots \wedge d\bar{z}_{i_q} \quad (\eta_{i_1, \ldots, i_q} \in E),$$

then

$$\iota(\eta) = \sum_{i_1 < \ldots < i_q} \left(\eta_{i_1, \ldots, i_q} \otimes (dz_1 \wedge \cdots \wedge dz_n)^{-1}\right) dz_1 \wedge \cdots \wedge dz_n \wedge d\bar{z}_{i_1} \wedge \cdots \wedge d\bar{z}_{i_q}.$$

First, we have the following proposition.

PROPOSITION 7.2. *The isomorphism $\iota : A^{0,q}(E) \to A^{n,q}(E \otimes \omega_X^{-1})$ has the following properties.*

(1) *The isomorphism ι is isometric.*
(2) $\bar{\partial} \circ \iota = \iota \circ \bar{\partial}$.
(3) *If X is compact, then $\bar{\partial}^* \circ \iota = \iota \circ \bar{\partial}^*$.*
(4) *If X is compact, then $\Delta_{\bar{\partial}} \circ \iota = \iota \circ \Delta_{\bar{\partial}}$.*

PROOF. (1) This is a local question. Let $\theta_1, \ldots, \theta_n$ be a C^∞ local orthonormal basis of $(1,0)$-from. For $\alpha, \beta \in A^{0,q}(E)$, we write

$$\begin{cases} \alpha = \sum_{i_1 < \cdots < i_q} \alpha_{i_1,\ldots,i_q} \bar{\theta}_{i_1} \wedge \cdots \wedge \bar{\theta}_{i_q}, \\ \beta = \sum_{i_1 < \cdots < i_q} \beta_{i_1,\ldots,i_q} \bar{\theta}_{i_1} \wedge \cdots \wedge \bar{\theta}_{i_q}, \end{cases}$$

then

$$\langle \alpha, \beta \rangle (x) = \sum_{i_1 < \cdots < i_q} h(\alpha_{i_1,\ldots,i_q}, \beta_{i_1,\ldots,i_q})(x).$$

On the other hand, as

$$\begin{cases} \iota(\alpha) = \sum_{i_1 < \cdots < i_q} \left(\alpha_{i_1,\ldots,i_q} \otimes (\theta_1 \wedge \cdots \theta_n)^{-1}\right) \theta_1 \wedge \cdots \wedge \theta_n \wedge \bar{\theta}_{i_1} \wedge \cdots \wedge \bar{\theta}_{i_q}, \\ \iota(\beta) = \sum_{i_1 < \cdots < i_q} \left(\beta_{i_1,\ldots,i_q} \otimes (\theta_1 \wedge \cdots \theta_n)^{-1}\right) \theta_1 \wedge \cdots \wedge \theta_n \wedge \bar{\theta}_{i_1} \wedge \cdots \wedge \bar{\theta}_{i_q}, \end{cases}$$

we have

$$\begin{aligned} \langle \iota(\alpha), \iota(\beta) \rangle (x) &= \sum_{i_1 < \cdots < i_q} \frac{h(\alpha_{i_1,\ldots,i_q}, \beta_{i_1,\ldots,i_q})(x)}{\langle \theta_1 \wedge \cdots \wedge \theta_n, \theta_1 \wedge \cdots \wedge \theta_n \rangle (x)} \\ &= \sum_{i_1 < \cdots < i_q} h(\alpha_{i_1,\ldots,i_q}, \beta_{i_1,\ldots,i_q})(x), \end{aligned}$$

as claimed.

(2) Locally we set $\alpha = \sum_{i_1 < \cdots < i_q} \alpha_{i_1,\ldots,i_q} d\bar{z}_{i_1} \wedge \cdots \wedge d\bar{z}_{i_q}$. Let $\omega_1, \ldots, \omega_r$ be a local holomorphic basis of E. Then there are local C^∞-functions $\alpha^1_{i_1,\ldots,i_q}, \ldots, \alpha^r_{i_1,\ldots,i_q}$ such that $\alpha_{i_1,\ldots,i_q} = \sum_{l=1}^r \alpha^l_{i_1,\ldots,i_q} \omega_l$, that is,

$$\alpha = \sum_{l=1}^r \omega_l \sum_{i_1 < \cdots < i_q} \alpha^l_{i_1,\ldots,i_q} d\bar{z}_{i_1} \wedge \cdots \wedge d\bar{z}_{i_q}.$$

Therefore, if we write $d\boldsymbol{z} = dz_1 \wedge \cdots \wedge dz_n$, then

$$\begin{aligned} \iota(\bar{\partial}(\alpha)) &= \iota\left(\sum_{l=1}^r \omega_l \sum_{i_1 < \cdots < i_q} \bar{\partial}(\alpha^l_{i_1,\ldots,i_q} d\bar{z}_{i_1} \wedge \cdots \wedge d\bar{z}_{i_q})\right) \\ &= \sum_{l=1}^r \omega_l \otimes d\boldsymbol{z}^{-1} \sum_{i_1 < \cdots < i_q} d\boldsymbol{z} \wedge \bar{\partial}(\alpha^l_{i_1,\ldots,i_q} d\bar{z}_{i_1} \wedge \cdots \wedge d\bar{z}_{i_q}) \\ &= \bar{\partial}\left(\sum_{l=1}^r \omega_l \otimes d\boldsymbol{z}^{-1} \sum_{i_1 < \cdots < i_q} d\boldsymbol{z} \wedge \alpha^l_{i_1,\ldots,i_q} d\bar{z}_{i_1} \wedge \cdots \wedge d\bar{z}_{i_q}\right) \\ &= \bar{\partial}(\iota(\alpha)). \end{aligned}$$

(3) Let $\alpha \in A^{0,q}(E)$ and $\beta \in A^{0,q+1}(E)$. By using (1) and (2),

$$\begin{aligned} (\iota(\alpha), \bar{\partial}^*(\iota(\beta))) &= (\bar{\partial}(\iota(\alpha)), \iota(\beta)) = (\iota(\bar{\partial}(\alpha)), \iota(\beta)) \\ &= (\bar{\partial}(\alpha), \beta) = (\alpha, \bar{\partial}^*(\beta)) = (\iota(\alpha), \iota(\bar{\partial}^*(\beta))), \end{aligned}$$

as required.

(4) is a consequence of (2) and (3). □

Let Ω be a $(1,1)$-form arising from the hermitian metric $ds^2 = \sum g_{\alpha\beta} dz_\alpha \otimes d\bar{z}_\beta$ on X, that is,

$$\Omega = \frac{\sqrt{-1}}{2} \sum g_{\alpha\beta} dz_\alpha \wedge d\bar{z}_\beta.$$

Let $L : A^{p,q}(E) \to A^{p+1,q+1}(E)$ be the operator given by $\eta \mapsto \Omega \wedge \eta$ and let Λ be the adjoint operator of L.

Let η be a real $(1,1)$-form on X, and let $\eta \wedge \Lambda : A^{n,1}(\omega_X^{-1}) \to A^{n,1}(\omega_X^{-1})$ be a homomorphism given by $(\eta \wedge \Lambda)(\xi) := \eta \wedge \Lambda(\xi)$. An operator $N(\eta) : A^{0,1} \to A^{0,1}$ is defined by a homomorphism which makes the following diagram commutative:

$$\begin{array}{ccc} A^{0,1}(\mathscr{O}_X) & \xrightarrow{N(\eta)} & A^{0,1}(\mathscr{O}_X) \\ {\scriptstyle \iota}\downarrow & & {\scriptstyle \iota}\downarrow \\ A^{n,1}(\omega_X^{-1}) & \xrightarrow{\eta\wedge\Lambda} & A^{n,1}(\omega_X^{-1}). \end{array}$$

For a C^∞-hermitian invertible sheaf $\overline{L}$ on X, $N(\overline{L})$ is defined by $N(2\pi c_1(\overline{L}))$, where $c_1(\overline{L})$ is the first Chern class of $\overline{L}$, that is, $2\pi c_1(\overline{L})$ is equal to

$$\sqrt{-1} \times (\text{the curvature form of } \overline{L}).$$

Moreover, for a C^∞-function ψ on X, we define $N(\psi)$ by

$$N(\psi) = N(\mathscr{O}_X, |\cdot| \exp(-\psi)) = N\left(2\sqrt{-1}\partial\bar{\partial}(\psi)\right).$$

Let $\theta_1, \ldots, \theta_n$ be a C^∞ local orthonormal basis of $(1,0)$-forms and

$$\eta = \frac{\sqrt{-1}}{2} \sum_{\alpha,\beta} h_{\alpha\beta} \theta_\alpha \wedge \bar{\theta}_\beta.$$

Then, by an easy calculation, we can check

$$N(\eta)(\bar{\theta}_\alpha) = \sum_\beta h_{\alpha\beta} \bar{\theta}_\beta. \tag{7.1}$$

Here we introduce several notations. Let (V, h) be a pair of finite-dimensional complex vector space and a hermitian inner product of V. A homomorphism $\phi : V \to V$ is called a hermitian transform if $h(\phi(x), y) = h(x, \phi(y))$ holds for all $x, y \in V$. We say that a *hermitian transform* ϕ is *positive semidefinite* if $h(\phi(x), x) \geq 0$ for all $x \in V$. For two hermitian transforms ϕ, ψ, $\phi \geq \psi$ means that $\phi - \psi$ is positive semidefinite.

As $(h_{\alpha\beta})$ in (7.1) is a hermitian matrix, $N(\eta)$ is a hermitian transform at each point. Moreover, if $\eta - \lambda$ is semipositive for real $(1,1)$-forms η and λ, then we denote it by $\eta \geq \lambda$. By (7.1), we can see that $\eta \geq \lambda$ implies $N(\eta) \geq N(\lambda)$.

PROPOSITION 7.3. *We assume that X is compact and Kähler. If there is a positive constant c such that*

$$N(\overline{L} \otimes \overline{\omega}_X^{-1}) \geq c\,\mathrm{id}$$

at all points of X, then

$$(\Delta_{\bar{\partial}}(\alpha), \alpha) \geq c(\alpha, \alpha)$$

for all $\alpha \in A^{0,q}(L)$, where a metric of ω_X is given by the Kähler metric of X.

PROOF. Since $N(\overline{L} \otimes \overline{\omega}_X^{-1}) \geq c\,\mathrm{id}$ at each point, for $\alpha \in A^{0,q}(L)$,

$$\langle N(\overline{L} \otimes \overline{\omega}_X^{-1})(\alpha), \alpha\rangle(x) \geq c\langle \alpha, \alpha\rangle(x)$$

at each point x of X. Therefore, integrating the above inequality, we have

$$(N(\overline{L} \otimes \overline{\omega}_X^{-1})(\alpha), \alpha) \geq c(\alpha, \alpha).$$

CLAIM 1 (Nakano's identity). *Let $\Theta = \Theta(\overline{L} \otimes \overline{\omega}_X^{-1})$ be the curvature form of $\overline{L} \otimes \overline{\omega}_X^{-1}$ and let $e(\Theta) : A^{p,q}(L \otimes \omega_X^{-1}) \to A^{p+1,q+1}(L \otimes \omega_X^{-1})$ be an operator given by $\eta \mapsto \Theta \wedge \eta$. Then*

$$\Delta_{\bar\partial} = \sqrt{-1}(e(\Theta)\Lambda - \Lambda e(\Theta)) + (D'D'^* + D'^*D')$$

on $A^{p,q}(L \otimes \omega_X^{-1})$, where D' is the $(1,0)$-part of the connection of $\overline{L} \otimes \overline{\omega}_X^{-1}$ and D'^* is the adjoint operator of D'.

PROOF. Let $\bar\partial^*$ be the adjoint operator of $\bar\partial$. Then, the following *Hodge identities*

$$\Lambda\bar\partial - \bar\partial\Lambda = -\sqrt{-1}D'^*, \quad \Lambda D' - D'\Lambda = -\sqrt{-1}\bar\partial^*$$

are well-known (for example, see [**37**]). On the other hand, since $e(\Theta) = \bar\partial D' + D'\bar\partial$,

$$\begin{aligned} e(\Theta)\Lambda - \Lambda e(\Theta) &= (\bar\partial D' + D'\bar\partial)\Lambda - \Lambda(\bar\partial D' + D'\bar\partial) \\ &= \bar\partial(D'\Lambda) + D'(\bar\partial\Lambda) - (\Lambda\bar\partial)D' - (\Lambda D')\bar\partial \\ &= \bar\partial(\Lambda D' - \sqrt{-1}\partial^*) + D'(\Lambda\bar\partial + \sqrt{-1}D'^*) \\ &\qquad - (\bar\partial\Lambda - \sqrt{-1}D'^*)D' - (D'\Lambda + \sqrt{-1}\bar\partial^*)\bar\partial \\ &= \sqrt{-1}(D'D'^* + D'^*D) - \sqrt{-1}(\bar\partial\bar\partial^* + \bar\partial^*\bar\partial), \end{aligned}$$

as claimed. □

As $2\pi c_1(\overline{L} \otimes \overline{\omega}_X^{-1}) = \sqrt{-1}\Theta$, using Nakano's identity and Proposition 7.2 for $\alpha \in A^{0,q}(L)$,

$$\begin{aligned} (\Delta_{\bar\partial}(\alpha), \alpha) &= (\iota(\Delta_{\bar\partial}(\alpha)), \iota(\alpha)) = (\Delta_{\bar\partial}(\iota(\alpha)), \iota(\alpha)) \\ &= \big((\sqrt{-1}e(\Theta)\Lambda)(\iota(\alpha)), \iota(\alpha)\big) + \|D'(\iota(\alpha))\|^2 + \|D'^*(\iota(\alpha))\|^2 \\ &\geq (\iota(N(\overline{L} \otimes \overline{\omega}_X^{-1})(\alpha)), \iota(\alpha)) = (N(\overline{L} \otimes \overline{\omega}_X^{-1})(\alpha), \alpha) \geq c(\alpha, \alpha). \end{aligned}$$ □

7.2. Bounded extension of holomorphic sections

Let us begin with the following fundamental lemma.

LEMMA 7.4. *Let X be a compact Kähler manifold and let $\overline{L}$ be a C^∞-hermitian invertible sheaf on X. We assume that there is a positive constant c such that*

$$N(\overline{L} \otimes \overline{\omega}_X^{-1}) \geq c\,\mathrm{id}$$

for all $x \in X$. Then, for any $w \in A^{0,1}(L)$ with $\bar\partial(w) = 0$, there is $u \in A^{0,0}(L)$ such that $\bar\partial(u) = w$ and $\|u\|^2 \leq c^{-1}\|w\|^2$.

PROOF. We set $\mathbf{H}_{\bar\partial}^{p,q}(L) = \{v \in A^{p,q}(L) \mid \Delta_{\bar\partial}(v) = 0\}$. Similarly as Theorem 1.33, $\mathbf{H}_{\bar\partial}^{p,q}(L)$ is a finite-dimensional complex vector space and there is the orthonormal projection $H_{\bar\partial} : A^{p,q}(L) \to \mathbf{H}_{\bar\partial}^{p,q}(L)$. Moreover, there is a Green operator $G_{\bar\partial} : A^{p,q}(L) \to A^{p,q}(L)$ uniquely such that

$$G_{\bar\partial}(\mathbf{H}_{\bar\partial}^{p,q}(L)) = \{0\}, \quad [\bar\partial, G_{\bar\partial}] = [\bar\partial^*, G_{\bar\partial}] = 0 \quad \text{and} \quad H_{\bar\partial} + \Delta_{\bar\partial}G_{\bar\partial} = \mathrm{id}_{A^{p,q}(L)}$$

(for example, see [**27**, Chapter 1, Section 2]).

By Proposition 7.3, $(\Delta_{\bar{\partial}}(v), v) \geq c(v, v)$ holds for $v \in A^{0,1}(L)$. In particular, $\mathbf{H}^{0,1}_{\bar{\partial}}(L) = \{0\}$. Thus, we have $G_{\bar{\partial}}\Delta_{\bar{\partial}} = \Delta_{\bar{\partial}} G_{\bar{\partial}} = \mathrm{id}_{A^{0,1}(L)}$.

Here let us see that

$$(G_{\bar{\partial}}(v), v) \leq c^{-1}\|v\|^2 \tag{7.2}$$

for $v \in A^{0,1}(L)$. There is a complete orthonormal system consisting of C^∞-eigenvectors of $\Delta_{\bar{\partial}}$ on the space of L^2-forms (for example, see [**27**, Chapter 0, Section 6]). Therefore, it is sufficient to see it in the case where v is a non-zero C^∞ eigenvector of $\Delta_{\bar{\partial}}$. Let λ be the eigenvalue of v. Then $\Delta_{\bar{\partial}}(v) = \lambda v$ and $\lambda G_{\bar{\partial}}(v) = v$. Thus $c(v, v) \leq (\Delta_{\bar{\partial}}(v), v) = \lambda(v, v)$, that is, $c \leq \lambda$. Therefore, $(G_{\bar{\partial}}(v), v) = \lambda^{-1}(v, v) \leq c^{-1}(v, v)$.

If we set $u = \bar{\partial}^* G_{\bar{\partial}} w$, then

$$w = \Delta_{\bar{\partial}} G_{\bar{\partial}} w = \bar{\partial}(\bar{\partial}^* G_{\bar{\partial}} w) = \bar{\partial}(u).$$

Moreover, by (7.2),

$$\|u\|^2 = (\bar{\partial}^* G_{\bar{\partial}} w, \bar{\partial}^* G_{\bar{\partial}} w) = (\Delta_{\bar{\partial}} G_{\bar{\partial}} w, G_{\bar{\partial}} w) = (w, G_{\bar{\partial}} w) \leq c^{-1}\|w\|^2. \qquad \square$$

LEMMA 7.5. *Let X be a compact Kähler manifold and let $\overline{L}$ be a C^∞-hermitian invertible sheaf on X. We assume that there is a sequence $\{\psi\}_{n=1}^{\infty}$ of C^∞ real-valued functions with the following properties:*

(1) $\psi_1 \geq \psi_2 \geq \cdots \geq \psi_i \geq \cdots$.

(2) *There is a positive constant c such that, for every i,*

$$(1/2)N(\psi_i) + N(\overline{L} \otimes \overline{\omega}_X^{-1}) \geq c\,\mathrm{id}$$

holds at any points of X.

For $\eta \in A^{0,1}(L)$, we define $\|\eta\|_\infty$ by

$$\|\eta\|_\infty = \lim_{i\to\infty} \|\eta \exp(-(1/2)\psi_i)\|,$$

where $\|\eta\|_\infty$ might be ∞. Then, for any $w \in A^{0,1}(L)$ with $\bar{\partial}(w) = 0$ and $\|w\|_\infty < \infty$, there is $u \in A^{0,0}(L)$ with $\bar{\partial}(u) = w$ and $\|u\|_\infty^2 \leq c^{-1}\|w\|_\infty^2$.

PROOF. We set $\overline{L}_i = \overline{L} \otimes (\mathscr{O}_X, |\cdot|\exp(-(1/2)\psi_i))$. A norm of $A^{0,1}(L)$ induced by the metric of $\overline{L}_i$ is denoted by $\|\cdot|_i$, that is, $\|\eta\|_i = \|\eta\exp(-(1/2)\psi_i)\|$. Then $\|\cdot\|_1 \leq \|\cdot\|_2 \leq \cdots \leq \|\cdot\|_i \leq \cdots \leq \|\cdot\|_\infty$ and $\|\eta\|_\infty = \lim_{i\to\infty}\|\eta\|_i$. Note that

$$N(\overline{L}_i \otimes \overline{\omega}_X^{-1}) = (1/2)N(\psi_i) + N(\overline{L} \otimes \overline{\omega}_X^{-1}).$$

Thus, by Lemma 7.4, there is $u_i \in A^{0,0}(L)$ with $\bar{\partial}(u_i) = w$ and $\|u_i\|_i^2 \leq c^{-1}\|w\|_i^2$. If we write $v_i = u_i - u_1$, then we have $v_i \in H^0(X, L)$ because $\bar{\partial}(v_i) = 0$. Let us see that

$$\|v_i\|_1 \leq \|u_1\|_1 + c^{-1/2}\|w\|_\infty.$$

Indeed,

$$c^{-1/2}\|w\|_\infty \geq c^{-1/2}\|w\|_i \geq \|u_i\|_i \geq \|u_i\|_1 = \|v_i + u_1\|_1 \geq \|v_i\|_1 - \|u_1\|_1.$$

Hence, $\{v_i \mid i = 1, 2, \ldots\}$ is a bounded set in $H^0(X, L)$. Therefore, as $H^0(X, L)$ is finite-dimensional, a subsequence $\{v_{i_k}\}$ of $\{v_i\}$ has the limit v. Here we set $u = u_1 + v$. Then $\bar{\partial}(u) = w$. Finally, we need to show an inequality:

$$\|u\|_\infty^2 \leq c^{-1}\|w\|_\infty^2.$$

For $i_k \geq i$, as
$$\|u_{i_k}\|_i^2 \leq \|u_{i_k}\|_{i_k}^2 \leq c^{-1}\|w\|_{i_k}^2 \leq c^{-1}\|w\|_\infty^2,$$
taking $k \to \infty$, we obtain $\|u\|_i^2 \leq c^{-1}\|w\|_\infty^2$. Therefore, taking $i \to \infty$, we have the inequality as desired. □

THEOREM 7.6. *Let X be a d-dimensional non-singular complex projective variety and let $\overline{L}$ be a positive C^∞-hermitian invertible sheaf on X. Let Y a reduced subscheme of X and let U be an open set of X with $Y \subseteq U$. Then there are positive constants n_0 and C such that, for any integer $n \geq n_0$ and any $l_U \in H^0(U, L^{\otimes n})$ with $\|l_U\|_{U,\mathrm{sup}} < \infty$, there is $l \in H^0(X, L^{\otimes n})$ such that $l|_Y = l_U|_Y$ and $\|l\|_{\mathrm{sup}} \leq Cn^{2d}\|l_U\|_{U,\mathrm{sup}}$, where $\|l_U\|_{U,\mathrm{sup}}$ is the supremum norm on U.*

PROOF. Let Φ be the volume form induced by the Kähler form on X. Let I be the defining ideal of Y, and let m be a positive integer m such that $I \otimes L^{\otimes m}$ is generated by global sections. Let $s_1, \ldots, s_N$ be a basis of $H^0(X, I \otimes L^{\otimes m})$ and
$$\rho = |s_1|^2 + \cdots + |s_N|^2.$$
Then we have the following.

CLAIM 2. (1) *$\rho(x) \geq 0$ for all $x \in X$ and*
$$\rho(x) = 0 \quad \Longleftrightarrow \quad x \in Y.$$
(2) *$\sqrt{-1}\partial\bar{\partial}(\log \rho) \geq -2\pi m c_1(\overline{L})$ on $X \setminus Y$.*
(3) *ρ^{-d} is not integrable on any neighborhood of Y.*

PROOF. (1) is obvious. Let us consider (2). If $x \in X \setminus Y$, then there is i with $s_i(x) \neq 0$. Thus, in a neighborhood of x, we can set $s_j = f_j s_i$ for some holomorphic function f_j in a neighborhood of x. Then
$$\rho = \frac{|s_1|^2 + \cdots + |s_N|^2}{|s_i|^2}|s_i|^2 = \left(|f_1|^2 + \cdots + |f_N|^2\right)|s_i|^2.$$
Thus
$$\sqrt{-1}\partial\bar{\partial}(\log \rho) = \sqrt{-1}\partial\bar{\partial}\log\left(|f_1|^2 + \cdots + |f_N|^2\right) - 2\pi m c_1(\overline{L}).$$
Here let us consider a map $f : U \to \mathbb{C}^{N-1}$ given by
$$f = (f_1, \ldots, f_{i-1}, f_{i+1}, \ldots, f_N).$$
Note that
$$\sqrt{-1}\partial\bar{\partial}\log\left(|f_1|^2 + \cdots + |f_N|^2\right) = f^*\left(\sqrt{-1}\partial\bar{\partial}\log\left(1 + |z_1|^2 + \cdots + |z_{N-1}|^2\right)\right)$$
and $\sqrt{-1}\partial\bar{\partial}\log\left(1 + |z_1|^2 + \cdots + |z_{N-1}|^2\right)$ is positive on $\mathbb{C}^{N-1}$. Thus we get (2).

Finally, we consider (3). Let $x \in Y$ be a non-singular point of Y. Then there is a local coordinate $(z_1, \ldots, z_d)$ of X around x such that $I_x = z_1\mathscr{O}_{X,x} + \cdots + z_{d'}\mathscr{O}_{X,x}$ $(d' \leq d)$. Let ω be a local basis of $L^{\otimes m}$ around x and $s_i = h_i\omega$ $(i = 1, \ldots, d)$. Then, as $h_i \in I_x$, we can write $h_i = \sum_{j=1}^{d'} g_{ij}z_j$. We take an open neighborhood W of x such that x_i, h_i, g_{ij} are defined on W. Let V be an open neighborhood of x with $\overline{V} \subseteq W$. If we set
$$M = \max\left\{\sup_{x\in V}|\omega|^2(x), \max_{ij}\{\sup_{x\in V}|g_{ij}(x)|\}\right\},$$

then

$$\rho = |\omega|^2 \sum_{i=1}^{N} |h_i|^2 \leq |\omega|^2 \sum_{i=1}^{N} \left(d' \sum_{j=1}^{d'} |g_{ij}|^2 |z_j|^2 \right)$$

$$\leq M^3 d' \sum_{i=1}^{N} \sum_{j=1}^{d'} |z_j|^2 = M^3 d' N \sum_{j=1}^{d'} |z_j|^2 \leq M^3 d' N \sum_{j=1}^{d} |z_j|^2$$

on V. Moreover, there is a positive constant C such that

$$\Phi \geq C(\sqrt{-1})^d dz_1 \wedge d\bar{z}_1 \wedge \cdots \wedge dz_d \wedge d\bar{z}_d$$

on V. Therefore,

$$\rho^{-d}\Phi \geq \frac{C}{(M^3 d' N)^d} \frac{(\sqrt{-1})^d dz_1 \wedge d\bar{z}_1 \wedge \cdots \wedge dz_d \wedge d\bar{z}_d}{(|z_1|^2 + \cdots + |z_d|^2)^d}.$$

It is easy to see that

$$\int_{\{z \in \mathbb{C} \mid |z| \leq r\}^d} \frac{(\sqrt{-1})^d dz_1 \wedge d\bar{z}_1 \wedge \cdots \wedge dz_d \wedge d\bar{z}_d}{(|z_1|^2 + \cdots + |z_d|^2)^d} = \infty$$

for any positive number r. Thus (3) follows. □

Let us choose a C^∞ function ϕ $\mathbb{R} \to \mathbb{R}$ such that $\phi(x) = x$ $(x \geq 0)$, $\phi(x) = -1/2$ for $(x \leq -1)$ and $\phi''(x) \geq 0$ $(x \in \mathbb{R})$. We set

$$\psi_i = \phi(\log(\rho^d) + i) - i \quad (i = 1, 2, 3, \ldots).$$

Clearly, ψ_i is C^∞-function. We have the following.

CLAIM 3. (1) *we have* $\psi_1 \geq \psi_2 \geq \cdots \geq \psi_i \geq \cdots$ *on* X. *Moreover,* $\psi_i \geq \log(\rho^d)$ *on* $X \setminus Y$.

(2) *For a non-negative continuous function* g *on* X, *the following equality holds:*

$$\lim_{i \to \infty} \int_X g \exp(-\psi_i)\Phi = \int_X g\rho^{-d}\Phi.$$

Note that each side might be ∞.

(3) $N(\psi_i) \geq -4\pi dm N(\overline{L})$ *on* X.

PROOF. (1) Since $0 \leq \phi'(t) \leq 1$ using the mean value theorem, for $i \leq j$,

$$\psi_j - \psi_i = \phi'(c)(j-i) - (j-i) = (j-i)(\phi'(c) - 1) \leq 0.$$

If we set $U_i = \{x \in X \mid \log(\rho(x)^d) + i > 0\}$, then $U_1 \subseteq U_2 \subseteq \cdots$ and $\bigcup_{i=1}^\infty U_i = X \setminus Y$. Therefore, for $x \in X \setminus Y$, if we choose j with $x \in U_j$ and $j > i$, then

$$\psi_i(x) \geq \psi_j(x) = \log(\rho(x)^d).$$

(2) We set $A_i = \displaystyle\int_X g\exp(-\psi_i)\Phi$ and $B_i = \displaystyle\int_{U_i} g\rho^{-d}\Phi$. Then

$$\begin{cases} A_1 \leq A_2 \leq \cdots \leq A_i \leq \cdots, \\ B_1 \leq B_2 \leq \cdots \leq B_i \leq \cdots, \\ B_i \leq A_i \quad (i = 1, 2, \ldots). \end{cases}$$

If $A = \lim_{i\to\infty} A_i < \infty$, then $B = \lim_{i\to\infty} B_i$ exists and $B \leq A$. This means that $\displaystyle\int_X g\rho^{-d}dx$ exists and it is equal to B. On the hand, by (1), $A_i \leq B$. Thus $A \leq B$.

Therefore $A = B$. Next we assume that $A = \infty$. Since $A_i \le B$ by (1), we have $B = \infty$.

(3) Our desired inequality is derived from the inequality

$$\sqrt{-1}\partial\bar{\partial}(\psi_k) \ge -2\pi d m c_1(\overline{L})$$

on X. Since $\sqrt{-1}\partial\bar{\partial}(\psi_k)$ and $c_1(\overline{L})$ are C^∞-forms, it is sufficient to show that $\sqrt{-1}\partial\bar{\partial}(\psi_k) \ge -2\pi d m c_1(\overline{L})$ holds on $X \setminus Y$. Note that

$$\frac{\partial^2 \psi_k}{\partial z_i \partial \bar{z}_j} = d^2 \cdot \phi''(\log(\rho^d) + k)\frac{\partial \log \rho}{\partial z_i}\frac{\partial \log \rho}{\partial \bar{z}_j} + d \cdot \phi'(\log(\rho^d) + k)\frac{\partial^2 \log \rho}{\partial z_i \partial \bar{z}_j}$$

and a matrix $(a_i \bar{a}_j)_{\substack{1 \le i \le d \\ 1 \le j \le d}}$ is semipositive for $a_1, \ldots, a_d \in \mathbb{C}$. Thus, by (2) of Claim 2,

$$\begin{aligned}\sqrt{-1}\partial\bar{\partial}(\psi_k) &\ge d \cdot \phi'(\log(\rho^d) + k)\sqrt{-1}\partial\bar{\partial}(\log \rho) \\ &\ge d \cdot \phi'(\log(\rho^d) + k)(-2\pi m c_1(\overline{L})) \ge -(2\pi m d)c_1(\overline{L})\end{aligned}$$

on $X \setminus Y$. □

Let us go back to the proof of the theorem. Since

$$(1/2)N(\psi_k) + nN(\overline{L}) + N(\overline{\omega}_X^{-1}) \ge (n - 2\pi m d)N(\overline{L}) + N(\overline{\omega}_X^{-1})$$

by (3) of Claim 3, and $\overline{L}$ is positive, there are a positive integer n_0 and a positive number c_1 such that, for $n \ge n_0$,

$$(1/2)N(\psi_k) + nN(\overline{L}) + N(\overline{\omega}_X^{-1}) \ge c_1 \,\mathrm{id}\,.$$

For $n \ge n_0$, let $l_U \in H^0(U, L^{\otimes n})$ with $\|l_U\|_{U,\sup} < \infty$ as in the theorem. Let θ be a C^∞ function on X such that θ is 0 outside of Y and 1 on a neighborhood U' of Y. We set $w = \bar{\partial}(\theta l_U)$. Then

$$\int_X |w|^2 \rho^{-d}\Phi \le \|l_U\|^2_{U,\sup}\int_{U\setminus U'} |\bar{\partial}(\theta)|^2 \rho^{-d}\Phi \tag{7.3}$$

because

$$\int_X |w|^2 \rho^{-d}\Phi = \int_U |\bar{\partial}(\theta)|^2 |l_U|^2 \rho^{-d}\Phi \le \|l_U\|^2_{U,\sup}\int_{U\setminus U'} |\bar{\partial}(\theta)|^2 \rho^{-d}\Phi.$$

In particular, $\|w\|_\infty < \infty$. Thus, by Lemma 7.5 and Claim 3, there is $l' \in A^{0,0}(L)$ with $\bar{\partial}(l') = w$ and

$$\int_X |l'|^2 \rho^{-d}\Phi \le c_1^{-1}\int_X |w|^2 \rho^{-d}\Phi.$$

On the other hand, if we write $c_2 = c_1^{-1}\int_{U\setminus U'} |\bar{\partial}(\theta)|^2 \rho^{-d}\Phi$, then by (7.3),

$$\int_X |l'|^2 \rho^{-d}\Phi \le c_2 \|l_U\|^2_{U,\sup}.$$

Here we set $l = \theta l_U - l'$. Then $l \in H^0(X, L^{\otimes n})$ because $\bar{\partial}(l) = 0$. Since

$$\int_X |l'|^2 \rho^{-d}\Phi < \infty,$$

by (3) in Claim 2, $l'|_Y = 0$, that is, $l|_Y = l_U|_Y$. Moreover, as

$$\begin{aligned}\int_X |l|^2\Phi &\leq 2\int_X |\theta l_U|^2\Phi + 2\int_X |l'|^2\Phi \\ &\leq 2\|\theta\|_{\sup}^2\|l_U\|_{U,\sup}^2 + 2\sup_{x\in X}\{\rho^d(x)\}\int_X |l'|^2\rho^{-d}\Phi,\end{aligned}$$

if we write $c_3 = 2\|\theta\|_{\sup}^2 + 2c_2\sup_{x\in X}\{\rho^d(x)\}$, then

$$\int_X |l|^2\Phi \leq c_3\|l_U\|_{U,\sup}^2$$

holds. Therefore, by virtue of Gromov's inequality (see Proposition 5.32), there is a positive constant c_4 with

$$\|l\|_{\sup}^2 \leq c_4 n^{2d}\int_X |l|^2\Phi,$$

and hence the theorem follows. □

Let us consider the main theorem of this section.

THEOREM 7.7. *Let X be a d-dimensional non-singular complex projective variety and let Y be a reduced subscheme of X. Let $\overline{L}$ be a C^∞-hermitian invertible sheaf on X such that L is ample and $c_1(\overline{L})$ is semipositive. For $l' \in H^0(Y, L|_Y)$ and a positive number ϵ, there is a positive integer n_1 such that, for any integer n with $n \geq n_1$, there is $l \in H^0(X, L^{\otimes n})$ with $l|_Y = l'^{\otimes n}$ and*

$$\|l\|_{X,\sup} \leq \exp(\epsilon n)\|l'\|_{Y,\sup}^n,$$

where $\|\cdot\|_{Y,\sup}$ is the supremum norm on Y.

PROOF. Let us begin with the following claim:

CLAIM 4. *We may assume that $c_1(\overline{L})$ is positive.*

PROOF. Let $|\cdot|$ be the metric of $\overline{L}$. As L is ample, we can choose a C^∞-hermitian metric $|\cdot|'$ of L such that $|\cdot| \leq |\cdot|'$ and $c_1(L, |\cdot|')$ is positive. Let us also choose a sufficient large integer m such that $1 \leq (|\cdot|'/|\cdot|)^{1/m} \leq \exp(\epsilon/2)$. If we set $|\cdot|'' = |\cdot|^{1-1/m}|\cdot|'^{1/m}$, then $1 \leq |\cdot|''/|\cdot| \leq \exp(\epsilon/2)$ and $c_1(L, |\cdot|'')$ is positive. We assume that the theorem holds for $(L, |\cdot|'')$. Then, for $l' \in H^0(Y, L|_Y)$, there is a positive integer n_1 such that, for all $n \geq n_1$, we can find $l \in H^0(X, nL)$ with

$$l|_Y = l'^{\otimes n} \quad \text{and} \quad \|l\|''_{X,\sup} \leq \exp((\epsilon/2)n)(\|l'\|''_{Y,\sup})^n,$$

where $\|\cdot\|''_{X,\sup}$ and $\|\cdot\|''_{Y,\sup}$ is the norm arising from $|\cdot|''$. Note that

$$\|l\|_{X,\sup} \leq \|l\|''_{X,\sup} \quad \text{and} \quad \|l'\|''_{Y,\sup} \leq \exp(\epsilon/2)\|l'\|_{Y,\sup}.$$

Thus

$$\|l\|_{X,\sup} \leq \exp(\epsilon n)(\|l'\|_{Y,\sup})^n$$

as required. □

From now on, we assume that $c_1(\overline{L})$ is positive. There is a positive integer n_0 such that, for $n \geq n_0$, $H^0(X, L^{\otimes n}) \to H^0(Y, L^{\otimes n}|_Y)$ is surjective. Thus, for any integer k with $n_0 \leq k < 2n_0$, there is $l_k \in H^0(X, L^{\otimes k})$ with $l_k|_Y = l'^{\otimes k}$. Let us choose an open neighborhood U of Y such that

$$\|l_k\|_{U,\sup} \leq \exp(\epsilon/2)\|l'\|_{Y,\sup}^k$$

holds for all k with $n_0 \leq k < 2n_0$.

For an integer n with $n \geq 2n_0$, there are positive integers q and r with $n = n_0 q + r$ ($n_0 \leq r < 2n_0$). Applying Theorem 7.6 to $l_{n_0}^{\otimes q} \otimes l_r\big|_U \in H^0(U, L^{\otimes n})$, for a sufficiently large n, there is $l \in H^0(X, L^{\otimes n})$ with $l|_Y = l'^{\otimes n}$ and

$$\|l\|_{X,\sup} \leq \exp(n\epsilon/2) \left\|l_{n_0}^{\otimes q} \otimes l_r\right\|_{U,\sup}.$$

On the other hand, since

$$\begin{aligned}\left\|l_{n_0}^{\otimes q} \otimes l_r\right\|_{U,\sup} &\leq \|l_{n_0}\|_{U,\sup}^{q} \|l_r\|_{U,\sup} \\ &\leq \exp(q\epsilon/2)\|l'\|_{Y,\sup}^{n_0 q} \exp(\epsilon/2)\|l'\|_{Y,\sup}^{r} \\ &\leq \exp(n\epsilon/2)\|l'\|_{Y,\sup}^{n},\end{aligned}$$

we obtain $\|l\|_{X,\sup} \leq \exp(n\epsilon)\|l'\|_{Y,\sup}^{n}$, as required. □

COROLLARY 7.8. *Let X' be a projective and smooth scheme over $\mathbb{R}$, Y' a reduced subscheme of X' and L' an ample invertible sheaf on X'. We set*

$$\begin{cases} X = X' \times_{\mathrm{Spec}(\mathbb{R})} \mathrm{Spec}(\mathbb{C}), \\ Y = Y' \times_{\mathrm{Spec}(\mathbb{R})} \mathrm{Spec}(\mathbb{C}) \quad (Y \textit{ is reduced by Proposition } 1.30), \\ L = L' \otimes_{\mathbb{R}} \mathbb{C}. \end{cases}$$

Let $|\cdot|$ be a C^∞-hermitian metric of L of real type such that $c_1(L, |\cdot|)$ is semipositive. For $l' \in H^0(Y', L'|_{Y'})$ and a positive number ϵ, there is a positive integer n_1 such that, for any integer n with $n \geq n_1$, there is $l \in H^0(X', L'^{\otimes n})$ with

$$l|_{Y'} = l'^{\otimes n} \quad \textit{and} \quad \|l\|_{X,\sup} \leq \exp(\epsilon n)\|l'\|_{Y,\sup}^{n}.$$

PROOF. We use notations in Section 5.1.2. By Theorem 7.7, there is a positive integer n_1 such that, for all $n \geq n_1$, there is

$$l_1 \in H^0(X, L^{\otimes n}) = H^0(X', L'^{\otimes n}) \otimes_{\mathbb{R}} \mathbb{C}$$

with

$$l_1|_Y = l'^{\otimes n} \quad \text{and} \quad \|l_1\|_{X,\sup} \leq \exp(\epsilon n)\|l'\|_{Y,\sup}^{n},$$

Here we set $l = (l_1 + F_\infty^*(l_1))/2$. Then $l \in H^0(X', L'^{\otimes n})$ and $l|_{Y'} = l'^{\otimes n}$. Moreover, using (5.3) in Section 5.1.2,

$$\|l\|_{\sup} \leq \frac{1}{2}(\|l_1\|_{\sup} + \|F_\infty^*(l_1)\|_{\sup}) = \|l_1\|_{\sup}.$$

Thus the corollary follows. □

7.3. Proof of Nakai-Moishezon's criterion on an arithmetic variety

The purpose of this section is to give the proof of Theorem 7.1. Let Y be a horizontal integral subscheme of X. First we assume that Y is vertical, that is, Y is contained in a fiber X_p of $X \to \mathrm{Spec}(\mathbb{Z})$ over a prime number p. Then

$$\widehat{\deg}\left(\widehat{c}_1(\overline{L})^{\cdot \dim Y} \cdot (Y, 0)\right) = \deg\left((L|_Y)^{\dim Y}\right) \log(p) > 0,$$

so that, by virtue of algebraic Nakai-Moishezon's criterion, we can see that L is ample on X_p for any prime number p. Therefore, L is ample on X.

In order to see the last assertion of Theorem 7.1, we apply Corollary 5.48 to $\overline{L}$. As L is ample, it is sufficient to see condition (3) of Corollary 5.48. From now on, we assume that Y is horizontal, that is, Y is flat over $\mathbb{Z}$. Since L is ample, by Corollary 6.14 and the assumptions of Theorem 7.1,

$$\widehat{\mathrm{vol}}\left(\overline{L}\big|_Y\right) \geq \widehat{\deg}\left(\widehat{c}_1\left(\overline{L}\big|_Y\right)^{\cdot \dim Y}\right) > 0,$$

and hence there are a positive integer n_0 and a non-zero $s' \in H^0(X|Y, n_0L)$ such that $\|s'\|_{Y,\mathrm{sup}} < 1$. Let ϵ_Y be a positive number with $\exp(\epsilon_Y)\|s'\|_{Y,\mathrm{sup}} < 1$. Note that $Y_{\mathbb{R}}$ and $Y_{\mathbb{C}}$ are reduced schemes by Proposition 1.30. Thus, by Corollary 7.8, there are a positive integer n_1 and $s \in H^0(X, n_1n_0L)_{\mathbb{R}}$ such that

$$s|_Y = s'^{\otimes n_1} \quad \text{and} \quad \|s\|_{X,\mathrm{sup}} \leq (\exp(\epsilon_Y)\|s'\|_{Y,\mathrm{sup}})^{n_1} < 1.$$

In particular, $\|s'^{\otimes n_1}\|_{X|Y,\mathrm{sup},\mathrm{quot}} < 1$, where $\|\cdot\|_{X|Y,\mathrm{sup},\mathrm{quot}}$ is the quotient norm of $H^0(X|Y, n_1n_0L)_{\mathbb{R}}$ induced by $\|\cdot\|_{X,\mathrm{sup}}$ of $H^0(X, n_1n_0L)_{\mathbb{R}}$. Therefore, the assertion follows.

7.4. Arithmetic Hilbert-Samuel formula

In this section, we consider the arithmetic Hilbert-Samuel formula for a nef C^∞-hermitian invertible sheaf due to Moriwaki [**51**] as an application of the continuity of arithmetic volumes and arithmetic Nakai-Moishezon's criterion. This is a generalization of Theorem 5.36 in some sense. Let us start with the following proposition.

PROPOSITION 7.9. *Let X be a generically smooth projective arithmetic variety, and let $\overline{L}$ and $\overline{A}$ be C^∞-hermitian invertible sheaves of real type on X. If $\overline{L}$ is nef and $\overline{A}$ is ample, then $\overline{L}+\overline{A}$ is ample*

PROOF. Let Z be an integral subscheme of X such that Z is flat over $\mathbb{Z}$. Then, by Proposition 5.39 and Theorem 6.15,

$$\widehat{\deg}\left(\widehat{c}_1\left(\overline{L}+\overline{A}\right)^{\dim Z}\cdot (Z,0)\right) > 0.$$

By virtue of arithmetic Nakai-Moishezon's criterion (Theorem 7.1), $\overline{L}+\overline{A}$ is ample. □

THEOREM 7.10 (Arithmetic Hilbert-Samuel formula). *Let $\overline{L}$ and $\overline{N}$ be C^∞-hermitian invertible sheaves of real type on an projective arithmetic variety X. If $\overline{L}$ is nef, then*

$$\hat{h}^0\left(H^0(X, mL+N), \|\cdot\|_{\mathrm{sup}}^{m\overline{L}+\overline{N}}\right) = \frac{\widehat{\deg}(\widehat{c}_1(\overline{L})^{\cdot d})}{d!}m^d + o(m^d) \quad (m \gg 1).$$

In particular, $\widehat{\mathrm{vol}}(\overline{L}) = \widehat{\deg}(\widehat{c}_1(\overline{L})^{\cdot d})$, and $\overline{L}$ is big if and only if

$$\widehat{\deg}(\widehat{c}_1(\overline{L})^{\cdot d}) > 0.$$

PROOF. It is sufficient to show the following:

$$\frac{\widehat{\deg}(\widehat{c}_1(\overline{L})^{\cdot d})}{d!} = \lim_{m\to\infty}\frac{\hat{h}^0(H^0(X, mL+N), \|\cdot\|_{\mathrm{sup}}^{m\overline{L}+\overline{N}})}{m^d}. \tag{7.4}$$

Let $\pi : X' \to X$ be a generic resolution of singularities of X. If the equality

$$\frac{\widehat{\deg}(\widehat{c}_1(\pi^*(\overline{L}))^{\cdot d})}{d!} = \lim_{m\to\infty}\frac{\hat{h}^0(H^0(X', \pi^*(mL+N)), \|\cdot\|_{\mathrm{sup}}^{\pi^*(m\overline{L}+\overline{N})})}{m^d}$$

holds, then by (2) in Theorem 6.4 and the projection formula, we have the same equality on X. Thus we may assume that X is generically smooth.

Let $\overline{A}$ be an ample C^∞-hermitian invertible sheaf of real type on X. Then, by Proposition 7.9, for a positive integer n, $\overline{L} + (1/n)\overline{A}$ is ample. Thus, by Proposition 5.41,

$$\widehat{\mathrm{vol}}(\overline{L} + (1/n)\overline{A}) = \widehat{\mathrm{vol}}_\flat(\overline{L} + (1/n)\overline{A}) = \widehat{\deg}\left(\left(\widehat{c}_1(\overline{L}) + (1/n)\widehat{c}_1(\overline{A})\right)^{\cdot d}\right).$$

Therefore, using the continuity of arithmetic volumes (Theorem 6.9), we obtain

$$\widehat{\mathrm{vol}}(\overline{L}) = \widehat{\mathrm{vol}}_\flat(\overline{L}) = \widehat{\deg}\left(\widehat{c}_1(\overline{L})^{\cdot d}\right).$$

Thus, by (1) in Theorem 6.4, (7.4) follows. □

CHAPTER 8

Arithmetic Bogomolov Inequality

Let E be a locally free coherent sheaf on a non-singular projective surface. If E is semistable with respect to an ample invertible sheaf H, that is, for any subsheaf F of E with $0 \subsetneq F \subsetneq E$, an inequality

$$\frac{\deg(c_1(F) \cdot c_1(H))}{\mathrm{rk}(F)} \leq \frac{\deg(c_1(E) \cdot c_1(H))}{\mathrm{rk}(E)}$$

holds, then we have the following inequality:

$$\deg\left(c_2(E) - \frac{\mathrm{rk}(E) - 1}{2\,\mathrm{rk}(E)} c_1(E)^2\right) \geq 0.$$

It is called the Bogomolov inequality and it has a lot of important application in the theory of algebraic surfaces. In this chapter, we consider its arithmetic analogue (Theorem 8.8). It was first proved by Miyaoka, but his paper was not published. Shortly afterward, Moriwaki [**46**] and Soulé [**65**] gave its proofs independently in published papers. This was also generalized to a higher dimensional arithmetic variety by Moriwaki [**48**]. This topic led the author to the study of Arakelov geometry. From the viewpoint of Arakelov geometry, the arithmetic Bogomolov inequality is an application of Generalized Hodge index theorem. However, we need a deep and fundamental theorem from differential geometry, which is the existence theorem of Hermite-Einstein metrics for stable locally free coherent sheaves. Its proof is beyond the scope of this book, so that we only refer to papers and books. Let us begin with the stability of locally free coherent sheaves on algebraic curves.

8.1. Semistable locally free coherent sheaves on algebraic curves

Let C be a non-singular and irreducible projective curve over an algebraically closed field k. Let E be a non-zero locally free coherent sheaf on C. The *slope* $\mu(E)$ of E is defined by $\mu(E) = \deg(E)/\mathrm{rk}(E)$. We say that E is *stable* (resp. *semistable*) if, for any subsheaf F of E with $0 \subsetneq F \subsetneq E$, $\mu(F) < \mu(E)$ (resp. $\mu(F) \leq \mu(E)$) holds. Note that E is stable (resp. semistable) if and only if, for any non-isomorphic surjective homomorphism $E \to Q$ ($Q \neq \{0\}$) of locally free coherent sheaves, we have $\mu(E) < \mu(Q)$ (resp. $\mu(E) \leq \mu(Q)$). We give the definition of another stability condition. Namely, E is said to be *polystable* if there are non-zero stable locally free coherent sheaves $E_1, \ldots, E_l$ on C such that $E \simeq E_1 \oplus \cdots \oplus E_l$ and $\mu(E_1) = \cdots = \mu(E_l)$. A polystable locally free coherent sheaf is semistable. Let us consider a criterion of semistability.

PROPOSITION 8.1. *We assume that the characteristic of k is zero. Let r be the rank of E. Let*

$$\pi : P = \mathrm{Proj}\left(\bigoplus\nolimits_{m\geq 0} \mathrm{Sym}^m(E)\right) \to C$$

be the projective bundle on E and $\mathscr{O}_P(1)$ the tautological invertible sheaf on P. Then we have the following.

(1) *Let $f : C' \to C$ be a surjective and finite morphism of non-singular irreducible projective curves over k. Then E is semistable if and only if $f^*(E)$ is semistable.*
(2) $(\mathscr{O}_P(r) - \pi^*(c_1(E)))^r = 0$.
(3) *The following are equivalent.*
   (3.1) *E is semistable.*
   (3.2) *$\mathscr{O}_P(r) - \pi^*(c_1(E))$ is nef.*

Proof. (1) Clearly, if $f^*(E)$ is semistable, then E is semistable. Conversely we assume that E is semistable. In order to show that $f^*(E)$ is semistable, we may assume that the function field $k(C')$ of C' is a Galois extension over the function field $k(C)$ of C. Here we assume that $f^*(E)$ is not semistable. Then there is a unique subsheaf F' of $f^*(E)$ such that $\mu(F')$ is maximal and $\operatorname{rk} F'$ is maximal among such F''s (for example, see [**37**, Chapter 5, Proposition 1.13]). In particular, $\mu(F') > \mu(f^*(E)) = \deg(f)\mu(E)$. For $\sigma \in \operatorname{Gal}(k(C')/k(C))$, it is easy to see that $\sigma^*(F') \subset f^*(E)$, $\deg(\sigma^*(F')) = \deg(F')$ and $\operatorname{rk} \sigma^*(F') = \operatorname{rk} F'$. Thus, using the uniqueness of F', we have $\sigma^*(F') = F'$. Therefore, F' descends to C, that is, there is a subsheaf F of E with $f^*(F) = F'$. In particular, $\mu(F) \le \mu(E)$, which implies that $\mu(F') \le \deg(f)\mu(E)$. This is a contradiction.

(2) Indeed,

$$\begin{aligned}(\mathscr{O}_P(r) - \pi^*(c_1(E)))^r &= r^r \mathscr{O}_P(1)^r - r^{r-1} r \mathscr{O}_P(1)^{r-1}\pi^*(c_1(E)) \\ &= r^r \deg(E) - r^r \deg(E) = 0.\end{aligned}$$

(3) We assume that E is semistable. Let D be an irreducible curve on P. If D lies in a fiber of π, then

$$((\mathscr{O}_P(r) - \pi^*(c_1(E))) \cdot D) = (\mathscr{O}_P(r) \cdot D) > 0$$

because $\mathscr{O}_P(r)$ is relatively ample with respect to π. Next we assume $\pi|_D : D \to C$ is surjective. Let C' be the normalization of D and the composition $C' \to D \to C$ is denoted by f . Let $\pi' : P' = \operatorname{Proj}\left(\bigoplus_{m\ge 0} \operatorname{Sym}^m(f^*(E))\right) \to C'$ be the projective bundle of $f^*(E)$ and $\mathscr{O}_{P'}(1)$ the tautological invertible sheaf on P'. Then there is a natural morphism $f' : P' \to P$ such that the following diagram is commutative:

$$\begin{array}{ccc} P' & \xrightarrow{f'} & P \\ {\scriptstyle \pi'}\big\downarrow & & \big\downarrow{\scriptstyle \pi} \\ C' & \xrightarrow{f} & C\,. \end{array}$$

Moreover, $f'^*(\mathscr{O}_P(r) - \pi^*(c_1(E))) = \mathscr{O}_{P'}(r) - \pi'^*(c_1(f^*(E)))$ and there is a section D' of $\pi' : P' \to C'$ with $f'_*(D') = D$. Therefore, as

$$(\mathscr{O}_{P'}(r) - \pi'^*(c_1(f^*(E))) \cdot D') = (\mathscr{O}_P(r) - \pi^*(c_1(E)) \cdot D),$$

it is sufficient to show $(\mathscr{O}_{P'}(r) - \pi'^*(c_1(f^*(E))) \cdot D') \ge 0$. Let Q' be the quotient invertible sheaf of $f^*(E)$ corresponding to D', that is, Q' is an invertible sheaf with $(\pi'|_{D'})^*(Q') = \mathscr{O}_{P'}(1)|_{D'}$. Then, since $f^*(E)$ is semistable by (1),

$$(\mathscr{O}_{P'}(r) - \pi'^*(c_1(f^*(E))) \cdot D') = r\deg(Q') - \deg(f^*(E)) \ge 0.$$

Next we assume that $\mathcal{O}_P(r) - \pi^*(c_1(E))$ is nef. Let Q be a quotient locally free coherent sheaf of E and $s = \operatorname{rk} Q$. Let Y be the subvariety of P corresponding to Q, that is, $Y = \operatorname{Proj}\left(\bigoplus_{m \geq 0} \operatorname{Sym}^m(Q)\right)$. Note that $\dim Y = s$. Thus

$$
\begin{aligned}
((\mathcal{O}_P(r) - \pi^*(c_1(E)))^s \cdot Y) &= r^s(\mathcal{O}_P(1)^s \cdot Y) - r^{s-1}s(\mathcal{O}_P(1)^{s-1}\pi^*(c_1(E)) \cdot Y) \\
&= r^s s\left(\frac{(\mathcal{O}_Y(1)^s)}{s} - \frac{(\mathcal{O}_Y(1)^{s-1} \cdot (\pi|_Y)^*(c_1(E)))}{r}\right) \\
&= r^s s\left(\frac{\deg(Q)}{s} - \frac{\deg(E)}{r}\right).
\end{aligned}
$$

Since $\mathcal{O}_P(r) - \pi^*(c_1(E))$ is nef, the above equation has a non-negative value, which shows the semistability of E. $\square$

PROPOSITION 8.2. *Let $k \subseteq K$ be algebraically closed fields. Let C be a non-singular irreducible projective curve over k and E a locally free coherent sheaf on C. We set $C_K = C \times_{\operatorname{Spec}(k)} \operatorname{Spec}(K)$ and $E_K = E \otimes_k K$. Then we have the following.*

(1) *If E is semistable on C, then E_K is semistable on C_K.*
(2) *If E is stable on C, then E_K is stable on C_K.*

PROOF. Let G be a subsheaf of E_K. Then there is an intermediate field F of $k \subseteq K$ such that G is defined over F and F is finitely generated over k. Let T be an algebraic variety over k such that the function field of T is F. Let $p : C \times T \to C$ and $q : C \times T \to T$ be the projections to the first factor and the second factor, respectively. Let η be the generic point of T and C_η the generic fiber of $q : C \times T \to T$, that is, $C_\eta = C \times_{\operatorname{Spec}(k)} \operatorname{Spec}(F)$. Then $p^*(E)|_{C_\eta} = E \otimes_k F$. Replacing T with an non-empty open set of T, if necessarily, we may assume that there is a locally free subsheaf $\mathcal{G}$ of $p^*(E)$ on $C \times T$ such that $\mathcal{G}|_{C_\eta} = G$ and, for all $t \in T$, the natural homomorphism $\mathcal{G}|_{q^{-1}(t)} \to p^*(E)|_{q^{-1}(t)}$ is injective. Let t_0 be a closed point of T and $G_0 = \mathcal{G}|_{C \times \{t_0\}}$. Then, as $q : C \times T \to T$ is flat, we can see that

$$
\deg(G) = \deg(G_0), \quad \operatorname{rk}(G) = \operatorname{rk}(G_0) \quad \text{and} \quad G_0 \subseteq E.
$$

In particular, $\mu(G) = \mu(G_0)$.

(1) We assume that E_K is not semistable on C_K. Then there is a subsheaf G of E_K with $\mu(G) > \mu(E_K)$. Let G_0 be a locally free coherent sheaf as constructed as above for this G. Then $\mu(G_0) = \mu(G) > \mu(E_K) = \mu(E)$. This contradicts the semistability of E.

(2) We assume that E_K is not semistable on C_K. Then there is a subsheaf G of E_K with $\mu(G) \geq \mu(E_K)$ and $0 < \operatorname{rk}(G) < \operatorname{rk}(E_K)$. Let us consider G_0 for this G. Then $\mu(G_0) = \mu(G) \geq \mu(E_K) = \mu(E)$. This contradicts the stability of E. $\square$

8.2. Hermite-Einstein metric and stability

Let C be a Riemann surface and Ω a Kähler form on C. Let $\overline{E} = (E, h)$ be a C^∞-hermitian locally free coherent sheaf on C. We say that h is a *Hermite-Einstein metric* with respect to Ω if there is a locally constant function c on C with $\Theta_{\overline{E}} = c\Omega \operatorname{id}_E$, where $\Theta_{\overline{E}}$ is the curvature form of $\overline{E}$.

The most important theorem for this chapter, which we do not prove here, is the following existence theorem of a Hermite-Einstein metric of a stable locally free sheaf on a connected compact Riemann surface.

THEOREM 8.3. *Let C be a connected compact Riemann surface and Ω a Kähler form on C. Let E be a locally free coherent sheaf on C. Then E has a Hermite-Einstein metric if and only if E is polystable.*

This theorem is first proved algebraically by Narasimhan-Seshadri [**56**], and Donaldson [**15**] gave a proof from the viewpoint of differential geometry. The generalization of the existence theorem of Hermite-Einstein metric on a higher dimensional Kähler manifold was called Kabayashi's conjecture. Donaldson [**16**] proved it on an algebraic surface and he extended it to a higher dimensional algebraic variety [**17**]. Afterward, a general result on a Kähler manifold was obtained by Uhlenbeck-Yau [**68**], [**69**].

Next let us recall Donaldson's Lagrangian. Let Ω be a Kähler form on a connected compact Riemann surface C. Let E be a locally free coherent sheaf on C, and let h and h' be C^∞-metrics of E. Let $\{h_t\}_{0\le t\le 1}$ be a C^∞-deformation of metrics of E with $h_0 = h'$ and $h_1 = h$. Then we set

$$\begin{cases} Q_1(E,h,h') = \log(\det(h)/\det(h')), \\ Q_2(E,h,h') = \sqrt{-1}\displaystyle\int_0^1 \operatorname{tr}(h_t^{-1}\partial_t(h_t)\Theta_{(E,h_t)})dt. \end{cases}$$

Donaldson's Lagrangian $L(h,h')$ is defined to be

$$L(h,h') = \int_C \left(Q_2(E,h,h') - 2\pi\mu(E)Q_1(E,h,h')\tilde{\Omega}\right),$$

where $\tilde{\Omega} = \left(\int_C \Omega\right)^{-1}\Omega$. By Proposition 5.29, we can see

$$L(h,h') = 2\pi\int_C \left(\mu(E)\tilde{c}_1(E,h,h')\tilde{\Omega} - \widetilde{\operatorname{ch}}_2(E,h,h')\right).$$

Next proposition gives basic properties of Donaldson's Lagrangian.

PROPOSITION 8.4. (1) *The following are equivalent:*

(1.1) *$L(h_0,h')$ gives the minimal value of the set*

$$\{L(h,h') \mid h \text{ is a } C^\infty\text{-metric of } E\}.$$

(1.2) *h_0 is a Hermite-Einstein metric with respect to Ω.*

(2) *We assume that E is stable. If h and k are Hermite-Einstein metrics of E with respect to Ω, then there is a positive constant c with $h = ck$.*

PROOF. For example, see [**37**, Chapter 6, Proposition (3.37)]. □

COROLLARY 8.5. *Let C be a compact Riemann surface (C is not necessarily connected) and let Ω be a Kähler form on C. Let E be a locally free coherent sheaf on C and let h and k be Hermite-Einstein metrics of E with respect to Ω. Then $h+k$ is also a Hermite-Einstein metric of E with respect to Ω.*

PROOF. Clearly we may assume that C is connected. Since E has a Hermite-Einstein metric, by Theorem 8.3, there are stable locally free coherent sheaves $E_1,\ldots,E_l$ such that $E = E_1 \oplus \cdots \oplus E_l$. If h' is a Hermite-Einstein metric with respect to Ω, then $h'_i = h'|_{E_i}$ is a Hermite-Einstein metric of E_i with respect to Ω and we have an orthogonal decomposition $(E,h') = (E_1,h'_1)\oplus\cdots\oplus(E_l,h'_l)$ (for example, see [**37**, Chapter 5, Theorem 2.5]). Therefore, we may assume that E is stable. Then, by (2) of Proposition 8.4, there is a positive constant c with $h = ck$. Therefore, $h+k = (1+c)h$ is a Hermite-Einstein metric. □

PROPOSITION 8.6. *Let X be a smooth projective curve over $\mathbb{R}$ and let Ω be a Kähler form on $X(\mathbb{C})$ with $F_\infty^*(\Omega) = -\Omega$, where $F_\infty : X(\mathbb{C}) \to X(\mathbb{C})$ is the complex conjugation map of $X(\mathbb{C})$ (for details, see Section 5.1.2). Let E be a locally free sheaf on X such that $E_\mathbb{C} = E \otimes_\mathbb{R} \mathbb{C}$ is polystable on each connected component of $X(\mathbb{C})$. Then $E_\mathbb{C}$ has a Hermite-Einstein metric h with respect to Ω such that $F_\infty^*(h) = \bar{h}$.*

PROOF. Since $E_\mathbb{C}$ is polystable on each connected component of $X(\mathbb{C})$, $E_\mathbb{C}$ has a Hermite-Einstein metric k with respect to Ω, that is, there is a locally constant function c on $X(\mathbb{C})$ with $\Theta_{(E_\mathbb{C},k)} = c\Omega \operatorname{id}_E$. Note that $d(F_\infty^*(\eta)) = F_\infty^*(d(\eta))$ for a C^∞ (p,q)-form η on $X(\mathbb{C})$. Thus

$$\begin{cases} \partial(F_\infty^*(\eta)) = F_\infty^*(\bar{\partial}(\eta)), \\ \bar{\partial}(F_\infty^*(\eta)) = F_\infty^*(\partial(\eta)). \end{cases}$$

Therefore, it is easy to see that $\Theta_{(E_\mathbb{C},\overline{F_\infty^*(k)})} = \overline{F_\infty^*(\Theta_{(E_\mathbb{C},k)})}$, and hence

$$\Theta_{(E_\mathbb{C},\overline{F_\infty^*(k)})} = \overline{F_\infty^*(c\Omega)} \operatorname{id}_E = -\overline{F_\infty^*(c)}\Omega \operatorname{id}_E,$$

which shows that $\overline{F_\infty^*(k)}$ is also a Hermite-Einstein metric with respect to Ω. Thus, if we set $h = k + \overline{F_\infty^*(k)}$, then $F_\infty^*(h) = \bar{h}$ and, by Corollary 8.5, h is a Hermite-Einstein metric with respect to Ω. □

PROPOSITION 8.7. *Let C be a Riemann surface, Ω a Kähler form on C and E a locally free coherent sheaf of rank r on C. Let*

$$\pi : P = \operatorname{Proj}\left(\bigoplus_{m\geq 0} \operatorname{Sym}^m(E)\right) \to C$$

be the projective bundle and $\mathscr{O}_P(1)$ the tautological invertible sheaf on P. Let h be a C^∞-metric of E and h'' the quotient metric of $\mathscr{O}_P(1)$ induced by the natural surjective homomorphism $\pi^*(E) \to \mathscr{O}_P(1)$ and the metric $\pi^*(h)$ of $\pi^*(E)$. Moreover, let $S = \operatorname{Ker}(\pi^*(E) \to \mathscr{O}_P(1))$, and h' the submetric of S induced by $\pi^*(h)$. Here we set

$$\overline{E} = (E,h), \quad \overline{S} = (S,h'), \quad \overline{\mathscr{O}}_P(1) = (\mathscr{O}_P(1),h'')$$

and

$$\overline{L} = \overline{\mathscr{O}}_P(1)^{\otimes r} \otimes \pi^*(\det(\overline{E}))^{\otimes -1}.$$

If h is a Hermite-Einstein metric with respect to Ω, then we have the following.

1. *$c_1(\overline{L})$ is semipositive.*
2. *Let us consider the following exact sequence of C^∞-hermitian locally free coherent sheaves:*

$$\mathcal{E}(\overline{E}) : 0 \to \overline{S} \to \pi^*(\overline{E}) \to \overline{\mathscr{O}}_P(1) \to 0.$$

   Then $c_1(\overline{L})\tilde{c}_r(\mathcal{E}(\overline{E}) \otimes \overline{\mathscr{O}}_P(-1)) = 0$.

PROOF. Since (1) and (2) are local with respect to C, we may assume

$$C = \{z \in \mathbb{C} \mid |z| \leq \epsilon\} \quad \text{and} \quad E \simeq \mathscr{O}_C^{\oplus r},$$

where ϵ is a sufficiently small positive number. We set

$$\overline{N} = (\mathscr{O}_C, \det(h)^{-1/r}) \quad \text{and} \quad \overline{E}' = \overline{E} \otimes \overline{N}.$$

Replacing $\overline{E}$ with $\overline{E}'$, we can define $\overline{\mathscr{O}}'_P(1)$, $\overline{S}'$, $\overline{L}'$ in the same way as before. Then

$$\overline{S}' = \overline{S} \otimes \pi^*(\overline{N}), \quad \overline{\mathscr{O}}'_P(1) = \overline{\mathscr{O}}(1) \otimes \pi^*(\overline{N}) \quad \text{and} \quad \overline{L}' = \overline{\mathscr{O}}'_P(1)^{\otimes r} \otimes \pi^*(\det(\overline{E}'))^{\otimes -1},$$

so that

$$\begin{aligned} \overline{L}' &= \overline{\mathscr{O}}'_P(1)^{\otimes r} \otimes \pi^*(\det(\overline{E}'))^{\otimes -1} \\ &= (\overline{\mathscr{O}}_P(1) \otimes \pi^*(\overline{N}))^{\otimes r} \otimes \pi^*(\det(\overline{E} \otimes \overline{N}))^{\otimes -1} \\ &= (\overline{\mathscr{O}}_P(1) \otimes \pi^*(\overline{N}))^{\otimes r} \otimes \pi^*(\det(\overline{E}) \otimes \overline{N}^{\otimes r}))^{\otimes -1} = \overline{L}. \end{aligned}$$

Moreover, since the natural exact sequence

$$\mathcal{E}(\overline{E}') : 0 \to \overline{S}' \to \pi^*(\overline{E}') \to \overline{\mathscr{O}}'_P(1) \to 0$$

arising from $\overline{E}'$ is equal to $\mathcal{E}(\overline{E}) \otimes \pi^*(\overline{N})$, we have $\mathcal{E}(\overline{E}) \otimes \overline{\mathscr{O}}_P(-1) = \mathcal{E}(\overline{E}') \otimes \overline{\mathscr{O}}'_P(-1)$. On the other hand, as $\overline{N}^{\otimes -r} = \det(\overline{E})$, if we set $\Theta_{\overline{E}} = c\Omega \operatorname{id}_E$, then

$$-r\Theta_{\overline{N}} = \Theta_{\overline{N}^{\otimes -r}} = \Theta_{\det(\overline{E})} = \operatorname{tr}(c\Omega \operatorname{id}_E) = rc\Omega.$$

Thus $\Theta_{\overline{N}} = -c\Omega$. Therefore,

$$\Theta_{\overline{E}'} = \Theta_{\overline{E}} + \Theta_{\overline{N}} \operatorname{id}_E = 0,$$

which means that $\overline{E}'$ is flat. By the above consideration, we may assume that $\overline{E}$ is flat in order to prove (1) and (2).

By virtue of [**37**, Chapter 1, Proposition 2.5], there are C^∞ local basis $s_1, \dots, s_r$ of E with $\nabla_{\overline{E}}(s_i) = 0$ $(i = 1, \dots, r)$. Taking the $(0,1)$-part of $\nabla_{\overline{E}}(s_i) = 0$, we have $\bar{\partial}(s_i) = 0$ $(i = 1, \dots, r)$, that is, $s_1, \dots, s_r$ are holomorphic sections of E. Moreover, as

$$d(h(s_i, s_j)) = h(\nabla_{\overline{E}}(s_i), s_i) + h(s_i, \nabla_{\overline{E}}(s_j)) = 0,$$

$h(s_i, s_j)$ is a constant for every i, j. Therefore, there are an r-dimensional complex vector space V and a hermitian inner product h_0 on V such that $\overline{E}$ is isometric to $(V, h_0) \otimes \mathscr{O}_C$. Moreover, taking an orthonormal basis of (V, h_0), we may further assume that $\overline{E}$ is isometric to $(\mathbb{C}^r, \langle\ ,\ \rangle) \otimes \mathscr{O}_C$, where $\langle\ ,\ \rangle$ is the standard inner product of $\mathbb{C}^r$. In this case, $P = C \times \mathbb{P}^{r-1}$ and $\pi : P \to C$ is the projection to the first factor. Let $\nu : P \to \mathbb{P}^{r-1}$ be the projection to the second factor. Then $\mathscr{O}_P(1) = \nu^*(\mathscr{O}_{\mathbb{P}^{r-1}}(1))$. Moreover, there is an exact sequence

$$\overline{\mathcal{E}}_0 : 0 \to \overline{S}_0 \to (\mathbb{C}^r, \langle , \rangle) \otimes \mathscr{O}_{\mathbb{P}^{r-1}} \to \overline{\mathscr{O}}^{FS}_{\mathbb{P}^{r-1}}(1) \to 0$$

on $\mathbb{P}^{r-1}$ such that $\nu^*(\overline{\mathcal{E}}_0) = \mathcal{E}(\overline{E})$, where the metric of $\overline{\mathscr{O}}^{FS}_{\mathbb{P}^{r-1}}(1)$ is the Fubini-Study metric. Therefore, (1) is obvious because $\overline{L} = \nu^*(\overline{\mathscr{O}}^{FS}_{\mathbb{P}^{r-1}}(1)^{\otimes r})$. (2) is also obvious because

$$c_1(\overline{L})\tilde{c}_r(\mathcal{E}(\overline{E}) \otimes \overline{\mathscr{O}}_P(-1)) = r\nu^*(c_1(\overline{\mathscr{O}}^{FS}_{\mathbb{P}^{r-1}}(1))\tilde{c}_r(\overline{\mathcal{E}}_0 \otimes \overline{\mathscr{O}}^{FS}_{\mathbb{P}^{r-1}}(-1)))$$

and $c_1(\overline{\mathscr{O}}^{FS}_{\mathbb{P}^{r-1}}(1))\tilde{c}_r(\overline{\mathcal{E}}_0 \otimes \overline{\mathscr{O}}^{FS}_{\mathbb{P}^{r-1}}(-1))$ is (r,r)-form on $\mathbb{P}^{r-1}$. $\square$

8.3. Arithmetic Bogomolov inequality and its proof

Let X be a regular projective arithmetic surface, that is, X is a two-dimensional, regular projective arithmetic variety. Let $X \to \operatorname{Spec}(O_K) \to \operatorname{Spec}(\mathbb{Z})$ be the Stein factorization of $X \to \operatorname{Spec}(\mathbb{Z})$, where K is an algebraic number field and O_K is the maximal order in K. The main theorem in this section is the following arithmetic Bogomolov inequality.

THEOREM 8.8 (Arithmetic Bogomolov inequality). *Let $\overline{E} = (E, h)$ be a C^∞-hermitian locally free coherent sheaf of rank r on X. If $E \otimes_{O_K} \overline{K}$ is semistable on $X_{\overline{K}} (= X \times_{\operatorname{Spec}(O_K)} \operatorname{Spec}(\overline{K}))$, then the inequality*

$$\widehat{\deg}\left(\widehat{c}_2(\overline{E}) - \frac{r-1}{2r}\widehat{c}_1(\overline{E})^2\right) \geq 0$$

holds, where $\overline{K}$ is an algebraic closure of K.

PROOF. Let $\pi : P = \operatorname{Proj}\left(\bigoplus_{m\geq 0} \operatorname{Sym}^m(E)\right) \to X$ be the projective bundle of E and $\mathcal{O}(1)$ the tautological invertible sheaf on P. The kernel of the natural surjective homomorphism $\pi^*(E) \to \mathcal{O}(1)$ is denoted by S. Note that $\pi^*(E)$ has a natural metric $\pi^*(h)$. Let h' be the submetric of S induced by $\pi^*(h)$ and h'' the quotient metric of $\mathcal{O}(1)$ induced by $\pi^*(h)$. We write $\overline{\mathcal{O}}(1) = (\mathcal{O}(1), h'')$ and $\overline{S} = (S, h')$. Then we have an exact sequence with metrics:

$$\mathcal{E}(\overline{E}) : 0 \to \overline{S} \to \pi^*(\overline{E}) \to \overline{\mathcal{O}}(1) \to 0.$$

Let us consider the following claim.

CLAIM 1. *If we set*

$$\theta = \int_{P(\mathbb{C})\to X(\mathbb{C})} (-1)^{r-1}\left(c_1(\overline{\mathcal{O}}(1)) - \frac{1}{r}\pi^*(c_1(\overline{E}))\right)\widetilde{c}_r(\mathcal{E}(\overline{E}) \otimes \overline{\mathcal{O}}(-1)),$$

then

$$\pi_*\left(\widehat{c}_1(\overline{\mathcal{O}}(1)) - \frac{1}{r}\pi^*(\widehat{c}_1(\overline{E}))\right)^{r+1} = \frac{r-1}{2r}\widehat{c}_1(\overline{E})^2 - \widehat{c}_2(\overline{E}) + (0, \theta)$$

PROOF. For $i \geq 1$, we write

$$\theta_i = (-1)^{r-1}\int_{P(\mathbb{C})\to X(\mathbb{C})} c_1(\overline{\mathcal{O}}(1))^{i-1}\widetilde{c}_r(\mathcal{E}(\overline{E}) \otimes \overline{\mathcal{O}}(-1)).$$

Then, by Proposition 5.30,

$$\begin{cases} \pi_*(\widehat{c}_1(\overline{\mathcal{O}}(1))^r) = \widehat{c}_1(\overline{E}) + (0, \theta_1), \\ \pi_*(\widehat{c}_1(\overline{\mathcal{O}}(1))^{r+1}) = \widehat{c}_1(\overline{E})^2 - \widehat{c}_2(\overline{E}) + (0, \theta_2) + \widehat{c}_1(\overline{E})(0, \theta_1) \end{cases}$$

and

$$\theta = \theta_2 - \frac{1}{r}c_1(\overline{E})\theta_1.$$

Therefore

$$\pi_*\left(\widehat{c}_1(\overline{\mathscr{O}}(1)) - \frac{1}{r}\pi^*(\widehat{c}_1(\overline{E}))\right)^{r+1}$$
$$= \pi_*(\widehat{c}_1(\overline{\mathscr{O}}(1))^{r+1}) - \frac{r+1}{r}\pi_*(\widehat{c}_1(\overline{\mathscr{O}}(1))^r)\widehat{c}_1(\overline{E}) + \frac{r+1}{2r}\widehat{c}_1(\overline{E})^2$$
$$= \frac{r-1}{2r}\widehat{c}_1(\overline{E})^2 - \widehat{c}_2(\overline{E}) + \left(0, \theta_2 - \frac{1}{r}c_1(\overline{E})\theta_1\right),$$

as required. □

Let Ω be a Kähler form on $X(\mathbb{C})$ with $F_\infty^*(\Omega) = -\Omega$.

CLAIM 2. *If h is a Hermite-Einstein metric with respect to Ω and h is of real type, the assertion of the theorem holds.*

PROOF. We set $\overline{L} = \overline{\mathscr{O}}(1)^{\otimes r} \otimes \pi^*(\det(\overline{E}))^{-1}$. By Claim 1 and Proposition 8.7, we obtain

$$\widehat{\deg}\left(\widehat{c}_1(\overline{L})^{r+1}\right) = r^{r+1}\widehat{\deg}\left(\frac{r-1}{2r}\widehat{c}_1(\overline{E})^2 - \widehat{c}_2(\overline{E})\right).$$

Moreover, $c_1(\overline{L})$ is semipositive by Proposition 8.7, L is nef on the generic fiber by Proposition 8.1, and L is not big on the generic fiber by Proposition 8.1. Therefore, if L has moderate growth of positive even cohomologies, then, by Theorem 6.12, we have $\widehat{\deg}\left(\widehat{c}_1(\overline{L})^{r+1}\right) \leq 0$, as claimed.

Let B be an ample invertible sheaf on X such that $A = \mathscr{O}_P(1) \otimes \pi^*(B)$ is ample on P. As $R^j\pi_*\mathscr{O}_P(l) = 0$ $(l \geq 0,\ j > 0)$, we obtain

$$\begin{aligned} H^i(P, (L^{\otimes n} \otimes A)^{\otimes m}) &= H^i(P, \mathscr{O}_P(mnr + m) \otimes \pi^*(B^{\otimes m} \otimes \det(E)^{\otimes -mn})) \\ &= H^i(X, \operatorname{Sym}^{mnr+m}(E) \otimes B^{\otimes m} \otimes \det(E)^{\otimes -mn}). \end{aligned}$$

In particular,

$$H^i(P, (L^{\otimes n} \otimes A)^{\otimes m}) = 0 \quad (i \geq 2),$$

which shows that L has moderate growth of positive even cohomologies. □

We define $\Delta(\overline{E})$ to be

$$\Delta(\overline{E}) = \widehat{\deg}\left(\widehat{c}_2(\overline{E}) - \frac{r-1}{2r}\widehat{c}_1(\overline{E})^2\right).$$

Then the following claim holds.

CLAIM 3. *$\Delta(\overline{E} \otimes \overline{M}) = \Delta(\overline{E})$ for a C^∞-hermitian invertible sheaf $\overline{M}$ on X.*

PROOF. Indeed, by using the axiom (3) of arithmetic characteristic classes, we have

$$\begin{cases} \widehat{c}_1(\overline{E} \otimes \overline{M}) = \widehat{c}_1(\overline{E}) + r\widehat{c}_1(\overline{M}), \\ \widehat{c}_2(\overline{E} \otimes \overline{M}) = \widehat{c}_2(\overline{E}) + (r-1)\widehat{c}_1(\overline{E})\widehat{c}_1(\overline{M}) + \dfrac{r(r-1)}{2}\widehat{c}_1(\overline{M})^2. \end{cases}$$

Thus, by an easy calculation, $\Delta(\overline{E} \otimes \overline{M}) = \Delta(\overline{E})$. □

CLAIM 4. *If $E_{\overline{K}}$ is polystable, then $\Delta(\overline{E}) \geq 0$.*

PROOF. For each $\sigma \in K(\mathbb{C})$, we set $X_\sigma = X \times^{\sigma}_{\mathrm{Spec}(K)} \mathrm{Spec}(\mathbb{C})$, where $\mathrm{Spec}(\mathbb{C})$ is viewed a scheme over $\mathrm{Spec}(K)$ via σ. Then $\{X_\sigma\}_{\sigma \in K(\mathbb{C})}$ is the set of all connected component of $X(\mathbb{C})$. For σ, if we fix an extension $\tilde{\sigma} : \overline{K} \to \mathbb{C}$ of σ, then $X_\sigma = X_{\overline{K}} \times^{\tilde{\sigma}}_{\mathrm{Spec}(\overline{K})} \mathrm{Spec}(\mathbb{C})$. Thus, by Proposition 8.2, E is polystable on each connected component of $X(\mathbb{C})$. Therefore, by Proposition 8.6, E has a Hermite-Einstein metric h_0 of real type. Then, by Claim 2,

$$\widehat{\deg}\left(\widehat{c}_2(E, h_0) - \frac{r-1}{2r}\widehat{c}_1(E, h_0)^2\right) \geq 0.$$

We set $\overline{M} = (\mathscr{O}_X, (\det(h_0)/\det(h))^{1/r})$. Then $\Delta(\overline{E} \otimes \overline{M}) = \Delta(\overline{E})$ and $\det(\overline{E} \otimes \overline{M}) = \det((E, h_0))$. Thus, in order to show the claim, we may assume that $\det(\overline{E}) = \det((E, h_0))$. Since

$$\begin{cases} \widehat{c}_1(\overline{E})^2 = 2(\widehat{\mathrm{ch}}_2(\overline{E}) + \widehat{c}_2(\overline{E})), \\ \widehat{c}_1((E, h_0))^2 = 2(\widehat{\mathrm{ch}}_2((E, h_0)) + \widehat{c}_2((E, h_0))), \end{cases}$$

by (5.18), we have

$$\widehat{c}_2((E, h_0)) - \widehat{c}_2(\overline{E}) = \widehat{\mathrm{ch}}_2(\overline{E}) - \widehat{\mathrm{ch}}_2((E, h_0)).$$

Therefore, using (5.17),

$$\begin{aligned} \Delta((E, h_0)) - \Delta(\overline{E}) &= \widehat{\deg}\left(\widehat{\mathrm{ch}}_2(\overline{E}) - \widehat{\mathrm{ch}}_2((E, h_0))\right) \\ &= \int_{X(\mathbb{C})} \widetilde{\mathrm{ch}}_2(h, h_0) = -(1/2\pi) L(h, h_0). \end{aligned}$$

On the other hand, by Proposition 8.4, $L(h, h_0) \geq L(h_0, h_0) = 0$. Hence, by Claim 2, $\Delta(\overline{E}) \geq \Delta((E, h_0)) \geq 0$. □

To proceed with further arguments, we need to show several facts.

CLAIM 5. (1) *Let $\mu : X' \to X$ be a generically finite morphism of regular projective arithmetic surfaces such that μ is smooth on the generic fibers. Then $\Delta(\mu^*(\overline{E})) = \deg(\mu)\Delta(\overline{E})$.*

(2) *Let $0 \to \overline{S} \to \overline{E} \to \overline{Q} \to 0$ be an exact sequence of C^∞-hermitian locally free coherent sheaves on X such that the metric $\overline{S}$ is the submetric induced by the metric of $\overline{E}$ and the metric of $\overline{Q}$ is the quotient metric induced by the metric of $\overline{E}$. Then $\Delta(\overline{E}) \geq \Delta(\overline{S} \oplus \overline{Q})$.*

(3) *Let M be a locally free coherent sheaf on X and N a subsheaf of M such that M/N is torsion free. Then there are a birational morphism of regular projective arithmetic varieties and a subsheaf N' of $\mu^*(M)$ such that N' coincides with N on X_K and that N' and $\mu^*(M)/N'$ are locally free.*

PROOF. (1) This is a consequence of the projection formula (Theorem 5.26).
(2) First, by Proposition 5.28,

$$\widehat{\deg}\left(\widehat{\mathrm{ch}}_2(\overline{S} \oplus \overline{Q}) - \widehat{\mathrm{ch}}_2(\overline{E})\right) = \int_{X(\mathbb{C})} \widetilde{\mathrm{ch}}_2(0 \to \overline{S} \to \overline{E} \to \overline{Q} \to 0) \geq 0.$$

Thus, as $\widehat{c}_1(\overline{E}) = \widehat{c}_1(\overline{S} \oplus \overline{Q})$, we obtain

$$\Delta(\overline{E}) - \Delta(\overline{S} \oplus \overline{Q}) = \widehat{\deg}\left(\widehat{c}_2(\overline{E}) - \widehat{c}_2(\overline{S} \oplus \overline{Q})\right) = \widehat{\deg}\left(\widehat{\mathrm{ch}}_2(\overline{S} \oplus \overline{Q}) - \widehat{\mathrm{ch}}_2(\overline{E})\right) \geq 0.$$

(3) By virtue of Raynaud's flattening theorem ([**62**, Theorem 1]), there is a birational morphism $\mu_1 : X_1 \to X$ of projective arithmetic varieties such that $\mu_1^*(M/N)/\{\text{torsion elements}\}$ is locally free. Moreover, by Lipman [**42**], there is a resolution of singularities $X' \to X_1$ of X_1. Thus we obtain a birational morphism of regular projective arithmetic varieties such that $\mu^*(M/N)/\{\text{torsion elements}\}$ is locally free. Therefore, if we set

$$N' = \operatorname{Ker}(\mu^*(M) \to \mu^*(M/N)/\{\text{torsion elements}\}),$$

then the assertion of (3) follows. □

Let us start of the proof of the theorem. Since $E_{\overline{K}}$ is semistable, there is a filtration

$$0 = F_0 \subsetneq F_1 \subsetneq \cdots \subsetneq F_{l-1} \subsetneq F_l = E_{\overline{K}}$$

of $E_{\overline{K}}$ on $X_{\overline{K}}$ such that F_i/F_{i-1} $(i = 1, \ldots, l)$ is a stable locally free coherent sheaf and $\mu(F_1/F_0) = \cdots = \mu(F_l/F_{l-1})$. Let K' be a finite extension of K such that F_i is defined over K' for every $i = 1, \ldots, l$. Let X' be a desingularization of $X \times_{\operatorname{Spec}(O_K)} \operatorname{Spec}(O_{K'})$ and $\mu : X' \to X$ the natural morphism. If $\Delta(\mu^*(\overline{E})) \geq 0$, then, by (1) in Claim 5, $\Delta(\overline{E}) \geq 0$. Therefore, we may assume that the filtration

$$0 = F_0 \subsetneq F_1 \subsetneq \cdots \subsetneq F_{l-1} \subsetneq F_l = E_{\overline{K}}$$

is defined over K. We set $E_i = \operatorname{Ker}(E \to E_K/F_i)$ $(i = 1, \ldots, l)$. Then

$$0 = E_0 \subsetneq E_1 \subsetneq \cdots \subsetneq E_{l-1} \subsetneq E_l = E,$$

E_i coincides with F_i on X_K, and E/E_i is torsion free. Moreover, using (1) and (3) in Claim 5, we may assume that E/E_i is locally free for all i. Let h_i be the submetric of E_i induced by h, and let q_i be the quotient metric of E_i/E_{i-1} induced by h_i and $E_i \to E_i/E_{i-1}$. Moreover, we write $\overline{E}_i = (E_i, h_i)$ and $\overline{Q}_i = (E_i/E_{i-1}, q_i)$. We consider an exact sequence

$$0 \to \overline{E}_{i-1} \to \overline{E}_i \oplus \overline{Q}_{i+1} \oplus \cdots \oplus \overline{Q}_l \to \overline{Q}_i \oplus \overline{Q}_{i+1} \oplus \cdots \oplus \overline{Q}_l \to 0.$$

By (2) in Claim 5, we have

$$\Delta(\overline{E}_i \oplus \overline{Q}_{i+1} \oplus \cdots \oplus \overline{Q}_l) \geq \Delta(\overline{E}_{i-1} \oplus \overline{Q}_i \oplus \overline{Q}_{i+1} \oplus \cdots \oplus \overline{Q}_l).$$

Therefore, we obtain $\Delta(\overline{E}) \geq \Delta(\overline{Q}_1 \oplus \cdots \oplus \overline{Q}_l)$. Here, since $\overline{Q}_1 \oplus \cdots \oplus \overline{Q}_l$ is polystable on $X_{\overline{K}}$, the theorem follows from Claim 4. □

REMARK 8.1. Let X be a d-dimensional regular projective arithmetic variety, and $X \to \operatorname{Spec}(O_K) \to \operatorname{Spec}(\mathbb{Z})$ the Stein factorization of $X \to \operatorname{Spec}(\mathbb{Z})$. Let $\overline{E} = (E, h)$ be a C^∞-hermitian locally free coherent sheaf on X. Then, as a generalization of Theorem 8.8, (1) and (2) below are known.

(1) We assume that $d = 2$. For $x \in \widehat{\operatorname{CH}}^1(X)_{\mathbb{Q}}$, the degree of the geometric part of x on X_K is denoted by $\deg_K(x)$. Then, by Moriwaki [**47**] and Naumann [**57**], the following holds: If $r \geq 2$ and $\Delta(\overline{E}) < 0$, then there is a non-zero saturated

subsheaf E' of E such that

$$\deg_K \left(\frac{\widehat{c}_1(\overline{E}')}{\operatorname{rk} E'} - \frac{\widehat{c}_1(\overline{E})}{r} \right) > 0 \quad \text{and}$$

$$\widehat{\deg} \left(\left(\frac{\widehat{c}_1(\overline{E}')}{\operatorname{rk} E'} - \frac{\widehat{c}_1(\overline{E})}{r} \right)^2 \right) \geq \frac{-2\Delta(\overline{E})}{r(r-1)},$$

where the metric of $\overline{E}'$ is the submetric of E' induced by h.

(2) Let $\overline{H}$ be an ample C^∞-hermitian invertible sheaf on X. We assume that $E_{\overline{K}}$ is semistable with respect to $H_{\overline{K}}$ on $X_{\overline{K}}$. Moriwaki [**48**] shows the inequality

$$\widehat{\deg} \left(\left(\widehat{c}_1(\overline{E}) - \frac{r-1}{2r} \widehat{c}_2(\overline{E})^2 \right) \cdot \widehat{c}_1(\overline{H})^{d-2} \right) \geq 0.$$

CHAPTER 9

Lang-Bogomolov Conjecture

In this chapter, we give the proof of the Lang-Bogomolov conjecture, which is a generalization of the following famous Lang conjecture (Faltings' theorem [**19**], [**21**]) :

> Let A be an abelian variety over a number field K and X a geometrically irreducible subvariety of A. If $X(K)$ is Zariski dense in X, then X is a translation of an abelian subvariety of A.

From the viewpoint of Arakelov geometry, the most important part is the proof of Bogomolov's conjecture. As we will show in this book, the Lang-Bogomolov conjecture is actually a consequence of the above Lang conjecture and Bogomolov's conjecture. Bogomolov's conjecture is the central topic of Arakelov geometry in the 90's. The equidistribution theorem, which was a crucial tool for Bogomolov's conjecture, was established by Szpiro-Ullmo-Zhang [**66**] and the conjecture itself was proved by Ullmo [**70**] in the case where curves and their Jacobian. Afterward, it was immediately solved by Zhang [**77**] in general. Moriwaki [**49**] also showed that it also holds over a finitely generated field over $\mathbb{Q}$. The Lang-Bogomolov conjecture was proved by Poonen [**59**] and Zhang [**78**]. It was also generalized to the case of a finitely generated field over $\mathbb{Q}$ by Moriwaki [**50**]. This generalization due to Moriwaki contains the absolute Lang conjecture. The contents of this chapter are based on [**49**], [**50**] and [**33**]. Let us begin with height functions.

9.1. Height function

First, we introduce a naive height of a $\overline{\mathbb{Q}}$-rational point of the projective space.

Let K be a number field, O_K the ring of integers in K and m-Spec(O_K) the set of all maximal ideals of O_K. For $P \in$ m-Spec(O_K) and $\sigma \in K(\mathbb{C})$, the absolute values $|\cdot|_P$ and $|\cdot|_\sigma$ associated with P and σ, respectively, are given by

$$\begin{cases} |x|_P := \#(O_K/P)^{-\operatorname{ord}_P(x)} & (x \in K), \\ |x|_\sigma = |\sigma(x)| & (x \in K). \end{cases}$$

For $x = (x_0, \ldots, x_n) \in K^{n+1} \setminus \{0\}$, we define $h_K(x_0, \ldots, x_n)$ to be

$$h_K(x_0, \ldots, x_n) := \frac{1}{[K:\mathbb{Q}]} \left(\sum_{P \in \text{m-Spec}(O_K)} \log \left(\max_{0 \le i \le n} \{|x_i|_P\} \right) \right.$$
$$\left. + \sum_{\sigma \in K(\mathbb{C})} \log \left(\max_{0 \le i \le n} \{|x_i|_\sigma\} \right) \right).$$

Then we have the following proposition.

PROPOSITION 9.1. (1) *If K'/K is a finite extension of algebraic number fields, then $h_{K'}(x) = h_K(x)$ for all $x \in K^{n+1} \setminus \{0\}$. Therefore, there is*

$$h : \overline{\mathbb{Q}}^{n+1} \setminus \{0\} \to \mathbb{R}$$

such that $h(x) = h_K(x)$ holds for all algebraic number fields K and $x \in K^{n+1} \setminus \{0\}$.

(2) $h(ax) = h(x)$ $(\forall a \in \overline{\mathbb{Q}}^{\times}, \forall x \in \overline{\mathbb{Q}}^{n+1} \setminus \{0\})$.

(3) $h(x_0, \ldots, x_n) = h(\sigma(x_0), \ldots, \sigma(x_n))$ $(\forall (x_0, \ldots, x_n) \in \overline{\mathbb{Q}}^{n+1} \setminus \{0\}, \forall \sigma \in \mathrm{Gal}(\overline{\mathbb{Q}}/\mathbb{Q}))$.

PROOF. (1) Let us begin with the following claim.

CLAIM 1. *Let $P \in \text{m-Spec}(O_K)$ and $PO_{K'} = P_1'^{e_1} \cdots P_r'^{e_r}$ the prime decomposition of $PO_{K'}$. If we set $f_i = [O_{K'}/P_i', O_K/P]$, then $[K' : K] = e_1 f_1 + \cdots + e_r f_r$.*

PROOF. Let t be a local parameter of $(O_K)_P$. Then, by (2) in Lemma 1.12, $\sum_{i=1}^{r} \mathrm{ord}_{P_i'}(t) f_i = \mathrm{ord}_P(N_{K'/K}(t))$. Note that

$$\mathrm{ord}_{P_i'}(t) = e_i \quad \text{and} \quad \mathrm{ord}_P(N_{K'/K}(t)) = [K' : K].$$

Thus the assertion follows. □

For $P \in \text{m-Spec}(O_K)$ and $\sigma \in K(\mathbb{C})$, we set

$$PO_{K'} = P_1'^{e_1} \cdots P_r'^{e_r} \quad \text{and} \quad K'(\mathbb{C})_\sigma = \{\sigma' \in K'(\mathbb{C}) \mid \sigma'|_K = \sigma\}.$$

Then, we have

$$|x|_{P_1'} \cdots |x|_{P_r'} = |x|_P^{[K':K]} \tag{9.1}$$

and

$$\prod_{\sigma' \in K'(\mathbb{C})_\sigma} |x|_{\sigma'} = |x|_\sigma^{[K':K]} \tag{9.2}$$

for $x \in K$. Indeed, as $\mathrm{ord}_{P_i'}(x) = e_i \,\mathrm{ord}_P(x)$, by using the above claim,

$$\begin{aligned} |x|_{P_1'} \cdots |x|_{P_r'} &= \#(O_{K'}/P_1')^{-\mathrm{ord}_{P_1'}(x)} \cdots \#(O_{K'}/P_r')^{-\mathrm{ord}_{P_r'}(x)} \\ &= \#(O_K/P)^{-(f_1 e_1 + \cdots + f_r e_r)\,\mathrm{ord}_P(x)} = |x|_P^{[K':K]}. \end{aligned}$$

Thus (9.1) follows. Moreover, as $\#K'(\mathbb{C})_\sigma = [K' : K]$, (9.2) follows. (1) is a consequence of (9.1) and (9.2).

(2) We choose an algebraic number field K with $a \in K$ and $x \in K^n \setminus \{0\}$. (2) follows from the product formula (Theorem 3.3).

(3) Let K be an algebraic number field such that $\sigma|_K \in \mathrm{Gal}(K/\mathbb{Q})$ and $(x_1, \ldots, x_n) \in K^n \setminus \{0\}$. Then it is easy to see that $h_K(x_1, \ldots, x_n) = h_K(\sigma(x_1), \ldots, \sigma(x_n))$, as required. □

DEFINITION 9.2. By (2) in Proposition 9.1, $h : \overline{\mathbb{Q}}^{n+1} \setminus \{0\} \to \mathbb{R}$ induces $\mathbb{P}^n(\overline{\mathbb{Q}}) \to \mathbb{R}$. It is called the *naive height function* on $\mathbb{P}^n(\mathbb{Q})$ and it is denoted by h_{nv}. For example, if $x \in \mathbb{P}^n(\mathbb{Q})$ with $x = (x_0 : \cdots : x_n)$ $(x_i \in \mathbb{Z}, x_0\mathbb{Z} + \cdots + x_n\mathbb{Z} = \mathbb{Z})$, then $h_{nv}(x) = \log \max\{|x_0|, \ldots, |x_n|\}$.

Let us define a height of a polynomial. Let F be a field and $|\cdot|_v$ an absolute value of F. For a polynomial $f = a_0 + a_1X + \cdots + a_nX^n \in F[X]$ over F, we define $\|f\|_v$ to be

$$\|f\|_v := \max_i\{|a_i|_v\}.$$

Let K be an algebraic number field and $f \in K[X] \setminus \{0\}$. The height $h_K(f)$ of f is defined by

$$h_K(f) := \frac{1}{[K:\mathbb{Q}]}\left(\sum_{P\in \text{m-Spec}(O_K)} \log(\|f\|_P) + \sum_{\sigma \in K(\mathbb{C})} \log(\|f\|_\sigma)\right),$$

that is, if we set $f = a_0 + a_1X + \cdots + a_nX^n$, then $h_K(f) = h_K(a_0, \ldots, a_n)$. Therefore, by (1) in Proposition 9.1, it extends to $h : \overline{\mathbb{Q}}[X] \setminus \{0\} \to \mathbb{R}$ naturally. Here we have the following.

PROPOSITION 9.3. *For $f, g \in \overline{\mathbb{Q}}[X] \setminus \{0\}$,*

$$h(fg) \leq h(f) + h(g) + \log(1 + \min\{\deg(f), \deg(g)\}).$$

PROOF. We may assume that $f, g \in K[X]$ for some algebraic number field K. Thus the proposition follows from the following lemma. □

LEMMA 9.4. *Let F be a field and $|\cdot|_v$ an absolute value of F. Then, for any $f, g \in F[X] \setminus \{0\}$,*

$$\|fg\|_v \leq (1 + \min\{\deg(f), \deg(g)\})\|f\|_v\|g\|_v.$$

Moreover, if $|\cdot|_v$ is discrete, then $\|fg\|_v = \|f\|_v\|g\|_v$ holds.

PROOF. The first part is easily checked by direct calculations. Let us consider the last part. We write

$$R = \{x \in F \mid |x|_v \leq 1\} \quad \text{and} \quad m = \{x \in F \mid |x|_v < 1\}.$$

As $|\cdot|_v$ is discrete, R is a 1-dimensional local integral domain such that m is the maximal ideal of R and there is $t \in R$ with $m = tR$. We set

$$f = \sum_i a_iX^i, \quad g = \sum_i b_iX^i \quad \text{and} \quad fg = \sum_i c_iX^i.$$

Moreover, let a, b and c be integers with

$$\sum_i Ra_i = Rt^a, \quad \sum_i Rb_i = Rt^b \quad \text{and} \quad \sum_i Rc_i = Rt^c.$$

Then $\|fg\|_v = \|f\|_v\|g\|_v$ is equivalent to, say, $a + b = c$. We write

$$f' = \sum_i t^{-a}a_iX^i, \quad g' = \sum_i t^{-b}b_iX^i \quad \text{and} \quad h' = \sum_i t^{-a-b}c_iX^i.$$

Then $h' = f'g'$, so that $h' \in R[X]$ because $f', g' \in R[X]$. Let J be the ideal generated by the coefficients of h'. We would like to show $J = R$. Otherwise, $J \subseteq m$. Let $\bar{f}'$, $\bar{g}'$ be the classes of f' and g' in $(R/m)[X]$, respectively. Then $\bar{f}'\bar{g}' = 0$. Since $(R/m)[X]$ is an integral domain, we have either $\bar{f}' = 0$ or $\bar{g}' = 0$. This is a contradiction because $\sum_i Rt^{-a}a_i = R$ and $\sum_i Rt^{-b}b_i = R$. Thus $J = R$, which means that $\sum_i Rt^{-a-b}c_i = R$. Therefore, $\sum_i Rc_i = Rt^{a+b}$, and hence $c = a+b$. □

By using Proposition 9.3, we have the following theorem.

THEOREM 9.5 (Northcott's theorem of naive height function). *For any positive number e and M, the set*

$$\{x \in \mathbb{P}^n(\overline{\mathbb{Q}}) \mid [\mathbb{Q}(x) : \mathbb{Q}] \leq e,\ h_{nv}(x) \leq M\}$$

is finite, where $\mathbb{Q}(x)$ is a field generated by the inhomogeneous coordinates over $\mathbb{Q}$, that is, if $x = (x_0 : \cdots : x_n)$ and $x_i \neq 0$, then $\mathbb{Q}(x) = \mathbb{Q}(x_0/x_i, \ldots, x_n/x_i)$.

PROOF. It is sufficient to see that, for any positive number e and M, the set

$$\left\{(a_1, \ldots, a_n) \in \overline{\mathbb{Q}}^n \mid [\mathbb{Q}(a_1, \ldots, a_n) : \mathbb{Q}] \leq e,\ h(1, a_1, \ldots, a_n) \leq M\right\}$$

is finite.

First we consider the case $n = 1$, that is, we show that the set

$$\{a \in \overline{\mathbb{Q}} \mid [\mathbb{Q}(a) : \mathbb{Q}] \leq e,\ h(1, a) \leq M\}$$

is finite. Let P be the minimal polynomial of a. If we set $d = \deg(P)$, then $d = [\mathbb{Q}(a) : \mathbb{Q}] \leq e$. Let $P = (X - \alpha_1) \cdots (X - \alpha_d)$ be the decomposition over $\overline{\mathbb{Q}}$. Then, by Proposition 9.3 and (3) in Proposition 9.1,

$$\begin{aligned} h(P) &\leq h(X - \alpha_1) + \cdots + h(X - \alpha_d) + d\log(2) \\ &= h(1, \alpha_1) + \cdots + h(1, \alpha_d) + d\log(2) \\ &= d(h(1, a) + \log(2)) \leq e(M + \log(2)). \end{aligned}$$

Here we write $P = X^d + c_{d-1}X^{d-1} + \cdots + c_0$ and find $b_0, \ldots, b_d \in \mathbb{Z}$ such that $c_i = b_i/b_d$ and $b_d\mathbb{Z} + b_{d-1}\mathbb{Z} + \cdots + b_0\mathbb{Z} = \mathbb{Z}$. Then, by (2) in Proposition 9.1,

$$h(P) = h(1, c_{d-1}, \ldots, c_0) = h(b_d, b_{d-1}, \ldots, b_0) = \log \max\{|b_d|, \ldots, |b_0|\}.$$

Therefore, there are only finitely many possibilities of P, so that we have the case $n = 1$.

Note that $h(1, a_i) \leq h(1, a_1, \ldots, a_n)$ and $\mathbb{Q}(a_i) \subseteq \mathbb{Q}(a_1, \ldots, a_n)$. Thus a general case follows from the case $n = 1$. □

Let K be an algebraic number field and X an integral projective scheme over K. Let L be an invertible sheaf on X. A pair $(\mathscr{X}, \overline{\mathscr{L}})$ is called a model of (X, L) if the following conditions are satisfied.

(1) $\mathscr{X}$ is a projective arithmetic variety with a natural morphism $\mathscr{X} \to \operatorname{Spec}(O_K)$, and the generic fiber of $\mathscr{X} \to \operatorname{Spec}(O_K)$ is X.
(2) $\overline{\mathscr{L}}$ is a C^∞-hermitian $\mathbb{Q}$-invertible sheaf on $\mathscr{X}$ such that $\mathscr{L}$ (forgotten the metric from $\overline{\mathscr{L}}$) is $\mathbb{Q}$-isomorphic to L over X, that is, there are a C^∞-hermitian invertible sheaf $\overline{\mathscr{M}}$ and a positive integer m such that $\mathscr{M}$ is isomorphic to mL over X and $\overline{\mathscr{L}} = (1/m)\overline{\mathscr{M}}$.

The existence of such a model is left to the reader. If $\overline{\mathscr{L}}$ is vertically nef (i.e. $\overline{\mathscr{M}}$ in (2) above is vertically nef), then $(\mathscr{X}, \overline{\mathscr{L}})$ is called a *vertically nef model.* If L is ample, then we can easily check that a vertical nef model of (X, L) exists.

We define $h_{(\mathscr{X}, \overline{\mathscr{L}})} : X(\overline{K}) \to \mathbb{R}$ as follows. For $x \in X(\overline{K})$, let Δ_x be the Zariski closure of the image of $\operatorname{Spec}(\overline{K}) \to X \to \mathscr{X}$. Then $h_{(\mathscr{X}, \overline{\mathscr{L}})}(x)$ is defined by

$$h_{(\mathscr{X}, \overline{\mathscr{L}})}(x) = \frac{\widehat{\deg}\left(\overline{\mathscr{L}}\big|_{\Delta_x}\right)}{[K(x) : \mathbb{Q}]} = \frac{\widehat{\deg}(\widehat{c}_1(\overline{\mathscr{L}}) \cdot (\Delta_x, 0))}{[K(x) : \mathbb{Q}]},$$

where $K(x)$ is the residue field at the image of $\mathrm{Spec}(\overline{K}) \to X$. More precisely, if we use $\overline{M}$ in the above (2), then

$$h_{(\mathscr{X},\overline{\mathscr{L}})}(x) = \frac{\widehat{\deg}\left(\overline{\mathscr{M}}|_{\Delta_x}\right)}{m[K(x):\mathbb{Q}]}$$

and it does not depend on the choice of $\overline{\mathscr{M}}$ and m.

LEMMA 9.6. *Let* $\mathrm{SBs}(L)$ *denote* $\bigcap_{n>0} \mathrm{Supp}(\mathrm{Coker}(H^0(X, nL) \otimes \mathscr{O}_X \to nL))$. *Then there is a constant* C *such that* $h_{(\mathscr{X},\overline{\mathscr{L}})}(x) \geq C$ *for all* $x \in (X \setminus \mathrm{SBs}(L))(\overline{K})$.

PROOF. For an invertible sheaf N on X, $\mathrm{Supp}(\mathrm{Coker}(H^0(X, N) \otimes \mathscr{O}_X \to N))$ is denoted by $\mathrm{Bs}(N)$. Note that $\mathrm{SBs}(N) = \bigcap_{n>0} \mathrm{Bs}(nN)$. Now we consider the following claim.

CLAIM 2. *There is a positive integer* a *with* $\mathrm{SBs}(N) = \mathrm{Bs}(aN)$.

PROOF. Since X is noetherian, it is sufficient to show that if

$$\mathrm{SBs}(N) \subsetneq \mathrm{Bs}(lN)$$

for a positive integer l, then there is a positive integer l' with

$$\mathrm{Bs}(ll'N) \subsetneq \mathrm{Bs}(lN).$$

We assume that $x \in \mathrm{Bs}(lN) \setminus \mathrm{SBs}(N)$. Then, since $x \notin \mathrm{SBs}(N)$, there is a positive integer l' with $x \notin \mathrm{Bs}(l'N)$. In particular, $x \notin \mathrm{Bs}(ll'N)$, so that $\mathrm{Bs}(ll'N) \subsetneq \mathrm{Bs}(lN)$. □

First we consider the case where $\overline{\mathscr{L}}$ is a C^∞-hermitian invertible sheaf in condition (2) of the above definition. By the claim, we can take a sufficiently large integer n with $\mathrm{SBs}(L) = \mathrm{Bs}(nL)$. Let $\{s_1, \ldots, s_r\}$ be a generators of $H^0(\mathscr{X}, n\mathscr{L})$ as an O_K-module. Then, $\{s_1, \ldots, s_r\}$ generates $H^0(X, nL)$ as a K-module. We set

$$D = \max\{\|s_1\|_{\sup}, \ldots, \|s_r\|_{\sup}\}.$$

For $x \in (X \setminus \mathrm{SBs}(L))(\overline{K})$, there is s_i with $s_i|_{\Delta_x} \neq 0$. Since $D^{-1}\|s_i\|_{\sup} \leq 1$, we have

$$\widehat{\deg}\left((n\mathscr{L}, D^{-1}|\cdot|_{n\overline{L}})|_{\Delta_x}\right) \geq 0.$$

On the other hand,

$$\widehat{\deg}\left((n\mathscr{L}, D^{-1}|\cdot|_{n\overline{L}})|_{\Delta_x}\right) = n\widehat{\deg}\left((\mathscr{L}, |\cdot|_{\overline{L}})|_{\Delta_x}\right) + \log(D)[K(x):\mathbb{Q}].$$

Therefore, the lemma follows in this case.

In a general case, there are a C^∞-hermitian invertible sheaf $\overline{\mathscr{M}}$ on $\mathscr{X}$ and a positive integer m such that $\mathscr{M}$ is isomorphic to mL over X and $\overline{\mathscr{L}} = (1/m)\overline{\mathscr{M}}$. Let M be the restriction of $\mathscr{M}$ to X. Then $M \simeq mL$ and, by the previous case, there is a constant C such that $h_{(\mathscr{X},\overline{\mathscr{M}})}(x) \geq C$ for all $x \in (X \setminus \mathrm{SBs}(M))(\overline{K})$. On the other hand, $\mathrm{SBs}(L) = \mathrm{SBs}(M)$. Indeed, by the claim,

$$\mathrm{SBs}(L) = \mathrm{Bs}(aL), \quad \mathrm{SBs}(M) = \mathrm{Bs}(bM)$$

for some positive integers a and b. Thus, as

$$\begin{cases} \mathrm{SBs}(L) \subseteq \mathrm{Bs}(abmL) \subseteq \mathrm{Bs}(aL), \\ \mathrm{SBs}(M) \subseteq \mathrm{Bs}(abM) \subseteq \mathrm{Bs}(bM), \end{cases}$$

we obtain $\mathrm{SBs}(L) = \mathrm{SBs}(M)$. Moreover, as $h_{(\mathscr{X},\overline{\mathscr{M}})}(x) = mh_{(\mathscr{X},\overline{\mathscr{L}})}(x)$, the general case follows. □

PROPOSITION 9.7. *Let $(\mathscr{X}', \overline{\mathscr{L}}')$ be another model of (X, L). Then there is a positive constant C such that*

$$\left| h_{(\mathscr{X}',\overline{\mathscr{L}}')}(x) - h_{(\mathscr{X},\overline{\mathscr{L}})}(x) \right| \leq C$$

for all $x \in X(\overline{K})$.

PROOF. Since $(\mathscr{X}', \overline{\mathscr{L}}')$ is a model of (X, L), there are a C^∞-hermitian invertible sheaf $\overline{\mathscr{M}}'$ on $\mathscr{X}'$ and a positive integer m' such that $\mathscr{M}'$ is isomorphic to $m'L$ on X and $\overline{\mathscr{L}}' = (1/m')\overline{\mathscr{M}}'$. Here we take a projective arithmetic variety $\mathscr{Y}$ with the following properties.

(1) There is a morphism $\mathscr{Y} \to \mathrm{Spec}(O_K)$ such that the generic fiber of $\mathscr{Y} \to \mathrm{Spec}(O_K)$ is X.
(2) There are birational morphisms $\mu : \mathscr{Y} \to \mathscr{X}$ and $\mu' : \mathscr{Y} \to \mathscr{X}'$ over $\mathrm{Spec}(O_K)$.

We set $\overline{\mathscr{A}} = m'\mu^*\overline{\mathscr{M}} - m\mu'^*\overline{\mathscr{M}}'$. Then $\mathscr{A}$ and $-\mathscr{A}$ is isomorphic to $\mathscr{O}_X$ on X. Thus, by applying the previous lemma to $\overline{\mathscr{A}}$ and $-\overline{\mathscr{A}}$, there is a positive constant C such that $|h_{(\mathscr{Y},\overline{\mathscr{A}})}(x)| \leq C$ for all $x \in X(\overline{K})$. On the other hand, by Lemma 3.18 and (1) in Lemma 3.20, we can see

$$\begin{aligned} h_{(\mathscr{Y},\overline{\mathscr{A}})}(x) &= m'h_{(\mathscr{Y},\mu^*\overline{\mathscr{M}})}(x) - mh_{(\mathscr{Y},\mu'^*\overline{\mathscr{M}}')}(x) = m'h_{(\mathscr{X},\overline{\mathscr{M}})}(x) - mh_{(\mathscr{X}',\overline{\mathscr{M}}')}(x) \\ &= mm'(h_{(\mathscr{X},\overline{\mathscr{L}})}(x) - h_{(\mathscr{X}',\overline{\mathscr{L}}')}(x)). \end{aligned}$$

Thus the proposition follows. □

DEFINITION 9.8. A map $h : X(\overline{K}) \to \mathbb{R}$ is called a *height function associated with L* if there is a model $(\mathscr{X}, \overline{\mathscr{L}})$ of (X, L) such that $h - h_{(\mathscr{X},\overline{\mathscr{L}})}$ is a bounded function on $X(\overline{K})$. A height function associated with L is not determined uniquely. However, by Proposition 9.7, it is unique modulo the set of bounded functions on $X(\overline{K})$.

PROPOSITION 9.9. *Let $f : X \to Y$ be a morphism of integral projective schemes over an algebraic number field K. Let L be an invertible sheaf on Y. Let $h_L : Y(\overline{K}) \to \mathbb{R}$ and $h_{f^*(L)} : X(\overline{K}) \to \mathbb{R}$ be height functions associated with L and $f^*(L)$, respectively. Then $h_{f^*(L)} - h_L \circ f$ is a bounded function.*

PROOF. Let $(\mathscr{Y}, \overline{\mathscr{L}})$ be a model of (Y, L). We choose a model $\mathscr{X}$ of X such that $f : X \to Y$ extends to $\tilde{f} : \mathscr{X} \to \mathscr{Y}$. Then $(\mathscr{X}, \tilde{f}^*(\overline{\mathscr{L}}))$ is a model of $(X, f^*(L))$. Thus, for $x \in X(\overline{K})$, by using the projection formula,

$$\widehat{\deg}\left(\tilde{f}^*(\overline{\mathscr{L}})\Big|_{\Delta_x} \right) = \deg(\Delta_x \to \Delta_{f(x)})\widehat{\deg}\left(\overline{\mathscr{L}}\big|_{\Delta_{f(x)}} \right).$$

Therefore, $h_{(\mathscr{X},\tilde{f}^*(\overline{\mathscr{L}}))}(x) = h_{(\mathscr{Y},\overline{\mathscr{L}})}(f(x))$, which shows the proposition. □

Let $\mathbb{P}^n_{\mathbb{Z}} = \mathrm{Proj}(\mathbb{Z}[X_0, \ldots, X_n])$ be the n-dimensional projective space over $\mathbb{Z}$. Note that $\dim \mathbb{P}^n_{\mathbb{Z}} = n + 1$. Let $\mathscr{O}_{\mathbb{P}^n}(1)$ be the tautological invertible sheaf on $\mathbb{P}^n_{\mathbb{Z}}$ and h_{FS} the Fubini-Study metric of $\mathscr{O}_{\mathbb{P}^n}(1)$, that is,

$$h_{FS}(X_i, X_j) = \frac{X_i \bar{X}_j}{|X_0|^2 + \cdots + |X_n|^2}.$$

Then we have the following.

PROPOSITION 9.10. *Let K be an algebraic number field. Then*

$$\widehat{\deg}(\,(\mathscr{O}_{\mathbb{P}^n}(1), h_{FS})|_{\Delta_x}) = \sum_{P \in \text{m-Spec}(O_K)} \log \max\{\|x_0\|_P, \ldots, \|x_n\|_P\} + \sum_{\sigma \in K(\mathbb{C})} \log \sqrt{|x_0|_\sigma^2 + \cdots + |x_n|_\sigma^2}$$

for $x = (x_0 : \cdots : x_n) \in \mathbb{P}^n_{\mathbb{Z}}(K)$. In particular, the naive height function $h_{nv} : \mathbb{P}^n(\overline{\mathbb{Q}}) \to \mathbb{R}$ in Definition 9.8 is a height function associated with $\mathscr{O}_{\mathbb{P}^n}(1)$.

PROOF. The morphism $\operatorname{Spec}(K) \to \mathbb{P}^n_{\mathbb{Z}}$ given by $x \in \mathbb{P}^n_{\mathbb{Z}}(K)$ extends to a morphism $\phi_x : \operatorname{Spec}(O_K) \to \mathbb{P}^n_{\mathbb{Z}}$ uniquely. Clearly,

$$\widehat{\deg}(\,(\mathscr{O}_{\mathbb{P}^n}(1), h_{FS})|_{\Delta_x}) = \widehat{\deg}(\phi_x^*(\mathscr{O}_{\mathbb{P}^n}(1), h_{FS})),$$

so that it is sufficient to see that

$$\widehat{\deg}(\phi_x^*(\mathscr{O}_{\mathbb{P}^n}(1), h_{FS})) = \sum_{P \in \text{m-Spec}(O_K)} \log \max\{\|x_0\|_P, \ldots, \|x_n\|_P\} + \sum_{\sigma \in K(\mathbb{C})} \log \sqrt{|x_0|_\sigma^2 + \cdots + |x_n|_\sigma^2}.$$

Without loss of generality, we may assume $x_0 \neq 0$. We write $\tilde{x}_i = \phi_x^*(X_i)$, $a_i = \tilde{x}_i/\tilde{x}_0 = x_i/x_0$. Then $a_i \in K$ and $\phi_x^*(\mathscr{O}_{\mathbb{P}^n}(1)) = O_K\tilde{x}_0 + \cdots + O_K\tilde{x}_n$. Thus,

$$\begin{aligned}
\frac{\phi_x^*(\mathscr{O}_{\mathbb{P}^n}(1))}{O_K\tilde{x}_0} &= \frac{O_K\tilde{x}_0 + \cdots + O_K\tilde{x}_n}{O_K\tilde{x}_0} \\
&= \frac{(O_K a_0 + O_K a_1 + \cdots + \mathscr{O}_K a_n)\tilde{x}_0}{O_K\tilde{x}_0} \\
&\simeq \frac{O_K a_0 + O_K a_1 + \cdots + \mathscr{O}_K a_n}{O_K}.
\end{aligned}$$

Therefore, we obtain

$$\begin{aligned}
\log \# \left(\frac{\phi_x^*(\mathscr{O}_{\mathbb{P}^n}(1))}{O_K\tilde{x}_0} \right) &= \log \# \left(\frac{O_K a_0 + O_K a_1 + \cdots + \mathscr{O}_K a_n}{O_K} \right) \\
&= \sum_{P \in \text{m-Spec}(O_K)} \log \# \left(\frac{(O_K a_0 + O_K a_1 + \cdots + \mathscr{O}_K a_n)_P}{(O_K)_P} \right).
\end{aligned}$$

Let t_P be a local parameter of $P(O_K)_P$. Then

$$(O_K a_0 + O_K a_1 + \cdots + \mathscr{O}_K a_n)_P = t_P^{\min\{\operatorname{ord}_P(a_i)\}}(O_K)_P,$$

and hence

$$\begin{aligned}
\log \# \left(\frac{(O_K a_0 + O_K a_1 + \cdots + \mathscr{O}_K a_n)_P}{(O_K)_P} \right) &= \log \left(\#(O_K/P)^{\max\{-\operatorname{ord}_P(a_i)\}} \right) \\
&= \log \left(\max\{\#(O_K/P)^{-\operatorname{ord}_P(a_i)}\} \right) \\
&= \log \left(\max\{\|a_i\|_P\} \right).
\end{aligned}$$

Therefore, we have

$$\log \# \left(\frac{\phi_x^*(\mathscr{O}_{\mathbb{P}}(1))}{O_K \tilde{x}_0} \right) = \sum_{P \in \text{m-Spec}(O_K)} \log \left(\max\{ \|x_i/x_0\|_P \} \right). \tag{9.3}$$

For each $\sigma \in K(\mathbb{C})$, $\phi_x^*(h_{FS})_\sigma(\tilde{x}_0 \otimes^\sigma 1, \tilde{x}_0 \otimes^\sigma 1)$ is the value of $h_{FS}(X_0, X_0)$ at the point $(\sigma(x_0) : \cdots : \sigma(x_n))$. Thus,

$$-(1/2)\log(\phi_x^*(h_{FS})_\sigma(\tilde{x}_0 \otimes^\sigma 1, \tilde{x}_0 \otimes^\sigma 1)) = \log \left(\frac{\sqrt{|x_0|_\sigma^2 + \cdots + |x_n|_\sigma^2}}{|x_0|_\sigma} \right). \tag{9.4}$$

Therefore, by (9.3), (9.4) and the product formula, the proposition follows. The last assertion follows from the inequalities:

$$\max\{|z_0|, \ldots, |z_n|\} \leq \sqrt{|z_0|^2 + \cdots + |z_n|^2} \leq \sqrt{n+1} \max\{|z_0|, \ldots, |z_n|\}$$

for $z_0, \ldots, z_n \in \mathbb{C}$. □

THEOREM 9.11 (Northcott's theorem). *Let X be an integral projective scheme over an algebraic number field K and L an ample invertible sheaf on X. Let $h_L : X(\overline{K}) \to \mathbb{R}$ be a height function associated with L. Then, for positive numbers e and M, the set*

$$\{x \in X(\overline{K}) \mid [K(x) : \mathbb{Q}] \leq e, \ h_L(x) \leq M\}$$

is finite.

PROOF. We may replace L with aL ($a \in \mathbb{Z}_{>0}$), so that we may assume that there is an embedding $\phi : X \hookrightarrow \mathbb{P}_K^n$ with $\phi^*(\mathscr{O}_{\mathbb{P}^n}(1)) = L$. Thus, by Proposition 9.9, $h_L - h_{nv} \circ \phi$ is a bounded function. Therefore, the theorem follows from Theorem 9.5. □

9.2. Height function on abelian variety

In this section, let us consider height functions on an abelian variety. The following result is a key lemma of this section.

LEMMA 9.12. *Let A be an abelian group and $h : A \to \mathbb{R}$ a map such that*

$$h(x+y+z) - h(x+y) - h(y+z) - h(z+x) + h(x) + h(y) + h(z)$$

is a bounded function on $A \times A \times A$. Then there are a unique symmetric bilinear map $b : A \times A \to \mathbb{R}$ and a unique linear map $l : A \to \mathbb{R}$ such that

$$h(x) - \frac{1}{2}b(x,x) - l(x)$$

is a bounded function on A.

PROOF. We set $\beta(x,y) = h(x+y) - h(x) - h(y)$. Clearly β is symmetric and, by our assumption, there is a constant C such that

$$|\beta(x+y,z) - \beta(x,z) - \beta(y,z)| \leq C$$

for all $x, y, z \in A$. Thus, as

$$\begin{aligned}
&\beta(2^{n+1}x, 2^{n+1}y) - 4\beta(2^n x, 2^n y) \\
&\qquad = \beta(2^n x + 2^n x, 2^{n+1}y) - \beta(2^n x, 2^{n+1}y) - \beta(2^n x, 2^{n+1}y) \\
&\qquad\qquad + 2\left(\beta(2^n y + 2^n y, 2^n x) - \beta(2^n y, 2^n x) - \beta(2^n y, 2^n x)\right),
\end{aligned}$$

we have

$$|\beta(2^{n+1}x, 2^{n+1}y) - 4\beta(2^n x, 2^n y)| \leq 3C,$$

that is,

$$|4^{-(n+1)}\beta(2^{n+1}x, 2^{n+1}y) - 4^{-n}\beta(2^n x, 2^n y)| \leq 3C4^{-(n+1)},$$

which means that a sequence $\{4^{-n}\beta(2^n x, 2^n y)\}_{n=1}^{\infty}$ is a Cauchy sequence. Therefore, there exists the limit $\lim_{n\to\infty} 4^{-n}\beta(2^n x, 2^n y)$. Here we define $b(x, y)$ to be

$$b(x, y) = \lim_{n\to\infty} 4^{-n}\beta(2^n x, 2^n y).$$

Then b is symmetric because β is symmetric. Moreover, since

$$|4^{-N}\beta(2^N x, 2^N y) - \beta(x, y)| \leq 3C \sum_{n=0}^{N-1} 4^{-(n+1)} \leq C$$

by the previous estimate, we obtain $|b - \beta| \leq C$. Further, b is bilinear because

$$|4^{-n}\beta(2^n(x+y), 2^n z) - 4^{-n}\beta(2^n x, 2^n z) - 4^{-n}\beta(2^n y, 2^n z)| \leq 4^{-n}C.$$

Next we set $\lambda(x) = h(x) - \frac{1}{2}b(x, x)$. Then, as

$$\lambda(x+y) - \lambda(x) - \lambda(y) = \beta(x, y) - b(x, y),$$

we have

$$|\lambda(x+y) - \lambda(x) - \lambda(y)| \leq C,$$

and hence

$$|2^{-(n+1)}\lambda(2^{n+1}x) - 2^{-n}\lambda(2^n x)| \leq 2^{-(n+1)}C,$$

so that we define $l(x)$ to be $l(x) = \lim_{n\to\infty} 2^{-n}\lambda(2^n x)$. We can easily see that $|l - \lambda| \leq C$. Moreover, in the same way as before, we can see that l is linear. Hence, by using b and l, we obtain

$$h(x) = \frac{1}{2}b(x, x) + l(x) + O(1).$$

Finally, we consider the uniqueness of b and l. Let b' and l' be another symmetric bilinear map and another linear map with $h = \frac{1}{2}b' + l' + O(1)$. Then, as $\beta = b' + O(1)$, we have $b = b' + O(1)$, that is, $|b - b'| \leq C'$ for some constant C'. In particular,

$$2^n|b(x, y) - b'(x, y)| = |b(2^n x, y) - b'(2^n x, y)| \leq C'.$$

Thus, taking n to ∞, we have $b = b'$, and hence $l = l' + O(1)$. Therefore, in the same way, we can see $l = l'$. □

THEOREM 9.13. *Let A be an abelian variety over $\overline{\mathbb{Q}}$. For any invertible sheaf L on A, there are a unique symmetric bilinear map $b_L : A(\overline{\mathbb{Q}}) \times A(\overline{\mathbb{Q}}) \to \mathbb{R}$ and a unique linear map $l_L : A(\overline{\mathbb{Q}}) \to \mathbb{R}$ such that*

$$h_L(x) = \frac{1}{2}b_L(x, x) + l_L(x) + O(1)$$

for $x \in A(\overline{\mathbb{Q}})$, where h_L is a height function associated with L. We define $\hat{h}_L(x)$ to be

$$\hat{h}_L(x) = \frac{1}{2}b_L(x, x) + l_L(x)$$

and it is called the canonical height *of L, or the* Néron-Tate height *of L. Moreover, the canonical height of L satisfies the following properties:*

(1) $\hat{h}_{L_1\otimes L_2} = \hat{h}_{L_1} + \hat{h}_{L_2}$ *for any invertible sheaves L_1 and L_2.*

(2) *Let* $f : A \to B$ *be a homomorphism of abelian varieties. Then* $\hat{h}_{f^*(L)} = f^*(\hat{h}_L)$.
(3) *Let* $[-1] : A \to A$ *be the endomorphism of* A *given by* $[-1](x) = -x$. *If* L *is* symmetric *(i.e.* $[-1]^*(L) \simeq L$*), then* $l_L = 0$. *Moreover, if* L *satisfies* $[-1]^*(L) = L^{-1}$, *then* $b_L = 0$.

Proof. To show the first half of the theorem, we need the following lemma (for details, see [**55**, p. 58, Corollary 2]).

Lemma 9.14 (Cubic theorem). *Let* X *be a projective variety and* A *an abelian variety. We assume that there are three morphisms* f, g, h *from* X *to* A. *Then, for any invertible sheaf* L *on* A, *we have*

$$(f+g+h)^*(L) \otimes (f+g)^*(L^{-1}) \otimes (g+h)^*(L^{-1}) \otimes (h+f)^*(L^{-1}) \otimes f^*(L) \otimes g^*(L) \otimes h^*(L) \simeq \mathcal{O}_X.$$

Let $p_i : A \times A \times A \to A$ be the projection to the i-th factor. Then, by Lemma 9.14,

$$(p_1+p_2+p_3)^*(L) \otimes (p_1+p_2)^*(L^{-1}) \otimes (p_2+p_3)^*(L^{-1}) \otimes (p_3+p_1)^*(L^{-1}) \otimes p_1^*(L) \otimes p_2^*(L) \otimes p_3^*(L) \simeq \mathcal{O}_A.$$

Thus we obtain

$$h_L(x+y+z) - h_L(x+y) - h_L(y+z) - h_L(z+x) + h_L(x) + h_L(y) + h_L(z) = O(1).$$

Therefore, the first half of the theorem follows from Lemma 9.12.

(1) and (2) are obvious.

(3) If $[-1]^*L \simeq L$, then, by (2), $\hat{h}_L(-x) = \hat{h}_L(x)$, which implies $l(-x) = l(x)$. Therefore, $l(x) = 0$. Moreover, if $[-1]^*L \simeq L^{-1}$, then, by (2), $\hat{h}_L(-x) = -\hat{h}_L(x)$, which implies $b(-x,-x) = -b(x,x)$. Thus $b(x,x) = 0$. Note that $2b(x,y) = b(x+y, x+y) - b(x,x) - b(y,y)$. Therefore, $b = 0$. □

Let A be an abelian variety over $\overline{\mathbb{Q}}$ and let L be a symmetric and ample invertible sheaf on A. From now on, $b_L(x,y)$ is often denoted by $\langle x, y\rangle_L$ and it is called *Néron-Tate height pairing.* By Theorem 9.13, $\hat{h}_L(x) = \frac{1}{2}\langle x, x\rangle_L$. We have the following theorem.

Theorem 9.15. (1) *The Néron-Tate height pairing* $\langle\ ,\ \rangle_L : A(\overline{\mathbb{Q}}) \times A(\overline{\mathbb{Q}}) \to \mathbb{R}$ *is positive definite modulo* $A(\overline{\mathbb{Q}})_{tor}$, *that is,* $\langle x, x\rangle_L \geq 0$ $(\forall x \in A(\overline{\mathbb{Q}}))$ *and* $\langle x, x\rangle_L = 0$ *if and only if* $x \in A(\overline{\mathbb{Q}})_{tor}$.
(2) $\langle\ ,\ \rangle_L$ *extends to an inner product of* $A(\overline{\mathbb{Q}}) \otimes \mathbb{R}$.

Proof. (1) As L is ample, nL is generated by global sections for a sufficiently large n. Thus, by Lemma 9.6, there is a constant C with $\hat{h}_{nL}(x) \geq C$, that is, $\hat{h}_L(x) \geq C/n$. Therefore,

$$m^2 \hat{h}_L(x) = \hat{h}_L(mx) \geq C/n.$$

Taking m to ∞, we have $\hat{h}_L(x) \geq 0$. To see that $\langle\ ,\ \rangle_L$ is positive definite, it is sufficient to show that if $\hat{h}_L(x) = 0$, then $x \in A(\overline{\mathbb{Q}})_{tor}$. We choose an algebraic number field K such that $x \in A(K)$. Let G be a cycle subgroup generated by x. For

$y = nx \in G$, $\hat{h}_L(y) = n^2\hat{h}_L(x) = 0$. Thus, by Northcott's theorem (Theorem 9.11), G is a finite group. Thus $x \in A(K)_{tor}$.

(2) We denote the natural homomorphism $A(\overline{\mathbb{Q}}) \to A(\overline{\mathbb{Q}}) \otimes \mathbb{R}$ by ι. Moreover, by abuse of notation, the extension of the symmetric $\mathbb{Z}$-bilinear map $\langle\ ,\ \rangle_L$ on $A(\overline{\mathbb{Q}})$ to $A(\overline{\mathbb{Q}}) \otimes \mathbb{R}$ is denoted by $\langle\ ,\ \rangle_L$, that is, $\langle x \otimes a, y \otimes b\rangle_L = ab\langle x, y\rangle_L$. We need to see that $\langle x, x\rangle_L \geq 0$ holds for $x \in A(\overline{\mathbb{Q}}) \otimes \mathbb{R}$ and that $\langle x, x\rangle_L = 0$ if and only if $x = 0$. We write

$$x = P_1 \otimes a_1 + \cdots + P_l \otimes a_l \quad (P_1, \ldots, P_l \in A(\overline{\mathbb{Q}}),\ a_1, \ldots, a_l \in \mathbb{R}).$$

Let M be a subgroup generated by $P_1, \ldots, P_l$ in $A(\overline{\mathbb{Q}})$ and V a vector subspace of $A(\overline{\mathbb{Q}}) \otimes \mathbb{R}$ generated by $\Lambda = \iota(M)$. Note that Λ is a lattice of V. Moreover, if we take an algebraic number field K with $P_1, \ldots, P_l \in A(K)$, then M is a subgroup of $A(K)$. For $v \in V$, we set $q(v) = \langle v, v\rangle_L$. Then, by (1), $q(\lambda) \geq 0$ for $\lambda \in \Lambda$ and $q(\lambda) = 0$ if and only if $\lambda = 0$. Moreover, by Northcott's theorem, for any positive number C, the set $\{\lambda \in \Lambda \mid q(\lambda) \leq C\}$ is finite. In particular,

$$\inf\{q(\lambda) \mid \lambda \in \Lambda \setminus \{0\}\} > 0.$$

Therefore, by Proposition 2.3, q is positive definite. Thus (2) follows. □

REMARK 9.1. Let L be a symmetric invertible sheaf on A and h_L a height function associated with L. Then, by the construction of Lemma 9.12,

$$\lim_{n\to\infty} \sup_{x \in A(\overline{\mathbb{Q}})} \left|\hat{h}_L(x) - 4^{-n} h_L(2^n x)\right| = 0.$$

REMARK 9.2. The Mordell-Weil theorem (for example, see [**55**]) says that $A(K)$ is finitely generated abelian group for an algebraic number field K. Thus

$$\langle\ ,\ \rangle_L : A(K) \times A(K) \to \mathbb{R}$$

yields an inner product on the finite-dimensional vector space $A(K) \otimes \mathbb{R}$.

REMARK 9.3. Let X be a regular projective arithmetic surface and let $X \to \operatorname{Spec}(O_K) \to \operatorname{Spec}(\mathbb{Z})$ be the Stein factorization of $X \to \operatorname{Spec}(\mathbb{Z})$, where K is an algebraic number field and O_K is the ring of integers in K. We assume that $X \to \operatorname{Spec}(O_K)$ is semistable. Let $\overline{L}$ be a C^∞-hermitian invertible sheaf on X such that $c_1(\overline{L}) = 0$ and $\deg(L|_F) = 0$ for all irreducible components of fibers of $X \to \operatorname{Spec}(O_K)$. The exact form of the arithmetic Hodge index theorem due to Faltings [**18**] and Hriljac [**31**] is the identity

$$-2[K : \mathbb{Q}]\hat{h}(L_K) = \widehat{\deg}(\widehat{c}_1(\overline{L})^2),$$

where $\hat{h}(L_K)$ is the Néron-Tate height viewed L_K as a point of $\operatorname{Pic}^0(X_K)$. In particular, $\widehat{\deg}(\widehat{c}_1(\overline{L})^2) \leq 0$.

9.3. Equidistribution theorem

In this section, we discuss the equidistribution theorem due to Szpiro-Ullmo-Zhang, which is a key result for Bogomolov's conjecture. First we consider the following theorem as an application of the generalized Hodge index theorem.

THEOREM 9.16. *Let $\mathscr{X}$ be a $(d+1)$-dimensional projective arithmetic variety and let $\overline{\mathscr{L}}$ be a C^∞-hermitian invertible sheaf of real type on $\mathscr{X}$. We assume that $\overline{\mathscr{L}}$ is vertically nef and $\deg(\mathscr{L}_{\mathbb{Q}}^d) > 0$. Then*

$$\inf_{x \in \mathscr{X}(\overline{\mathbb{Q}})} h_{(\mathscr{X},\overline{\mathscr{L}})}(x) \leq \frac{\widehat{\deg}(\widehat{c}_1(\overline{\mathscr{L}})^{d+1})}{(d+1)\deg(\mathscr{L}_{\mathbb{Q}}^d)} \leq \sup_{Y \subsetneq \mathscr{X}_{\mathbb{Q}}} \left\{ \inf_{x \in (\mathscr{X}_{\mathbb{Q}} \setminus Y)(\overline{\mathbb{Q}})} h_{(\mathscr{X},\overline{\mathscr{L}})}(x) \right\}$$

holds, where Y's run over all proper subvarieties of $\mathscr{X}_{\mathbb{Q}}$.

PROOF. First we have the following claim.

CLAIM 3. *Under the assumption of the theorem, if $\widehat{\deg}(\widehat{c}_1(\overline{\mathscr{L}})^{d+1}) > 0$, then*

$$\sup_{Y \subsetneq \mathscr{X}_{\mathbb{Q}}} \left\{ \inf_{x \in (\mathscr{X}_{\mathbb{Q}} \setminus Y)(\overline{\mathbb{Q}})} h_{(\mathscr{X},\overline{\mathscr{L}})}(x) \right\} \geq 0.$$

PROOF. By Theorem 6.12, replacing $\overline{\mathscr{L}}$ by $\overline{\mathscr{L}}^{\otimes n}$ $(n > 0)$ if necessary, we may assume that there is a small section s of $\mathscr{L}$. Note that

$$h_{(\mathscr{X},\overline{\mathscr{L}})}(x) = \frac{1}{\deg(\Delta_x \to B)} \widehat{\deg}\left(\widehat{c}_1(\overline{\mathscr{L}}) \cdot (\Delta_x, 0)\right)$$

for any $x \in \mathscr{X}_{\mathbb{Q}}(\overline{\mathbb{Q}})$. Thus, if $x \notin \operatorname{Supp}(\operatorname{div}(s))_{\mathbb{Q}}(\overline{\mathbb{Q}})$, then $h_{(\mathscr{X},\overline{\mathscr{L}})}(x) \geq 0$, which proves the claim. $\square$

Fix a real number c with $0 < c < 1$. Let $\overline{A}$ be a hermitian $\mathbb{Z}$-module given by $\overline{A} = (\mathscr{O}_{\operatorname{Spec}(\mathbb{Z})}, c|\cdot|)$. First we consider the right inequality. Let us fix an arbitrary rational number λ with

$$\lambda < \frac{\widehat{\deg}(\widehat{c}_1(\overline{\mathscr{L}})^{d+1})}{(d+1)\deg(\mathscr{L}_{\mathbb{Q}}^d)\widehat{\deg}(\overline{A})}.$$

It is easy to see

$$\widehat{\deg}\left(\left(\widehat{c}_1(\overline{\mathscr{L}}) - \lambda\widehat{c}_1(\pi^*(\overline{A}))\right)^{d+1}\right) > 0,$$

where $\pi : \mathscr{X} \to \operatorname{Spec}(\mathbb{Z})$ is the structure morphism. As $\overline{\mathscr{L}} - \lambda\pi^*(\overline{A})$ is vertically nef and it is big on $\mathscr{X}_{\mathbb{Q}}$, the above claim implies

$$\sup_{Y \subsetneq \mathscr{X}_{\mathbb{Q}}} \left\{ \inf_{x \in (\mathscr{X}_{\mathbb{Q}} \setminus Y)(\overline{\mathbb{Q}})} h_{(\mathscr{X},\overline{\mathscr{L}}-\lambda\pi^*(\overline{A}))}(x) \right\} \geq 0,$$

and hence

$$\sup_{Y \subsetneq \mathscr{X}_{\mathbb{Q}}} \left\{ \inf_{x \in (\mathscr{X}_{\mathbb{Q}} \setminus Y)(\overline{\mathbb{Q}})} h_{(\mathscr{X},\overline{\mathscr{L}})}(x) \right\} \geq \lambda\widehat{\deg}(\overline{A}).$$

Therefore, we have the right inequality.

Next let us see the left inequality. We fix an arbitrary rational number μ with

$$\mu \leq \frac{\inf_{x \in \mathscr{X}(\overline{\mathbb{Q}})} h_{(\mathscr{X},\overline{\mathscr{L}})}(x)}{\widehat{\deg}(\overline{A})},$$

so that

$$\inf_{x \in \mathscr{X}(\overline{\mathbb{Q}})} h_{(\mathscr{X},\overline{\mathscr{L}}-\mu\pi^*(\overline{A}))}(x) \geq 0,$$

which shows that $\overline{\mathscr{L}} - \mu\pi^*(\overline{A})$ is nef. Therefore, by Theorem 6.15, we obtain

$$\widehat{\deg}\left(\left(\widehat{c}_1(\overline{\mathscr{L}}) - \mu\widehat{c}_1(\pi^*(\overline{A}))\right)^{d+1}\right) \geq 0,$$

that is,

$$\widehat{\deg}\left(\widehat{c}_1(\overline{\mathscr{L}})^{d+1}\right) \geq \mu(d+1)\deg(\mathscr{L}_{\mathbb{Q}}^d)\widehat{\deg}(\overline{A}),$$

and hence the left inequality follows. □

Let K be an algebraic number field and let X be a geometrically irreducible projective variety over K. We denote the structure morphism $X \to \operatorname{Spec}(K)$ by p. Let $X(\mathbb{C})$ be the set of all $\mathbb{C}$-value points of X as a scheme over $\mathbb{Q}$. Let us fix a $\mathbb{C}$-valued point $\sigma : \operatorname{Spec}(\mathbb{C}) \to \operatorname{Spec}(K)$ of $\operatorname{Spec}(K)$, that is, an embedding $K \to \mathbb{C}$. Let $X_\sigma = X \times^{\sigma}_{\operatorname{Spec}(K)} \operatorname{Spec}(\mathbb{C})$ be the base change with respect to σ. Then a $\mathbb{C}$-valued point of X_σ as a scheme over $\mathbb{C}$ is nothing more than $x \in X(\mathbb{C})$ such that the following diagram is commutative:

$$\begin{array}{ccc} \operatorname{Spec}(\mathbb{C}) & \xrightarrow{\ x\ } & X \\ {\scriptstyle\sigma}\downarrow & \swarrow{\scriptstyle p} & \\ \operatorname{Spec}(K). & & \end{array}$$

Moreover, $X_\sigma(\mathbb{C})$ is one of the connected components of $X(\mathbb{C})$. We choose $\tilde{\sigma} : \operatorname{Spec}(\mathbb{C}) \to \operatorname{Spec}(\overline{K})$ such that the composition

$$\operatorname{Spec}(\mathbb{C}) \xrightarrow{\tilde{\sigma}} \operatorname{Spec}(\overline{K}) \to \operatorname{Spec}(K)$$

coincides with σ. Then, for $x \in X(\overline{K})$, the composition $\operatorname{Spec}(\mathbb{C}) \xrightarrow{\tilde{\sigma}} \operatorname{Spec}(\overline{K}) \xrightarrow{x} X$ gives rise to an element of $X_\sigma(\mathbb{C})$. Namely, we have a natural map $\iota_{\tilde{\sigma}} : X(\overline{K}) \to X_\sigma(\mathbb{C})$. Notice that $\operatorname{Gal}(\overline{K}/K)$ acts on $X(\overline{K})$ naturally. For $x \in X(\overline{K})$, the image of the $\operatorname{Gal}(\overline{K}/K)$-orbit of x by $\iota_{\tilde{\sigma}}$ does not depend on the choice of x, so that it is denoted by $O_\sigma(x)$.

Let $\{x_m\}_{m=1}^\infty$ be a sequence in $X(\overline{K})$. If any subsequence of $\{x_m\}_{m=1}^\infty$ is Zariski dense in X, then the sequence $\{x_m\}_{m=1}^\infty$ is said to be *generic*.

The following theorem, which is called the *equidistribution theorem*, is due to Szpiro-Ullmo-Zhang ([**66**], [**70**] and [**77**]).

THEOREM 9.17 (Equidistribution theorem). *Let L be an ample invertible sheaf on X. Let $h : X(\overline{X}) \to \mathbb{R}$ be a height function associated with L. Let $\left\{(\mathscr{X}_n, \overline{\mathscr{L}}_n)\right\}_{n=1}^\infty$ be a sequence of vertically nef models of (X, L) (i.e. $\overline{\mathscr{L}}_n$ is vertically nef). Let $\{x_m\}_{m=1}^\infty$ be a generic sequence in $X(\overline{K})$. Here we assume the following:*

(1) *If we set $\omega = c_1(\overline{\mathscr{L}}_1)$, then ω is positive and $c_1(\overline{\mathscr{L}}_n) = \omega$ $(\forall n)$.*
(2) *$h(x) \geq 0$ for $x \in X(\overline{K})$.*
(3) *$\lim_{n\to\infty} \sup_{x \in X(\overline{K})} |h(x) - h_{(\mathscr{X}_n, \overline{\mathscr{L}}_n)}(x)| = 0$.*
(4) *$\lim_{m\to\infty} h(x_m) = 0$.*
(5) *$\overline{\mathscr{L}}_n$ is of real type.*

If we set $\omega_\sigma = \omega|_{X_\sigma(\mathbb{C})}$, then we have the weak convergence

$$\lim_{m\to\infty}\left(\frac{1}{\#O_\sigma(x_m)} \sum_{z \in O_\sigma(x_m)} \delta_z\right) = \frac{\omega_\sigma^{\dim X}}{\deg(L^{\dim X})}$$

as currents on $X_\sigma(\mathbb{C})$. Namely, for any C^∞-function f on $X_\sigma(\mathbb{C})$,

$$\lim_{m\to\infty}\left(\frac{1}{\#O_\sigma(x_m)}\sum_{z\in O_\sigma(x_m)} f(z)\right) = \int_{X_\sigma} \frac{f\omega_\sigma^{\dim X}}{\deg(L^{\dim X})}.$$

PROOF. Let f be a C^∞-function on $X_\sigma(\mathbb{C})$. Considering the real part and the imaginary part of f, we may assume that f is real-valued. For non-negative real number λ, we define $\overline{\mathscr{L}}_{n,\lambda}$ as follows: Let $|\cdot|_{\overline{\mathscr{L}}_n}$ denote the metric of $\overline{\mathscr{L}}_n$. The metric of $\overline{\mathscr{L}}_{n,\lambda}$ on $X_\tau(\mathbb{C})$ is given by

$$\begin{cases} \exp\left(-\lambda(f+F_\infty^*(f))/2\right)|\cdot|_{\overline{\mathscr{L}}_n} & \text{if } \tau=\sigma \text{ and } \sigma=\bar\sigma,\\ \exp\left(-\lambda f/2\right)|\cdot|_{\overline{\mathscr{L}}_n} & \text{if } \tau=\sigma \text{ and } \sigma\neq\bar\sigma,\\ \exp\left(-\lambda F_\infty^*(f)/2\right)|\cdot|_{\overline{\mathscr{L}}_n} & \text{if } \tau=\bar\sigma \text{ and } \sigma\neq\bar\sigma,\\ |\cdot|_{\overline{\mathscr{L}}_n} & \text{otherwise,}\end{cases}$$

where $F_\infty : X(\mathbb{C}) \to X(\mathbb{C})$ is the complex conjugation map. Since ω_τ is positive and $X_\tau(\mathbb{C})$ is compact, there is a positive constant λ_0 such that λ_0 does not depend on n and that $\overline{\mathscr{L}}_{n,\lambda}$ is vertically nef for all λ with $0 \le \lambda \le \lambda_0$. Therefore, by Theorem 9.16,

$$\begin{aligned} &\sup_{Y\subsetneq X}\left\{\inf_{x\in(X\setminus Y)(\overline{K})} h_{(\mathscr{X}_n,\overline{\mathscr{L}}_{n,\lambda})}(x)\right\} \\ &\qquad \ge \frac{\widehat{\deg}(\widehat{c}_1(\overline{\mathscr{L}}_{n,\lambda})^{d+1})}{(d+1)\deg(L^d)} \ge \inf_{x\in X(\overline{K})} h_{(\mathscr{X}_n,\overline{\mathscr{L}}_{n,\lambda})}(x), \end{aligned} \tag{9.5}$$

where $d=\dim X$. In particular, considering the case where $\lambda=0$, we obtain

$$\frac{\widehat{\deg}(\widehat{c}_1(\overline{\mathscr{L}}_n)^{d+1})}{(d+1)\deg(L^d)} \ge \inf_{x\in X(\overline{K})} h_{(\mathscr{X}_n,\overline{\mathscr{L}}_n)}(x).$$

On the other hand, for any $\epsilon>0$, there is n_0 such that, for $n \ge n_0$,

$$h(x)-\epsilon \le h_{(\mathscr{X}_n,\overline{\mathscr{L}}_n)}(x) \le h(x)+\epsilon$$

for all $x\in X(\overline{K})$, and hence,

$$\frac{\widehat{\deg}(\widehat{c}_1(\overline{\mathscr{L}}_n)^{d+1})}{(d+1)\deg(L^d)} \ge \inf_{x\in X(\overline{K})} h(x)-\epsilon \ge -\epsilon.$$

Note that

$$\int_{X_{\bar\sigma}} F_\infty^*(f)\omega_{\bar\sigma}^d = \int_{X_\sigma} f\omega_\sigma^d$$

(cf. Subsection 5.1.2), so that, by an easy calculation, we can see

$$\frac{\widehat{\deg}(\widehat{c}_1(\overline{\mathscr{L}}_{n,\lambda})^{d+1})}{(d+1)\deg(L^d)} = \frac{\widehat{\deg}(\widehat{c}_1(\overline{\mathscr{L}}_n)^{d+1})}{(d+1)\deg(L^d)} + \frac{\lambda}{\deg(L^d)}\int_{X_\sigma} f\omega_\sigma^d + O(\lambda^2)$$

and the term $O(\lambda^2)$ does not depend on n. Therefore, by (9.5), if $n \ge n_0$, then

$$\sup_{Y\subsetneq X}\left\{\inf_{x\in(X\setminus Y)(\overline{K})} h_{(\mathscr{X}_n,\overline{\mathscr{L}}_{n,\lambda})}(x)\right\} \ge -\epsilon + \frac{\lambda}{\deg(L^d)}\int_{X_\sigma} f\omega_\sigma^d + O(\lambda^2).$$

Hence, using the generic sequence $\{x_m\}$, we get

$$\liminf_{m\to\infty} h_{(\mathscr{X}_n,\overline{\mathscr{L}}_{n,\lambda})}(x_m) \geq -\epsilon + \frac{\lambda}{\deg(L^d)}\int_{X_\sigma} f\omega_\sigma^d + O(\lambda^2). \tag{9.6}$$

On the other hand, as

$$h_{(\mathscr{X}_n,\overline{\mathscr{L}}_{n,\lambda})}(x) = h_{(\mathscr{X}_n,\overline{\mathscr{L}}_n)}(x) + \frac{\lambda}{\#O_\sigma(x)}\sum_{z\in O_\sigma(x)} f(z),$$

if $n \geq n_0$, then

$$\begin{aligned}\liminf_{m\to\infty} h_{(\mathscr{X}_n,\overline{\mathscr{L}}_{n,\lambda})}(x_m) &\leq \liminf_{m\to\infty}\left(\epsilon + h(x_m) + \frac{\lambda}{\#O_\sigma(x_m)}\sum_{z\in O_\sigma(x_m)} f(z)\right)\\ &= \epsilon + \liminf_{m\to\infty}\left(\frac{\lambda}{\#O_\sigma(x_m)}\sum_{z\in O_\sigma(x_m)} f(z)\right).\end{aligned} \tag{9.7}$$

Thus, by (9.6) and (9.7), we have

$$2\epsilon + \liminf_{m\to\infty}\left(\frac{\lambda}{\#O_\sigma(x_m)}\sum_{z\in O_\sigma(x_m)} f(z)\right) \geq \frac{\lambda}{\deg(L^d)}\int_{X_\sigma} f\omega_\sigma^d + O(\lambda^2).$$

Here ϵ is an arbitrary positive number, and hence the above inequality implies

$$\lambda\liminf_{m\to\infty}\left(\frac{1}{\#O_\sigma(x_m)}\sum_{z\in O_\sigma(x_m)} f(z)\right) \geq \frac{\lambda}{\deg(L^d)}\int_{X_\sigma} f\omega_\sigma^d + O(\lambda^2).$$

Thus, taking $\lambda \searrow 0$ with $\lambda > 0$, we obtain

$$\liminf_{m\to\infty}\left(\frac{1}{\#O_\sigma(x_m)}\sum_{z\in O_\sigma(x_m)} f(z)\right) \geq \frac{1}{\deg(L^d)}\int_{X_\sigma} f\omega_\sigma^d.$$

This inequality still holds even if we replace f by $-f$. Thus we have

$$\limsup_{m\to\infty}\left(\frac{1}{\#O_\sigma(x_m)}\sum_{z\in O_\sigma(x_m)} f(z)\right) \leq \frac{1}{\deg(L^d)}\int_{X_\sigma} f\omega_\sigma^d.$$

Therefore the theorem follows. □

9.4. Cubic metric on complex abelian variety

Let A be a complex abelian variety. For $a \in A$, let $T_a : A \to A$ denote the translation by a, that is, $T_a(x) = x + a$. A form η on A is called an *invariant form* if

$$T_a^*(\eta) = \eta \qquad (\forall a \in A).$$

The group $\mathbb{H}^{1,1}(A,\mathbb{R})$ of invariant real $(1,1)$-forms on A is canonically isomorphic to the $(1,1)$-part $H^{1,1}(A,\mathbb{R})$ of the cohomology group $H^2(A,\mathbb{R})$. Let $\mathbb{C}^g \to A$ be the universal covering of A and $(z_1,\ldots,z_g)$ a coordinate of $\mathbb{C}^g$. Then an element η of $\mathbb{H}^{1,1}(A,\mathbb{R})$ can be written by

$$\eta = \sqrt{-1}\sum_{1\leq i,j\leq g} h_{ij}dz_i\wedge d\bar{z}_j$$

for some $g \times g$ hermitian matrix $H = (h_{ij})$.

Let L be an invertible sheaf on A. A C^∞-metric h of L is called a *cubic metric* if the first Chern form $c_1(L, h)$ is an element of $\mathbb{H}^{1,1}(A, \mathbb{R})$. Then we have the following.

PROPOSITION 9.18. (1) *Every invertible sheaf L on A has a cubic metric.*

(2) *If h and h' are cubic metrics of L, then there is a positive constant c with $h' = ch$.*

(3) *If h and h' are cubic metrics of L, then $h + h'$ is also a cubic metric.*

(4) *If L is ample and h is a cubic metric, then (L, h) is positive.*

PROOF. (1) Let h_0 be a C^∞-metric of L. Then, by $\partial\bar{\partial}$-Lemma (Lemma 1.36), there are $\eta \in \mathbb{H}^{1,1}(A, \mathbb{R})$ and a C^∞ real-valued function f with

$$c_1(L, h_0) = \eta + \frac{\sqrt{-1}}{2\pi}\partial\bar{\partial}(f).$$

Thus, if we set $h = \exp(f)h_0$, then $c_1(L, h) = \eta$.

(2) Since $c_1(L, h)$ and $c_1(L, h')$ are the same cohomology class, we have $c_1(L, h) = c_1(L, h')$. We choose C^∞ real-valued function f with $h' = \exp(f)h$. Then $\partial\bar{\partial}(f) = 0$, which means that f is harmonic, so that f is a constant.

(3) follows from (2).

(4) Replacing (L, h) with $(L^{\otimes n}, h^{\otimes n})$ $(n > 0)$, we may assume that L is very ample. Thus there is a non-zero element $s \in H^0(A, L)$ such that $B := \operatorname{div}(s)$ is non-singular. Let $\mathbb{C}^g \to A$ be the universal covering of A. If we choose a coordinate $(z_1, \ldots, z_g)$ of $\mathbb{C}^g$ suitably, then $c_1(L, h)$ can be written by

$$c_1(L, h) = \sqrt{-1}\sum_{i=1}^{g} \lambda_i dz_i \wedge d\bar{z}_i,$$

where $\lambda_1, \ldots, \lambda_g$ are real numbers. Here we set

$$\Omega = \sqrt{-1}\sum\nolimits_{i=1}^{g} dz_i \wedge d\bar{z}_i \quad \text{and} \quad \Omega_i = \Omega - \sqrt{-1}dz_i \wedge d\bar{z}_i.$$

Then $c_1(L, h)\Omega_i^{g-1} = (\lambda_i/g)\Omega^g$. As Ω_i is semipositive and $d(\Omega_i) = 0$, by the Poincaré-Lelong formula (Theorem 1.39),

$$0 \leq \int_B \Omega_i^{g-1} = \int_A c_1(L, h)\Omega_i^{g-1} = (\lambda_i/g)\int_A \Omega^g,$$

that is $\lambda_i \geq 0$ $(\forall i)$. On the other hand, $c_1(L, h)^g = \lambda_1 \cdots \lambda_g \Omega^g$ and L is ample. Thus

$$0 < \int_A c_1(L, h)^g = \lambda_1 \cdots \lambda_g \int_A \Omega^g,$$

and hence $\lambda_1 \cdots \lambda_g > 0$, so that $\lambda_i > 0$ $(\forall i)$. □

Let $\bar{\ }^a : \operatorname{Spec}(\mathbb{C}) \to \operatorname{Spec}(\mathbb{C})$ be a scheme morphism induced by the complex conjugation map. We set $A' = A \times_{\operatorname{Spec}(\mathbb{C})}^{\bar{\ }^a} \operatorname{Spec}(\mathbb{C})$. For $x \in A(\mathbb{C})$, that is, a scheme morphism $x : \operatorname{Spec}(\mathbb{C}) \to A$ over $\mathbb{C}$, the composition $\operatorname{Spec}(\mathbb{C}) \xrightarrow{\bar{\ }^a} \operatorname{Spec}(\mathbb{C}) \xrightarrow{x} A$ is not a morphism over $\mathbb{C}$. Let $\bar{x} : \operatorname{Spec}(\mathbb{C}) \to A'$ be a morphism arising from the following commutative diagram:

$$\begin{array}{ccc} A & \xleftarrow{x\circ -^a} & \mathrm{Spec}(\mathbb{C}) \\ \big\downarrow & & \big\| \\ \mathrm{Spec}(\mathbb{C}) & \xleftarrow{-^a} & \mathrm{Spec}(\mathbb{C})\,. \end{array}$$

Notice that $\bar{x}$ is a morphism over $\mathbb{C}$ that is, $\bar{x} \in A'(\mathbb{C})$. A map $A(\mathbb{C}) \to A'(\mathbb{C})$ given by $x \mapsto \bar{x}$ is denoted by F_∞. As $(A')'$ is naturally isomorphic to A over $\mathbb{C}$, we can also define $A'(\mathbb{C}) \to A(\mathbb{C})$ as before. By abuse of notation, it is also denoted by F_∞. Then $F_\infty^2 = \mathrm{id}$. Moreover, the group structure of A yields a group structure of A' and F_∞ is a homomorphism. In particular, for $a \in A(\mathbb{C})$, $T_a \circ F_\infty = F_\infty \circ T_{\bar{a}}$ holds. Further, F_∞ is a C^∞-map.

Let L be an invertible sheaf on A and $L' = F_\infty^*(L) = L \otimes_{\mathbb{C}}^{-} \mathbb{C}$, where $\otimes^-$ means the tensor product viewing $\mathbb{C}$ as a $\mathbb{C}$-module via the complex conjugation. Then L' is an invertible sheaf on A'. Let h be a C^∞-metric of L. We define h' to be

$$h'(s \otimes^- 1, t \otimes^- 1)(\bar{x}) = \overline{h(s,t)(x)} \qquad (x \in A(\mathbb{C}),\ s, t \in L_x),$$

that is, $h' = \overline{F_\infty^*(h)}$, which is a C^∞-metric of L'. Then we have the following.

PROPOSITION 9.19. *If h is a cubic metric of L, then h' is also a cubic metric of L'.*

PROOF. Let s be a non-zero rational section of L. We set

$$f = -\log h(s,s) \quad \text{and} \quad g = -\log h'(s \otimes^- 1, s \otimes^- 1).$$

Then $g = F_\infty^*(f)$. On the other hand, as $d \circ F_\infty^* = F_\infty^* \circ d$, we obtain $\partial \circ F_\infty^* = F_\infty^* \circ \bar{\partial}$ and $\bar{\partial} \circ F_\infty^* = F_\infty^* \circ \partial$. Therefore,

$$c_1(L', h') = \frac{\sqrt{-1}}{2\pi}\partial\bar{\partial}(g) = -F_\infty^*\left(\frac{\sqrt{-1}}{2\pi}\partial\bar{\partial}(f)\right) = -F_\infty^*(c_1(L,h)).$$

Thus, for $a \in A(\mathbb{C})$, we have

$$\begin{aligned} T_{\bar{a}}^*(c_1(L', h')) &= -T_{\bar{a}}^*(F_\infty^*(c_1(L,h)) = -F_\infty^*(T_a^*(c_1(L,h))) \\ &= -F_\infty^*(c_1(L,h)) = c_1(L', h'), \end{aligned}$$

which shows that h' is a cubic metric. □

9.5. Bogomolov's conjecture

Let A be an abelian variety over an algebraic number field K and let L be a symmetric and ample invertible sheaf on A. In addition, let X be an irreducible subvariety of $A_{\overline{K}}$ $(:= A \times_{\mathrm{Spec}(K)} \mathrm{Spec}(\overline{K}))$. Bogomolov's conjecture proved by Ullmo [**70**] and Zhang [**77**] is the following theorem.

THEOREM 9.20 (Bogomolov's conjecture). *Let $\hat{h}_L$ be the Néron-Tate height function given by L. If the set $\{x \in X(\overline{K}) \mid \hat{h}_L(x) \leq \epsilon\}$ is Zariski dense in X for any positive number $\epsilon > 0$, then there exist an abelian subvariety B of $A_{\overline{K}}$ and a torsion point b of $A(\overline{K})$ such that $X = B + b$.*

PROOF. Let $G(X)$ be an algebraic subgroup of $A_{\overline{K}}$ given by

$$G(X) = \{a \in A_{\overline{K}} \mid a + X = X\}.$$

First, we have the following lemma.

LEMMA 9.21. *For an integer m with $m \geq 2$, let $\alpha_m : A_{\overline{K}}^m \to A_{\overline{K}}^{m-1}$ be a morphism given by*

$$\alpha_m(x_1, \ldots, x_m) = (x_1 - x_2, \ldots, x_{m-1} - x_m).$$

If $G(X) = \{0\}$, then $\alpha_m|_{X^m} : X^m \to \alpha_m(X^m)$ is birational for a sufficiently large m.

PROOF. For $x_1, \ldots, x_m \in X$, we set

$$G(x_1, \ldots, x_m) = \{a \in A_{\overline{K}} \mid a + x_1, \ldots, a + x_m \in X\},$$

which is an algebraic subgroup of $A_{\overline{K}}$. Here let us check the following:

$$(\alpha_m|_{X^m})^{-1}(\alpha_m(x_1, \ldots, x_m)) = \{(x_1 + a, \ldots, x_m + a) \mid a \in G(x_1, \ldots, x_m)\}.$$

Indeed, if we assume that $(y_1, \ldots, y_m) \in (\alpha_m|_{X^m})^{-1}(\alpha_m(x_1, \ldots, x_m))$, then we have $y_1 - x_1 = \cdots = y_m - x_m$ because $x_i - x_{i+1} = y_i - y_{i+1}$ for all $1 \leq i < m$. Thus there is $a \in G(x_1, \ldots, x_m)$ with $y_i = x_i + a$ for all i. Therefore,

$$(\alpha_m|_{X^m})^{-1}(\alpha_m(x_1, \ldots, x_m)) \subseteq \{(x_1 + a, \ldots, x_m + a) \mid a \in G(x_1, \ldots, x_m)\}.$$

The converse inclusion is obvious.

Note that $G(X) = \{a \in A_{\overline{K}} \mid a + x \in X \ (\forall x \in X)\}$. If m is sufficiently large, then there is a Zariski open set U of X^m such that $G(x_1, \ldots, x_m) = 0$ for all $(x_1, \ldots, x_m) \in U$. Thus the lemma follows from the above claim. □

Let us start the proof of the theorem. First we assume that $G(X) = \{0\}$. It is sufficient to show that $\dim X = 0$. Indeed, if we write $X = \{x\}$, then $0 \leq \hat{h}_L(x) \leq \epsilon$ for every positive number ϵ. Thus $\hat{h}_L(x) = 0$, which means that $x \in A(\overline{K})_{tor}$, and hence, the theorem follows.

Let us see $\dim X = 0$ under the assumption $G(X) = \{0\}$. We assume the contrary, that is, $\dim X > 0$. Since $G(X) = \{0\}$, by Lemma 9.21, for a sufficiently large m, $\alpha_m|_{X^m} : X^m \to \alpha_m(X^m)$ is birational. For the proof of the theorem, we may replace K by a finite extension of K. Thus we may assume that X is defined over K and $X^m \to \alpha_m(X^m)$ is a birational map over K.

Let l be a positive integer and $p_i : A^l \to A$ the projection to the i-th factor. We set $L_l = p_1^*(L) + \cdots + p_l^*(L)$ on A^l. Let $\hat{h}_{L_l}$ be the Néron-Tate height function given by L_l on A^l. Then

$$\hat{h}_{L_l}(a_1, \ldots, a_l) = \hat{h}_L(a_1) + \cdots + \hat{h}_L(a_l)$$

for $(a_1, \ldots, a_l) \in A^l(\overline{K})$. Here let us consider the following claim.

CLAIM 4. *Let Y be a proper subscheme of X^l. Then, for any $\epsilon > 0$, there is $x \in X^l \setminus Y$ with $\hat{h}_{L_l}(x) \leq \epsilon$.*

PROOF. We prove it by induction on l. If $l = 1$, then the assertion is obvious by the assumption of the theorem. We assume $l > 1$. Let $q : X^l \to X^{l-1}$ be a morphism given by $q(x_1, \ldots, x_l) = (x_2, \ldots, x_l)$. Then there is a non-empty Zariski open set W of X^{l-1} such that $q^{-1}(x'') \cap Y$ is a proper subscheme of $q^{-1}(x'')$ for all $x'' \in W$. By the hypothesis of induction, there is $x'' \in W$ such that $\hat{h}_{L_{l-1}}(x'') \leq \epsilon/2$. On the other hand, by the case $l = 1$, there is $x' \in q^{-1}(x'') \setminus q^{-1}(x'') \cap Y$ with $\hat{h}_L(x') \leq \epsilon/2$. If we set $x = (x', x'')$, then $x \in X^l \setminus Y$ and $\hat{h}_{L_l}(x) \leq \epsilon$, as required. □

From now on , we assume that $l = m$. By abuse of notation, $\alpha_m|_{X^m}$ is denoted by α_m. As $\alpha_m : X^m \to \alpha_m(X^m)$ is birational over K, there is a Zariski open U over K such that $\alpha_m : \alpha_m^{-1}(U) \xrightarrow{\sim} U$. Since X^m has only countably many subvarieties over $\overline{K}$, let $\{Z_i\}_{i=1}^{\infty}$ be the set of all proper subvarieties of X^m over $\overline{K}$. By the above claim, for any positive integer n, we can find $x_n \in X^m(\overline{K})$ such that

$$x_n \notin \bigcup_{i=1}^{n} Z_i \cup (X^m \setminus \alpha_m^{-1}(U))$$

and $\hat{h}_{L_m}(x_n) \leq 1/n$. Then the sequence $\{x_n\}$ is generic in X^m and $\lim_{n\to\infty} \hat{h}_{L_m}(x_n) = 0$. Moreover, as $x_n \in \alpha_m^{-1}(U)$, $\{\alpha_m(x_n)\}$ is also generic. Notice that $\hat{h}_L^{1/2}$ gives a seminorm on $A(\overline{K})$. Thus we can see $\lim_{n\to\infty} \hat{h}_{L_{m-1}}(\alpha_m(x_n)) = 0$.

Let $(\mathscr{A}, \overline{\mathscr{L}})$ be a model of (A, L) such that $\overline{\mathscr{L}}$ is vertically nef and of real type, $c_1(\overline{\mathscr{L}})$ is positive and the metric of $\overline{\mathscr{L}}$ on $A_\sigma(\mathbb{C})$ is a cubic metric for all $\sigma \in K(\mathbb{C})$ (the existence of the above cubic metric of real type follows from Proposition 9.18 and Proposition 9.19). Let $\mathscr{A}_n$ be a model of A such that the multiplication map $[2^n] : A \to A$ by 2^n extends to $f_n : \mathscr{A}_n \to \mathscr{A}$. If we set $\overline{\mathscr{L}}_n = 4^{-n} f_n^*(\overline{\mathscr{L}})$, then $(\mathscr{A}_n, \overline{\mathscr{L}}_n)$ is a model of (A, L) such that $\overline{\mathscr{L}}_n$ is vertically nef and $c_1(\overline{\mathscr{L}}_n) = c_1(\overline{\mathscr{L}})$ because the metric of $\overline{\mathscr{L}}$ is cubic. Moreover, for $x \in A(\overline{K})$,

$$h_{(\mathscr{A}_n, \overline{\mathscr{L}}_n)}(x) = 4^{-n} h_{(\mathscr{A}, \overline{\mathscr{L}})}(2^n x).$$

Thus, by Remark 9.1,

$$\lim_{n\to\infty} \sup_{x \in A(\overline{K})} |h_{(\mathscr{A}_n, \overline{\mathscr{L}}_n)}(x) - \hat{h}_L(x)| = 0.$$

Let $\mathscr{X}_{m,n}$ (resp. $\mathscr{Y}_{m,n}$) denote the Zariski closure of X^m (resp. $\alpha_m(X^m)$) in

$$\overbrace{\mathscr{A}_n \times_{O_K} \cdots \times_{O_K} \mathscr{A}_n}^{m \text{ times}} \qquad (\text{resp. } \overbrace{\mathscr{A}_n \times_{O_K} \cdots \times_{O_K} \mathscr{A}_n}^{m-1 \text{ times}}).$$

Moreover,

$$(p_1^*(\overline{\mathscr{L}}_n) + \cdots + p_m^*(\overline{\mathscr{L}}_n))\big|_{\mathscr{X}_{m,n}} \quad \text{and} \quad (p_1^*(\overline{\mathscr{L}}_n) + \cdots + p_{m-1}^*(\overline{\mathscr{L}}_n))\big|_{\mathscr{Y}_{m,n}}$$

are denoted by $\overline{\mathscr{M}}_{m,n}$ and $\overline{\mathscr{N}}_{m,n}$, respectively, where p_i is the projection to the i-th factor. Then $\overline{\mathscr{M}}_{m,n}$ and $\overline{\mathscr{N}}_{m,n}$ is vertically nef and $c_1(\overline{\mathscr{M}}_{m,n})$ and $c_1(\overline{\mathscr{N}}_{m,n})$ coincides with $c_1(\overline{\mathscr{M}}_{m,1})$ and $c_1(\overline{\mathscr{N}}_{m,1})$, respectively. Further,

$$\lim_{n\to\infty} \sup_{x \in X(\overline{K})} |h_{(\mathscr{X}_{m,n}, \overline{\mathscr{M}}_{m,n})}(x) - \hat{h}_{L_m}(x)| = 0$$

and

$$\lim_{n\to\infty} \sup_{x \in \alpha_m(X^m)(\overline{K})} |h_{(\mathscr{Y}_{m,n}, \overline{\mathscr{N}}_{m,n})}(x) - \hat{h}_{L_{m-1}}(x)| = 0,$$

so that we can apply Theorem 9.17 on X^m and $\alpha_m(X^m)$. We fix $\sigma \in \operatorname{Spec}(K)(\mathbb{C})$ and write $d = \dim X^m = \dim \alpha_m(X^m)$. By Theorem 9.17, we have

$$\lim_{n\to\infty} \left(\frac{1}{\#O_\sigma(x_n)} \sum_{z \in O_\sigma(x_n)} \delta_z \right) = \frac{\left(c_1(\overline{L}_{m\sigma})|_{X_\sigma^m(\mathbb{C})}\right)^{\wedge d}}{\deg\left((L_m|_{X^m})^d\right)}$$

on $X^m_\sigma(\mathbb{C})$ and

$$\lim_{n\to\infty}\left(\frac{1}{\#O_\sigma(\alpha_m(x_n))}\sum_{z\in O_\sigma(\alpha_m(x_n))}\delta_z\right) = \frac{\left(c_1(\overline{L}_{m-1\sigma})\big|_{\alpha_m(X^m)_\sigma(\mathbb{C})}\right)^{\wedge d}}{\deg\left(\left(L_{m-1}\big|_{\alpha_m(X^m)}\right)^d\right)}$$

on $\alpha_m(X^m)_\sigma$. Therefore, the above convergences also hold on $\alpha_m^{-1}(U)_\sigma(\mathbb{C})$ and on $U_\sigma(\mathbb{C})$. Note that $\alpha_m^{-1}(U)_\sigma(\mathbb{C}) \simeq U_\sigma(\mathbb{C})$. Thus we have

$$\frac{\left(c_1(\overline{L}_{m\sigma})\big|_{X^m_\sigma(\mathbb{C})}\right)^{\wedge d}}{\deg\left(\left(L_m\big|_{X^m}\right)^d\right)} = \alpha_m^*\left(\frac{\left(c_1(\overline{L}_{m-1\sigma})\big|_{\alpha_m(X^m)_\sigma(\mathbb{C})}\right)^{\wedge d}}{\deg\left(\left(L_{m-1}\big|_{\alpha_m(X^m)}\right)^d\right)}\right)$$

on $\alpha_m^{-1}(U)_\sigma(\mathbb{C})$. Since each side of the above equation is C^∞-forms, the equation holds on X^m_σ. On the other hand, $\alpha_m : X^m \to \alpha_m(X^m)$ is not an isomorphism on the diagonal $\{(x,\dots,x) \mid x \in X\}$ by the construction of α_m. Thus the right side form is not positive on the diagonal. This is a contradiction because the left side form is positive.

Finally, we have a general case. Let $A' := A_{\overline{K}}/G(X)$, $\pi : A_{\overline{K}} \to A'$ the natural homomorphism and $X' := \pi(X)$. It is easy to see that $G(X') = \{0\}$ and $\pi^{-1}(X') = X$. Let L' be a symmetric and ample invertible sheaf on A'. We choose a positive integer a such that $L^{\otimes a} \otimes \pi^*(L')^{\otimes -1}$ is ample. Then $\pi^*(\hat{h}_{L'}) \le a\hat{h}_L$. For any positive number ϵ, $\{x' \in X'(\overline{K}) \mid \hat{h}_{L'}(x') \le \epsilon\}$ is Zariski dense because

$$\pi\left(\{x \in X(\overline{K}) \mid \hat{h}_L(x) \le \epsilon/a\}\right) \subseteq \{x' \in X'(\overline{K}) \mid \hat{h}_{L'}(x') \le \epsilon\}.$$

Therefore, by the previous case, there is $x' \in A'(\overline{K})_{tor}$ with $X' = \{x'\}$, so that $X = \pi^{-1}(x')$. In particular, $G(X)$ is an abelian subvariety. On the other hand, by the following lemma, $\pi(A(\overline{K})_{tor}) = A'(\overline{K})_{tor}$. Thus the assertion of the theorem follows. □

LEMMA 9.22. *Let A be an abelian variety over an algebraically closed field k and let B be an abelian subvariety of A. Let $A' = A/B$ and let $\pi : A \to A'$ be the natural homomorphism. Let Γ be a subgroup of $A(k)$ and $\Gamma' = \pi(\Gamma)$. Moreover, we set*

$$\begin{cases} \Gamma_{\text{div}} = \{x \in A(k) \mid \textit{there is a positive integer } n \textit{ with } nx \in \Gamma\}, \\ \Gamma'_{\text{div}} = \{x' \in A'(k) \mid \textit{there is a positive integer } n \textit{ with } nx' \in \Gamma'\}. \end{cases}$$

Then $\pi(\Gamma_{\text{div}}) = \Gamma'_{\text{div}}$.

PROOF. Clearly $\pi(\Gamma_{\text{div}}) \subseteq \Gamma'_{\text{div}}$. Conversely we assume $x' \in \Gamma'_{\text{div}}$. As $\pi : A(k) \to A'(k)$ is surjective, there is $x_1 \in A(k)$ with $\pi(x_1) = x'$. We choose a positive integer n with $nx' \in \Gamma'$. Thus $nx_1 \in B + \Gamma$. Note that the multiplication map $[n] : A \to A$ by n is surjective, there are $b \in B(k)$ and $\gamma \in \Gamma_{\text{div}}$ with $nx_1 = nb + n\gamma$. Therefore, $x_1 - b - \gamma \in A(k)_{tor}$. Thus, if we set $x = x_1 - b$, then $\pi(x) = x'$ and

$$x \in \Gamma_{\text{div}} + A(k)_{tor} \subseteq \Gamma_{\text{div}}.$$

□

9.6. The Lang-Bogomolov Conjecture

In this section, we introduce the Lang-Bogomolov conjecture, which is the mixed version of Lang's conjecture and Bogomolov's conjecture. This was proved by Poonen [**59**] and Zhang [**78**]. Afterward, it was generalized to the case of finitely generated fields over $\mathbb{Q}$ by Moriwaki [**50**]. This generalization contains the following absolute version of Lang's conjecture.

THEOREM 9.23 (Absolute version of Lang's conjecture). *Let A be an abelian variety over $\mathbb{C}$ and Γ a subgroup of $A(\mathbb{C})$ with $\dim(\Gamma \otimes_{\mathbb{Z}} \mathbb{Q}) < \infty$. Let X be an irreducible subscheme of A. Then there are abelian subvarieties $C_1, \ldots, C_n$ of A and $\gamma_1, \ldots, \gamma_n \in \Gamma$ such that*

$$\overline{X(\mathbb{C}) \cap \Gamma} = \bigcup_{i=1}^{n} (C_i(\mathbb{C}) + \gamma_i) \quad \text{and} \quad X(\mathbb{C}) \cap \Gamma = \bigcup_{i=1}^{n} (C_i(\mathbb{C}) + \gamma_i) \cap \Gamma.$$

Let A be an abelian variety over an algebraic number field K and let L be a symmetric and ample invertible sheaf on A. Let $\widehat{h}_L : A(\overline{K}) \to \mathbb{R}$ be the Néron-Tate height function arising from L. Let $\langle\ ,\ \rangle_L$ be the induced symmetric bilinear form on $A(\overline{K})$, that is,

$$\langle x, y \rangle_L = \frac{1}{2} \left(\widehat{h}_L(x+y) - \widehat{h}_L(x) - \widehat{h}_L(y) \right)$$

for $x, y \in A(\overline{K})$. Note that $\langle\ ,\ \rangle_L : A(\overline{K}) \times A(\overline{K}) \to \mathbb{R}$ is positive semidefinite. By Theorem 9.15, $\langle\ ,\ \rangle_L$ induces an inner product on $A(\overline{K}) \otimes \mathbb{R}$. We put

$$\delta_L(x_1, \ldots, x_l) = \det\left(\langle x_i, x_j \rangle_L \right)$$

for $x_1, \ldots, x_l \in A(\overline{K})$.

Let Γ be a subgroup of $A(\overline{K})$ such that Γ is of finite rank, that is, $\dim_{\mathbb{Q}}(\Gamma \otimes \mathbb{Q}) < \infty$. Let Γ_{div} be the saturation of Γ in $A(\overline{K})$, that is,

$$\Gamma_{\mathrm{div}} = \{ x \in A(\overline{K}) \mid \text{there is a positive integer } n \text{ with } nx \in \Gamma \}.$$

Then Γ_{div} is also a subgroup of finite rank.

The following theorem is the Lang-Bogomolov conjecture due to Poonen [**59**] and Zhang [**78**].

THEOREM 9.24 (Lang-Bogomolov conjecture). *Let X be a subvariety of $A_{\overline{K}}$. Let $\gamma_1, \ldots, \gamma_k$ be elements of Γ such that $\gamma_1, \ldots, \gamma_k$ form a basis of $\Gamma \otimes \mathbb{Q}$. If*

$$\{ x \in X(\overline{K}) \;\; \delta_L(\gamma_1, \ldots, \gamma_k, x) \leq \epsilon \}$$

is Zariski dense in X for any positive number ϵ, then X is a translation of an abelian subvariety of $A_{\overline{K}}$ by an element of Γ_{div}.

In the case where $\Gamma = 0$, this theorem is nothing more than Bogomolov's conjecture (Theorem 9.20). In addition, Theorem 9.24 is a generalization of the following Lang's conjecture (the case where $\Gamma = A(K)$) which was proved by Faltings (cf. [**19**] and [**21**]).

THEOREM 9.25 (Lang's conjecture over an algebraic number field). *Let A be an abelian variety over an algebraic number field K and X a geometrically irreducible subvariety of A. If $X(K)$ is Zariski dense in X, then X is a translation of an abelian subvariety of $A_{\overline{K}}$.*

In the following sections, under the assumption of Faltings' theorem (Lang's conjecture), we give the proof of Theorem 9.24 according to Moriwaki's paper [**50**].

9.7. Small points with respect to a subgroup of finite rank

The contexts of this section follow from Moriwaki's paper [**50**] and the article [**33**].

Let A be an abelian variety over an algebraic number field K and let L be a symmetric and ample invertible sheaf on A. Let Γ be a subgroup of $A(\overline{K})$ of finite rank. In this section, we assume that there is a subgroup Γ_0 of Γ such that $\Gamma_0 \subseteq A(K)$ and $\Gamma_0 \otimes \mathbb{Q} = \Gamma \otimes \mathbb{Q}$. This assumption always holds if we replace K by a finite extension of K.

A non-empty subset S of $A(\overline{K})$ is said to be *small with respect to* Γ if, for each $s \in S$, we can find a decomposition $s = \gamma(s) + z(s)$ satisfying the following properties:

(i) $\gamma(s) \in \Gamma$ for all $s \in S$.
(ii) For any positive number $\epsilon > 0$, there exists a non-empty subset $S(\epsilon)$ of S such that $S \setminus S(\epsilon)$ is finite set and that $\langle z(s), z(s)\rangle_L \leq \epsilon$ for $s \in S(\epsilon)$.

In addition, S is said to be *small* if S is small with respect to $\{0\}$.

Let $A_{\overline{K}} := A \times_{\mathrm{Spec}(K)} \mathrm{Spec}(\overline{K})$ and let $f : A_{\overline{K}} \to A'$ be a homomorphism of abelian varieties over $\overline{K}$. Let L' be a symmetric and ample invertible sheaf on A'. As L is ample, there is a positive integer a such that $L^{\otimes a} \otimes f^*(L')^{-1}$ is generated by global sections. Thus, by Lemma 9.6,

$$\langle f(x), f(x)\rangle_{L'} = \langle x, x\rangle_{f^*(L')} \leq a\langle x, x\rangle_L$$

for $x \in A(\overline{K})$. In particular, in the case where $A' = A_{\overline{K}}$ and $f = \mathrm{id}$, the above inequality means that the smallness with respect to Γ does not depend on the choice of the symmetric and ample invertible sheaf.

In terms of smallness, Bogomolov's conjecture (Theorem 9.20) can be rewritten as follows.

THEOREM 9.26. *Let S be a non-empty small set of $A(\overline{K})$. Then there exist abelian subvarieties $C_1, \ldots, C_r$ of $A_{\overline{K}}$, torsion points $c_1, \ldots, c_r$ and non-torsion points $b_1, \ldots, b_m$ such that*

$$\text{the Zariski closure of } S = \bigcup_{i=1}^{r} (C_i + c_i) \cup \{b_1, \ldots, b_m\}.$$

PROOF. Let X be a positive-dimensional irreducible component of the Zariski closure of S. If we set $S' = S \cap X(\overline{K})$, then the Zariski closure of S' is X and S' is small, so that, by Theorem 9.20, X is a translation of an abelian subvariety by a torsion point. Therefore, the assertion follows. □

Let F be a finite extension field of K. From now on, every finite extension field of K is viewed as a subfield of $\overline{K}$. For $x \in A(\overline{K})$, we define $O_F(x)$ to be

$$O_F(x) = \left\{\sigma(x) \in A(\overline{K}) \mid \sigma \in \mathrm{Gal}(\overline{K}/F)\right\}.$$

Moreover, for $n \geq 2$, a homomorphism $\beta_n : A^n_{\overline{K}} \to A^{n-1}_{\overline{K}}$ is defined by

$$\beta_n(x_1, \ldots, x_n) = (x_2 - x_1, x_3 - x_1, \ldots, x_n - x_1).$$

For $T \subseteq A(\overline{K})$, a subset $\mathscr{D}_n(T, F)$ of $A(\overline{K})^{n-1}$ is defined by

$$\mathscr{D}_n(T, F) = \bigcup_{s \in T} \beta_n\left(O_F(s)^n\right).$$

Moreover, the Zariski closure of $\mathscr{D}_n(T, F)$ in $A_{\overline{K}}^{n-1}$ is denoted by $\overline{\mathscr{D}}_n(T, F)$.

Let $\langle\ ,\ \rangle_{L_n}$ be the Néron-Tate height pair on A^n defined by $L_n = \bigotimes_{i=1}^n p_i^*(L)$, where $p_i : A^n \to A$ is the projection to the i-th factor. Note that $\langle x, x\rangle_{L_n}$ is given by

$$\langle x, x\rangle_{L_n} = \langle x_1, x_1\rangle_L + \cdots + \langle x_n, x_n\rangle_L$$

for $x = (x_1, \ldots, x_n) \in A(\overline{K})^n$. By abuse of notation, $\langle\ ,\ \rangle_{L_n}$ is denote by $\langle\ ,\ \rangle_L$. We have the following proposition.

PROPOSITION 9.27. *Let S be a non-empty subset of $A(\overline{K})$ such that S is small with respect to Γ. Let $f : A_{\overline{K}} \to A'$ be a homomorphism of abelian varieties over $\overline{K}$ and let $f^{n-1} : A_{\overline{K}}^{n-1} \to A'^{n-1}$ be a homomorphism given by $(x_1, \ldots, x_{n-1}) \mapsto (f(x_1), \ldots, f(x_{n-1}))$. For a finite extension field F of K, we have the following:*

1. *$f^{n-1}(\mathscr{D}_n(S, F))$ is small.*
2. *If $b_1, \ldots, b_l$ are non-torsion points of $f^{n-1}(\mathscr{D}_n(S, F))$, then there is a non-empty subset S' of S such that $S \setminus S'$ is finite and that $b_i \notin f^{n-1}(\mathscr{D}_n(S', F))$ for all i.*

PROOF. First, note that we assume that there is a subgroup Γ_0 of Γ with $\Gamma_0 \subseteq A(K)$ and $\Gamma_0 \otimes \mathbb{Q} = \Gamma \otimes \mathbb{Q}$. For simplicity, we denote $\sqrt{\langle\cdot,\cdot\rangle_L}$ by $\|\cdot\|_L$. For $s \in S$, let $s = \gamma(s) + z(s)$ be the decomposition as in the definition of smallness. Then there is a positive integer m with $m\gamma(s) \in \Gamma_0 \subseteq A(K)$. Thus $\sigma(\gamma(s)) - \tau(\gamma(s))$ is a torsion point for $\sigma, \tau \in \mathrm{Gal}(\overline{K}/F)$, so that

$$\|\sigma(s) - \tau(s)\|_L = \|\sigma(z(s)) - \tau(z(s))\|_L \leq 2\|z(s)\|_L.$$

Therefore, for any $x \in O_F(s)^n$, $\|\beta_n(x)\|_L \leq 2\sqrt{n-1}\|z(s)\|_L$. Let L' be a symmetric and ample invertible sheaf on A'. We can find a positive integer a with $\langle f(x), f(x)\rangle_{L'} \leq a\langle x, x\rangle_L$, so that

$$\|f^{n-1}(\beta_n(x))\|_{L'} \leq 2\sqrt{a(n-1)}\|z(s)\|_L \qquad (\forall x \in O_F(s)^n). \tag{9.8}$$

(1) Since S is small with respect to Γ, for any $\epsilon > 0$, there is a non-empty subset $S(\epsilon)$ of S such that $S \setminus S(\epsilon)$ is finite and

$$\|z(s)\|_L \leq \frac{\epsilon}{2\sqrt{a(n-1)}} \qquad (\forall s \in S(\epsilon)).$$

Then, by (9.8), $\|y\|_{L'} \leq \epsilon$ for any $y \in f^{n-1}(\mathscr{D}_n(S(\epsilon), F))$. On the other hand, as

$$f^{n-1}(\mathscr{D}_n(S, F)) \setminus f^{n-1}(\mathscr{D}_n(S(\epsilon), F)) \subseteq f^{n-1}(\mathscr{D}_n(S \setminus S(\epsilon), F)),$$

$f^{n-1}(\mathscr{D}_n(S, F)) \setminus f^{n-1}(\mathscr{D}_n(S(\epsilon), F))$ is finite.

(2) If we set $\mu = \min_{i=1,\ldots,l}\{\|b_i\|_{L'}\}$, then there is a non-empty subset S' of S such that $S \setminus S'$ is finite and

$$\|z(s)\|_L < \frac{\mu}{2\sqrt{a(n-1)}} \qquad (\forall s \in S'),$$

and hence, by (9.8), we obtain

$$\|y\|_{L'} < \mu \qquad (\forall y \in f^{n-1}(\mathscr{D}_n(S', F))),$$

that is, $b_i \notin f^{n-1}(\mathscr{D}_n(S', F))$ for all i. □

If T' is a subset of T and F' is a finite extension of F, then $\mathscr{D}_n(T', F') \subseteq \mathscr{D}_n(T, F)$, so that $\overline{\mathscr{D}}_n(T', F') \subseteq \overline{\mathscr{D}}_n(T, F)$. Let $[N] : A \to A$ be the multiplication map given by a positive integer N. Then, as $[N](\mathscr{D}_n(T, F)) = \mathscr{D}_n([N](T), F)$, we have $\overline{\mathscr{D}}_n([N](T), F) = [N](\overline{\mathscr{D}}_n(T, F))$. A pair (T, F) is said to be *n-minimized* if the following conditions are satisfied.

(a) T is an infinite set.
(b) $\overline{\mathscr{D}}_n(T', F') = \overline{\mathscr{D}}_n(T, F)$ holds for any infinite subset T' of T and any finite extension field F' of F.
(c) $[N](\overline{\mathscr{D}}_n(T, F)) = \overline{\mathscr{D}}_n(T, F)$ holds for any positive integer N.

PROPOSITION 9.28. *Let S be an infinite set of $A(\overline{K})$ such that S is small with respect to Γ. Then there are an infinite subset T of S, a finite extension field F of K and a positive integer N such that $([N](T), F)$ is n-minimized.*

PROOF. Since the space $A^{n-1}_{\overline{K}}$ is noetherian, the following set

$$\{\overline{\mathscr{D}}_n(T, F) \mid T \text{ is an infinite subset of } S \text{ and } F \text{ is a finite extension field of } K\}$$

has the minimal element. Let us take (T, F) such that $\overline{\mathscr{D}}_n(T, F)$ is minimal in the above set.

By (1) in Proposition 9.27, $\mathscr{D}_n(T, F)$ is small. Thus, by Theorem 9.26, there exist abelian subvarieties $C_1, \ldots, C_r$, torsion points $c_1, \ldots, c_r$ and non-torsion points $b_1, \ldots, b_m$ such that

$$\overline{\mathscr{D}}_n(T, F) = \bigcup_{i=1}^{r} (C_i + c_i) \cup \{b_1, \ldots, b_m\}.$$

By (2) in Proposition 9.27, there is a finite subset M of T with

$$\mathscr{D}_n(T \setminus M, F) \subseteq \bigcup_{i=1}^{r} (C_i + c_i),$$

so that $\overline{\mathscr{D}}_n(T \setminus M, F) \subseteq \bigcup_{i=1}^{r} (C_i + c_i)$. Therefore, by the minimality of $\overline{\mathscr{D}}_n(T, F)$, we obtain

$$\overline{\mathscr{D}}_n(T, F) = \bigcup_{i=1}^{r} (C_i + c_i). \tag{9.9}$$

Let us choose a positive integer N with $Nc_i = 0$ $(i = 1, \ldots, r)$. Let us see that $([N](T), F)$ is n-minimized. First, $[N](T)$ is an infinite set because T is an infinite set and $[N]$ is a finite morphism. Let T_1 be an infinite subset of $[N](T)$ and F' a finite extension field of F. If we choose a subset T' of T with $[N](T') = T_1$, then

$$\begin{aligned} \overline{\mathscr{D}}_n(T_1, F') = \overline{\mathscr{D}}_n([N](T'), F') &= [N](\overline{\mathscr{D}}_n(T', F')) \\ &= [N](\overline{\mathscr{D}}_n(T, F)) = \overline{\mathscr{D}}_n([N](T), F). \end{aligned}$$

Moreover, as $\overline{\mathscr{D}}_n([N](T), F) = \bigcup_{i=1}^{r} C_i$, for any positive integer N',

$$\begin{aligned} [N'](\overline{\mathscr{D}}_n([N](T), F)) &= [N'] \left(\bigcup_{i=1}^{r} C_i \right) \\ &= \bigcup_{i=1}^{r} C_i = \overline{\mathscr{D}}_n([N](T), F), \end{aligned}$$

so that $([N](T), F)$ is n-minimized. □

The following theorem is a major result for the proof of Theorem 9.24.

THEOREM 9.29 ([**59**], [**50**]). *Let S be an infinite subset of $A(\overline{K})$ such that S is small with respect to Γ. Let F be a finite extension field of K. If (S, F) is 2-minimized, there is an abelian subvariety C of $A_{\overline{K}}$ such that $\overline{\mathscr{D}}_n(S, F) = C^{n-1}$ for all $n \geq 2$.*

PROOF. First, we have the following key claim for the proof of the theorem:

CLAIM 5. *Let S' be an infinite subset of $A(\overline{K})$ such that S' is small with respect to Γ. Let F' be a finite extension field of K. Let B be an abelian subscheme of $A_{F'}$ over F'. For some positive integer e, we assume the following property: for any $s' \in S'$, we can find a subset $T(s')$ of $O_{F'}(s') \times O_{F'}(s')$ such that $\beta_2(T(s')) \subseteq B(\overline{K})$ and $\#T(s') \geq \#(O_{F'}(s') \times O_{F'}(s'))/e$. Then there are a finite subset M of S' and a positive integer N such that $\overline{\mathscr{D}}_2([N](S' \setminus M), F') \subseteq B(\overline{K})$.*

PROOF. Let $\pi : A_{F'} \to A_{F'}/B$ be the natural homomorphism over F'. We fix an arbitrary point s' of S'. Let F'' be a finite extension of $F'(s')$ such that F''/F' is a Galois extension. Let $\phi : G := \mathrm{Gal}(F''/F') \to O_{F'}(s')$ be a map given by $\sigma \mapsto \sigma(s')$. Let $G_{\pi(s')}$ be the stabilizer at $\pi(s')$ by the action of G. If we set

$$R = \{(\sigma, \tau) \in G \times G \mid \sigma(\pi(s')) = \tau(\pi(s'))\},$$

then $\#R = \#G_{\pi(s')} \cdot \#G$. In addition, note that

$$\begin{aligned} &(\phi \times \phi)^{-1}(T(s')) \subseteq R, \ \#(\phi \times \phi)^{-1}(T(s')) = (\#G_{s'})^2 \cdot \#T(s') \text{ and} \\ &\#G = \#G_{s'} \cdot \#O_{F'}(s'), \end{aligned}$$

where $G_{s'}$ be the stabilizer at s' by the action of G, so that

$$\#R \geq \#(G \times G)/e$$

because $\#T(s') \geq \#(O_{F'}(s') \times O_{F'}(s'))/e$. Therefore, we have $[G : G_{\pi(s')}] \leq e$, that is,

$$[F'(\pi(s')) : F'] \leq e. \tag{9.10}$$

By Proposition 9.27, $\pi(\mathscr{D}_2(S', F'))$ is small. Thus, by (9.10) together with Northcott's theorem (Theorem 9.11), $\pi(\mathscr{D}_2(S', F'))$ is a finite set, so that, by using Proposition 9.27 again, there is a finite subset M of S' such that

$$[N](\pi(\mathscr{D}_2(S' \setminus M, F'))) = \{0\}$$

for some positive integer N. Thus $\mathscr{D}_2([N](S' \setminus M), F') \subseteq B(\overline{K})$, which implies

$$\overline{\mathscr{D}}_2([N](S' \setminus M), F') \subseteq B(\overline{K}),$$

as required. □

As (S, F) is 2-minimized, $\overline{\mathscr{D}}_2(S, F)$ is invariant under the endomorphism $[N]$ for any positive integer N, so that, in the same way as (9.9), there are abelian subvarieties $C_1, \ldots, C_e$ such that

$$\overline{\mathscr{D}}_2(S, F) = \bigcup_{i=1}^{e} C_i.$$

Let us see $e = 1$. For this purpose, it suffices to find C_i, an infinite subset S_1 of S, a positive integer N_1 and a finite extension field F_1 of F such that

$$\overline{\mathscr{D}}_2([N_1](S_1), F_1) \subseteq C_i(\overline{K}).$$

Indeed, let F_1 be a finite extension field of F such that all C_i's are defined over F_1. For each $s \in S$, we set $T_i(s) = \{x \in O_{F_1}(s) \times O_{F_1}(s) \mid \beta_2(x) \subseteq C_i(\overline{K})\}$. By the pigeonhole principle, we can find i and an infinite subset S' of S such that $\#T_i(s) \geq \#(O_{F_1}(s) \times O_{F_1}(s))/e$ for all $s \in S'$. Thus, by Claim 5, there is an infinite subset S_1 of S' and a positive integer N_1 with $\overline{\mathscr{D}}_2([N_1](S_1), F_1) \subseteq C_i(\overline{K})$.

From now on, we denote C_i by C, so that $\overline{\mathscr{D}}_2(S, F) = C$. Finally, we have $\overline{\mathscr{D}}_n(S, F) = C^{n-1}$ for all $n \geq 2$. Obviously $\overline{\mathscr{D}}_n(S, F) \subseteq C^{n-1}$, so that it is sufficient to find a positive integer N_2, an infinite subset S_2 of S and a finite extension field F_2 of F such that with

$$\overline{\mathscr{D}}_n([N_2](S_2), F_2) = C^{n-1}.$$

By Proposition 9.28, there are a positive integer N_2, an infinite subset S_2 of S and a finite extension field F_2 of F such that $([N_2](S_2), F_2)$ is n-minimized. In the same way as above, we can find abelian subvarieties $B_1, \ldots, B_l$ such that $\overline{\mathscr{D}}_n([N_2](S_2), F_2) = \bigcup_{j=1}^{l} B_j$. In addition, as $([N_2](S_2), F_2)$ is n-minimized, replacing F_2 by a finite extension field if necessarily, we may assume that $C, B_1, \ldots, B_l$ are defined over F_2. Let us check $\overline{\mathscr{D}}_n([N_2](S_2), F_2) = C^{n-1}$.

As before, there are B_j and an infinite subset S'' of $[N_2](S_2)$ such that, for all $s \in S''$, we can find a subset $U(s)$ of $O_{F_2}(s)^n$ with $\#U(s) \geq \#(O_{F_2}(s)^n)/l$ and $\beta_n(U(s)) \subseteq B_j(\overline{K})$. In the following, B_j is denoted by B. Let $C^{(q)}$ be the q-axis of C^{n-1}, that is,

$$C^{(q)} = 0 \times \cdots \times \overset{q\text{-th}}{C} \times \cdots \times 0.$$

Moreover, we set $B^{(q)} = B \cap C^{(q)}$. Then we have the following claim.

CLAIM 6. *$B^{(q)} = C^{(q)}$ holds for all $1 \leq q \leq n-1$.*

PROOF. We fix an arbitrary $s \in S''$. For $t = (t_1, \ldots, t_q, t_{q+2}, \ldots, t_{n-1}) \in O_{F_2}(s)^{n-1}$, we set

$$L_t(s) = \{x \in O_{F_2}(s) \mid (t_1, \ldots, t_q, x, t_{q+2}, \ldots, t_{n-1}) \in U(s)\}.$$

Pick up $t' = (t'_1, \ldots, t'_{n-1})$ of $O_{F_2}(s)^{n-1}$ such that $\#L_{t'}(s)$ is maximal among $\{\#L_t(s)\}_{t \in O_{F_2}(s)^{n-1}}$. As

$$\#L_{t'}(s) \cdot \#O_{F_2}(s)^{n-1} \geq \#U(s) \geq \#(O_{F_2}(s)^n)/l,$$

we have $\#L_{t'}(s) \geq \#O_{F_2}(s)/l$. If we set

$$T(s) := L_{t'}(s) \times L_{t'}(s) \subseteq O_{F_2}(s) \times O_{F_2}(s),$$

then

$$\begin{aligned} &\beta_n\left(0, \ldots, 0, y - x, 0, \ldots, 0\right) \\ &\quad = \beta_n\left(t'_1, \ldots, t'_q, y, t'_{q+2}, \ldots, t'_{n-1}\right) - \beta_n\left(t'_1, \ldots, t'_q, x, t'_{q+2}, \ldots, t'_{n-1}\right) \in B(\overline{K}) \end{aligned}$$

for $(x, y) \in T(s)$, which means that $\beta_2(x, y) \in B^{(q)}(\overline{K})$, where we view $B^{(q)}$ as a subscheme of A. Thus, as

$$\beta_2(T(s)) \subseteq B^{(q)}(\overline{K}) \quad \text{and} \quad \#T(s) \geq \#(O_{F_2}(s) \times O_{F_2}(s))/l^2,$$

by Claim 5, there are an infinite subset S_3 of S'' and a positive integer N'' with $\overline{\mathscr{D}}_2([N''](S_3), F_2) \subseteq B^{(q)}(\overline{K})$. Note that

$$[N''](S_3) \subseteq [N''](S'') \subseteq [N'']([N_2](S_2)) \subseteq [N'' N_2](S).$$

Thus if we take an infinite subset T_1 of S with $[N''](S_3) = [N''N_2](T_1)$, then

$$\overline{\mathscr{D}}_2([N''](S_3), F_2) = [N''N_2](\overline{\mathscr{D}}_2(T_1, F_2)) = [N''N_2](C) = C.$$

Therefore, $B^{(q)} = C^{(q)}$ holds. □

By Claim 6 above,

$$C^{n-1} = C^{(1)} + \cdots + C^{(n-1)} = B^{(1)} + \cdots + B^{(n-1)} \subseteq B \subseteq \overline{\mathscr{D}}_n([N_2](S_2), F_2) \subseteq C^{n-1},$$

as required. □

9.8. The proof of Theorem 9.24

First let us prepare several facts from linear algebra. Let V be a real vector space and $\langle\ ,\ \rangle$ an inner product of V. Let $\Lambda = \{v_1, \ldots, v_k\}$ be a subset of V such that $v_1, \ldots, v_k$ are linearly independent over $\mathbb{R}$. We define

$$\Delta_\Lambda : V \times V \to \mathbb{R}$$

to be

$$\Delta_\Lambda(x, y) = \det \begin{pmatrix} \langle v_1, v_1\rangle & \cdots & \langle v_1, v_k\rangle & \langle v_1, y\rangle \\ \vdots & \ddots & \vdots & \vdots \\ \langle v_k, v_1\rangle & \cdots & \langle v_k, v_k\rangle & \langle v_k, y\rangle \\ \langle x, v_1\rangle & \cdots & \langle x, v_k\rangle & \langle x, y\rangle \end{pmatrix},$$

which is a symmetric and positive semidefinite bilinear form. Moreover, it is easy to see that $\Delta_\Lambda(v, x) = 0$ for $v \in \mathrm{Span}(\Lambda)$ and $x \in V$. Here let us see the following proposition.

PROPOSITION 9.30. *Let $f : V \to V'$ be a linear map of real vector spaces. Let $\langle\ ,\ \rangle$ and $\langle\ ,\ \rangle'$ be inner products of V and V', respectively. We assume that there is a positive constant a such that $\langle f(x), f(x)\rangle' \leq a\langle x, x\rangle$ for all $x \in V$. Let $\Lambda = \{v_1, \ldots, v_k\}$ be a finite set consisting of linear independent vectors in V and let $\Lambda' = \{v'_1, \ldots, v'_{k'}\}$ be a basis of $f(\mathrm{Span}(\Lambda))$. Then*

$$\Delta_{\Lambda'}(f(x), f(x)) \leq a\frac{\det(\Lambda')}{\det(\Lambda)}\Delta_\Lambda(x, x)$$

for all $x \in V$, where $\det(\Lambda) = \det(\langle v_i, v_j\rangle)$ and $\det(\Lambda') = \det(\langle v'_{i'}, v'_{j'}\rangle')$.

PROOF. Let us begin with the following claim.

CLAIM 7. (1) *Let $\mathrm{Span}(\Lambda)^\perp$ be the subspace consisting of vectors perpendicular to $\mathrm{Span}(\Lambda)$, that is,*

$$\mathrm{Span}(\Lambda)^\perp = \{x \in V \mid \langle x, v\rangle = 0 \text{ for all } v \in \mathrm{Span}(\Lambda)\}.$$

Then there exist linear maps $p_\Lambda : V \to \mathrm{Span}(\Lambda)$ and $q_\Lambda : V \to \mathrm{Span}(\Lambda)^\perp$ such that $x = p_\Lambda(x) + q_\Lambda(x)$ for $x \in V$.

(2) *For all $x \in V$, $\Delta_\Lambda(x, x) = \det(\Lambda)\langle q_\Lambda(x), q_\Lambda(x)\rangle$. In particular, $\Delta_\Lambda(x, x) \leq \det(\Lambda)\langle x, x\rangle$ and the equality holds if and only if $x \in \mathrm{Span}(\Lambda)^\perp$.*

PROOF. (1) For $x \in V$, we can find a unique solution $(\lambda_1, \ldots, \lambda_k) \in \mathbb{R}^k$ satisfying equations

$$\sum_{j=1}^{k} \lambda_j\langle v_j, v_i\rangle = \langle x, v_i\rangle \quad (\forall\ i = 1, \ldots, n),$$

which means that there is a unique vector $v \in \operatorname{Span}(\Lambda)$ such that $x - v$ is perpendicular to $\operatorname{Span}(\Lambda)$. We denote v by $p_\Lambda(x)$ and $x - v$ by $q_\Lambda(x)$, respectively. Then (1) follows.

(2) By using (1), we obtain

$$\begin{aligned}\Delta_\Lambda(x,x) &= \Delta_\Lambda(p_\Lambda(x), p_\Lambda(x)) + 2\Delta_\Lambda(p_\Lambda(x), q_\Lambda(x)) + \Delta_\Lambda(q_\Lambda(x), q_\Lambda(x)) \\ &= \Delta_\Lambda(q_\Lambda(x), q_\Lambda(x)) = \det(\Lambda)\langle q_\Lambda(x), q_\Lambda(x)\rangle,\end{aligned}$$

as required. □

Let x be an element of V. By virtue of (1) in the above claim, we can decompose x in such a way that

$$x = v + y \quad (v \in \operatorname{Span}(\Lambda), y \in \operatorname{Span}(\Lambda)^\perp).$$

Thus, applying (2) in the above claim to Λ',

$$\begin{aligned}a\frac{\det(\Lambda')}{\det(\Lambda)}\Delta_\Lambda(x,x) &= \det(\Lambda')a\langle y, y\rangle \\ &\geq \det(\Lambda')\langle f(y), f(y)\rangle' \\ &\geq \Delta_{\Lambda'}(f(y), f(y)).\end{aligned}$$

Moreover, as $f(v) \in \operatorname{Span}(\Lambda')$ and $\Delta_{\Lambda'}(v', x') = 0$ for $v' \in \operatorname{Span}(\Lambda')$ and $x' \in V'$,

$$\Delta_{\Lambda'}(f(x), f(x)) = \Delta_{\Lambda'}(f(y), f(y)).$$

Thus the proposition follows. □

In addition, we need one more lemma for the proof of Theorem 9.24.

LEMMA 9.31. *Let $f : A \to A'$ be a homomorphism of abelian varieties over an algebraically closed field such that f is surjective and* $\operatorname{Ker}(f)$ *is finite. For an irreducible subscheme X of A, we set*

$$\begin{cases} G(X) = \{a \in A \mid a + X = X\}, \\ G(f(X)) = \{a' \in A' \mid a' + f(X) = f(X)\}. \end{cases}$$

Then $f(G(X)) = G(f(X))$.

PROOF. It is obvious that $f(G(X)) \subseteq G(f(X))$, so that it is sufficient to see that $G(f(X)) \subseteq f(G(X))$. Let $a' \in G(f(X))$. We choose $a \in A$ with $f(a) = a'$. Then, as $f(a + x) = a' + f(x) \in f(X)$ for $x \in X$, there is $x' \in X$ with $f(a + x) = f(x')$. Thus we can find $\xi(x) \in \operatorname{Ker}(f)$ with $a + \xi(x) + x \in X$. We assume that $\dim(X) = 0$, that is, $X = \{x\}$. Then $a + \xi(x) \in G(X)$. Thus $a' = f(a + \xi(x)) \in f(G(X))$. Next we assume $\dim(X) > 0$. Then we consider a map $X \to \operatorname{Ker}(f)$ given by $x \mapsto \xi(x)$. Since $\dim(X) > 0$ and $\operatorname{Ker}(f)$ is finite, one of the fibers of the map $X \to \operatorname{Ker}(f)$ is Zariski dense in X, that is, there are $\xi_0 \in \operatorname{Ker}(f)$ and a Zariski dense subset S of X with $a + \xi_0 + S \subseteq X$. Thus, taking the Zariski closures of both sides, we obtain $a + \xi_0 + X \subseteq X$, which means that $a + \xi_0 \in G(X)$, so that $a' = f(a + \xi_0) \in f(G(X))$. □

PROOF OF THEOREM 9.24. In order to show Theorem 9.24, we may replace K by a finite extension of K. Therefore, we may assume that X is defined over K and there is a subgroup Γ_0 of Γ with $\Gamma_0 \subseteq A(K)$ and $\Gamma_0 \otimes \mathbb{Q} = \Gamma \otimes \mathbb{Q}$.

As in Lemma 9.31, let $G(X)$ be the stabilizer of X, that is,

$$G(X) = \{a \in A_{\overline{K}} \mid a + X = X\}.$$

First we consider the following claim.

CLAIM 8. *If $G(X)$ is finite, then there is $P \in \Gamma_{\text{div}}$ with $X = \{P\}$.*

PROOF. If $\dim X = 0$, then $X = \{P\}$ for some $P \in A(K)$. Since

$$\delta_L(\gamma_1, \dots, \gamma_k, P) = 0,$$

we have $P \in \Gamma_{\text{div}}$, so that it is sufficient to show $\dim(X) = 0$. We assume $\dim(X) > 0$. Then there is a sequence $\{s_l\}_{l=1}^{\infty}$ in $X(\overline{K})$ with the following properties.

(a) $\{s_l\}_{l=1}^{\infty}$ consists of distinct points.
(b) Any infinite subset of $S = \{s_l \mid l = 1, 2, \dots\}$ is a Zariski dense in X.
(c) $\delta_L(\gamma_1, \dots, \gamma_k, s_l) \leq 1/l$ for all l.

By (c), S is small with respect to Γ_{div}. By Lemma 9.31, $[N](G(X)) = G([N](X))$ for all positive integer N. Therefore, by Proposition 9.28, replacing K, X and S with a finite extension of K, $[N](X)$ and $[N](S)$, respectively, if necessarily, we can take an infinite subset S of $X(\overline{K})$ with the following properties.

(i) S is small with respect to Γ_{div}.
(ii) S is a Zariski dense in X.
(iii) (S, K) is 2-minimized.

Then, by Theorem 9.29, there is an abelian subvariety C of $A_{\overline{K}}$ such that $\overline{\mathscr{D}}_n(S, K) = C^{n-1}$ for all $n \geq 2$.

If $\dim(C) = 0$, then $S \subseteq A(K)$. Thus, by Lang's conjecture over an algebraic number field (Theorem 9.25), X is a translation of an abelian subvariety B of A. In particular, $G(X) = B$, which contradicts the facts that $\dim(X) > 0$ and $G(X)$ are finite.

Otherwise (that is, $\dim(C) > 0$), we choose a positive integer n with $n > 2\dim(A)$. Let $\pi : A \to A/C$ be the natural homomorphism and $T = \pi(X)$. Let X_T^n be the fiber product of X over T. Let us consider a morphism $\beta_n : X_T^n \to A^{n-1}$ given by $\beta_n(x_1, \dots, x_n) = (x_2 - x_1, \dots, x_n - x_1)$. Note that $O_K(s)^n \subseteq X_T^n$ for $s \in S$. Let Y be the Zariski closure of $\bigcup_{s \in S} O_K(s)^n$ in X_T^n. Since

$$\beta_n(Y) = \overline{\beta_n(Y)} \supseteq \overline{\bigcup_{s \in S} \beta_n(O_K(s)^n)} = C^{n-1},$$

we have

$$\dim(X_T^n) \geq \dim(Y) \geq \dim(C^{n-1}).$$

On the other hand, as $G(X)$ is finite, $\dim(X/T) \leq \dim(C) - 1$, so that

$$\begin{aligned}
\dim(X_T^n) - \dim(C^{n-1}) &= (n \dim(X/T) + \dim(T)) - (n-1)\dim(C) \\
&\leq (n(\dim(C) - 1) + \dim(T)) - (n-1)\dim(C) \\
&= \dim(C) + \dim(T) - n \\
&\leq 2\dim(A) - n < 0.
\end{aligned}$$

This is a contradiction. □

Let us start the proof of the theorem. Let $A' = A_{\overline{K}}/G(X)$ and $\pi : A_{\overline{K}} \to A'$ be the natural homomorphism. We set $X' = \pi(X)$ and $\Gamma' = \pi(\Gamma)$. It is easy to see that $\pi^{-1}(X') = X$ and $G(X') = 0$. Let L' be an ample and symmetric invertible sheaf on A'. We can take a positive integer a such that $L^{\otimes a} \otimes \pi^*(L')^{-1}$ is ample, so that $a\hat{h}_L(x,x) \geq \hat{h}_{L'}(\pi(x), \pi(x))$ $(x \in A(\overline{K}))$. Therefore, if $\gamma_1, \ldots, \gamma_k$ is a basis of $\Gamma \otimes \mathbb{Q}$ and $\gamma'_1, \ldots, \gamma'_{k'}$ is a basis of $\Gamma' \otimes \mathbb{Q}$, then, by Proposition 9.30, there is a positive constant C_1 such that

$$\delta_{L'}(\gamma'_1, \ldots, \gamma'_{k'}, \pi(x)) \leq C_1 \delta_L(\gamma_1, \ldots, \gamma_k, x)$$

holds for all $x \in A(\overline{K})$. Thus, for any $\epsilon > 0$,

$$\pi\left(\{x \in X(\overline{K}) \mid \delta_L(\gamma_1, \ldots, \gamma_k, x) \leq \epsilon/C_1\}\right) \subseteq \{x' \in X'(\overline{K}) \mid \delta_{L'}(\gamma'_1, \ldots, \gamma'_{k'}, x') \leq \epsilon\}.$$

Hence the set $\{x' \in X'(\overline{K}) \mid \delta_{L'}(\gamma'_1, \ldots, \gamma'_{k'}, x') \leq \epsilon\}$ is Zariski dense in X'. Thus, by the previous claim, $X' = \{P'\}$ for some $P' \in \Gamma'_{\text{div}}$. In particular, X is a translation of $G(X)$ and $G(X)$ is an abelian subvariety. By Lemma 9.22, $\pi : \Gamma_{\text{div}} \to \Gamma'_{\text{div}}$ is surjective. Namely, X is a translation of $G(X)$ by an element of Γ_{div}. $\square$

Bibliography

[1] A. Abbes and T. Bouche, Théorème de Hilbert-Samuel "arithmétique", Ann. Inst. Fourier(Grenoble) **45** (1995), 375–401.

[2] S. J. Arakelov, Intersection theory for divisors on an arithmetic surface, Izv. Akad. Nauk SSSR Ser. Mat. **38** (1974), 1179–1192; AMS Translation **8**, (1974), 1167–1180.

[3] M. F. Atiyah and I. G. Macdonald, *Introduction to Commutative Algebra*, Addison-Wesley (1969).

[4] T. Aubin, *Nonlinear Analysis on Manifolds. Monge-Ampère Equations*, Grundlehren der Mathematischen Wissenschaften **252**, Springer (1982).

[5] J.-M. Bismut, Quillen metrics and singular fibres in arbitrary relative dimension, J. Algebraic Geom. **6** (1997), 19–149.

[6] J.-M. Bismut, H. Gillet and C. Soulé, Analytic torsion and holomorphic determinant bundles. I. Bott-Chern forms and analytic torsion, Comm. Math. Phys. **115** (1988), 49–78.

[7] J.-M. Bismut and G. Lebeau, Complex immersions and Quillen metrics, Inst. Hautes Études Sci. Publ. Math. **74** (1991), 1–297.

[8] J.-M. Bismut and É. Vasserot, The asymptotics of the Ray-Singer analytic torsion associated with high powers of a positive line bundle, Comm. Math. Phys. **125** (1989), 355–367.

[9] R. Bott and S. S. Chern, Hermitian vector bundles and the equidistribution of the zeros of their holomorphic sections, Acta Math. **114** (1965), 71-112.

[10] T. Bouche, Convergence de la métrique de Fubini-Study d'un fibré linéaire positif, Ann Inst. Fourier (Grenoble), **40** (1990), 117-130.

[11] N. Bourbaki, *Éléments de Mathématique, Algèbre commutative*, Springer-Verlag (2006).

[12] H. Chen, Positive degree and arithmetic bigness, preprint (arXiv:0803.2583).

[13] H. Chen, Arithmetic Fujita approximation, Ann. Sci. École Norm. Sup., **43** (2010), 555–578.

[14] P. Deligne, Le déterminant de la cohomologie, Comtemporary Mathematics **67** (1987), 93–177.

[15] S. K. Donaldson, A new proof of a theorem of Narasimhan and Seshadri, J. Diff. Geometry **18** (1983), 269-277.

[16] S. K. Donaldson, Anti self-dual Yang-Mills connections over complex algebraic surfaces and stable vector bundles, Proc. London Math. Soc. **50** (1985), 1-26.

[17] S. K. Donaldson, Infinite determinants, stable bundles and curvature, Duke Math. J. **54** (1987), 231–247.

[18] G. Faltings, Calculus on arithmetic surfaces, Ann. of Math. **119** (1984), 387–424.

[19] G. Faltings, Diophantine approximation on abelian varieties, Ann. of Math. **133** (1991), 549–576.

[20] G. Faltings, *Lectures on the arithmetic Riemann-Roch theorem*, notes taken by S. Zhang, Princeton Univ. Press (1992).

[21] G. Faltings, The general case of S. Lang's conjecture in *Barsotti Symposium in Algebraic Geometry (Abano Terme, 1991)*, 175–182, Academic Press, (1994).

[22] W. Fulton, *Intersection Theory*, Ergebnisse der Mathematik und ihrer Grenzgebiete, Springer-Verlag (1984).

[23] H. Gillet and C. Soulé, Arithmetic intersection theory, Inst. Hautes Études Sci. Publ. Math. **72** (1990), 93–174.

[24] H. Gillet and C. Soulé, Characteristic classes for algebraic vector bundles with Hermitian metric. I, II, Ann. of Math. **131** (1990), 163–203; ibid **131** (1990), 205–238.

[25] H. Gillet and C. Soulé, On the number of lattice points in convex symmetric bodies and their duals, Israel J. Math. **74** (1991), 347-357.

[26] H. Gillet and C. Soulé, An arithmetic Riemann-Roch theorem, Invent. Math. **110** (1992), 473–543.

[27] P. Griffiths and J. Harris, *Principles of algebraic geometry*, Pure and Applied Mathematics. Wiley-Intersci., New York, (1978).

[28] P. M. Gruber and C. G. Lekkerkerker, *Geometry of numbers*, Second edition, North-Holland, Amsterdam, (1987).

[29] R. Hartshorne, *Algebraic geometry*, Graduate Texts in Mathematics, No. **52**. Springer, New York, (1977).

[30] H. Hironaka, Resolution of singularities of an algebraic variety over a field of characteristic zero. I, II, Ann. of Math. **79** (1964), 109–203; ibid. **79** (1964), 205–326.

[31] P. Hriljac, Height and Arakerov's intersection theory, Amer. J. Math., **107** (1985), 23–38.

[32] S. Kawaguchi and A. Moriwaki, Inequalities for semistable families for arithmetic varieties, J. Math. Kyoto Univ. **41** (2001), 97–182.

[33] S. Kawaguchi, A. Moriwaki and K. Yamaki, Introudcution to Arakelov Geometry, Proceedings of the Symposium on Algebraic geometry in East Asia (2002), 1-74.

[34] A. Khovanskii, Newton polyhedron, Hilbert polynomial and sums of finite sets, Funct. Anal. Appl. **26** (1993), 276–281.

[35] S. L. Kleiman, Toward a numerical theory of ampleness, Ann. of Math. **84** (1966), 293–344.

[36] F. F. Knudsen and D. Mumford, The projectivity of the moduli space of stable curves. I. Preliminaries on "det" and "Div", Math. Scand. **39** (1976), 19–55.

[37] S. Kobayashi, *Differential geometry of complex vector bundles*, Publications of the Mathematical Society of Japan, **15**, Iwanami Shoten, Publishers and Princeton University Press (1987).

[38] S. Lang, *Fundamentals of Diophantine geometry*, Springer, New York, (1983).

[39] S. Lang, *Introduction to Arakelov Theory*, Springer-Verlag, (1988).

[40] R. Lazarsfeld, *Positivity in Algebraic Geometry I, II*, Springer-Verlag (2004).

[41] R. Lazarsfeld and M. Mustaţă, Convex bodies associated to linear series, preprint (arXiv:0805.4559)

[42] J. Lipman, Desingularization of two-dimensional schemes, Ann. of Math., **107** (1978), 151-207.

[43] Q. Liu, *Algebraic Geometry and Arithmetic Curves*, Oxford Graduate Texts in Mathematics **6**, Oxford University Press (2002).

[44] H. Matsumura, *Commutative Algebra* (2nd edition), Mathematics Lecture Note Series, **56**, The Benjamin/Cummings Publishing Company, Inc., (1980).

[45] H. Matsumura, Commutative Ring Theory, Cambridge Studies in Advanced Mathematics **8**, Cambridge University Press (1989).

[46] A. Moriwaki, Inequality of Bogomolov-Gieseker type on arithmetic surfaces, Duke Math. J. **74** (1994), 713–761.

[47] A. Moriwaki, Bogomolov unstability on arithmetic surfaces, Math. Research Letters **1** (1994), 601-611.

[48] A. Moriwaki, Arithmetic Bogomolov-Gieseker's inequality, Amer. J. Math. **117** (1995), 1325-1347.

[49] A. Moriwaki, Arithmetic height functions over finitely generated fields, Invent. Math. **140** (2000), 101–142.

[50] A. Moriwaki, A generalization of Bogomolov and Lang over finitely generated fields, Duke Math. J. **107** (2001), 85–102.

[51] A. Moriwaki, Continuity of volumes on arithmetic varieties, J. Algebraic Geom. **18** (2009), 407-457.

[52] A. Moriwaki, Continuous extension of arithmetic volumes, International Mathematics Research Notices (2009), 3598-3638.

[53] A. Moriwaki, Estimation of arithmetic linear series, Kyoto J. of Math., **50** (Memorial issue for the late Professor Nagata) (2010), 685–725.

[54] A. Moriwaki, Free basis consisting of strictly small sections, International Mathematics Research Notices (2011), 1245-1267.

[55] D. Mumford, *Abelian Varieties*, Oxford University Press (1974).

[56] M. S. Narasimhan and C. S. Seshadri, Stable and unitary vector bundles on compact Riemann surfaces, Ann. of Math. **82** (1965), 540-567.

[57] N. Naumann, A quantitative sharpening of Moriwaki's arithmetic Bogomolov inequality, Math. Research Letters **12** (2005), 877-884.

[58] J. Noguchi and T. Ochiai, *Geometric Function Theory in Several Complex Variables*, Translations of Mathematical Monographs **80**, AMS (1990).

[59] B. Poonen, Mordell-Lang plus Bogomolov, Invent. Math. **137** (1999), 413–425.

[60] D. Quillen, Determinants of Cauchy-Riemann operators over a Riemann surface, Funct. Anal. Appl. (1985), 31–34.

[61] D. B. Ray and I. M. Singer, Analytic torsion for complex manifolds, Ann. of Math. **98** (1973), 154–177.

[62] M. Raynaud, Flat modules in algebraic geometry, Compositio Math. **24** (1972), 11–31.

[63] W. Rudin, *Real and Complex Analysis*, 3rd edition, McGraw-Hill (1987).

[64] C. Soulé, *Lectures on Arakelov Geometry*, with the collaboration of D. Abramovich, J.-F. Burnol and J. Kramer. Cambridge Univ. Press, (1992).

[65] C. Soulé, A vanishing theorem on arithmetic surfaces, Invent. Math. **116** (1994), 577–599.

[66] L. Szpiro, E. Ullmo and S. Zhang, Équirépartition des petits points, Invent. Math. **127** (1997), 337–347.

[67] G. Tian, On a set of polarized Kähler metrics on algebraic manifolds, J. Diff. Geometry, **32** (1990), 99-130.

[68] K. Uhlenbeck and S.-Y. Yau, On the existence of Hermitian Yang-Mills connections in stable vector bundles, Comm. Pure Appl. Math. **39** (1986), 257-293.

[69] K. Uhlenbeck and S.-Y. Yau, A note on our previous paper: On the existence of Hermitian Yang-Mills connections in stable vector bundles, Comm. Pure Appl. Math. **42** (1989), 703–707.

[70] E. Ullmo, Positivité et discrétion des points algébriques des courbes, Ann. of Math. **147** (1998), 167–179.

[71] P. Vojta, Siegel's theorem in the compact case, Ann. of Math., **133** (1991), 509–548

[72] R. Webster, *Convexity*, Oxford University Press, (1994).

[73] X. Yuan, Big line bundles over arithmetic varieties, Invent. Math. **173** (2008), 603–649.

[74] X. Yuan, On volumes of arithmetic line bundles, Compositio Mathematicae 145 (2009), 1447-1464.

[75] S. Zhang, Positive line bundles on arithmetic surfaces, Ann. of Math. **136** (1992), 569–587.

[76] S. Zhang, Positive line bundles on arithmetic varieties, J. Amer. Math. Soc. **8** (1995), 187–221.

[77] S. Zhang, Equidistribution of small points on abelian varieties, Ann. of Math. **147** (1998), 159–165.

[78] S. Zhang, Distribution of almost division points, Duke Math. J. **103** (2000), 39–46.

Index